Explorations in
College Algebra

Explorations in College Algebra

Linda Almgren Kime
Judy Clark

University of Massachusetts, Boston

In Collaboration With

Norma M. Agras
Miami Dade Community College

Robert F. Almgren
University of Chicago

Linda Falstein
University of Massachusetts, Boston

Meg Hickey
Massachusetts College of Art

John A. Lutts
University of Massachusetts, Boston

Beverly K. Michael
University of Pittsburgh

Jeremiah V. Russell
University of Massachusetts, Boston
Boston Public Schools

Software Developed by
Hubert Hohn
Massachusetts College of Art

Funded by a National Science Foundation Grant

JOHN WILEY & SONS, INC.
New York • Chichester • Weinheim • Brisbane • Toronto • Singapore

EXECUTIVE EDITOR   Ruth Baruth
DEVELOPMENTAL EDITOR   Madalyn Stone
SENIOR MARKETING MANAGER   Leslie Hines
PRODUCTION EDITOR   Tracey Kuehn
DESIGNER   Harry Nolan
PHOTO EDITOR   Kim Khatchatourian
ILLUSTRATION EDITOR   Sigmund Malinowski
ELECTRONIC ILLUSTRATIONS   Radiant
ASSISTANT EDITOR   Barbara Bredenko
CD-ROM   Diamelle
COVER PHOTOGRAPH   © David Doubilet

This project was supported, in part, by the

National Science Foundation

Opinions expressed are those of the authors
and not necessarily those of the Foundation

This material is based upon work supported by the National Science Foundation under Grant No. USE-9254117.

This book was set in Times Roman by Progressive Information Technologies and was printed and bound by Quebecor/Kingsport. The cover was printed by Lehigh Press.

This book is printed on acid-free paper ☺.

The paper in this book was manufactured by a mill whose forest management programs include sustained yield harvesting of its timberlands. Sustained yield harvesting principles ensure that the numbers of trees cut each year does not exceed the amount of new growth.

Library of Congress Cataloging-in-Publication Data
Kime, Linda Almgren.
    Explorations in college algebra / Linda Almgren Kime, Judy Clark.
        p.   cm.
    Includes index.
    ISBN 0-471-10698-4 (pbk. : alk. paper)
    1. Algebra.   I. Clark, Judy, 1945–   .   II. Title.
QA152.2.K55   1998
512.9—dc21

                                                                97–9515
                                                                CIP

Printed in the United States of America

10 9 8 7 6 5 4 3 2 1

To all of our students, who have been thoughtful critics
and enthusiastic participants in the development of these materials

and

to our parents and other members of our families
who provided the inspiration for us to pursue mathematics.

To the Students

FROM A STUDENT

Have you ever thought of math as something that you like to do? No, I'm not crazy. You might think of math as merely a bunch of numbers and variables and long hours of monotonous repetition, but this class makes math fun and interesting.

Math will come alive right in front of you. This class, unlike many other math courses, is different in that it shows you practical applications of math to real life situations and gives meaning to math. The numbers and variables are symbolic representations of things in the world. Math will finally make sense.

This class is also different in that there will be lecture, but much of the class time will be spent *doing* math. I found this to be very helpful in that you're actually doing the math right there and then and not trying to decipher and make sense of your lecture notes at home where it's too late to ask questions.

At first things will take a little time to get used to. But watch out. Once you see the big picture of this class it can be very addictive. All you have to do is give it a chance to show you what it's all about. This class opens up your mind and helps you think critically and thoughtfully about math as you never have before. I hope you enjoy it as much as I have.

Sincerely,
Philip Wan
Nursing/Pre-Med Major, UMass/Boston

FROM THE AUTHORS

Mathematics offers a way of looking at the world. Just as musical training increases awareness of sounds, or knowledge of history gives a deeper perspective on the present, mathematics heightens your perception of underlying order and systematic patterns in the world around you.

Our approach to learning algebra starts with questions from the physical and social sciences. You will learn to use algebra in your search for answers. You will also be asked to *read* and analyze mathematical arguments from a variety of different sources and viewpoints. You will be asked to *reason* through your own quantitative arguments, using both traditional algebraic tools and contemporary technological ones. And you will be asked to *write* about your conclusions.

The classes may not look much like the ones you're used to. The professor might lecture a lot less, and you may find yourself working with your classmates in small groups. An exploratory approach results in a lot of questions and discussions. This process of exploring is as important as memorizing the "facts."

We hope that you learn mathematics in a way that will prove useful throughout your life.

With warm regards,
Linda and Judy
Cambridge, Massachusetts, June 1997

Preface

This book grew from a desire to reshape the standard college algebra course. We wanted to make algebra interesting and relevant to our students. Thus we designed the materials to shift the focus from learning a set of discrete mechanical rules to exploring how algebra is used to answer questions about the physical and social world around us.

The materials are in the spirit of the reform movement. The content and approach reflect the standards established by the National Council of Teachers of Mathematics (NCTM) and the American Mathematical Association of Two-Year Colleges (AMATYC), and the recommendations of the Mathematical Association of America (MAA). Our goal is to adequately prepare students for future mathematics or other quantitatively based courses. This text, while designed for college algebra, has been used successfully for other types of courses including pre-calculus for nonmath majors and quantitative reasoning courses.

OUR PROCESS OF DEVELOPMENT

The materials were first used in an experimental course, Explorations in College Algebra, funded in part by a grant from the National Science Foundation, and taught at the University of Massachusetts, Boston. The text and accompanying software have evolved from the ongoing collaboration of faculty from a broad consortium of schools. The Rough Draft was tested at fourteen beta sites. Their feedback shaped the Preliminary Edition, which was in turn critiqued by additional mathematics faculty, teaching assistants, and students. We also consulted with a wide range of discipline specialists, such as economists, physicists, bankers, biologists, political scientists, and astronomers, in order to understand what kinds of questions and answers were relevant to these fields. We wanted to know how they actually *used* mathematics, not so much in classrooms as in their work.

We conducted an extensive survey of faculty members teaching college algebra courses. The results helped us decide which topics to include in the text, but also convinced us that there is no such thing as a generic college algebra course.

Our students graciously allowed us to experiment, revise, and experiment again. They enthusiastically provided honest and insightful advice, and suggestions for improvement. Their input, perhaps more than any others, shaped the content and ideas of the course.

GENERAL PRINCIPLES

The following general principles served as our guide.

- Algebra is a powerful tool that can help answer questions that arise in the social, physical, and life sciences. Algebraic procedures and concepts can be developed from the investigation of practical problems.

- Mathematical ideas can be represented and understood in multiple ways—through words, numbers, graphs, and symbols.

- Students can make connections among various representations and understand concepts more deeply by becoming actively engaged in doing mathematics, as well as by communicating their ideas to others.

- Materials should be flexible. They should accommodate a broad range of teaching and learning styles in a variety of settings.

- Technology should be used when appropriate. While technology is not required to teach this course, its use can greatly enhance student understanding.

- Conceptual understanding should be accompanied by sufficient practice in building skills.

ORGANIZATION

The text includes more material than college algebra courses usually include in one semester. Instructors will need to make choices and may wish to cover the chapters in a different order. It is recommended, however, that students become familiar with the material covered in Chapters 1, 2, 3, and 4 before attempting to do the other chapters. The rest of the chapters may be covered in any order, with two exceptions: the material on exponents in Chapter 7 should be covered prior to material in the other chapters in Part II; and Chapter 12, "Logarithmic Links: Logarithmic, Exponential, and Power Functions" assumes Sec. 7.7, "Logarithms Base 10," and Chapter 8, "Growth and Decay." Entering students should have had elementary high school algebra and be familiar with the use of symbols to represent unknown numerical quantities.

Part I focuses on algebraic applications in the social sciences. Chapter 1 introduces the basic issues in working with and writing about data. This chapter is the least like a traditional algebra chapter, but perhaps the most relevant to students' everyday lives. Students can be introduced to the basics of using a graphing calculator or computer. The amount of time spent on this chapter (from 2 days to 3 weeks) will vary a great deal from instructor to instructor. The time spent is probably dictated by what mathematics courses most of the students would take next and how long it takes for students to become familiar with a particular technology.

Chapter 2 introduces the fundamental concept of functions and their representations in words, tables, graphs, and equations. In Chapter 3, students encounter the notion of the *average rate of change,* which becomes a central thread throughout the course.

Chapter 4 is the first of three chapters dedicated to *linear functions* and their applications. It is motivated by looking at phenomena in which the rate of change is constant and focuses on constructing, graphing, and interpreting linear functions. Chapter 5 uses a case study approach to introduce a major tool of the social sciences—fitting lines to data. Students use real U.S. Census data on 1000 families to describe and analyze the relationship between education and income. Chapter 6 deals with *systems of linear equations* and their uses in economics, including a comparison of the costs of different heating systems and the benefits of different income tax plans.

Part II opens with Chapter 7, "Deep Time and Deep Space." Studying the age and size of objects in the known universe provides a context for learning *scientific notation* and strategies for comparing objects of widely differing sizes. Students learn to estimate answers, a critical skill in this age of calculators and computers. The basic laws of manipulating *exponents* are applied to different bases and powers. *Logarithms* base 10 are used to scale graphs and to construct measurement systems for numbers with different orders of magnitude.

In Chapter 8, the growth of *E. coli* bacteria motivates the study of *exponential functions* of the form $y = Ca^x$. Models of exponential growth or decay are applied to human populations, radioactive decay, and Medicare costs. Semi-log plots are introduced to identify exponential patterns in data. *Power functions* are introduced in Chapter 9 through studying the relationship between size and shape in biological organisms. Students learn about general *polynomial functions* as the sum of power functions with positive integer exponents.

Chapters 10 and 11 cover the family of *quadratic functions*. A case study of a classic free-fall experiment in Chapter 10 gives students some insights into how Galileo uncovered the basic laws of motion. In Chapter 11, "Parabolic Reflections," the properties of quadratic functions and their graphs are examined.

Logarithmic functions are motivated in Chapter 12 by the need to solve exponential equations studied in Chapter 8. Chapter 12 extends the discussion in Section 7.7 of logarithms as numbers to logarithms as functions. It introduces *e,* the natural logarithm, and exponential functions of the form $y = Ce^{rx}$. Data are graphed on linear, semi-log, and log-log plots to find an appropriate function model.

SPECIAL FEATURES AND THEIR USES

In addition to the usual collection of chapters, chapter overviews and summaries, exercises, and answer sections, these materials contain many special features. These features represent an array of resources from which the instructor and student can pick and choose. Suggestions for using each resource component appear at the appropriate point in the text and in the Instructor's Manual. The following list contains an overview of the features and some suggested ways to use them.

Explorations

Explorations, located at the end of each chapter, allow students to experiment with ideas. The Explorations can be used with students working in small groups or individually, in or out of the classroom, or assigned as special projects. They are generally more open ended than exercises, and are designed to be used in parallel with reading the text. An Exploration icon indicates when a particular Exploration could be started, rather than waiting until the end of the chapter.

CD-ROM

The text comes with a CD-ROM that contains multiple additional resources for the course. The course software, data files, and Graphing Calculator Workbook are all on the CD-ROM and are discussed in the following section on technology use. The CD also contains the Student's Solution Manual, additional exercises, readings, and Explorations. A separate Instructor's Resource CD can be ordered from Wiley. It contains the materials on the text CD as well as the instructor's Manual and Instructor's Solution Manual. The CD icon appears in the text when a resource is available on CD-ROM.

Anthology of Readings

The *Anthology of Readings* is in the Appendix with additional readings on the CD-ROM. The Anthology contains a variety of articles from newspapers, government publications, scientific and popular magazines, etc. Some readings highlight a controversy about a topic discussed in a chapter, others offer expanded coverage of a mathematical topic only mentioned in the text. Each reading is referenced with an icon at an appropriate point in the text, exercises, or Explorations. Instructors, however, may wish to find new and creative ways to use these or other readings.

Something to think about

Short inserts entitled *Something to think about* pose provocative questions that could be used as the basis of a class discussion or an assignment or for students to ponder on their own.

Writing assignments and class presentations

Many of the exercises or Explorations contain a writing component. Chapter 1 includes suggestions for writing strategies. The notion of a "60 Second Summary" is introduced as a way for students to present key ideas clearly and concisely. Some of the Explorations also ask students to present their results to the class.

TECHNOLOGY COMPONENTS

There are many Explorations and exercises built around graphing calculators and computers, but there are no specific hardware requirements. Some schools use only graphing calculators, others use only computers with the course software, and some use a combination of both. The course can also be taught entirely in a computer lab using a spreadsheet program, a simple function graphing program, and the course software. The accompanying CD-ROM includes several resources to make using technology easier. A CD icon indicates when one of the electronic resources might prove useful. The CD can be used with either a PC (with Windows) or Mac and includes:

Custom Software

These programs offer an easy-to-use interactive environment that helps students visualize specific mathematical concepts, and provides practice in basic skills. The software can be used for classroom demonstrations or in a lab as the basis for several of the Explorations, or at home by the student.

Graph Link Files

The CD-ROM contains numerous graph link files for the TI-82 or TI-83 graphing calculators and instructions for file downloading. These short programs contain all the major data sets used in the course. They can all be transferred at once to every student's calculator at the beginning of the course.

Graphing Calculator Workbook

Basic instructions for using TI-82 and TI-83 calculators are included in the Graphing Calculator Workbook. The instructions are linked directly with the chapters, and introduce calculator skills as needed. The CD-ROM contains additional worksheets with problems that provide additional practice in basic skills. If students are using graphing calculators for the first time, we strongly suggest they read through the workbook instructions before doing the Explorations or exercises for each chapter. The workbook may be ordered in hard copy, as well.

Spreadsheet Files

All the major data sets (including the entire FAM 1000 U.S. Census data set used in the case study in Chapter 5) exist as Excel files on the CD-ROM.

SKILL-BUILDING COMPONENTS

In addition to doing selected end of chapter exercises, students can build skills in manipulating algebraic expressions through:

Algebra Aerobics

Algebra Aerobics are short collections of problems integrated throughout the text, usually at the end of sections, that provide basic practice in skills either recently introduced or assumed as prerequisites. Answers to these drill problems are provided in the Appendices. Students are encouraged to read the text with pencil and paper in hand, and to stop and work through these skill problems and check their answers before moving on.

Custom Software

Certain programs in the custom software provide skill-building practice. Instructors can assign these programs for homework, or students on their own may wish to use them to practice.

Worksheets on the CD-ROM

The CD-ROM contains additional worksheets and answers to reinforce basic manipulative skills. Instructors are free to print out and xerox the worksheets as handouts for an entire class, or just for students who need extra help in a particular topic. Students may also wish to print them out and use them for practice on their own.

OTHER TEXT SUPPLEMENTS

Instructors can order printed versions of the Graphing Calculator Workbook or the Student's Solution Manual that appear on the CD-ROM packaged with the text. The Instructor's Manual and Instructor's Solution Manual exist on a separate Instructor's Resource CD and in print versions.

MAJOR CHANGES FROM THE PRELIMINARY TO THE FIRST EDITION

- A new Chapter 12 introduces logarithmic functions and exponential functions base e.
- The text has been refined based on feedback from numerous test sites for increased clarity, and additional problems and examples have been added where needed.

- The Algebra Aerobics sections, for skill building, have been increased to provide additional practice with drill exercises.
- The CD-ROM, packaged with every text, provides students with course software and data sets, along with a getting-started Graphing Calculator Manual for the TI-82 and 83. Other resources on the disk: additional skill building material, readings, worksheets, and exercises. The Student's Solution Manual, with answers to the odd exercises found in the text, is also included on the CD.
- The Instructor's Resource CD-ROM contains the Instructor's Manual, Instructor's Solution Manual, and all features contained on the student CD.
- The separate Graphing Calculator Workbook replaces the graphing calculator appendix.

KEEPING IN TOUCH

You may want to visit our Web site at *http://www.wiley.com/college/math/mathem/kimeclark* to look for (or suggest) new ideas for Explorations, readings, and exercises, to leave mail for the authors, or to join the ongoing discussion of the College Algebra Consortium on reforming college algebra. The authors would be delighted to receive reactions to the materials. Please contact us at the University of Massachusetts, Boston, 100 Morrissey Blvd, Boston, MA 02125.

Readers interested in further information about the materials should contact John Wiley & Sons, Inc., by mail at 605 Third Avenue, New York, NY 10158 or by e-mail at *math@wiley.com.*

Acknowledgments

We wish to express our appreciation to all those who have helped and supported us in this collaborative endeavor. We are grateful for the support of the National Science Foundation, whose funding made this project possible, and for the generous help from our program officers, Elizabeth Teles and Marjorie Enneking. We wish to thank the members of our Advisory Board: Deborah Hughes Hallett, Philip and Phylis Morrison, Nicholas Rubino and Ethan Bolker for providing encouragement and invaluable advice in each phase of our work.

The text could not have been produced without the generous and on-going support we received from students, faculty, staff and administrators at the University of Massachusetts, Boston. Our thanks especially to the administrators who provided resources for many aspects of the project: Chancellor Sherry Penney, Provosts Louis Esposito and Fuad Safwat, Associate Provost Theresa Mortimer, and Deans Patricia Davidson, Christine Armett-Kibel, Ellie Kutz, William Dandridge, and Martin Quitt. Special thanks to Paul O'Keefe, Mike Larsen, and Carol DeSouza for helping us with the endless paperwork and procedures that are an inevitable part of obtaining and running a grant.

We are especially grateful to the following colleagues at the University of Massachusetts, Boston, and elsewhere, for their contributions to the text: Bob Seeley, George Lukas, Rachel Skvirsky, Ron Etter, Lowell Schwartz, Max West, Tony Roman, Joe Check, Suzy Groden, Joan Lukas, Bernice Auslander, Karen Callaghan, Lou Ferleger, Art MacEwan, Randy Albelda, Rosanne Donahue, Gourish Hosangady, Bob Lee, Ken Kustin, John Looney, Brenda Cherry, Peg Cronin, Bob Morris, Margaret Zaleskas, Debra Borkovitz, Paul Foster, Elizabeth Cavicchi, Eric Entemann, Bob Martin, David Hruby, Steve Rodi, David Ames, Sergio Schirato, David Harrison, Max Knight, Holly Handlin, Brian Butler, Jie Chen, John Murphy, and Rachel Dyal. Particular thanks are owed to Barry Bluestone from whom we borrowed the concept of the FAM 1000 data set used in Chapter 5, "Looking for Links."

A project cannot function without an efficient and forgiving Project Administrator. In Theresa Fougere we found a woman with a divine combination of attributes: common sense, dedication, attention to detail, and a sense of humor. Our special thanks to her and others who have shared administrative duties: Clare Crawford, Marie Coleman, Aurora Alamariu, Val Goktuk, Jonathan Rose, Wendy Sanders, Matt Gunderson, Karen Sullivan, Eric Dunlap, Jan McLeod, Matt Smith, Lynne Bowen, Patricia Dognazzi, John Harper, Angie Hunter, Sandy Mabbit, and David Wilson.

A text designed around the application of real world data would have been impossible if not for the long and selfless hours put in by Myrna Kustin and David Wilson researching and acquiring copyright permissions from around the globe.

We also wish to thank our terrific student teaching assistants: John Koveos, Tony Beckwith, Irene Blach, Philip Wan, Kristen Demopoulos, Tony Horne, Cathy Briggs, and Arlene Russo. They excelled as students and as teachers.

We are deeply indebted to Peter Renz, Dick Cluster and Kristen Clark for their gracious and supportive editorial help.

Kudos are due our publisher, John Wiley & Sons. We are grateful that William

Pesce gave us a chance to present our ideas to the Wiley staff. He asked good questions and we are still searching for some of the answers. Special thanks are due to Madalyn Stone, our delightful and patient developmental editor; Tracey Kuehn, our gentle and long-suffering production editor; Harry Nolan, our creative graphic designer; and Sigmund Malinowski, who has worked long and hard on the artwork. Many others at Wiley, including Wayne Anderson, Mary Johenk, and Eileen Navagh have been extraordinarily helpful. Particular thanks go to our wonderful editor, Ruth Baruth. After our first two hour al fresco lunch in Harvard Square, we knew that we would have a pleasant long-term relationship. Ruth has provided the gentle guidance necessary for first-time authors. She has generously given us many hours of practical advice and has cheerfully helped us with the endless details of publishing a book.

We are especially grateful to the fourteen beta sites scattered across the United States. One of the joys of this project has been working with so many dedicated faculty who are searching for new ways to reach out to students. These faculty, teaching assistants, and students all offered incredible support and encouragement, and a wealth of helpful suggestions. Some became codevelopers of the materials, and many became good friends. Our heartful thanks to:

Sandi Athanassiou and Mark Yannotta, Erika Kwiatkowski, Whitney Latimer, Sergei Kosakovsky, and all the other wonderful TA's at University of Missouri-Columbia
Peg McPartland, Golden Gate University
Peggy Tibbs and John Watson, Arkansas Tech University
Josie Hamer, Robert Hoburg, and Bruce King, Western Connecticut State University
Judy Stubblefield, Garden City Community College
Bob Lee, Boston Public Schools and University of Massachusetts, Boston
Jean Prendergast and Keith Desrosiers, Bridgewater State College
Lida McDowell, Jan Davis, and Jeff Stuart, University of Southern Mississippi
Russell Reich, Sierra Nevada College
Christopher Olsen, George Washington High School
Judy Jones and Beverly Taylor, Valencia Community College

Tina Bond from Pensacola Junior College and Curtis Card from Black Hills State University were also invaluable critical readers of the materials.

Our families couldn't help but become caught up in this time consuming endeavor. Linda's husband, Milford, generously helped subsidize a sabbatical year, and her son Kristian willingly spent his vacations constructing our Web site and helping with the endless list of production details. Judy's husband, Gerry, became our Consortium lawyer, and her daughters, Kristin, Rachel and Caroline, provided support, understanding, laughter, and "whatever." All our family members ran errands, made dinners, listened to our concerns, and gave us the time and space to work on the text. Our love and thanks.

Finally, we wish to thank all of our students. Without them, this book would not have been written.

Contents

Exponential Functions

Power Functions

Quadratic and Other Polynomial Functions

APPENDICES

PART

Algebra in the Social Sciences

INTRODUCTION TO PART I

Former Secretary of Labor Robert Reich pronounced "learning as the key to earning" in a speech on the eve of Labor Day 1994. He asserted that "The fundamental fault line running through today's workforce is based on education and skills. . . . As recently as 1979 a male college graduate earned 49 percent more than a similar man with only a high school diploma. . . . By 1992, however, the average male college graduate was earning 83 percent more than his high-school graduate counterpart, and the notion of common prospects had faded considerably."

Presidential hopefuls battle over the virtues of a flat income tax. Republican analyst Kevin Phillips claims that the public fascination is based on the misconception that "the flat tax, by closing loopholes, will make the rich pay more." Yet, according to Steve Forbes, "Everyone gets a tax break with the flat tax."

Life expectancy at birth for both sexes has increased dramatically in the last century for Americans. In 1900 the average life expectancy was slightly over 49 years. By 1990 the average life expectancy had risen to 75.8 years. This is due in large part to the fact that the United States is the world's richest nation and spends far more in combined public and private funds on health care than does any other nation. Yet according to the United Nations' 1994 Human Development Report, people in other countries live longer and get more care.

Political and personal decisions often depend on how data are analyzed and interpreted. What are the benefits of an education? Who will gain from a flat tax? How will changes in life expectancy affect social policy? The social sciences, such as political science, economics, history, demography, and sociology, try to answer questions like these. Unlike physical scientists, who try to discover fundamental laws of the universe, social scientists do not expect to discover exact principles that govern society. However important social trends can be uncovered by collecting numerical data and using algebraic strategies to search for patterns. The conclusions from such research often form the bases for decisions about how our society should function.

Part I approaches the standard topics in college algebra from the perspective of the social sciences. The tools required to deal with real data are developed: verbal and visual methods of data presentation; algebraic methods of single- and two-variable data analysis, including numerical data summaries, rate of change calculations, linear functions, and linear regression models; and systems of equations. We encourage an attitude of "reasoned skepticism," constantly asking questions about the process of generating and drawing conclusions from data.

Making Sense of Data

Overview

In all sciences answering questions requires the collection of data. A physicist or chemist generates data using a controlled laboratory experiment; an economist or sociologist collects data using society as a laboratory. As citizens, it is essential for us to understand the issues involved in collecting, representing, and interpreting data. We will often have to make decisions based on data presented in the news media, academic reports, or market surveys.

Chapter 1 deals with the issues encountered in using real data. It introduces the mathematical and technological skills needed to organize and analyze data. You learn how to reason about data, interpret results, and communicate your ideas clearly to others.

The Explorations give you an opportunity to apply these tools to data collected on your class or on the U.S. population. They also provide experience with any technology (graphing calculators or computers) you might be using in this course.

After reading this chapter you should be able to:

- read and interpret data tables
- visualize data
- understand the differences among the terms mean, median, and mode
- construct a "60 second summary"
- ask critical questions about data and conclusions drawn from data

1.1 AN INTRODUCTION TO VARIABLES AND DATA

Exploration 1.1 provides an opportunity to collect your own data and to think about issues related to classifying and interpreting data.

This course starts with you. How would you describe yourself to others? Are you a 5′ 6″, black, 26-year-old, female engineer? Or perhaps you are a 5′ 10″, Chinese, 18-year-old, male flute player, sharing an apartment with two friends. In statistical terms, characteristics such as height, race, age, and major that vary from person to person are called *variables*. Information collected about a variable is called *data*.[1]

Some variables that you might use to describe yourself, such as age, height, or number of people in your household, can be represented by a number and a unit of measure (such as 18 years, 6 feet, or 3 people). These are called *quantitative variables*. For other variables, such as gender or college major, we use categories (such as male and female, or physics, English, and psychology) to classify information. These are called *qualitative variables*.

Many of the controversies in the social sciences have centered on how particular variables are defined and measured. For nearly two centuries, the categories used by the U.S. Census Bureau to classify race and ethnicity have been the subject of debate. For example, Hispanic used to be considered a racial classification. It is now considered an ethnic classification, since Hispanics can be black, or white, or any other race. Orlando Patterson, a sociologist at Harvard University, argues that the U.S. Census should eliminate racial categories altogether because they have no scientific basis and are racially divisive.

Data can be represented by tables, graphs, and numerical descriptors. We begin by seeing how tables and graphs can reveal patterns in data. We look at a small data set collected by students, and then examine a large data set of the ages of the United States population. By comparing present age distribution to future projections, we can ask how the change in distribution will affect such issues as Social Security.

The search for ways to identify, summarize, and describe patterns in the data can be thought of as a search for a "60 second summary." If you had to describe, in 60 seconds, the main idea you have drawn from the data, what would you say? Creating a one-minute summary may take quite a bit of time. Constructing a brief, succinct description is much more difficult than constructing a lengthy one. Keep in mind that the process is as important as the product. You will need to ask and answer a series of probing questions, focus in on a key idea, and then put your conclusions into writing. The first few sections of this chapter prepare you to construct 60 second summaries of current and projected age data for the United States.

Something to think about

Describe in your own words the difference between quantitative and qualitative variables. Give examples of each. Now look at the class questionnaire in Exploration 1.1 (at the end of the chapter exercises) and identify each variable as quantitative or qualitative.

1.2 READING AND INTERPRETING DATA TABLES

A first step in organizing raw data is to create a table. We start with a simple data table and look for ways to summarize the numbers it contains.

Table 1.1 displays data collected in a survey of one section of college algebra at the University of Massachusetts at Boston. The first column lists values from 18 years to 46 years for the variable "age." The second column gives the number of people for each age and represents the *frequency* or *frequency count* for each age. For example, in this sample, there are one 18-year-old, one 19-year-old, three 20-year-olds, etc.

---

[1] *Data* is the plural of the Latin word *datum* (meaning something given). Hence one datum, two data.

The information on students' ages in Table 1.1 can be condensed by compressing the data into intervals of 2 or more years. The first step is to decide on the interval size and where the first interval starts. We can then count the number of people that fall into each of the intervals. For example, in Table 1.2, we constructed 3-year intervals and started with 18- to 20-year-olds. There are 5 people in the 18- to 20-year interval; 5 people in the 21- to 23-year interval; 1 person in the 24- to 26-year interval, etc. Note that the interval from 18 to 20 years includes those people who were 18, 19, or 20 years old.

Table 1.2

Ages of Students in Three-Year Intervals

Age Interval	Frequency Count
18–20	5
21–23	5
24–26	1
27–29	3
30–32	2
33–35	1
36–38	1
39–41	2
42–44	0
45–47	1
Total	**21**

Table 1.1

Ages of Students in UMass Class

Age (years)	Frequency Count
18	1
19	1
20	3
21	1
22	1
23	3
24	0
25	0
26	1
27	1
28	1
29	1
30	2
31	0
32	0
33	1
34	0
35	0
36	1
37	0
38	0
39	1
40	0
41	1
42	0
43	0
44	0
45	0
46	1
Total	**21**

Converting Frequency Counts into Percentages

In order to get a better sense of the relative number of students in each age interval, we can compare the frequency count in each interval to the total frequency count for the sample. Such a comparison is called a *relative frequency* and is expressed as either a decimal or a percentage.

> The *relative frequency* of any value or interval of values is the fraction (often expressed as a percentage) of all the data in the sample having that value or lying in that interval.

The relative frequency for each age interval in our data set can be found by dividing the number of people in each age interval by the total number of people in the class. For example, there were 5 people in the first age interval of 18 to 20 years and the total number of students in the class was 21. How do we express the relative frequency of 18- to 20-year-olds as a percentage of the students in the class?

divide 5 by 21	$5/21 = 0.238$
round to two places	$= 0.24$
convert to % by multiplying by 100%	$(0.24)(100\%) = 24\%$

Table 1.3 is the age data table with an additional column for the relative frequency (expressed as a percentage) of students in each age interval.

Table 1.3		
Student Ages with Relative Frequency		
Age Interval	Frequency Count	Relative Frequency (%)*
18–20	5	24
21–23	5	24
24–26	1	5
27–29	3	14
30–32	2	10
33–35	1	5
36–38	1	5
39–41	2	10
42–44	0	0
45–47	1	5
Total	**21**	**100**

* Since we rounded off each of the percents, the percentages actually add to 102%. But 21 students represents 100% of our class.

Relative frequencies are important since they can be used to make comparisons among categories of data. For example, if we want to compare this mathematics class to another group, especially if the group is not the same size, we must consider the percentages of students in each particular age interval, not just the total numbers. Why? Suppose you know that 7 students in one section were age 18 and that 1235 students in the college were age 18. This doesn't tell you a great deal because you don't know the total number of students in the class or the college. But if you know that 7 out of 35 students (or 20%) of the class are 18, and 1235 out of 4231 students (or about 29%) of all the students in the college are 18, then the numbers are seen relative to the size of the class and the population of the college. We say that the relative frequency of 18-year-olds in the class is 20%, and the relative frequency of 18-year-olds in the college as a whole is 29%. We now know that the class has a smaller number of 18-year-olds compared to the college as a whole.

Summarizing Data to Reveal Patterns

Stop and look at Table 1.4 for a minute. It displays information collected by the U.S. Census Bureau on the age and sex of people in the United States in 1991. Note that the counts are all in thousands. For example, the total population of the United States is listed as approximately 252,177 thousand or, equivalently, 252,177,000. The table presents frequency counts in different ways. There are frequency counts of the total population and of males and females separately, both in 1-year intervals and in 5-year intervals. The frequency counts for the 5-year intervals are in bold type. In a large data table like this, it can be difficult to recognize patterns. But by grouping the data and condensing the details, some overall patterns start to emerge, and we have taken the first step toward a 60 second summary.

Table 1.4

Resident US Population, by Sex and Age: 1991 (In thousands, except as indicated. As of July1.)

Age	Total	Male	Female	Age	Total	Male	Female
Total	**252,177**	**122,979**	**129,198**				
Under 5 years old	**19,222**	**9,836**	**9,386**	**45 to 49 years old**	**14,094**	**6,907**	**7,188**
Under 1 yr	4,011	2,052	1,959	45 years old	2,832	1,392	1,440
1 year old	3,969	2,030	1,938	46 years old	2,821	1,384	1,437
2 years old	3,806	1,949	1,857	47 years old	2,848	1,392	1,456
3 years old	3,718	1,902	1,816	48 years old	2,856	1,396	1,459
4 years old	3,717	1,902	1,815	49 years old	2,737	1,342	1,395
5 to 9 years old	**18,237**	**9,337**	**8,900**	**50 to 54 years old**	**11,645**	**5,656**	**5,989**
5 years old	3,702	1,897	1,806	50 years old	2,528	1,234	1,294
6 years old	3,681	1,884	1,797	51 years old	2,340	1,139	1,200
7 years old	3,575	1,829	1,746	52 years old	2,298	1,116	1,182
8 years old	3,512	1,797	1,715	53 years old	2,280	1,105	1,175
9 years old	3,767	1,930	1,836	54 years old	2,200	1,063	1,137
10 to 14 years old	**17,671**	**9,051**	**8,620**	**55 to 59 years old**	**10,423**	**4,987**	**5,436**
10 years old	3,703	1,899	1,804	55 years old	2,129	1,023	1,106
11 years old	3,662	1,875	1,786	56 years old	2,195	1,054	1,141
12 years old	3,484	1,783	1,701	57 years old	2,068	991	1,077
13 years old	3,414	1,746	1,668	58 years old	1,946	927	1,019
14 years old	3,409	1,747	1,661	59 years old	2,085	992	1,093
15 to 19 years old	**17,205**	**8,834**	**8,371**	**60 to 64 years old**	**10,582**	**4,945**	**5,637**
15 years old	3,293	1,690	1,603	60 years old	2,124	994	1,129
16 years old	3,362	1,732	1,630	61 years old	2,100	994	1,106
17 years old	3,360	1,733	1,627	62 years old	2,076	971	1,105
18 years old	3,383	1,733	1,650	63 years old	2,145	1,004	1,141
19 years old	3,808	1,948	1,860	64 years old	2,138	982	1,156
20 to 24 years old	**19,194**	**9,775**	**9,419**	**65 to 69 years old**	**10,037**	**4,491**	**5,546**
20 years old	4,080	2,087	1,992	65 years old	2,072	939	1,132
21 years old	3,969	2,029	1,940	66 years old	2,069	933	1,136
22 years old	3,732	1,902	1,829	67 years old	2,026	910	1,116
23 years old	3,633	1,845	1,788	68 years old	1,911	848	1,063
24 years old	3,781	1,912	1,869	69 years old	1,960	862	1,099
25 to 29 years old	**20,718**	**10,393**	**10,325**	**70 to 74 years old**	**8,242**	**3,531**	**4,712**
25 years old	3,837	1,935	1,902	70 years old	1,886	828	1,058
26 years old	4,042	2,026	2,016	71 years old	1,731	750	981
27 years old	4,217	2,116	2,101	72 years old	1,649	710	939
28 years old	4,063	2,033	2,030	73 years old	1,518	637	881
29 years old	4,559	2,283	2,276	74 years old	1,458	606	852
30 to 34 years old	**22,159**	**11,034**	**11,125**	**75 to 79 years old**	**6,279**	**2,482**	**3,797**
30 years old	4,482	2,234	2,247	75 years old	1,405	576	829
31 years old	4,414	2,200	2,214	76 years old	1,334	540	795
32 years old	4,371	2,173	2,197	77 years old	1,244	491	753
33 years old	4,372	2,169	2,203	78 years old	1,204	462	742
34 years old	4,520	2,258	2,263	79 years old	1,091	414	677
35 to 39 years old	**20,518**	**10,174**	**10,344**	**80 to 84 years old**	**4,035**	**1,406**	**2,629**
35 years old	4,312	2,142	2,170	80 years old	970	352	618
36 years old	4,189	2,079	2,109	81 years old	883	314	568
37 years old	4,120	2,040	2,080	82 years old	799	279	521
38 years old	3,778	1,867	1,912	83 years old	739	249	489
39 years old	4,119	2,047	2,072	84 years old	644	211	433
40 to 44 years old	**18,754**	**9,258**	**9,496**	**85 to 89 years old**	**2,090**	**625**	**1,465**
40 years old	3,829	1,891	1,938	**90 to 94 years old**	**812**	**201**	**611**
41 years old	3,716	1,833	1,883	**95 to 99 years old**	**214**	**45**	**169**
42 years old	3,629	1,788	1,841	**100 years and over**	**44**	**10**	**34**
43 years old	3,585	1,761	1,824	**Median age (yrs)**	**33.1**	**31.9**	**34.3**
44 years old	3,996	1,986	2,010				

Source: U.S. Bureau of the Census, Current Population Reports, P25–1095, in *The American Almanac 1993–1994: Statistical Abstract of the United States* (Austin, Texas: The Reference Press).

* See Excel file USAGE1A5. A modified version of Table 1.4 (containing ages for 1-year intervals only) appears in the Excel and the graph link file USAGES1.

Exploration 1.4 follows an alternate path to a 60 second summary and compares the distribution of ages for men versus women.

Table 1.5 shows only the data on 5-year age intervals for the total population from Table 1.4. By focusing on one aspect of the data, and then simplifying details, some patterns become recognizable.

In Table 1.5 the first and third columns display age intervals of 5 years and the second and fourth columns display the frequency counts for those intervals. For example, the interval of 5 to 9 years of age includes the end points of 5 years and 9 years; thus the interval includes all those people who are 5, 6, 7, 8, or 9 years of age. There are approximately 18,237,000 people in the 5- to 9-year interval.

Examine the table closely. What 5-year interval has the highest frequency count? As we look at the age intervals from youngest to oldest, when is the frequency count increasing? When is it decreasing?

We could further condense the data by compressing them into 10-year intervals and calculating the relative frequencies as shown in Table 1.6. What patterns can now be seen? What information has been lost? Is there any advantage to using these larger intervals?

Table 1.5

Ages of 1991 U.S. Population in 5-Year Intervals

Age	Frequency Count in 1000s	Age	Frequency Count in 1000s
Under 5	19,222	55 to 59	10,423
5 to 9	18,237	60 to 64	10,582
10 to 14	17,671	65 to 69	10,037
15 to 19	17,205	70 to 74	8,242
20 to 24	19,194	75 to 79	6,279
25 to 29	20,718	80 to 84	4,035
30 to 34	22,159	85 to 89	2,090
35 to 39	20,518	90 to 94	812
40 to 44	18,754	95 to 99	214
45 to 49	14,094	100 & over	44
50 to 54	11,645	**Total**	**252,177**

Source: U.S. Bureau of the Census, Current Population Reports, P25–1095, in *The American Almanac 1993–1994: Statistical Abstract of the United States* (Austin, Texas: The Reference Press).
Notes: Includes armed forces abroad. Frequency count may not be exact because of rounding.

Table 1.6

Ages of 1991 U.S. Population in 10-year Intervals

Age	Frequency Count in 1000s	Relative Frequency (%)*
Under 10	37,459	15
10 to19	34,876	14
20 to 29	39,912	16
30 to 39	42,677	17
40 to 49	32,848	13
50 to 59	22,068	9
60 to 69	20,619	8
70 to 79	14,521	6
80 and over	7,195	3
Total	**252,177**	**100**

Based on data from: U.S. Bureau of the Census, Current Population Reports, P25–1095, in *The American Almanac 1993–1994: Statistical Abstract of the United States* (Austin, Texas: The Reference Press).
Note: Includes armed forces abroad.

* Frequency count and relative frequency total may not be exact because of rounding.

AN INTRODUCTION TO ALGEBRA AEROBICS

At the end of most sections in the text there are short "Algebra Aerobics" workouts with answers in the back of the book. They are intended to give you practice with the mechanical skills introduced in the prior section, and to review other skills we assume you have learned in previous courses. In general the problems shouldn't take you more than a few minutes to do. We recommend reading the text with pencil in hand, working out these practice problems, and then checking your solutions in the back of the book. The Algebra Aerobics are numbered according to the section number of the chapter in which they occur.

Algebra Aerobics 1.2

1. In Table 1.3, what percentage of students is under 30 years old?
2. Using Table 1.5, calculate the percentage of the population that is 50 to 64 years old.
3. Fill in Table 1.7.

Table 1.7		
Age	Frequency Count	Relative Frequency (%)
1–20	52	
21–40	35	
41–60	28	
61–80		
Total	**137**	

1.3 VISUALIZING DATA

Humans are visual creatures. By converting a data table into a picture, we are able to recognize patterns that may be hard to discern from a list of numbers. There are many different ways to display data. The type of graph or chart that is used influences which aspects of the data emerge and which remain hidden. In this chapter we focus on *histograms,* which show how data for quantitative variables are distributed over a range of values. A histogram is a graph on which the horizontal axis represents a section of the number line. The horizontal axis is usually marked off in equal-sized intervals to facilitate comparisons among intervals. The histograms in Figures 1.1 through 1.4 have been constructed from the previous tables of data on U.S. population. They all show relative frequencies, but each uses a different interval size on the horizontal axis. In Figures 1.1 to 1.4, each bar of the histogram includes the left end point of the interval and excludes the right end point of the interval. For example, the second bar in Figure 1.2 represents the percentage of people who are 5, 6, 7, and 8 years old but does not include 10-year-olds.

The program "F1: Histograms" in FAM 1000 Census Data *allows you easily to construct histograms for several different variables using 1996 U.S. Census data.*

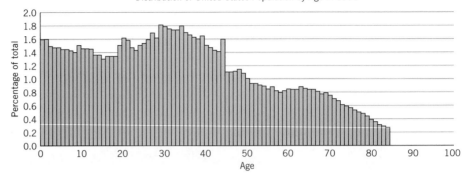

Figure 1.1   1991 population—interval size 1 (Data in one year intervals not available for 85 yrs and older.)

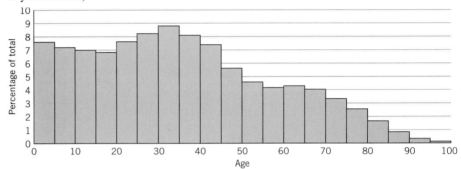

Figure 1.2   1991 population—interval size 5.

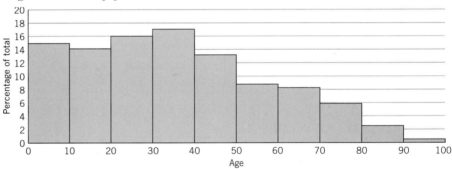

Figure 1.3   1991 population—interval size 10.

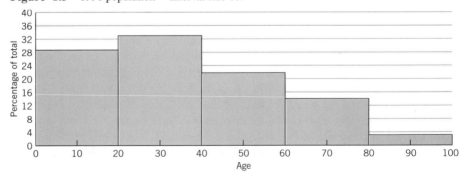

Figure 1.4   1991 population—interval size 20.

Source: U.S. Bureau of the Census, Current Population Reports, P25-1095, in *The American Almanac 1993–1994: Statistical Abstract of the United States* (Austin, Texas: The Reference Press).

Trade-offs in Representing Data

The size of the interval affects what type of information is revealed or hidden in a histogram of the data. When condensing data into larger intervals, subjective decisions must be made. With a large number of small intervals, the complexity of the image can be confusing. With a small number of large intervals, meaningful information may be lost. For example, the slight rise in the percentage of people in the 60 to 64 group compared to the 55 to 59 group in Figure 1.2 is lost in the histogram with 10-year intervals in Figure 1.3. The decline in the percentage of people in the 10 to 19 group compared to the 0 to 9 group in Figure 1.3 is lost using 20-year intervals in Figure 1.4.

The type of graph we choose to use can also affect what aspects of the data are highlighted. The pie chart (Figure 1.5) is a common graph used in newspapers and magazine articles to display relative frequencies.

An Important Aside: What a Good Graph Should Include

When you encounter a graph in a newspaper or you produce one for class, there are three elements that should always be present:

1. An informative title that succinctly describes the graph
2. Clearly labeled axes (or legend) including the units of measurement. (For example, whether age is being measured in years or months.)
3. The source of the data, cited in either the data table, in the text, or on the graph.

Jottings for a 60 Second Summary

When data are summarized, information is lost but certain patterns may be revealed. Thus, it is helpful to weave back and forth between tables and graphs in the search for patterns in the data. We start by recording some observations, focusing on patterns that may be useful for constructing a 60 second summary.

Questions and observations:

What can we notice about all four histograms?

- They all show a general overall decline in the percentage of people, interrupted by a large bump occurring between about 20 and 45 years (the baby boom?).

What can we say about this bump?

- On the 1-year interval histogram there seem to be few peaks, with the highest peaks around 30 and 35 years. Checking back in Table 1.4, we can determine that there are more people 29 years old (about 4,559,000 people) than any other age.
- On the 5-year interval histogram, the peak occurs in the 30- to 34-year interval.
- On the 10-year interval histogram, the peak occurs in the 30- to 39-year interval.

Any other observations?

- On the 1-year interval histogram, there is a sharp drop at about age 45.

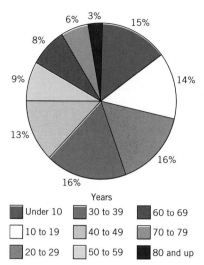

Figure 1.5 Percentage distribution of ages (by 10-year intervals) for the U.S. population in 1991.

Based on data from: U.S. Bureau of the Census, Current Population Reports, P25-1095, in *The American Almanac 1993–1994: Statistical Abstract of the United States* (Austin, Texas: The Reference Press).

Something to think about

In what ways is the pie chart easier to read and interpret than the histogram for 10-year age intervals? What type of information is easiest to see in the histogram? What are some trade-offs in using pie charts versus histograms?

- On the 5-year interval histogram, between 0 and 19, the percentage of people in each successive age interval decreases; between 15 and 34, the percentage of people in each successive age interval starts increasing, reaching a peak in the 30- to 34-year interval; after 34 years, there is a steady decline in each interval except for the 60- to 64-year interval.

- On the 10-year interval table, the biggest drops (each of about 10,000,000 people) seem to occur between those in their thirties and forties, and those in their forties and fifties.

Take a moment to jot down in your class notebook any patterns you find in the graphs and tables that might be useful for a 60 second summary.

As we examine the graphs and tables, questions about the data begin to emerge. Sometimes questions may require additional calculations. For instance, you may be interested in the percentage of the population that is of working age or the percentage that is retired. Some questions may require more data or research. For example, how will the percentage breakdown between working age and retired shift over the next 50 years? Interesting questions should be noted in your jottings.

Algebra Aerobics 1.3

1. Use Table 1.8 to create a histogram and a pie chart.

Table 1.8

Age	Frequency Count	Relative Frequency (%)
1–20	52	38
21–40	35	26
41–60	28	20
61–80	22	16
Total	**137**	**100**

mate the relative frequencies from the graph and then calculate the frequency count in each interval.)

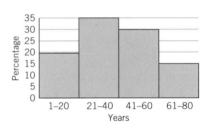

Figure 1.6   Distribution of ages (in years).

2. From the histogram in Figure 1.6, create a frequency distribution table. Assume that the total number of people represented by the histogram is 1352. (Hint: Estimate the relative frequencies from the graph and then calculate the frequency count in each interval.)

3. Use Figure 1.4 to estimate the percentage of people in the U.S. in 1991 who were between 20 and 59 years old.

1.4 WHAT IS AVERAGE ANYWAY?

In addition to graphs and tables, numerical descriptors are often used to summarize data. Three numerical descriptors, called the *mean, median,* and *mode,* offer different ways to describe and compare data sets. Each descriptor can be thought of as a "one second summary" of data. Although all of these terms are sometimes referred to as *average,* technically only the mean should be called the average.[2]

Numerical Descriptors

The mean   The *mean* is the arithmetic average, with which you are probably already familiar.

> The *mean* of a list of numbers is their sum divided by the number of terms in the list.

The following set of data extracted from Table 1.1 contains the ages (in years) of students in a college algebra class.

18, 19, 20, 20, 20, 21, 22, 23, 23, 23, 26, 27, 28, 29, 30, 30, 33, 36, 39, 41, 46

The mean age can be found by adding together all of the 21 observed ages and dividing the resulting sum by 21, the total number of terms in the list.

$$(18 + 19 + 20 + 20 + 20 + 21 + 22 + 23 + 23 + 23 + 26 + 27 + 28 + 29 + 30 + 30 + 33 + 36 + 39 + 41 + 46) = 574$$

So the mean is

$$574/21 = 27.33$$

According to an article in the electronic magazine CHANCE News, *the mean is not always correctly used in the popular press.*

To describe this process in general, we can represent the age of each student by a letter with a numerical subscript. Thus, for a class with n students, the ages can be represented as $a_1, a_2, a_3, \ldots, a_n$.

The mean of n values, $a_1, a_2, a_3, \ldots, a_n$, is found by adding these values and dividing the sum by n, the number of values.

$$\text{mean} = \frac{(a_1 + a_2 + a_3 \cdots + a_n)}{n}$$

The Greek letter Σ (called sigma) is used to represent the sum of all of the terms of a certain group. Thus $a_1 + a_2 + a_3 \cdots + a_n$ can be written as

$$\sum_{i=1}^{n} a_i$$

which means to add together all of the values of a_i from a_1 to a_n.

$$\sum_{i=1}^{n} a_i = a_1 + a_2 \cdots + a_n$$

The mean can be expressed as

$$\text{mean} = \frac{(a_1 + a_2 + a_3 \cdots + a_n)}{n} = \frac{\sum_{i=1}^{n} a_i}{n}$$

---

[2] The word "average" has an interesting derivation according to Klein's etymological dictionary. It comes originally from the Arabic word "awariyan" which means merchandise damaged by seawater. The idea being debated was that if your ships arrived with water-damaged merchandise, should you have to bear all the losses yourself, or should they be spread around or "averaged" among all the other merchants? The words *"avería"* in Spanish, *"avaría"* in Italian, *"avarie"* in French still mean "damage."

The median   The *median* locates the "middle" of a numerically ordered list. More formally:

> The *median* separates a numerically ordered list of numbers into two parts, with half the numbers at or below the median, and half at or above the median. If the number of observations is odd, the median is the middle number in the ordered list. If the number of observations is even, the median is the mean of the two middle numbers in the ordered list.

In the previous ordered list of ages, there are 21 numbers. Since the list has an odd number of terms, the median is the middle or 11th number, which is 26 years.

18, 19, 20, 20, 20, 21, 22, 23, 23, 23, **26**, 27, 28, 29, 30, 30, 33, 36, 39, 41, 46

median = 26

See the reading "The Median Isn't the Message," to find out how an understanding of the median gave renewed hope to the renowned scientist, Stephen J. Gould, when he was diagnosed with cancer.

If one of the 23-year-old people dropped out of the class, the list would contain the ages of only 20 students. Since that list now has an even number of terms, the median would be the mean of the middle two numbers, in this case the 10th and 11th numbers. So the median would be the mean of 26 and 27, which equals (26 + 27)/2 or 26.5

18, 19, 20, 20, 20, 21, 22, 23, 23, **26, 27,** 28, 29, 30, 30, 33, 36, 39, 41, 46

median = (26 + 27)/2 = 26.5

Neither the mean nor the median needs to be one of the numbers in the original list.

The mode   The value that occurs most frequently is called the *mode* and therefore, it is one of the numbers contained in the list. In the above class list, without one of the 23-year-olds, the mode is 20. If there is more than one such value, the data set is *multimodal*. In the original class data of 21 students, there are 3 people who are 20, and 3 people who are 23; thus both 20 and 23 are modes. This data set is said to be *multimodal* or in this particular case where there are two modes, *bimodal*. Of the three measures, mean, median, and mode, the mode is the least used.

> The *mode* is the number that occurs with the greatest frequency. If data are grouped into intervals, then the interval with the greatest frequency count is called the *modal interval.*

Something to think about

1. If someone tells you that in his town "all of the children are above average," you might be skeptical. (This is called the "Lake Wobegon effect.") But could most (more than half) of the children be above average? Explain.

2. Why do you think most researchers use median rather than mean income when studying "typical" households?

Since the word "average" is sometimes used incorrectly to represent any one of these three terms, it is often not clear what is meant by a statement such as "the *average* salary for an American is $23,000." Is $23,000 the mean—the result of adding together the salaries of all Americans and then dividing by the total number of people whose salaries were counted? Does it mean that $23,000 is the median—that half of Americans make less than $23,000 and half make more? Or is $23,000 the mode—if you rounded off salaries to the nearest thousand, would more Americans make $23,000 than any other salary? Since the term "average" is used so loosely, it is important to ask how it is being used.

Significance of the Mean, Median, and Mode

Mean, median, and mode are all useful in different ways. The mean and median are ways to describe the "center" of the data and are called *measures of central tendency.*

The *mode* is the most common value in a set of data. The term "mode" can be thought of as the most popular data value, or the one that is most in fashion. We can look at the course of the mode as it moves through time like the peak of a wave. One example is the "baby boom," the explosion in the number of babies that occurred just after World War II. College admissions officers scrutinize population histograms to plan for advancing waves of students. As a middle-aged population peak advances into old age, society needs to plan to be able to meet the housing, health care, and other needs of this group of people. A moving peak also has a negative side: If a country has built additional schools for an incoming wave of high-school and college students, what do they do with the facilities when the crest has passed by, and the student population falls? Understanding the mode and how it might change is important in both short- and long-range planning.

The special significance of the *median* is that it divides the number of entries in the data set into two equal halves. If the median age in a large urban housing project is 17, then half the population is 17 or under. Hence issues of day care, recreation, supervision of minors, and kindergarten through 12th grade education should be very high priorities with the management. The other half of the population is, of course, 17 or older. If the adults in that same project included a fairly large proportion of elders, the *mean* age could be 35; by itself this number would give no indication that a majority of the residents were minors.

The major disadvantage of the median is that it is unchanged by the redistribution of values above and below the median. For example, as long as the median income is larger than the poverty level, it will remain the same even if all poor people suddenly increase their incomes up to that level and everyone else's income remains the same.

The *mean* is the most commonly cited statistic in the news media. One advantage of the mean is that it can be used for calculations relating to the whole data set. Suppose a corporation wants to open a new factory similar to its other factories. If the managers know the mean cost of wages and benefits for an employee, they can make an estimate of what it will cost to employ the number of workers needed to run the new factory.

total employee cost = (mean cost for employees) · (number of employees)

The major disadvantage of the mean is that it is affected by *outliers,* extreme values in the data set. John Schwartz comments in a *Washington Post* article ("Mean statistics: When is the average best?" December 6, 1995, p. H7) that if Bill Gates, billionaire chief executive of Microsoft, were to move into a town with 10,000 penniless people, the mean income would be more than a million dollars. This might suggest that the town is full of millionaires, which is obviously far from the truth. The median income, however, is still $0.

The program "F4: Measures of Central Tendency" in the FAM 1000 Census Graphs *can help you understand mean, median, and mode and their relationships to histograms.*

Algebra Aerobics 1.4

1. Calculate the mean, median, and mode for the following data:

    a. $475, $250, $300, $450, $275, $300, $6000, $400, $300

    b. 0.4, 0.3, 0.3, 0.7, 1.2, 0.5, 0.9, 0.4

2. Explain why the mean may be a misleading numerical summary of the data in 1(a).

3. Seven members of an eight-person track team reported the following times (in seconds) for the 100 meter dash: 14.5, 13.1, 15.2, 16.0, 14.7, 12.5, and 13.3.

    If the team's mean time was 14.2 seconds, what was the time of the eighth runner?

1.5 WRITING ABOUT DATA

Whether you are constructing a 60 second summary to present to your class, writing an executive summary for a report to your boss, or simply trying to understand what is going on in your business, you can help your ideas take shape by putting them down on paper. At first your notes may be mere sketches, incomplete and perhaps with errors. As you think through your ideas they will become clearer, stronger, and more focused. Writing a clear summary is difficult. Although the end product may be short, the time invested in creating it can be quite lengthy. The essential steps are: to get started no matter how tentatively, and then to polish and refine your ideas. We offer here one, though certainly not the only, strategy for constructing short, clear summaries.

Constructing a 60 Second Summary of U.S. Age Data

How would you go about constructing a paragraph describing the distribution of ages in the United States? Many different types of summaries are possible. For a one paragraph summary, it is usually best to have a single key point and then to weave together evidence to support that point. The following steps illustrate one process that can be used.

1. *Collect relevant data tables and construct graphs.*
2. *Make jottings.*

    Start by writing a list of initial questions and observations from tables, graphs and numerical descriptors. We have included below some of the previous jottings and have added numerical descriptors.

    What were the general patterns in the data?
    *   a general steady decline in percentage of people in each age interval with a large bump roughly between ages 20 and 45.
    *   on the five-year interval histogram in Figure 1.2, between 0 and 19, the percentage of people in each successive age interval decreased; between 15 and

34, the percentage of people in each successive age interval started increasing, reaching a peak in the 30- to 34-year interval; after 34 years, there was steady decline in each interval (except for the 60- to 64-year interval.)

Any other observations?

- there was a sharp drop at about age 45.

What are the mean, median, and mode for the U.S. population?

- median age is 33.1 years (given at the bottom of Table 1.4, the original U.S. population table on age).

- mode is 29 years, so there are more people 29 years old (about 4,559,000 people) than any other age.

- mean age is not given in Census data in Table 1.4. (Exercise 23 asks you to use Table 1.6 to estimate the mean age for the U.S. population.)

3. *Pose additional questions.*

Exploring the data may raise additional questions such as:

- What percentage of the population are children? What percentage elderly?

- What percentage of the population is in the work force?

- Where does the baby boom that occurred right after World War II show up in the data?

- What caused the sharp drop at age 45?

4. *Identify the key idea.*

The next step is to identify a key idea (out of possibly many ideas). This forces some kind of summary statement about the data. The key idea is often used in the *topic* or *opening* sentence. For example:

> *In 1991 there was a general decline in the percentage of the population in each successive age interval with the exception of a large increase between the ages of 20 and 45 years.*

5. *Provide additional supporting evidence.*

Collect evidence that supports, expands, or modifies the key idea. Our jottings may give enough information or, as in the following, we may have to go back to tables and graphs to find more information.

- about 17% of the population was under 20

- about 13% was 65 or over

- the "baby boomers" were born roughly between 1946 and 1960. Since the data are for 1991, those born between 1946 and 1960 would be between 31 and 45 years old in the data set. So if we think of reading the individual year histogram (Figure 1.1) from right to left, we see that there is a big surge in 45-year-olds (as compared to 46-year-olds). This may signal the beginning of the baby boom. Continuing to read from right to left, the boom seems to peak with the 30-year-olds, perhaps signaling the end of the baby boom. Thus the baby boom may be the cause of the bump, and (reading from left to right) the sharp drop at age 45.

6. *Put ideas together.*

Construct a paragraph defining and defending the key idea. You may wish to start with a topic sentence containing your key idea, followed by supporting evidence that may include a graph or table. You will probably want to weave back and forth among the steps in order to organize your thoughts, then pick and choose the evidence you want to include. You could finish with a concluding sentence. The following is one example of a 60 second summary.

Patterns in the 1991 U.S. Population

In 1991, there was a general decline in the percentage of the population in each successive 5-year interval that was interrupted by a large surge, roughly between the ages of 20 and 45. We can see in the accompanying histogram (Figure 1.7) that from newborns to teenagers, the percentage of people in each succes-

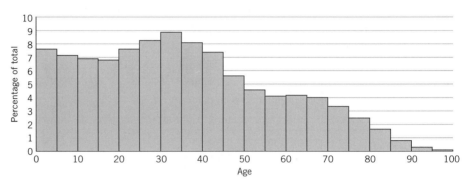

Figure 1.7   1991 U.S. population in 5-year intervals.

sive 5-year interval decreased. Between the ages of 15 and 34, the percentage of people in each successive interval increased, reaching a maximum number of almost 9% for those between 30 and 34 years. The median age of 33.1 is located in this peak. After age 34, there is a return to a steady decline in the number of people for each age group with the exception of a slight rise in the 60- to 64-year age interval. This overall wave pattern may be caused in part by the "baby boom," the surge of births after World War II.

Questions That Require More Research

The previous paragraph offered a description of the age distribution of the U.S. population. Analysis often exposes limitations of the data and generates new questions. Some of the questions raised in the analysis of age data would require more data, such as information about death rates and birth rates, to answer them. Other questions may require research in other disciplines such as history, sociology, medicine, or political science.

- Is the baby boom really the cause of the sharp drop in the population percentage after 45 years of age?

  People 45 years and older in 1991 were born in 1946 or before. Does the drop reflect an increase in the birth rate starting in 1946? Or does it perhaps reflect a sharp increase in the death rate for people at age 46? Could it reflect some historical event such as the Vietnam War that killed many who would have been 45 years old in 1991? Was there an influx in immigrants who are now 45 and under? The number of people at any given age is determined by a number of factors, including the birth rate, the death rate, and immigration patterns.

- How will the distribution of ages in the United States change over time? What will be the impact on America's future as those in the 20- to 45-year population bulge get older?

- How does the distribution of ages in the United States compare with those of other industrialized nations such as Japan or Germany?

- How does the distribution of ages in the United States compare to those of developing countries such as China? (Exercise 25 includes data to help make this comparison.)

These are the kinds of questions asked by social scientists.

Starting With a Question: What Lies Ahead?

Data are usually collected and explored for a specific purpose. Often data analysis is prompted by the need to answer a particular question. One question raised in the previous discussion was:

How will the distribution of ages in the United States change over time?

To create a 60 second summary answering this question, we need estimates for the distribution of ages in the future. In addition to collecting current demographic data, the U.S. Census Bureau makes predictions about the future. Table 1.9 and Figure 1.8 show

Source: Based on data from: U.S. Bureau of the Census, Current Population Reports, P25–1095 and P25–1104, in *The American Almanac 1993–1994 and 1996–1997: Statistical Abstract of the United States* (Austin, Texas: The Reference Press). Note: Includes armed forces abroad.

Relative frequency total may not be exact because of rounding.

Table 1.9
Distribution of Ages of U.S. Population in 1991 and 2050 (Projected)

	Pop. in 1991		Pop. in 2050			Pop. in 1991		Pop. in 2050	
Age	Total (in 1000s)	%	Total (in 1000s)	%	Age	Total (in 1000s)	%	Total (in 1000s)	%
Under 5	19,222	8	25,382	6	60 to 64	10,582	4	20,553	5
5 to 9	18,237	7	25,222	6	65 to 69	10,037	4	18,859	5
10 to 14	17,671	7	25,650	7	70 to 74	8,242	3	15,769	4
15 to 19	17,205	7	25,897	7	75 to 79	6,279	2	14,510	4
20 to 24	19,194	8	25,313	6	80 to 84	4,035	2	12,078	3
25 to 29	20,718	8	24,659	6	85 to 89	2,090	1	9,283	2
30 to 34	22,159	9	24,803	6	90 to 94	812	0	5,779	1
35 to 39	20,518	8	24,316	6	95 to 99	214	0	2,623	1
40 to 44	18,754	7	23,423	6	100 and over	44	0	1,208	0
45 to 49	14,094	6	22,266	6	**Total**	**252,177**	**100**	**392,031**	**100**
50 to 54	11,645	5	22,071	6	**Median age**	**33**		**39**	
55 to 59	10,423	4	22,367	6					

* See Excel or graph link file PROJAGES.

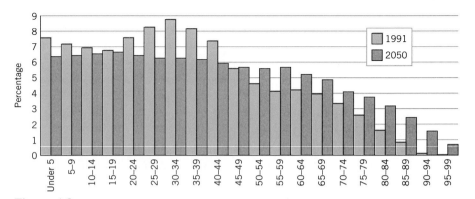

Figure 1.8   Distribution of ages of U.S. population in 1991 and 2050 (projected).

Source: U.S. Bureau of the Census, Current Population Reports, P25-1095 and P25-1104, in *The American Almanac 1993–1994 and 1996–1997: Statistical Abstract of the United States* (Austin, Texas: The Reference Press).

the Bureau's projections for the distribution of ages in the U.S. population in the year 2050.

The first step is to examine the table and graph for patterns in data. The next steps are to make some jottings, identify the key idea, collect supporting evidence, and then start the process of constructing a paragraph that summarizes the findings. The following is a summary written by a student.

Comparison of Age Distribution Between the Year 1991 and the Year 2050

Between 1991 and 2050 the distribution of ages in the U.S. population is expected to shift from a heavy emphasis on the young to a more even spread among the ages. The aging of one of the United States' largest population groups, the baby boomers, will push the average age of the population upward from its original youth-heavy distribution. In 1991 people between the ages of 0 and 44 years made up 69% of the population. By 2050 the same age group will contain only 56%. Those 45 and over in 1991 were 31% of the population, while in 2050 that age bracket is expected to contain 44%. We can see from the graph (Figure 1.8) that in 2050 the peaks of 5-year intervals for the ages of 0 to 70 years form a steady line at around 6% of the total population. The 1991 interval data varies from 3% to 9%, with the largest percentage occurring between the ages of 30 and 34 years. In the 1991 group, the distribution of ages varies dramatically, whereas the 2050 data follow a steady downward curve.

The Search for Answers

Chet Raymo's article "True Nature of Math Remains, in Sum, a Mystery" gives a view of the power of mathematics in understanding the world.

Mathematics provides the tools to describe patterns in populations. It is the job of social scientists to interpret what these patterns mean. Social scientists use these tools to form and test hypotheses, and to reach (perhaps controversial) conclusions. For example, since the U.S. Census Bureau projections indicate that there will be fewer people working, and more people retired, some policy advisers have warned that the United

States is facing a Social Security funding crisis.[3] In a speech to the U.S. Senate, Bill Bradley produced a chart that showed that in 1950 one Social Security beneficiary was supported by 16.5 workers, in 1980 one beneficiary was supported by 3.3 workers, and by 2030 there may be only 2 workers supporting each beneficiary. In *Too Many Promises: The Uncertain Future of Social Security,* Michael Boskin, chief economic adviser to President Bush, argued that the well-off should rely on private pension programs rather than Social Security benefits.

Other social scientists are not so pessimistic. Although everyone seems to agree on the inevitability of an increasingly older population, some point out that the population will shrink at the bottom while it grows at the top. They argue that we should be looking at the total dependent population, which consists of the young as well as the elderly. In 1991, the sum of those under 20 and 65 or over was approximately 41%. In 2050, the corresponding sum is expected to be about 46%. So the percentage of the population requiring support may only increase by about 5%. Federal expenditures for children, however, are roughly one-sixth of the amount spent on adults. Most of the financial support for children comes from their parents. To shift resources from children to the elderly would require drastic and probably unpopular political decisions. The harsh reality is that most adults are far more willing to pay the expenses for their own children than to be taxed to support an elderly population.

1.6 DOES THE WHOLE PROCESS MAKE SENSE?

Whether reading a quantitative argument in the newspaper or constructing your own 60 second summary, the fundamental question to keep in mind is "Does it all make sense?" Any conclusion that claims to be drawn from the analysis of data should be approached with reasoned skepticism. We need constantly to ask questions about how much we trust the data and the conclusions drawn from them.

How Good Are the Data?

- **What definitions were used?**

    During the Great Depression five reputable agencies published five very different unemployment estimates for the same month (Table 1.10). The variance was due primarily to differences in the definitions of unemployment. Some estimates included people leaving school and seeking employment for the first time and some did not; some measured only blue-collar unemployment, whereas others included unemployed professionals.

- **Was the choice of definition biased by the goal? Would another definition have been more useful?**

    Different groups collecting data on the same issue may be gathering information for different purposes. The different definitions used by the above groups are tied to the group's concerns and objectives. Clearly, different arguments about the seri-

---

[3] The following analysis relies heavily on arguments found in a section entitled "Will You Still Feed Me?" in Mark Maier's *The Data Game: Controversies in Social Science Statistics,*" (Armonk, New York: M. E. Sharpe, 1991).

Table 1.10

**Estimates of Unemployment for the Month of November 1935
According to Five Reporting Agencies**

Agency Preparing Estimates	Estimate of Number Unemployed
The National Industrial Conference Board	9,177,000
Government Committee on Economic Security	10,913,000
The American Federation of Labor	10,077,000
National Research League	14,173,000
Labor Research Association	17,029,000

Source: Jerome B. Cohen, "The Misuse of Statistics," *Journal of the American Statistical Association,* XXXIII, No. 204, (1938), p. 657. Reprinted with permission from *Journal of the American Statistical Association.* Copyright ©, 1938 by the American Statistical Association.

ousness of the unemployment problem can be made depending on which estimates you choose to use for the number of unemployed people.

- **What is being measured? Is it relevant?**

  In the nineteenth century, the "science" of craniology equated intelligence with brain size, and hence skull size. It is possible to measure skull size with extreme accuracy, but is it reasonable to believe that head size is related to intelligence? Today, we attempt to measure intelligence using the Intelligence Quotient and Scholastic Aptitude Tests. Many questions have been raised about what these tests really measure. Some people claim that they are twentieth century versions of craniology.

- **Are the measurements accurate and reliable?**

  In May 1993, *The New York Times* published an article entitled "Job Loss in Recession: Scratch Those Figures" explaining measurement errors in employment figures.

  > *Having declared last June that the recent recession had eliminated many more jobs than anyone had realized, the Labor Department is now canceling that view. The department says it overstated by 540,000 the number of jobs that were created in the late 1980's. And then it overstated how many jobs had disappeared in the recession.*[4]

  The measurements used to count jobs were faulty. Among other things, the U.S. Department of Labor counted paychecks instead of the people receiving paychecks. Any person receiving both a regular check and a separate check for overtime was counted twice. Look at Figure 1.9 (page 23) from *The New York Times* article showing the original count and what purports to be a more accurate count.

- **How big is the sample? How was the sample chosen? Is the sample representative of the total population?**

  In the decennial census, the U.S. Census Bureau has attempted to collect data on everyone in the United States. Starting in the year 2000, however, the Bureau will

---

[4] "Job Loss in Recession: Scratch Those Figures." *The New York Times,* Friday, May 7, 1993, p. D1.

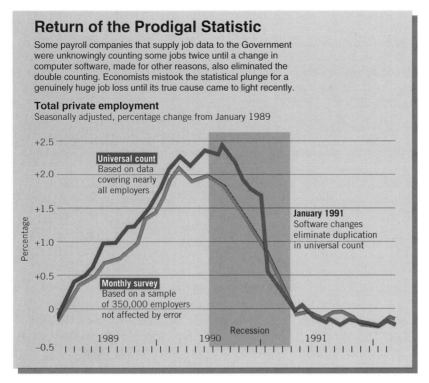

Figure 1.9 Graph from *The New York Times* on federal miscounting of employment numbers.

Source: Bureau of Labor Statistics. Copyright © 1993 by the New York Times Company. Reprinted by permission.

collect data on 90% of the population, and by sampling the remaining 10% extrapolate information on the rest.

In his book *Flaws and Fallacies in Statistical Thinking,* Stephen Campbell cites the classic case of a poorly chosen sample. *The Literary Digest* during the 1936 Presidential election

. . . *mailed out 10,000,000 ballots and had 2,300,000 returned. On the basis of this unusually large sample, it confidently predicted that Alfred M. Landon would win by a comfortable margin. As it turned out, however, Franklin D. Roosevelt received 60 percent of the votes cast, a proportion representing one of the largest majorities in American presidential history. . . . [It turns out that the] ballots were sent primarily to upper-income types . . . as a result of the magazine's having selected names from lists of its own subscribers, and telephone and automobile owners. . . . In this election, the presence of bias rendered the sample, despite its enormous size, inadequate as a representative subset of the population.*[5]

[5] Stephen Campbell, *Flaws and Fallacies in Statistical Thinking,* p. 148, copyright © 1974, Reprinted by permission of Prentice Hall, Inc., Upper Saddle River, New Jersey.

How Good Are the Conclusions?

Edward R. Tufte in *Data Analysis for Politics and Policy* points out that

> *Almost all efforts at data analysis seek, at some point, to generalize the results and extend the reach of the conclusions beyond a particular set of data. The inferential leap may be from past experiences to future ones, from a sample of a population to the whole population, or from a narrow range of a variable to a wider range. The real difficulty is in deciding when the extrapolation beyond the range of the variables is warranted and when it is merely naive. As usual, it is largely a matter of substantive judgment. . . .*

When we read conclusions drawn from data that are, often a matter of judgment, we should ask:

- **Is the analysis accurate?**

  > *Boston researchers disclosed today [Oct. 6, 1993] that they had made a mathematical error in a major study published in July that seemed to provide reassuring news about the risk of breast cancer in women with a family history of the disease. . . . The study of nearly 130,000 women published in the July 21st issue of the* Journal of the American Medical Association *found that only 2.5 percent of all breast cancers occurred in women with a family history of the disease. When corrected, the percentage turned out to be 6 percent, much closer to previous estimates.*[6]

  Apparently the error occurred when a researcher carelessly transcribed numbers from two different tables and used them in a formula. Because the erroneous percentage was so much lower than previous estimates, the report was front page news when it was published. Many women with family histories of breast cancer who had been cheered by the erroneous results were disappointed by the corrected findings.

- **Would alternative conclusions better fit the data?**
  When alternative plausible conclusions are derived from the same data, both conclusions may be wrong or they might both possess some truth. Stephen Campbell argues in his book, *Flaws and Fallacies in Statistical Thinking,* that:

  > *When equally plausible alternative conclusions can be reached from exactly the same statistical evidence, the logical link between evidence and conclusion offered is probably rather weak.*
  > *Fact: "Fifteen individual nations have improved their infant mortality rates more than the United States since 1950."*
  > *Conclusion offered: The state of health care in the United States is inadequate. Equally plausible alternative conclusion: The state of health care in several other countries—probably underdeveloped countries—has improved markedly since 1950, at least with respect to infant mortality.*[7]

---

[6] "Cancer study error found. Key finding unchanged on breast disease risk," *The Boston Globe,* Wednesday, October 6, 1993, p. 1.

[7] Stephen Campbell, *Flaws and Fallacies in Statistical Thinking,* p. 111, copyright © 1974, Reprinted by permission of Prentice Hall, Inc., Upper Saddle River, New Jersey.

- **Are the data selectively reported?**

  In his book *We're Number ONE: Where America Stands—and Falls—In the New World Order,* Andrew Shapiro challenges the assumption that America is "Number One" by showing how "statistics can be found to tell any story." One section is devoted to debunking America's Number One status on spending on education. He claims:

  > *We're Number One in private spending on education.*
  > *But . . . We're Number 17 in public spending on education.*

  > *Including public and private spending, the United States spends about 7 percent of its gross domestic product (GDP) on education annually, or around $330 billion. By this measure, we do fairly well; only a few developed nations spend more overall relative to the size of their economies. But as with health care expenditures, the spending comparisons change dramatically when we distinguish between public and private sources. The United States is first in private spending on education among the nineteen major industrial nations, but only seventeenth in public spending as a percentage of GDP.*[8]

  In most cases you will not have access to the original data used, or have the time and energy to reconstruct the analysis. But you should always ask the kind of questions that we have raised here and critically examine the arguments presented. Keep asking yourself:

  > *Does it all make sense?*

CHAPTER SUMMARY

In statistical terms, characteristics such as height, race, age, and college major that vary from person to person are called *variables*. Information collected about variables is called *data*.

Variables can be classified as *quantitative variables* or *qualitative variables*. Quantitative variables, such as age, height, or number of people in your household can be represented by a number and a unit of measurement (for example, 18 years, 6 feet, or 3 people). Qualitative variables, such as gender or college major, require us to construct categories (such as male or female; physics, English, or psychology).

The *relative frequency* of any value or interval of values is the fraction (often expressed as a percentage) of all the data in the sample having that value or lying in that interval.

Data can be described using tables, graphs (such as *histograms* and *pie charts*), or numbers (such as the *mean, median,* or *mode*).

The *mean* of a list of numbers is their sum divided by the number of terms in the list. The mean is often referred to as the "average."

The *median* separates a numerically ordered list of numbers into two parts, with half the numbers at or below the median and half at or above the median. If the number of observations is odd, the median is the middle number in the ordered list. If the number of observations is even, the median is the mean of the two middle numbers in the ordered list.

[8] Andrew Shapiro, *We're Number ONE: Where America Stands—and Falls—In the New World Order* (New York: Random House, 1992).

The *mode* is the number that occurs with the greatest frequency. If data are grouped into intervals, then the interval with the greatest frequency count is called the *modal interval*. Modes indicate peaks in the data. If there are two distinct peaks the data are said to be *bimodal*. In general, data with more than one distinct peak are called *multimodal*.

Outliers are extreme, nontypical values in a data set.

A good graph contains:

1. An informative title
2. Clearly labeled axes (or legend) including the units of measurement
3. The source of the data

A "60 second summary" gives a concise and focused description of patterns found in a data set.

EXERCISES

1. If you are using technology in your classroom, you need to become familiar with the features of your particular machine. This may mean completing a computer tutorial on the general features of an IBM or Mac, or working assignments on a graphing calculator. If appropriate, you may want to refer to the basic set of instructions for the TI-82 and TI-83 contained in the Graphing Calculator Workbook. The instructions cover, chapter by chapter, the minimal TI-82 or TI-83 skills needed to do some of the chapter exercises and Explorations. Your instructor will be able to guide you.

2. The following bar chart shows the predictions of the U.S. Census Bureau about the future racial composition of American society.

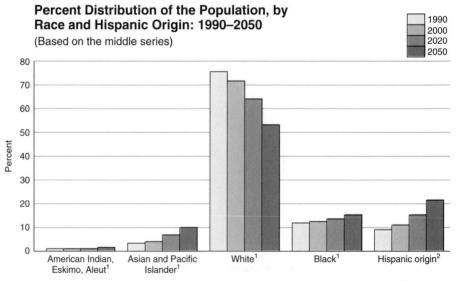

Percent Distribution of the Population, by Race and Hispanic Origin: 1990–2050
(Based on the middle series)

[1]Non-Hispanic   [2]Persons of Hispanic origin may be of any race.

Source: U.S. Bureau of the Census, Current Population Reports, P25-1095, in *The American Almanac 1993–1994: Statistical Abstract of the United States* (Austin, Texas: The Reference Press).

a. Estimate the following percentages:

American Indian, Eskimo, Aleutian in the year 2050

Combined white and black population in the year 2020

Non-Hispanic population in the year 1990

b. Write a topic sentence describing the overall trend.

3. This pie chart of America's spending patterns appeared in *The New York Times,* Thursday, Jan. 20, 1994.

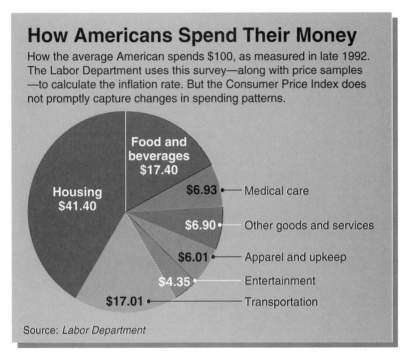

How Americans Spend Their Money

How the average American spends $100, as measured in late 1992. The Labor Department uses this survey—along with price samples —to calculate the inflation rate. But the Consumer Price Index does not promptly capture changes in spending patterns.

Food and beverages $17.40

Housing $41.40

$6.93 — Medical care

$6.90 — Other goods and services

$6.01 — Apparel and upkeep

$4.35 — Entertainment

$17.01 — Transportation

Source: *Labor Department*

Source: Copyright © 1994 by The New York Times Company. Reprinted by permission.

a. In what single category do Americans spend the largest percentage of their income? What is this percentage?

b. If an American family had an income of $30,000, according to this chart how much of it would be spent on food and beverages?

c. If you were to write a newspaper article to accompany this pie chart, what would your opening topic sentence be?

4. Shown here is a table of salaries taken from a survey of recent graduates (with bachelor degrees) from a well-known university in Pittsburgh.

a. How many graduates were surveyed?

b. Is this quantitative or qualitative data? Explain why.

c. What is the relative frequency of people having a salary between $26,000 and $30,000?

d. Draw a histogram of the data.

Salary (in thousands)	Number
11–15	2
16–20	5
21–25	20
26–30	10
31–35	6
36–40	1

5. Population pyramids are a type of graph used to depict the overall age structure of a society.

   a. Using the accompanying population pyramids estimate the percent of:

   Males between the ages 35 and 39 in year 2000

   Females between the ages 55 and 59 in year 2000

   Males 85 and over in year 2050; females 85 and over in year 2050

   All males and females between the ages of 0 and 9 in year 2050

   b. The largest proportions of people are indicated by the greatest bulges in the pyramid. Briefly describe what these bulges suggest about trends occurring in American society.

Age Distribution of the U.S. Population, by Sex: 2000

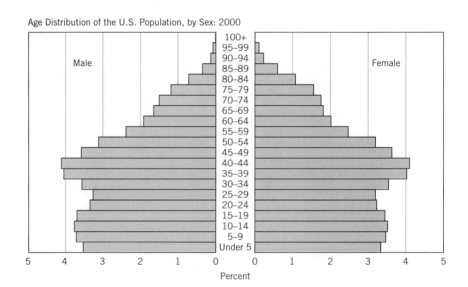

Age Distribution of the U.S. Population, by Sex: 2050

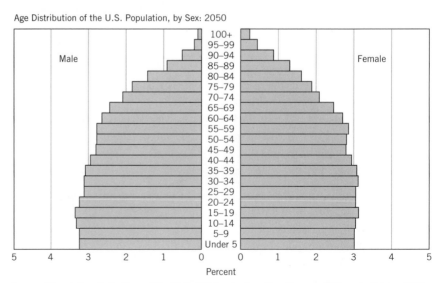

Source: "Population Projections of the United States, by Age, Sex, Race, and Hispanic Origin: 1992 to 2050," in U. S. Bureau of the Census Current Population Reports, P25-1092, November 1992.

6. This bar graph shows the distribution of tornadoes in Illinois by month for the years 1870–1910. Write a paragraph that summarizes the information in the graph.

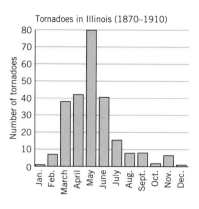

Tornadoes in Illinois (1870–1910)

7. The diagram below shows the percentage of jet crashes occuring in each phase of a flight.

    a. Does the chart include information about all worldwide jet crashes during 1959–1993? How do you know?

    b. During what phase of a flight is a crash most likely to occur? Least likely to occur?

    c. Can you tell from the diagram how many crashes occurred worldwide from 1959 to 1993? Why or why not?

    d. List 2 or 3 features other than those mentioned in parts (b) and (c) that the diagram does or doesn't tell you about airline crashes.

When Crashes Occur

Percentage of jet crashes worldwide occuring in each phase of a flight. Statistics are for 1959–93, and exclude crashes caused by sabotage or military action.

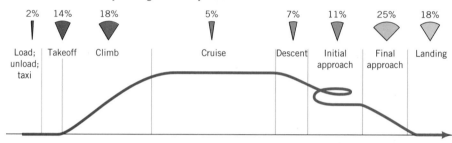

Source: Boeing; rpt. in "When Crashes Occur," *The New York Times,* Dec. 18, 1994: 36. Copyright © 1994 by the New York Times Company. Reprinted by permission.

8. According to *The Arkansas Democrat-Gazette,* Dec. 2, 1995 these school districts reported the following college remediation rates (the percentage of college students from each district taking remedial courses).

Alma: 52.9%	Eudora: 86.7%	Little Rock: 61.8%
Arkadelphis: 56.1%	Fountain Hill: 100%	Maynard: 15%
Calico Rock: 15.2%	Hamburg: 69.8%	Pea Ridge: 33.3%
Deer: 25%	Lead Hill: 50%	Russellville: 36%

Construct a histogram illustrating these percentages using intervals of 25 (i.e., 1–25%; 26–50%; 51–75%; 76–100%).

9. Try looking for some interesting graphs of data in the newspaper or in textbooks from other courses and bring them into class. Why did you find them interesting? What do the graphs tell you about the data? What questions are left unanswered by the graphs?

10. Compute the mean, median, and mode for the following sets of numbers:

    a. 6, 1, -4, 7, 10, 1, 13

    b. Use numbers in part (a) but replace 13 with 1014.

    c. 9.0, -2.3, 9.0, 4.5, 5.3, 4.5

11. Herb Caen, who was a Pulitzer Prize–winning columnist for the *San Francisco Chronicle,* remarked that a person moving from state A to state B could raise the average IQ in both states. Is he right? Explain.

12. Suppose you are waiting in the checkout line at the grocery store, and you suddenly wonder if you have enough cash. You count $15 in your pocket and 7 items in your basket. Estimate the largest "mean" item price you will be able to afford. (Sometimes, just by running your eyes over your basket, you can quickly guess whether the mean price of your selections is above or below the affordable value without adding up all the prices.)

13. **a.** Compute the mean, median, and mode for the list: 5, 18, 22, 46, 80, 105, 110.

    b. Now change one of the entries in the list so that the median stays the same and the mean increases.

14. Suppose that a church congregation has 100 members, each of whom tithes (donates) 10% of his or her income to the church. The church collected $250,000 last year from its members.

    a. What was the mean contribution of its members?

    b. What was the mean income of its members?

    c. Can you predict the median or the mode income of its members?

15. Read Stephen Jay Gould's article "The Median Isn't the Message" and explain how an understanding of statistics brought hope to a cancer victim. (P.S.— Stephen Jay Gould is alive and well today.)

16. **a.** On the first quiz (worth 25 points) given in a section of college algebra, 1 person received a score of 16, 2 people got 18, 1 got 21, 3 got 22, 1 got 23 and 1 got 25. What were the mean, median, and mode of the quiz scores for this group of students?

    b. On the second quiz (again worth 25 points), the scores for 8 of the students were: 16, 17, 18, 20, 22, 23, 25, and 25.

       i. If the mean of the scores for the 9 students was 21, then what was the missing score?

       ii. If the median of the scores was 22, then what are possible scores for the missing ninth student?

       iii. If the data were bimodal and one mode was 25, then what are possible scores for the ninth student?

17. **a.** Using Σ notation, write an algebraic expression for the mean of the 5 numbers: x_1, x_2, x_3, x_4, x_5.

    b. Using Σ notation, write an algebraic expression for the mean of n numbers: $t_1, t_2, t_3, \ldots t_n$.

18. Consider the data given in the table on page 31 describing wolves captured and radio-collared in Montana and southern British Columbia, Canada, in 1993.

    a. Look at the variables: sex, weight, and age. Identify each variable as quantitative or qualitative.

    b. Calculate the mean, median, and mode weights of female wolves, of male wolves, and of pups.

    c. Discuss which measurement of central tendency (mean, median, or mode) might be more appropriate.

d. Estimate the mean age of the females, of the males, and of the total population.

e. Write a 60 second summary of this wolf population.

Pack	Capture Date	Wolf Number	Eartag Number	Sex	Weight (Pounds)	Age (Years)
Spruce Creek	6/14	9318	102103	M	100	2–3
	6/16	9381	104105	F	73	1–2
North Camas	6/2	9375	none	F	77	4–5
	6/4	9378	100101	F	72	3–4
	6/4	9376	9596	F	66	2–3
	6/4	9377	9798	F	63	1–2
	6/9	9379	none	F	59	?
	6/11	9380	none	F	59	1
South Camas	5/27	9474	9091	F	61	1
	5/29	9317	9293	M	95	2–3
	5/30	8756	2627	F	77	6
Murphy Lake	6/22	1718	1718	F	73	3
	6/23	2627	2627	F	21	pup
	6/23	3637	3637	M	96	4–5
	10/15	2223	2223	F	62	pup
Sawtooth	2/26	8808	8808	M	122	5
	9/22	2829	2829	F	50	pup
	9/25	3031	3031	F	53	pup
Ninemile	8/27	4243	4243	F	73	1

Source: 1993 Annual Report of the Montana Interagency Wolf Working Group, prepared by the U.S. Fish and Wildlife Service.

19. The accompanying table (and Excel and graph link file MEDICAL) show the number of community hospital beds and personnel per 1000 people for states in the Northeast and the Midwest in 1993. (Community hospitals comprise the majority of all hospitals in the United States. This category excludes long-term, psychiatric, tuberculosis, and federal hospitals.)

Region and State	Beds (per 1000)	Occupancy Rate (%)	Personnel (per 1000)	Region and State	Beds (per 1000)	Occupancy Rate (%)	Personnel (per 1000)
Northeast				Midwest			
Maine	4.4	68.0	18.5	Ohio	41.1	60.5	176.2
New Hampshire	3.4	63.7	13.8	Indiana	21.3	58.7	90.6
Vermont	1.9	64.2	7.0	Illinois	44.1	63.5	180.0
Massachusetts	21.1	71.5	107.8	Michigan	30.9	64.7	140.9
Rhode Island	3.0	73.3	14.7	Wisconsin	17.7	63.4	65.5
Connecticut	9.2	74.4	44.8	Minnesota	18.4	66.0	55.0
New York	77.4	82.8	328.7	Iowa	13.4	57.9	44.1
New Jersey	31.1	77.0	121.0	Missouri	23.6	58.9	95.9
Pennsylvania	53.4	72.6	230.3	North Dakota	4.4	64.2	12.2
Mean	—	**71.9**	**98.5**	South Dakota	4.3	60.6	11.4
Median	—	**72.3**	**71.7**	Nebraska	8.4	55.2	25.7
				Kansas	11.3	54.2	36.3
				Mean	—	**60.7**	**77.8**
				Median	—	**60.6**	**60.3**

Source: U.S. Bureau of the Census, Current Population Reports, P25-1104, in *The American Almanac 1996–1997: Statistical Abstract of the United States* (Austin, Texas: The Reference Press).

a. What is the mean and median for the number of hospital beds per 1000 people for the Northeast and Midwest?

b. Write a summary paragraph that describes some aspect of the information in the table.

 20. Read the *CHANCE News* article and explain why the author was horrified.

21. According to the 1992 U.S. Census, the median net worth of American families was $52,200. The mean net worth was $220,300. How could there be such a wide discrepancy?

22. The chart to the left gives the ages of students in a mathematics class.

a. Use this information to estimate the mean age of the students in the class. Show your work.

b. What is the largest value the mean could have? The smallest? Why?

23. (Graphing calculator or spreadsheet recommended) Use the following table to generate an estimate of the mean age of the U.S. population. Show your work.

Age Interval	Frequency Count
15–19	2
20–24	8
25–29	4
30–34	3
35–39	2
40–44	1
45–49	1
Total	**21**

Ages of United States Population in 1991 in 10-Year Intervals

Age	Total (thousands)
Under 10	37,459
10–19	34,876
20–29	39,912
30–39	42,677
40–49	32,848
50–59	22,068
60–69	20,620
70–79	14,521
80 and over	7,195
Total	**252,177**

Based on data from U.S. Bureau of the Census, Current Population Reports, P25-1095, in *The American Almanac 1993–1994: Statistical Abstract of the United States*, 1993.

Note: Includes Armed Forces abroad. Frequency counts may not be exact because of rounding.

 24. (Requires computer and software on course CD-ROM) Open up the program "F1: Histograms" in the *FAM1000 Census Graphs* in the course software. 1996 U.S. Census data about 1000 randomly selected United States individuals and their families are imbedded in this program. You can use it to create histograms for education, age, and different measures of income. Try using different interval sizes to see what patterns emerge. Decide on one variable (say education) and then compare the histograms of this variable for different groups of people. For example, you could compare education histograms for men and women, or for people living in two different regions of the country, etc. Pick a comparison that you think is interesting and, if possible, print out your histograms. Then write a 60 second summary describing your observations.

25. The following table and graphs (and related Excel and graph link file USCHINA) contain information about the populations of the United States and China. Write a 60 second summary comparing the two populations.

Age Distribution of the 1991 Population (in 1000s) of China and the United States

Age Interval	China Totals	China %	United States Totals	United States %
0–4	116,624	10.30	19,222	7.62
5–9	99,440	8.79	18,237	7.23
10–14	97,456	8.61	17,671	7.01
15–19	120,403	10.64	17,205	6.82
20–24	125,877	11.12	19,194	7.61
25–29	104,268	9.21	20,718	8.22
30–34	83,805	7.40	22,159	8.79
35–39	86,315	7.63	20,518	8.14
40–44	63,844	5.64	18,754	7.44
45–49	49,181	4.35	14,094	5.59
50–54	45,666	4.03	11,645	4.62
55–59	41,753	3.69	10,423	4.13
60–64	34,057	3.01	10,582	4.20
65–69	26,395	2.33	10,037	3.98
70–74	18,119	1.60	8,242	3.27
75–79	10,971	0.97	6,279	2.49
80–84	5,373	0.47	4,035	1.60
85 and over	2,338	0.21	3,160	1.25
Total	**1,131,885**	**100.00**	**252,177**	**100.00**
Median age	**25**		**33**	
Mean age	**28**		**36**	

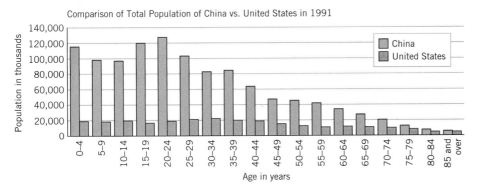

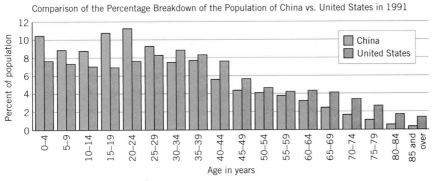

Source: U.S. Bureau of the Census, Current Population Reports, P25-1095 and P25-1104, in *The American Almanac 1993–1994 and 1996–1997: Statistical Abstract of the United States* (Austin, Texas: The Reference Press). Chinese population/statistical information reprinted with permission from the State Statistical Bureau of the People's Republic of China from *The China Statistical Yearbook,* copyright © 1991. Counts may not be exact because of rounding.

26. A New York Times/CBS News poll asked a question about the level of spending for "welfare" and 23% of the respondents said the nation was spending "too little" on welfare. The question was reworded with one small change: "assistance to the poor" was substituted for "welfare" and 64% of the same respondents said the nation was spending "too little."

Are these results contradictory? Explain your reasoning.

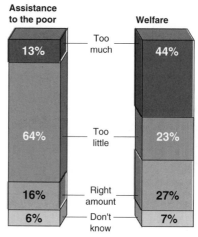

The Answer Depends on the Question

Are we spending too much, too little, or about the right amount on:

Assistance to the poor / **Welfare**

	Assistance to the poor	Welfare
Too much	13%	44%
Too little	64%	23%
Right amount	16%	27%
Don't know	6%	7%

Numbers do not add to 100 because of rounding.

Source: From a nationwide *New York Times*/CBS News Poll conducted May 6–8, 1992, results published in *The New York Times,* July 5, 1992. Copyright © 1992 by The New York Times Company. Reprinted by permission.

27. Read *The New York Times* OP-ED article, "A Fragmented War on Cancer" by Hamilton Jordan, President Jimmy Carter's chief of staff. Jordan claims that we are on the verge of a cancer epidemic. What questions are raised about the evidence he uses to support this claim?

a. Use what you have learned about the distribution of ages over time in the United States to refute his claim.

b. Read *The New York Times* letter to the editor by William M. London, Director of Public Health, American Council on Science and Health. London argues that Hamilton Jordan's assertions are misleading. What questions are raised by the arguments of William London? What additional data would you need to evaluate his arguments?

EXPLORATIONS

The Explorations are intended primarily for in-class use. However, most can be done outside of class if you have access to a computer or graphing calculator. The first three Explorations can be done in sequence using the class data collected in the first Exploration or each can stand alone. You should bring a notebook to class to keep a record of your results and observations.

Data that can be used with the Explorations are on graph link and spreadsheet files on the CD-ROM that accompanies the text. Basic instructions for using the TI-82 and TI-83 are included in the Graphing Calculator Workbook.

EXPLORATION 1.1

Collecting Data

Objectives

- explore some of the issues related to collecting data
- collect data about students in order to describe the class

Materials/Equipment

- class questionnaire
- measuring tapes in centimeters and inches
- optional measuring devices (Descriptions are included in "Health Measurements," a reading on the CD-ROM)

  eye chart

  flexibility tester constructed from cardboard box and yardstick

  measuring device for blood pressure and person who knows how to use it

Related Reading

"U.S. Government Definitions of Census Terms"

Procedure

In a Small Group or with a Partner

1. Pick (or your instructor will assign you) one of the undefined variables in parts 7–10 of the questionnaire. Spend about 15 minutes coming up with a workable definition of that vari-

able. Write the definition on the board. Be sure there is a way in which responses can be answered by a number or single letter.

2. Consult the reading "Health Measurements" on the CD-ROM if you decide to collect health data.

Discussion/Analysis

Class Discussion

After all the definitions are recorded on the board, discuss your definition with the class. Is it clear? Does everyone in the class fall into one of the categories of your definition? Can anyone think of someone who might not fit into any of the categories? Modify the definition until all can agree on the wording.

Are there any other variables or questions you would like to add to the questionnaire? Propose them to the class and, if the class agrees, pencil them in at the bottom of the questionnaire.

As a class, decide on the final version of the questionnaire. Write down the final definitions in your class notebook so you can refer to them in later classes.

Summary

In a Small Group or with a Partner

Fill out the entire questionnaire, and help each other when necessary to take measurements.

Questionnaires remain anonymous, and you can leave blank any question you can't or don't want to answer.

Hand in your questionnaire to your instructor by the end of class.

Exploration-Linked Homework

Read "U.S. Government Definitions of Census Terms" for a glimpse into some of the federal government's definitions of the variables you have defined in class. How do the "class" definitions and the "official" ones differ?

ALGEBRA CLASS QUESTIONNAIRE

(You may leave any category blank.)

☐   1.   Age (in years)

☐   2.   Sex (female = 1, male = 2)

☐   3.   Your height (in inches)

☐   4.   Distance from your navel to the floor (in centimeters)

☐   5.   Estimate your average travel time to school (in minutes)

☐   6.   What is your most frequent mode of transportation to school?
       (F =by foot, C = car, P = public transportation, B = bike, O = other)

The following variables will be defined in class. We will discuss ways of coding possible responses and then use the results to record our personal data.

☐   7.   The number of people in your household

☐   8.   Your employment status

☐   9.   Your ethnic classification

☐   10.   Your attitude toward mathematics

Health data

☐   11.   Your pulse rate before jumping (beats per minute)

☐   12.   Your pulse rate after jumping for 1 minute (beats per minute)

☐   13.   Blood pressure: systolic (mm Hg)

☐   14.   Blood pressure: diastolic (mm Hg)

☐   15.   Flexibility (in inches)

☐   16.   Vision, left eye

☐   17.   Vision, right eye

Other Data

☐

☐

EXPLORATION 1.2

Organizing and Visualizing Data

Objectives

- learn techniques for organizing and graphing data using a computer (with a spreadsheet program) or a graphing calculator
- describe and analyze the overall shape of single-variable data using frequency and relative frequency histograms

Materials/Equipment

- computer with spreadsheet program or graphing calculator with list-sorting and statistical plotting features
- data from class questionnaire or other small data set in both electronic and hard copy form. If using class data, copy of questionnaire decided upon in Exploration 1.1.
- overhead projector and projection panel for computer or graphing calculator
- transparencies for printing graphs (optional). Could also use transparencies with special colored pens to draw graphs.

Related Reading

"Presentation of Mathematical Papers at Joint Mathematics Meetings and Mathfests"

Procedure

Class Demonstration

1. If you haven't used a spreadsheet or graphing calculator before, you'll need a basic technical introduction. If you are using a TI-82 or TI-83 graphing calculator there are basic instructions in the Graphing Calculator Workbook.

    If you're using a spreadsheet,
    - copy the column with the data onto a new spreadsheet.
    - graph the data. What does this graph tell you about the data?
    - sort the data and graph it again. Is this graph any better at conveying information about the data?

    If you're using a graphing calculator,
    - discuss window sizes, changing interval sizes, and statistical plot procedures.
2. Choose a variable, select an interval size and construct a frequency and then a relative frequency histogram. If possible, label one of these carefully and print it out. If you have access to a laser printer, you can print onto an overhead transparency.

In a Small Group or with a Partner

Choose another variable from your data. Pick an interval size and then generate both a frequency and a relative frequency histogram. If possible, make copies of the histograms for both your partner and yourself.

Discussion/Analysis

With your partner(s), analyze and jot down patterns that emerge from the data. What are some limitations of the data? What other questions are raised and how may they be resolved? In your notebook, record jottings for a 60 second summary that describes your results.

Exploration-Linked Homework

A Verbal Report of Your Results

With your partner(s), prepare a verbal 60 second summary to give to the class. If possible use an overhead projector with a transparency of your histogram or a projector linked to your graphing calculator. If not, bring in a paper copy of your histogram. (The short pamphlet "Presentation of Mathematical Papers at Joint Mathematics Meetings and Mathfests" will show you that even mathematical professionals need very basic advice about talking before an audience.)

A Written Report of Your Results

Construct a written 60 second summary. (Section 1.5 of this text offers some suggestions for writing a 60 second summary.)

EXPLORATION 1.3

Describing Data Using Numbers and Graphs

Objectives

- describe and analyze data using numbers and graphs
- explore how to make comparisons using numbers and graphs

Materials/Equipment

- computer with spreadsheet program **or** graphing calculator
- data from class questionnaire or other small data set, either as spreadsheet or graph link file
- overhead projector and projection panel for computer or graphing calculator
- transparencies for printing or drawing graphs for overhead projector (optional)

Related Software

"F4: Measures of Central Tendency" in *FAM 1000 Census Graphs*

Procedure

Class Demonstration

1. Calculate mean, median, and mode using software functions or graphing calculator functions.

2. Using two columns of data, do a double sort (for example, using sex and ages, sort on sex and then on age.) Now compare the mean, median, and mode for different subsets (e.g., ages of men and ages of women).

Working in Small Groups or with a Partner

Work on a 60 second summary comparing at least two variables from your small data set. Decide on two variables and what approach you will take.

Think of ways of dividing up the work. By the end of the next class, your small group should be prepared to do a short class presentation of your findings, using overhead transparencies, enlarged Xeroxes, or projections from your graphing calculator.

Discussion/Analysis

Possible research topics and possible questions you may want to address are:

Pulse rates before and after exercising

What does a histogram of resting pulse rates look like? A histogram of pulse rates after exercise? What is an "average" pulse rate before exercising? After exercising? Are there any conclusions you can draw?

Heights of men vs. women

What does a histogram of women's heights look like? A histogram of men's heights? What is an "average" height for women? For men? Are there any generalizations you can draw?

Commuting time versus mode of transportation

What is the "average" commuting time for the whole class? What percentage of the class commutes by foot? By car? By public transportation? What is the "average" commuting time for each type of transportation?

There are many different ways of comparing two sets of data and there is no one right way of making comparisons. You can generate numbers such as averages, create graphs such as histograms, or try out other ways of visualizing data. You just need to be able to describe your results.

Be creative. It may be interesting to compare blood pressure of students with different commuting times, the flexibility of males and females, or the vision of students of different ages.

Exploration-Linked Homework

A Verbal Report of Your Results:

With your partner(s) prepare a 60 second summary of your results. See Section 1.5 and 1.6 of the text for guidelines. If possible use a transparency or graphing calculator projector with your talk.

A Written Report of Your Results:

Write your own 60 second summary of your results. Include graphs and numerical descriptors in summarizing your data.

EXPLORATION 1.4

Comparing the Ages of Men and Women in the United States

Objectives

- describe and analyze ages of men and women in the United States.
- explore how to make comparisons using numbers and graphs.

Materials/Equipment

- computer with spreadsheet program or graphing calculator (basic instructions for using the TI-82 and TI-83 with this Exploration are in the Graphing Calculator Workbook).
- data files and hard copy of ages of men and women in the United States. See Table 1.4 and its electronic versions USAGE1A5 (Excel file only) and USAGES1 (Excel and graph link versions), which contains ages in 1-year intervals only.
- overhead projector and projection panel for computer or graphing calculator
- transparencies for printing or drawing graphs for overhead projector (optional)

Procedure

Working with a Partner

1. Make predictions as to differences or similarities between the ages of men and women in the United States. For example, do you think the number of men in relation to the number of women varies by age?

2. Using information from Table 1.4 find the mean and median age of men and of women in the United States in 1991.

3. Compress the age data using the same interval size for men and women. Find the relative frequency for each age interval for men and women. Construct frequency or relative frequency histograms showing the distribution of ages for men and women.

4. In your class notebook, jot down your observations about the age data. Do they suggest any differences between men and women? Compare and contrast the age data using numbers such as mean and median and information from your graphs.

Discussion/Analysis

Discuss your findings with your partner(s). What do the graphs and numbers reveal about the ages of men and women? What type of arguments can be made? What are the limitations of the data? What questions have been raised by your analysis?

Exploration-Linked Homework

With your partner(s) prepare a short class presentation of your results. See Sections 1.5 and 1.6 of text for guidelines. If possible use an overhead projector with a transparency or graphing calculator projector to demonstrate your results.

Write your own 60 second summary, including graphs and numbers.

An Introduction to Functions

Overview

A central theme in mathematics and its applications is the search for patterns. Mathematics offers powerful tools both for recognizing patterns and for describing them succinctly. In Chapter 1 we saw how images such as histograms and pie charts, or numbers such as mean and median, are used to reveal patterns in single-variable data. In this chapter we explore how to describe the relationship between two variables. We focus on patterns of change. Does change in one variable affect change in another?

If one variable depends on another, we speak informally of one variable being a function of another. We talk, for instance, of how one's political views are often a function of one's upbringing. In mathematics the term *function* is used more rigorously to describe a specific type of relationship between two variables. Mathematical functions can be represented by words, tables, graphs, and equations. The Explorations will help you develop an intuitive understanding of functions and describe patterns in data tables and graphs.

After reading this chapter you should be able to:

- recognize from a graph when one variable with respect to another is increasing or decreasing and when it reaches a maximum or minimum value
- define a function
- identify the independent and dependent variable, and the domain and range of a function
- represent functions with words, tables, graphs, and equations

2.1 RELATIONSHIPS BETWEEN TWO VARIABLES

We can obtain important but static snapshots of patterns in single-variable data from images such as histograms and pie charts, or from numbers such as mean and median. By looking at two-variable data, we gain information about relationships between variables. We can ask how change in one variable affects change in another. How does the weight of a child determine the amount of medication prescribed by a pediatrician? How does median age or income change over time? Mathematics offers powerful tools both for recognizing patterns of change and for describing them succinctly.

One of the most common and important cases of two-variable data is the *time series*. In a time series, one of the variables is time which is measured in units such as minutes, months, years, or decades. The other variable may be anything that can be measured at different times, for example, median age, a child's weight, or number of voters. Some estimates claim that over 80% of the graphs in the popular press are time series.

Example   1

Table 2.1 shows a time series for median age. The first column contains the year, and the second the corresponding median age for the U.S. population.

Table 2.1	
Median Age of U.S. Population 1850–2050 **(data for 1995–2050 are projected)**	
Year	Median Age
1850	18.9
1860	19.4
1870	20.2
1880	20.9
1890	22.0
1900	22.9
1910	24.1
1920	25.3
1930	26.4
1940	29.0
1950	30.2
1960	29.5
1970	28.0
1980	30.0
1990	32.8
1991	33.1
1995	34.0
2000	35.5
2025	38.1
2050	39.0

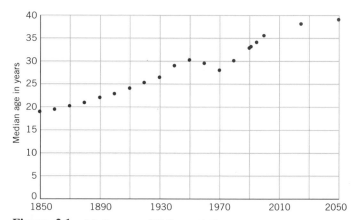

Figure 2.1   Median age of U.S. population.

Source: U.S. Bureau of the Census, U.S. Census of Population, Current Population Reports, in *The American Almanac 1995–1996: Statistical Abstract of the United States,* 1995.

We can think of the rows in Table 2.1 as ordered pairs. The first row, for example, corresponds to the ordered pair (1850, 18.9). The second row corresponds to (1860, 19.4). We can graph these ordered pairs using the convention of plotting the first coordinate (year) on the horizontal axis, and the second coordinate (median age) on the vertical axis. The resulting graph in Figure 2.1 is called a *scatter plot*.

It is apparent from both Table 2.1 and Figure 2.1 that, in general, as the years increase, the median age increases as well. This is not surprising, since we saw in Chapter 1 (and repeated in Table 2.1) that the Bureau of the Census predicted the median age of 33.1 in 1991 would rise to 39.0 by the year 2050. In Table 2.1 the *minimum* median age of 18.9 occurs in 1850 and the *maximum* median age is projected to be 39.0 in 2050.

The median age, however, does decrease between 1950 and 1970; it drops from 30.2 in 1950, to 29.5 in 1960, down to 28.0 in 1970. The corresponding dip is easy to see in Figure 2.1.

Something to think about

1. What might have caused a lowering of the median age between 1950 and 1970? More babies? Fewer old people? More young immigrants?
2. What are some of the trade-offs in using the median instead of the mean age?

Table 2.2 and Figure 2.2 show the annual federal budget surplus (+) or deficit (−) since World War II.

Example 2

Table 2.2

Federal Budget Surplus (+) or Deficit (−)

Year	Billions of Dollars	Year	Billions of Dollars
1945	−47.6	1981	−79.0
1950	−3.1	1982	−128.0
1955	−3.0	1983	−207.8
1960	0.3	1984	−185.4
1965	−1.4	1985*	−212.3
1970	−2.8	1986	−221.2
1971	−23.0	1987	−149.8
1972	−23.4	1988	−155.2
1973	−14.9	1989	−152.5
1974	−6.1	1990	−221.4
1975	−53.2	1991	−269.4
1976	−73.7	1992	−290.4
1977	−53.7	1993	−255.1
1978	−59.2	1994	−203.2
1979	−40.7	1995	−163.9
1980	−73.8	1996 (est.)	−145.6

* The Balanced Budget and Emergency Deficit Control Act of 1985 put all the previously off-budget federal entities into the budget and moved Social Security off-budget.

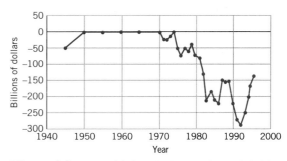

Figure 2.2 Annual federal budget surplus or deficit in billions of dollars.

Source: U.S. Office of Management and Budget, "Budget of the United States Government," annual, in *The American Almanac 1995–1996 and 1996–1997: Statistical Abstract of the United States,* 1995 and 1996.

Figure 2.2 indicates that the federal budget was fairly balanced (with little deficit or surplus) until about 1970. Since then, the federal budget has been running an annual deficit, which has been generally getting larger until 1992. If we look at the values in Table 2.2, we see that the graph reaches a maximum in 1960 when there was actually a surplus of about 0.3 billion or 300 million and a minimum in 1992 when the annual deficit was almost 300 *billion dollars*. Table 2.2 shows only the *annual* surpluses or

Something to think about

If the annual deficit returns to zero, does that mean that the federal debt is zero?

deficits. In order to calculate the total federal debt we need to take into account the cumulative effect of all the deficits and surpluses for each year together with any interest or payback of principal. Exercise 6 in the problem set at the end of the chapter contains data on the total federal debt. The size of the annual deficit and total debt in future years is a matter of intense national debate. Will the graph return to a downward path, or will it continue the upward trend started in 1992 and eventually reach zero?

Example   3

We can describe mathematical relationships between two abstract variables x and y using equations like:

$$y = x^2 + 2x - 3$$

Think of the equation $y = x^2 + 2x - 3$ as a mathematical sentence. Some combinations of values of x and y make the sentence true, such as $x = 0$ and $y = -3$. Other combinations make it false, for example, $x = 0$ and $y = 5$. Combinations of values for x and y that make the sentence true are called *solutions* to the equation. We can express these solutions as ordered pairs of the form (x, y).

> The *solutions* to an equation in two variables x and y are all the ordered pairs (x, y) that make the equation a true statement.

By substituting various values for x and finding the associated values for y, we can generate a table of solutions. There are infinitely many possible solutions since we can substitute any number for x and find a corresponding value for y. Table 2.3 shows a finite number of solutions. We can plot the points by hand or use technology to generate a graph of the relationship. Each point on the graph in Figure 2.3 corresponds to an ordered pair of the form (x, y) that makes the equation a true statement.

> The *graph* of an equation in two variables is the set of points corresponding to the ordered pairs of numbers that make the equation a true statement.

Table 2.3

x	y
−4	5
−3	0
−2	−3
−1	−4
0	−3
1	0
2	5
3	12

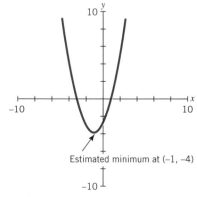

Estimated minimum at (−1, −4)

Figure 2.3   Graph of $y = x^2 + 2x - 3$

We can "read" the *graph* from left to right as we did the time series graphs and ask what effect changes in x have on y. In Figure 2.3, initially as x increases, y decreases until it reaches an estimated minimum value of -4 when x is about -1. (In later chapters we'll learn how to determine this minimum value exactly.) When $x > -1$, as x increases, y now increases. Figure 2.3 shows only a small section of the graph of all the solutions to the equation. Both "arms" of the graph extend infinitely upward, so there is no maximum value for y.

Algebra Aerobics 2.1

Table 2.4

Year	Median Family Income in Constant 1990 $	Year	Median Family Income in Constant 1990 $
1973	35,474	1982	31,738
1974	34,205	1983	32,378
1975	33,328	1984	33,251
1976	34,359	1985	33,689
1977	34,528	1986	35,129
1978	35,361	1987	35,632
1979	35,262	1988	35,565
1980	33,346	1989	36,062
1981	32,190	1990	35,353

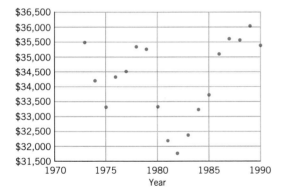

Figure 2.4   Median family income in 1990 dollars.

Source: U.S. Bureau of the Census, Current Population Reports, P-60 Series, in *The American Almanac 1993–1994: Statistical Abstract of the United States,* 1993.

1. Table 2.4 and the scatter plot in Figure 2.4 show median family income over time in constant 1990 dollars.[1]

   a. What is the maximum value for the median family income during this period? In what year does this occur? What are the coordinates of this point in Figure 2.4?

   b. What is the minimum value for the median family income? In what year does this occur? What are the coordinates of this point?

   c. What is the longest time period during which the median income was increasing? Decreasing?

2. a. Describe in your own words how to compute the value for y given a value for x using the following equation:

$$y = -x^2 + 1$$

   b. Which of the following ordered pairs represent solutions to the equation?

   $(0, 0), (0,1), (1, 0), (-1, 2), (-2, 3), (-1, 0)$

   c. Generate a small table of values that represent solutions to the equation.

3. Repeat the directions in Problem 2(a), (b) and (c) using the equation $y = (x - 1)^2$.

[1] "Constant dollars" is a measure used by economists to compare incomes and other variables in terms of purchasing power, eliminating the effects of inflation. To say the median income in 1973 was $35,474 in "constant 1990 dollars" means that the median income in 1973 could buy an amount of goods and services that would cost $35,474 to buy in 1990. The actual median income in 1973 (measured in what economists call "current dollars") was much lower. Income corrected for inflation is sometimes called "real" income.

2.2 FUNCTIONS

The Definition of a Function

In Section 2.1 we used ordered pairs to describe the relationships between time and real social variables (median age, the federal debt, and median family income), as well as the relationship between two abstract variables x and y. We think of the value of the first variable in the ordered pair as determining the value of the second variable. In our examples, each year determines one particular value of median age, debt or income, and each value of x determines one and only one value for y. We also think of age, debt, and income as being functions of time, and we think of y as a function of x.

When we speak informally of one thing being a function of some other thing, we mean that one depends on the other. How well a car runs is a function of how well it is maintained. One's political views are often a function of one's upbringing. The response of a patient to medication is a function of the dose he or she is given.

With mathematical functions, the *independent variable* is the one that may be changed at will or that, like time, governs a process. The other variable, which responds, is the *dependent variable*. The dependent variable is a *function* of the independent variable if each value of the independent variable determines one and only one value of the dependent variable.

> A variable y is a *function* of a variable x if each value of x determines one and only one value of y. We say y is the *value of the function* or the *dependent variable* and x is the *independent variable*.

Exploration 2.1 will help you develop an intuitive sense of functions.

Hat size is a function of the distance around your head. What you pay at the gas station is a function of the number of gallons of gas you pump into your car.

Each example in Section 2.1 represents a function. For instance, in Table 2.1 and Figure 2.1 median age is a function of time, since each year determines a single value for median age. The equation $y = x^2 + 2x - 3$ describes y as a function of x, since each value of x determines a single corresponding value of y.

One variable is not a function of another if the value of the first does not uniquely determine the value of the second. Hat size is not a function of how much you ate for lunch, since people who ate the same lunch could have different hat sizes. In the following set of ordered pairs in the form (x, y), y is not a function of x.

$$(1, 2), (1, 3), (2, 4), (3, 5)$$

When x equals 1, y can be 2 or 3. Thus, there is *not* one and only one value of y for each value of x.

Representing Functions with Words, Tables, Graphs, and Equations

We can think of a function as defining a "rule" that determines one and only one value of y for each value of x. The rule can be described using words, tables, graphs, or equations.

- Using words, the rule may say "y is five percent of x."
- Using a table, the rule may say "find the value of x in the left column of the table and read the corresponding value of y from the right column."
- Using a graph, it may say "go along the horizontal axis a distance x, then go up or down until you meet a point on the curve, and read y from the point's vertical coordinate."
- Using an equation, the rule may be written as $y = 0.05x$.

Example   1

The sales tax in Illinois is 8%; that is, for each dollar you spend in a store, the law says that you must pay a tax of 8 cents or 0.08 dollars. (We have rounded off the sales tax to simplify calculations.) We may write this as an equation,

$$T = 0.08P$$

where T is the amount of sales tax, P is the price of your purchase and both are measured in dollars. This equation is a rule for determining a value for T given a value for P. It says, "Take the given value of P and multiply it by 0.08; the result is the value of T."

P is the independent variable and T the dependent variable; the sales tax depends upon the purchase price. The equation represents T as a function of P, since for each value of P the equation determines a unique value of T. A function that is used to describe a real world phenomenon is called a *mathematical model*.

Of course, we could also use this formula to make a table of values for T determined by the many different values of P (Table 2.5). Such tables are indeed posted next to many cash registers. We can plot the ordered pairs in the table to create a graph of the function (Figure 2.5). *By convention the independent variable is plotted on the horizontal axis, and the dependent variable on the vertical axis.* The words, equation, table, and graph are all representations of an 8% sales tax as a function of purchase price.

$$T = 0.08P$$

Figure 2.5   Illinois sales tax.

Table 2.5	
Illinois Sales Tax	
P (purchase price in $)	*T* (sales tax in $)
0	0.00
1	0.08
2	0.16
3	0.24
4	0.32
5	0.40
6	0.48
7	0.56
8	0.64
9	0.72
10	0.80

We selected a few values for P (integers between 0 and 10) in order to generate Table 2.5 and Figure 2.5. But what are all the possible values for P for which it makes sense to use the formula $T = 0.08P$? Since P represents purchase price, negative values for P are meaningless. We need to restrict P to possible dollar amounts such that $P \geq 0$. In theory there is no upper limit on prices, so we assume P has no maximum amount. We call the set of possible values for the independent variable the *domain*. So in this example:

$$\text{domain} = \text{ all possible dollar amounts } P \text{ greater than or equal to } 0$$

What are the corresponding values for the tax T? The values for T in our model will also always be nonnegative. Assuming that there is no maximum value for P, then there will be no maximum value for T. So T could assume any dollar amount such that $T \geq 0$. We call the set of possible values for the dependent variable the *range*. So here

$$\text{range} = \text{ all dollar amounts } T \text{ greater than or equal to } 0$$

> The *domain* of a function is the set of possible values of the independent variable. The *range* is the set of corresponding values of the dependent variable.

Table 2.5 only shows the sales tax for a few selected purchase prices, but we assume that we could have used any positive dollar amount for P. We have "connected the dots" on the scatter plot to suggest the many possible intermediate price values. For example, if $P = \$2.50$, then $T = \$0.20$.

Algebra Aerobics 2.2a

1. **a.** Write an equation for computing a 15% tip in a restaurant. Does your equation represent a function? If so, what are your choices for the independent and dependent variables? What are reasonable choices for the domain and range?

   b. How much would the equation suggest you tip for an $8 meal?

   c. Compute a 15% tip on a total check of $26.42.

2. If we let D stand for ampicillin dosage expressed in milligrams and W stand for a child's weight in kilograms, then the equation

   $$D = 50W$$

   gives a rule for finding the safe maximum daily drug dosage of ampicillin (used to treat respiratory infections) for children who weigh less than 10 kilograms (about 22 pounds).[2]

   a. What are logical choices for the independent and dependent variables?

   b. Does the equation represent a function? Why?

   c. What is a reasonable domain? Range?

   d. Generate a small table and graph of the function.

[2] Information extracted from Anna M. Curren and Laurie D. Munday, *Math for Meds: Dosages and Solutions,* 6th ed. (San Diego: W. I. Publications, 1990), p. 198.

3. Table 2.6 lists the pulse rate (P) in beats/min and respiration rate (R) in breaths/min taken at different times for one individual. Based on this table, is this person's respiration rate a function of pulse rate?

Table 2.6	
P	R
75	12
95	17
75	14
130	20
110	19

Function Notation

In the Illinois sales tax example, we found that tax T is a function of price P. We can give a letter name to the function; f is a common, obvious choice. To indicate T is a function of P, we write

$$f(P) = T$$

which is read "f of P equals T." Sometimes we call the independent variable (in this case P) the *input*, and the dependent variable (in this case T) the *output*. Note that the use of parentheses in $f(P)$ does *not* signify multiplication. The symbol $f(P)$ means f evaluated at P, not f times P.

A function, f, is a rule that tells us what to do to a value of an independent variable in order to produce a unique value of the dependent variable. The rule could be defined using an equation, words, graphs, or tables. In this case, f is defined by the equation $T = 0.08P$. We say

$$f(P) = T \qquad \text{where} \qquad T = 0.08P$$

We must now be careful to distinguish among three different symbols for:

the independent variable, P

the dependent variable, T

the function name, f

The $f(x)$ [or in this case $f(P)$] notation is particularly useful when a function is being evaluated at a specific point. For example, instead of saying "the value for T when $P = 10$," we write simply "$f(10)$." Then we have

$$f(10) = (0.08)(10)$$
$$= 0.8$$

Similarly

$$f(0.5) = (0.08)(0.5) \qquad f(7/4) = (0.08)(7/4) \qquad f(6.5) = (0.08)(6.5)$$
$$= 0.04 \qquad\qquad\quad = 0.14 \qquad\qquad\quad = 0.52$$

The function notation $f(x)$ emphasizes the choice of one variable as an independent input, and the other as a dependent output. We encounter this notation regularly in mathematics texts. The $f(x)$ notation is used less frequently in social science and science texts. This may stem from either the reluctance to classify one variable as depending upon another, or the desire to be able to retain the flexibility of an equation that can be rewritten in many different forms.

Algebra Aerobics 2.2b

1. If $f(x) = 5x$, evaluate $f(1), f(10)$, and $f(-2)$.

2. If $g(x) = 2x^2 - x + 4$, evaluate $g(0), g(1)$, and $g(-1)$.

3. From the Table 2.7, find $f(0)$ and $f(20)$. Find two values of x for which $f(x) = 10$.

4. From the graph in Figure 2.6, estimate $f(-4), f(-1)$, $f(0)$, and $f(3)$. Approximate two values for x such that $f(x) = 0$.

Table 2.7

x	$f(x)$
0	20
10	10
20	0
30	10
40	20

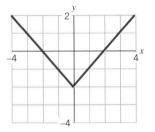

Figure 2.6   Graph of $f(x)$.

The Tax Equation in the Abstract

Let's look again at the function $f(P) = T$, where

$$T = 0.08P$$

and this time divorce it from the context of Illinois taxes. We are no longer thinking of the function as a real world model, but as describing a relationship between two abstract variables P and T. T is still a function of P. Since P no longer represents purchase price, P is not measured in dollars and cents. The values for P, the independent variable, are all real numbers, positive, negative, or zero.[3] So now

$$\text{domain} = \text{all real numbers}$$

Values for the dependent variable T can also be any real number. So

$$\text{range} = \text{all real numbers}$$

We can generate a small table of values (including negative numbers) (Table 2.8) and

---

[3] Recall that the real numbers are all numbers that can be written as signed decimal expressions, for example, 1, -3, 20.12, $\pi = 3.14159$ . . . , $1/3 = 0.33333$. . . . We often visualize the real numbers as points on a line that has been marked off with a zero, and divided into equally spaced unit lengths.

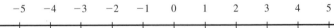

The real numbers split into two distinct categories: rational and irrational. Rational numbers (which include all integers and fractions) are numbers whose decimal expansions eventually terminate or contain a fixed pattern that infinitely repeats. For example, 7, $1/4 = 0.25$, $2/3 = 0.66666$ . . . , and $-2/11 = -0.1818181818$ . . . are all rational numbers. Irrational numbers are those whose decimal patterns do not contain an infinitely repeating pattern and do not terminate. Some examples are $\sqrt{2} = 1.414213$ . . . and $\pi = 3.14159$. . . .

an associated graph (Figure 2.7). The line now extends indefinitely in two directions. P can now assume values that wouldn't have made sense as prices, such as negative values like -100, or irrational values like π or $\sqrt{2}$.

Table 2.8	
P	T
-10	-0.80
-5	-0.40
0	0.00
5	0.40
10	0.80

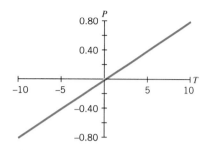

Figure 2.7   Graph of $T = 0.08P$.

A Note About Functions and Their Domains

An implicit part of the definition of a function is its domain; functions and their domains go hand in hand. We need to be careful when we use the same functional notation $f(P) = 0.08P$ to denote both the Illinois sales tax function (where the domain was restricted to values of $P \geq 0$) and its abstraction (where the domain was all the real numbers). By changing the domain, we have changed the function. For example, if f represents the functional rule defining Illinois sales taxes, then $f(-1)$ is undefined. If we take f to represent an abstract relationship, then $f(-1) = (0.08)(-1)$ or -0.08. Any time we use a function, we must consider the function paired with its domain.

Algebra Aerobics 2.2c

1. Think of the function for generating 15% tips as an abstract mathematical function having nothing to do with meal prices. What is the domain and range?

2. Think of the function $D = 50W$ for ampicillin dosage in Algebra Aerobics 2.2(a) as an abstract mathematical equation. What is the domain? The range? How could the table and graph reflect these changes? Rewrite the function using f to represent the rule relating the abstract variables W and D. What is $f(15)$? $f(-15)$? $f(15000)$?

2.3 EQUATIONS THAT REPRESENT FUNCTIONS

Putting Equations into "Function Form"

A solution to an equation such as

$$3x^2 - 2y = 5x \qquad (1)$$

is a pair of values for x and y that make the equation true. For example, $x = 1$ and $y = -1$ are a solution for Equation (1) since they make it a true statement.

given	$3x^2 - 2y = 5x$
if $x = 1$ and $y = -1$, then	$3(1)^2 - 2(-1) = 5(1)$
	$3 + 2 = 5$
	$5 = 5$

There are in fact an infinite number of solutions for this equation. But it is often difficult to find solutions to equations in a form such as $3x^2 - 2y = 5x$ by anything other than trial and error.

Not all equations in two variables represent functions. But when an equation is a function, it is useful to solve the equation for the dependent variable in terms of the independent variable. This is called putting the equation in *function form*. Once an equation is in function form, we can easily generate many solutions by picking different values for the independent variable and using the functional rule to compute the corresponding values of the dependent variable. Many graphing calculators and computer graphing programs accept only equations in function form as input.

To put an equation into function form, we first need to identify the independent and the dependent variables. Sometimes the choice may be obvious, other times not. There may be more than one possible correct choice. For now, in order to practice the mechanics of solving for one variable in terms of another, we adopt the mathematical convention that x represents the independent and y the dependent variable. Solutions for functions are often written as ordered pairs in the form

(independent variable, dependent variable)

In order to put each of the following equations into function form, we need to represent y, the dependent variable, in terms of x, the independent variable. We want

$$y = \text{(some expression involving } x\text{)}$$

We start by putting all the terms involving y on one side of the equation and everything else on the other side.

Examples of Equations That Represent Functions

Example   1

Analyze the equation $4x - 3y = 6$. Can it be put in function form? What do the symbols tell us to do? Generate a table and graph of solutions to the equation. Decide whether or not the equation represents a function. If it does, what are the domain and range?

SOLUTION
First, try to put the equation in function form.

given the equation	$4x - 3y = 6$
subtract $4x$ from both sides	$-3y = 6 - 4x$
divide both sides by -3	$(-3y)/(-3) = (6 - 4x)/(-3)$
simplify	$y = \dfrac{6}{(-3)} - \dfrac{4x}{(-3)}$
simplify and rearrange terms	$y = (4/3)x - 2$

We now have an expression for y in terms of x. We can generate a small table of solutions (Table 2.9) to the equation, and graph the results (Figure 2.8).

Table 2.9	
x	$f(x)$
-6	-10
-3	-6
0	-2
3	2
6	6

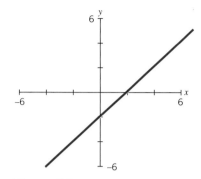

Figure 2.8   Graph of $y = (4/3)x - 2$.

Does the equation $y = (4/3)x - 2$ represent y as a function of x? That is, given a particular value for x, does the equation determine one and only one value for y? This is not always an easy question to answer. What do the symbols in this equation tell us to do? They say:

"To find y, take the value of x, multiply it by (4/3), and then add -2."

Starting with a particular value for x, each step in this process determines one and only one number, resulting in a unique corresponding value for y. So y is indeed a function of x. If we use f to represent the function, we could write $y = f(x)$ where $f(x) = (4/3)x - 2$.

What is the domain, the set of possible values for x? In most cases throughout this text, we implicitly assume that the domain (and range) are restricted to real numbers. Are there any other constraints on the values which x may assume? No. In this case we could use any real number for x. So,

$$\text{domain} = \text{all real numbers}$$

What is the range, the corresponding set of possible values for $f(x)$? Since the line in Figure 2.8 extends indefinitely in both directions, there are no maximum or minimum values for $f(x)$. This implies that

$$\text{range} = \text{all real numbers}$$

Throughout the graph, as x increases, $f(x)$ increases. We call $f(x)$ an *increasing function*.

Analyze the equation $xy = 1$. Can it be put in function form? What do the symbols tell us to do? Generate a table and graph of solutions to the equation. Decide whether or not the equation represents a function. If it does, what are the domain and range?

Example   2

SOLUTION
First, put the equation into function form.

given the equation	$xy = 1$
divide both sides by x (assumes $x \neq 0$)	$y = 1/x$

Note that when we divided by x, we assumed that x was not zero, since division by zero is undefined. By doing so, have we placed a constraint on x that didn't exist in the original equation, $xy = 1$? No. Since the product of x and y equals 1, neither x (nor y) could ever be zero.

Table 2.10 and Figure 2.9 show some solutions to the equation $y = 1/x$.

Table 2.10	
x	y
-5	$-1/5$
-1	-1
$-1/10$	-10
$-1/100$	-100
0	not defined
$1/100$	100
$1/10$	10
1	1
5	$1/5$

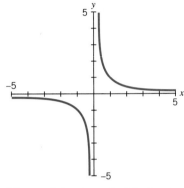

Figure 2.9   A graph of the function $y = 1/x$.

Does the equation $y = 1/x$ describe a function? For any nonzero value for x, does the equation determine a unique value for y? Yes. The symbols say given any x, take its reciprocal. This will give us one and only one value for y.

What is the domain of this function? We know that x cannot equal 0. Are there any other limitations on x? If we choose any nonzero value for x, is $1/x$ a real number? Yes. Then

$$\text{domain} = \text{all real numbers except } 0$$

It's important to look carefully at the behavior of a graph near points or intervals that are missing from the domain. If you're using technology to graph the function, you may want to change your window size several times in order to gain a better idea of the function's shape near the y-axis where $x = 0$.

We can see that the function appears to "blow up" when x is near zero, splitting the graph into two pieces: one when $x > 0$ and one when $x < 0$.

When $x > 0$, $y > 0$. As x increases, the values for y decrease, approaching but never reaching zero. For example, the following points are solutions to the equation $y = 1/x$ and hence lie on the graph: $(1, 1)$, $(10, 1/10)$, $(100, 1/100)$.

As x approaches 0 from the right, the solutions $(1, 1)$, $(1/10, 10)$, $(1/100, 100)$ suggest that the values for y get larger and larger. As x stays positive, and gets closer and closer to 0, the y values get larger and larger. The function $y = 1/x$ has no maximum or minimum value. The range is all real numbers except 0.

Something to think about

Can you describe what happens to the graph of $y = \dfrac{1}{x}$, when $x < 0$?

Not All Equations Represent Functions

Example   3

Analyze the equation $y^2 - x = 0$. Can it be put in function form? Generate a table and graph of solutions to the equation. Decide whether or not the equation represents a function.

long

SOLUTION
Put the equation in function form.

Given the equation $y^2 - x = 0$

add x to both sides $y^2 = x$

In order to solve this equation, we need to take the *square root* of both sides. An equation such as

$$y^2 = 4$$

has two solutions,

$$y = \sqrt{4} \text{ or } 2 \qquad \text{and} \qquad y = -\sqrt{4} \text{ or } -2$$

since both $2^2 = 4$ and $(-2)^2 = 4$. We can combine these two solutions in the single expression

$$y = \pm \sqrt{4} \text{ or } \pm 2$$

So given

$$y^2 = x$$

taking the square root of both sides gives us two solutions

$$y = \pm \sqrt{x}$$

If we limit ourselves to values of x and y that are real, then in order to take the square root of x we must assume that x is not negative. There is no real number whose square is a negative number. Hence we require

$$x \geq 0$$

Any positive value for x is associated with two different y values, namely $+\sqrt{x}$ and $-\sqrt{x}$. So the equation $y^2 = x$ does *not* represent y as a function of x, unless we restrict ourselves to a trivial domain of only 0.

We can generate a table (Table 2.11) and graph of solutions to the equation $y^2 - x = 0$, where $x \geq 0$.

We can see on the graph in Figure 2.10 that the equation $y^2 = x$ does not represent y as a function of x since, in particular, when x is 4, y may be $+2$ or -2. y is a function of x if there is only one value for y for each value of x, but here there are two.

Something to think about

Can you think of an easy visual test to determine whether or not a graph is a function?

Table 2.11	
x	*y*
0	0
1	1 or −1
2	$\sqrt{2}$ or $-\sqrt{2}$
4	2 or −2
9	3 or −3

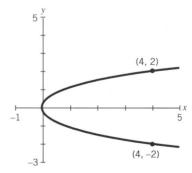

Figure 2.10 Graph of the equation $y^2 = x$.

Algebra Aerobics 2.3a

1. Consider the function $f(x) = x^2 - 5x + 6$. Find $f(-3), f(0)$, and $f(1)$.

In problems 2–5:

Solve for y in terms of x. Determine if y is a function of x. If it is, rewrite using $f(x)$ notation. If y is a function of x, determine the domain.

2. $2(x - 1) - 3(y + 5) = 10$

3. $x^2 + 2x - 3y + 4 = 0$

4. $\dfrac{x}{2} + \dfrac{y}{3} = 1$

5. $x + \sqrt{x - 2} + y = 4$

When Is a Relationship Not a Function?

A two-variable table, graph, or equation is not a function if you can find even one value for the independent variable that determines more than one corresponding value for the dependent variable. That means that the relationship does not define a clear unambiguous way of getting from any value of x to one and only one associated value for y. Graphs often offer the easiest way to recognize a relationship between two variables that is not a function.

The vertical line test   Graphs do not represent functions if they fail the *vertical line test;* that is, if you can draw a vertical line that crosses the graph two or more times. Any two points of the relationship that lie on the line would share the same first coordinate, but have two different second coordinates. So the relationship would match two different values of the dependent variable with one value of the independent variable. In other words, for that value of x, there would not be a unique value of y.

For example, Figure 2.11 is a graph of the equation $y^2 = x$. We verified in the previous section that the equation is not a function. But the graph offers simple visual evidence. One can draw a vertical line (an infinite number in fact) which would cross the graph twice. One such line is sketched in, and it crosses the graph at (4, 2) and (4, −2). That means that the value $x = 4$ does not determine one and only one value of y. It corresponds to y values of both 2 and −2.

Figure 2.12 shows that the graph of the function $y^2 = x$ is actually a composite of two separate pieces each of which individually *is* a function. The upper half is the function $y = +\sqrt{x}$ and the lower half (dotted on the graph) is the function $y = -\sqrt{x}$.

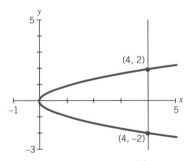

Figure 2.11   Graph of $y^2 = x$.

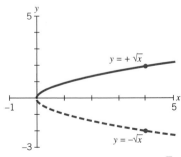

Figure 2.12   Graph of $y = +\sqrt{x}$ and $y = -\sqrt{x}$.

If y is a function of x, is x also a function of y? **Answer: Not necessarily.**

The Pensacola tides are a function of the time of day. Figure 2.13 shows the water level over a 24-hour period in Pensacola, Florida. A water level of 0 represents the mean water level.

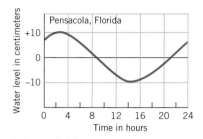

Figure 2.13 Diurnal tides in a 24-hour period in Pensacola, Florida.*

The independent variable is time, and the dependent variable is water level. The domain is from 0 to 24 hours, and the range is from about -10 to $+10$ centimeters. From 0 to 2 hours the water level appears to be steadily increasing until it reaches a maximum of about $+10$ centimeters. After 2 hours the water level steadily decreases until it reaches a minimum level of about -10 centimeters at 14 hours. After 14 hours the water level starts increasing again, continuing the tidal cycle. This graph represents a function since it passes the vertical line test.

Is the time of day a function of the Pensacola tides? Is each water level in Pensacola associated with one and only one time of day?

No. The same water level may be reached more than once during the same day. Look at Figure 2.14, which shows the Pensacola tides, now with water level on the horizontal axis and time on the vertical axis. The graph fails the vertical line test. For instance, if you draw a vertical line corresponding to a water level of 0 centimeters (see the black line on the graph) the line crosses the graph in two different places (at about 8 and 22 hours). There is not one and only one hour in the 24-hour period associated with a water level of 0 centimeters. Therefore, the time of day is not a function of the water level. There is something unusual about Pensacola's tides. Normally there is roughly a 6-hour time difference between high and low tide, creating two high and two low tides a day. What is the time difference between high and low tide in Pensacola? Such a tide is called *diurnal*.

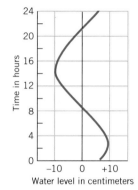

Figure 2.14 The time of day is not a function of the tide.

*Graph adapted from Fig. 8-2, in *Oceanography: An Introduction to the Planet Oceanus* by Paul R. Pinet. Copyright ©1992 by West Publishing Company, St. Paul, Minnesota. All rights reserved.

Algebra Aerobics 2.3b

1. Describe the graph in Figure 2.15.
   a. Is y a function of x?
   b. When is the function increasing? Decreasing?
   c. What seems to happen to the graph when the value of x approaches 0?

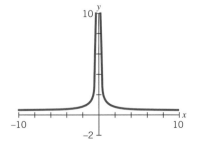

Figure 2.15 Graph of an abstract relationship between x and y.

2. Which of the following graphs represent functions and which do not? Why?

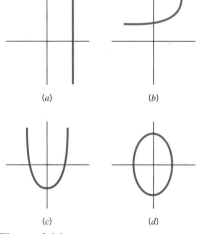

(a)   (b)

(c)   (d)

Figure 2.16

3. Consider the following chart (Figure 2.17). Is weight a function of height? Is height a function of weight?

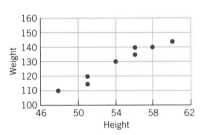

Figure 2.17 Graph of weight versus height.

4. Consider Table 2.12:

Table 2.12	
Y	*D*
1992	$2.5
1993	$2.7
1994	$2.4
1995	−$0.5
1996	$0.7
1997	$2.7

a. Is *D* a function of *Y*?

b. Is *Y* a function of *D*?

2.4 FITTING FUNCTIONS TO DATA

Exploration 2.2 will give you practice in finding functions to describe patterns in data tables.

So far we have not tried to determine an equation or formula to describe a functional relationship. In all previous examples, either the equation was given (as in the Illinois tax example) or was unknown (as in the median income example). But often we want to model a set of data with an equation that describes one variable as a function of another.

Consider the following experimental data on growth of a colony of *Sordaria finicola* fungus. The area of the colony (growing in a petri dish) is measured at a series of times. We denote by *t* the time in hours since the introduction of the sample into the dish, and by *A* the area of the colony in square centimeters. Table 2.13 shows the results.

Our first step is to plot the data (Figure 2.18). We have drawn straight lines between successive data points to guide our eyes.

Table 2.13	
Growth of Fungus	
Time *t* (hours)	**Area *A*** (sq cm)
0	0.0
5	0.6
10	2.5
15	5.6
20	10.0
25	15.6

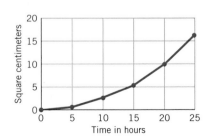

Figure 2.18 Area of *Sordaria finicola* colony.

In this case, the area certainly is a function of time: there is only one area measurement at each time.

Let us try to guess an algebraic expression for this function.

Our first guess might be a simple proportionality as in the Illinois tax example; that is, the area might be able to be expressed as the product of a constant number and the variable time. Since $A = 0$ at $t = 0$, and since substituting $t = 20$ should give us $A = 10$, we may guess the formula

$A = \frac{1}{2}t$ or $t/2$ (matches the data points $t = 0, A = 0$ and $t = 20, A = 10$)

But this formula does not fit any other data point. For example, at $t = 10$, this formula predicts $A = 10/2 = 5$, but the measured value was $A = 2.5$. Table 2.14 and Figure 2.19 show the predicted area (using the formula $A = t/2$) and the actual area. The fit doesn't look very good.

Table 2.14

Predicted versus Actual Fungal Growth

Time t (hours)	Predicted Area A $A = t/2$ (sq cm)	Actual Area A (sq cm)
0	$(0)/2 = 0.0$	0.0
5	$(5)/2 = 2.5$	0.6
10	$(10)/2 = 5.0$	2.5
15	$(15)/2 = 7.5$	5.6
20	$(20)/2 = 10.0$	10.0
25	$(25)/2 = 12.5$	15.6

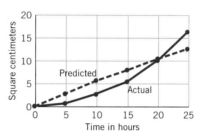

Figure 2.19 Predicted versus the actual area growth of *Sordaria finicola* colony.

As a second guess, we may try a function formula involving a square. If we again try to match the data point $t = 20, A = 10$, we now may write the formula

$$A = t^2/40$$

Carrying out the instructions on the right side, we can set $t = 20$ and compute $A = t^2/40 = 20^2/40 = 400/40 = 10$. Again, we check whether this formula correctly matches the other data points (see Table 2.15 and Figure 2.20). This time the fit is very close! The formula $A = t^2/40$ is a model that summarizes all the information in the original table of data values.

Table 2.15

New Predicted versus Actual Fungal Growth

Time t (hours)	Predicted Area A (sq cm)	Actual Area A (sq cm)
0	$(0^2)/40 = 0.0$	0.0
5	$(5^2)/40 = 0.6$	0.6
10	$(10^2)/40 = 2.5$	2.5
15	$(15^2)/40 = 5.6$	5.6
20	$(20^2)/40 = 10.0$	10.0
25	$(25^2)/40 = 15.6$	15.6

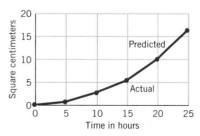

Figure 2.20 Predicted and actual area coincide.

This formula describes growth of the fungus colony for values of t between 0 and 25 hours. It may be true for times somewhat beyond 25 hours, but we do not know how far; perhaps at 26 hours the colony met the edge of the petri dish and stopped growing completely. The domain of the function is the values of t from 0 to 25 hours.

We should remember that determining this formula does not tell us *why* the formula is true. It is simply a way of summarizing the information we have measured. However, knowing the functional expression for the data can suggest various ideas for the reason. In this case, we might guess that the colony has the shape of a circle whose radius is increasing at a constant rate.

Algebra Aerobics 2.4

For each of the following tables find a function which takes the x values and produces the y values.

1.

x	y
0	0
1	3
2	6
3	9
4	12

2.

x	y
0	-2
1	1
2	4
3	7
4	10

3.

x	y
0	0
1	-1
2	-4
3	-9
4	-16

Hint: Table 2 is related to Table 1.

A Messier Example

Scatter plots that come from real data are rarely as clean as the fungus data. Real data are usually messy. As in Figure 2.21, there may be many values of the dependent vari-

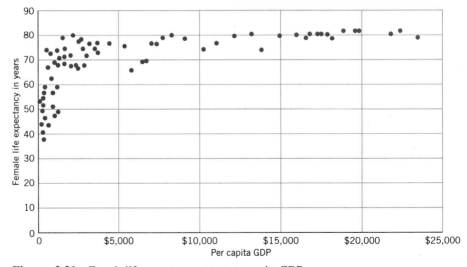

Figure 2.21   Female life expectancy versus per capita GDP.

*See the Excel or graph link
file NATIONS.*

able associated with a single value of the independent variable. This particular graph shows a scatter plot of the relationship between per capita gross domestic product (GDP) measured in U.S. dollars, and female life expectancy for 91 different countries. (The raw data can be found on the CD-ROM.) The gross domestic product for a country is the total output of goods and services produced in that country and valued at market prices. "Per capita" is a Latin phrase that means *per person*.

Clearly we could not find a function that would be a perfect fit to the data. Yet the graph does seem to suggest a relationship between per capita GDP and life expectancy. If the GDP is less than $5000 per person, an increase in per capita GDP seems to be associated with an increase in female life expectancy. If the per capita GDP is greater than $5000, the scatter plot is fairly flat, suggesting that an additional increase in per capita GDP does not significantly increase female life expectancy. There seems to be a "ceiling" for life expectancy of about 80 years. We can imagine penciling in a function that would roughly approximate the shape of the data (Figure 2.22).

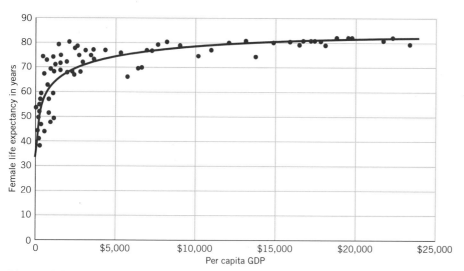

Figure 2.22   Possible model for female life expectancy as a function of per capita GDP.

If we could find an equation for this function, we might have a reasonable way of describing the relationship between per capita GDP and life expectancy. This function would be only an approximation of the data, a simplification of reality. But it might provide a model that could help us to describe and understand what is happening. We can then go back to the raw data and explore in more detail countries with a per capita GDP below and above $5000 or those countries which seem to fall outside the general pattern. We can ask whether the same relationship holds for male life expectancy. We can study the relationship between life expectancy and other variables such as the number of doctors, or pollution levels.

Mathematical models can provide tools for understanding *what* is happening, but they do not tell us *why* it is happening. Why does female life expectancy remain fairly stable if per capita GDP is greater than $5000? Why does it increase as the per capita GDP increases, but remains less than $5000? We can posit plausible explanations. We could hypothesize that up to a certain point greater per person national wealth implies access to better health care and hence longer life expectancy. However we should be aware that we have now taken a leap beyond the mathematical description of the data.

What Lies Ahead

The rest of this text is dedicated to building a library of functions that can serve as mathematical models for real world data sets like this one. We will study the properties of these functions to enable us to better understand how they can be used to describe the world around us.

CHAPTER SUMMARY

A variable y is a *function* of variable x if each value of x determines one and only one value of y. In such cases, y is the *value of the function* or the *dependent variable* and x is the *independent variable*.

We can give a letter name to the function; f is a common obvious choice. We write

$$y = f(x)$$

to indicate that y is a function of x. A function, f, is a rule that tells us what to do to a value of an independent variable in order to produce one and only one value of the dependent variable. The rule could be defined using an equation, words, a graph, or a table.

The $f(x)$ notation is particularly useful when a function is being evaluated at a specific point. For example, if $f(P) = T$ is defined by the equation $T = 0.08P$, instead of saying "the value for T when $P = 10$," we simply write "$f(10)$." Then we have

$$f(10) = (0.08)(10)$$
$$= 0.8$$

The *domain* of a function is the set of all possible values of the independent variable. The *range* is the set of corresponding values of the dependent variable. Unless otherwise specified, the domain is assumed to be the set of all real numbers x such that the corresponding y values in the range are real numbers.

The *solutions* to an equation in two variables x and y are the set of ordered pairs of the form (x, y) that make the equation a true statement. The *graph* of an equation is the set of points corresponding to the ordered-pair solutions where the first coordinate corresponds to values on the horizontal axis and the second to values on the vertical axis.

If an equation represents a function, then solving the equation for the dependent variable in terms of the independent variable is called putting the equation in *function form*. The ordered pair solutions of the function are conventionally listed with the value of the independent variable first, and the dependent variable second.

(independent variable, dependent variable)

The graph of a function by convention has the independent variable on the horizontal axis and the dependent variable on the vertical axis.

A graph does not represent a function if it fails the *vertical line test*, that is, if you can draw a vertical line that crosses the graph two or more times.

A formula or equation used to describe a real world phenomenon is called a *mathematical model*.

EXERCISES

1. Consider the following graph of U.S. military sales to foreign governments.
   a. During what years did sales increase?
   b. During what years did sales decrease?
   c. Estimate the maximum value for sales.
   d. Estimate the minimum value for sales.

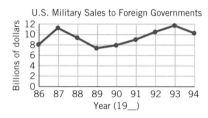

U.S. Military Sales to Foreign Governments

Source: U.S. Bureau of the Census, in *The American Almanac 1996–1997: Statistical Abstract of the United States.* 1996.

2. The following graph shows the 24-hour temperature cycle of a normal man. The man was confined to bed in order to minimize temperature fluctuations caused by activity.
   a. Estimate the man's maximum temperature.
   b. Estimate his minimum temperature.
   c. Give a short general description of the 24-hour temperature cycle.

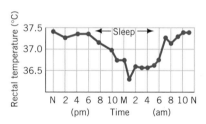

Source: V.B. Mountcastle, *Medical Physiology*, vol. 2, ed. 14. St. Louis, Mosby-Year Book, Inc.

3. Sketch a plausible graph for each of the following and label the axes:
   a. The amount of snow on your backyard each day from December 1 to March 1.
   b. The temperature during a 24-hour period in your home town during one day in July.
   c. The amount of water inside your fishing boat if your boat leaks a little and your fishing partner bails out water every once in a while.
   d. The total hours of daylight each day of the year.
   e. The temperature of a cup of hot coffee left to stand.
   f. The temperature of an ice-cold drink left to stand.

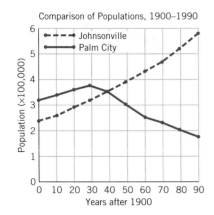

Comparison of Populations, 1900–1990

4. Look at the graph that shows the populations of two towns.
    a. What is the range of population size for Johnsonville? For Palm City?
    b. During what years did the population of Palm City increase?
    c. During what years did the population of Palm City decrease?
    d. When were the populations equal?

5. Given the four points $(-1, 3)$, $(1, 0)$, $(2, 3)$, and $(1, 2)$, for each of the following equations, identify which points are solutions for that equation.

    a. $y = 2x + 5$ c. $y = x^2 - x + 1$

    b. $y = x^2 - 1$ d. $y = \dfrac{4}{x + 1}$

6. In each of the following three examples:
    a. identify logical choices for the independent and dependent variable
    b. verify whether or not the example represents a function

    If the example is a function, then:
    c. generate a table of values (if not already given)
    d. generate a graph (if not already given)
    e. identify the intervals on which the function is increasing or decreasing
    f. identify any maximum or minimum values of the function

Example 1

Table 2.2 and Figure 2.2 showed the annual budget deficit. The following table and graph show the accumulated gross federal debt. Data is available in Excel or graph link file FEDDEBT.

Accumulated Gross Federal Debt

Year	Billions of $	Year	Billions of $
1945	260	1981	994
1950	257	1982	1137
1955	274	1983	1371
1960	291	1984	1564
1965	322	1985	1827
1970	361	1986	2120
1971	408	1987	2346
1972	436	1988	2601
1973	466	1989	2868
1974	484	1990	3266
1975	541	1991	3599
1976	629	1992	4082
1977	706	1993	4436
1978	777	1994	4721
1979	829	1995	4961
1980	909	(est)	

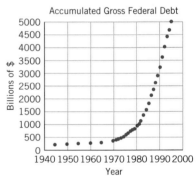

Accumulated Gross Federal Debt

Source: U.S. Office of Management and Budget, "Budget of the United States Government," annual, in *The American Almanac 1995–1996 and 1996–1997: Statistical Abstract of the United States,* 1995 and 1996.

Example 2
The formula

$$A = 25W$$

where A = ampicillin dosage in milligrams, and W = child's weight in kilograms, represents the *minimum* effective pediatric daily drug dosage of ampicillin as a function of a child's weight. You may recall that $D = 50W$ represented the *maximum* recommended dosage in Algebra Aerobics 2.2(a). Both formulas apply only for children weighing up to 10 kilograms.

Example 3

The following table shows the progress of national regulations in controlling carbon monoxide emissions.

National Ambient Air Pollutant Concentrations: 1985 to 1994 [Air quality standard is 0 parts per million (ppm)]									
	1985	1987	1988	1989	1990	1991	1992	1993	1994
Carbon monoxide (ppm)	6.97	6.69	6.38	6.34	5.87	5.55	5.18	4.88	5.57

Source: U.S. Environmental Protection Agency, "National Air Quality and Emissions Trends Report" annual, as taken from *The American Almanac 1995–1996: Statistical Abstract of the United States,* 1995.

7. Consider the following chart of juvenile arrests for murder.
    a. During what intervals did the number of arrests show a slight decrease?
    b. Approximate the number of juvenile arrests for murder in 1988 and 1993.
    c. Estimate the ratio of the number of arrests in 1993 to the number of arrests in 1988.
    d. What can be said generally about the number of juveniles arrested for murder based on this chart?

Source: U.S. Bureau of the Census, in *The American Almanac 1996–1997: Statistical Abstract of the United States,* 1996.

8. Assume that for persons who earn less than $20,000 a year, income tax is 16% of their income.
    a. Generate a formula that describes income tax in terms of income for people earning less than $20,000 a year.
    b. What are you treating as the independent variable? The dependent variable?
    c. Does your formula represent a function? Explain.
    d. If it is a function, what is the domain? The range?

9. The price of gasoline is $1.24 per gallon.
    a. Generate a formula that describes the cost of buying gas as a function of the amount of gasoline purchased.
    b. What is the independent variable? The dependent variable?

    c. Does your formula represent a function? Explain.

    d. If it is a function, what is a suitable domain? Range?

    e. Generate a small table of values and a graph.

10. Consider the following table, listing the weight (W) and height (H) of five individuals. Based on this table, is height a function of weight? Is weight a function of height?

Weight (pounds) W	Height (inches) H
120	54
120	55
125	58
130	60
135	56

11. Consider the following table.

Y	P
1990	$1.4
1991	$2.3
1992	$0
1993	$-$0.5
1994	$1.4
1995	$1.2

    a. Is P a function of Y?

    b. What is the domain? What is the range?

    c. What is the maximum value of P? In what year did this occur?

    d. During what intervals was P increasing? Decreasing?

    e. Now consider P as the independent variable and Y as the dependent variable. Is Y a function of P?

12. Look at the following table.

    a. Find $p(-4)$, $p(5)$ and $p(1)$.

    b. For what value(s) of n does $p(n) = 2$?

n	$p(n)$
-4	0.063
-3	0.125
-2	0.25
-1	0.5
0	1
1	2
2	4
3	8
4	16
5	32

13. Consider the function graphed in the figure to the right.

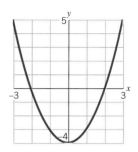

  a. Find $f(-3)$, $f(0)$, and $f(1)$

  b. Find two values of x such that $f(x) = 0$

14. For each equation below, write an equivalent equation that expresses z in terms of t. Is z a function of t? Why or why not?

  a. $3t - 5z = 10$

  b. $(2t - 3)(t + 4) - 2z = 0$

  c. $t + z^2 + 10 = 0$

15. Find the implied domains for each of the following functions:

  a. $f(x) = 300.4 + 3.2x$

  b. $g(x) = \sqrt{x + 2}$

  c. $g(x) = \dfrac{5 - 2x}{2}$

  d. $j(x) = \dfrac{1}{x + 1}$

  e. $k(x) = 3$

  f. $l(x) = \dfrac{x}{4 - x^2}$

16. Determine whether y is a function of x in each of the following equations. If the equation is not a function, find a value of x that is associated with 2 different y values.

  a. $y = x^2 + 1$

  b. $y = 3x - 2$

  c. $y = 5$

  d. $x^2 + y^2 = 25$

  e. $y = \sqrt{x}$

  f. $y^2 = x$

17. Given $f(x) = 1 - 0.5x$ and $g(x) = x^2 + 1$, evaluate

  a. $f(0) + g(0)$

  b. $f(-2) - g(-3)$

  c. $g(2)/f(1)$

  d. $f(3) \cdot g(3)$

18. Which of the following graphs describe functions?

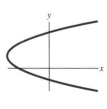

(a)

(b)

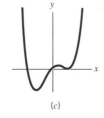

(c)

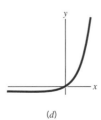

(d)

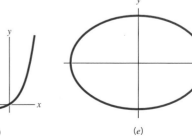

(e)

EXPLORATION 2.1

Picturing Functions

Objective

- develop an intuitive understanding of functions.

Materials/Equipment

None required

Procedure

PART I

Class Discussion

Bridget, the 6-year-old daughter of a professor at the University of Pittsburgh, loves playing with her rubber duckie in the bath at night. Her mother drew the following graph for her math class. It shows the water level in Bridget's tub as a function of time.

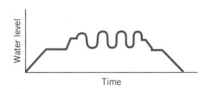

Pick out the time period during which:

- the tub is being filled
- Bridget is entering the tub
- she is playing with her rubber duckie
- she leaves the tub
- the tub is being drained

With a Partner

Create your own graph of a function that tells a story. Be as inventive as possible. (Some students have drawn functions that showed the decibel levels during a phone conversation of a boy and girl friend, number of hours spent doing homework during one week, amount of money in their pocket during the week.)

Class Discussion

Draw your graph on the blackboard and tell its story to the class.

PART II

With a Partner

Generate a plausible graph for each of the following:

1. Time spent driving to work as a function of the amount of snow on the road. (Note: the first inch or so may not make any difference; the domain may be only up to about a foot of snow since after that you may not be able to get to work.)
2. The hours of daylight as a function of the time of year.
3. The temperature of the coffee in your cup as a function of time.
4. The distance that a cannonball (or javelin or baseball) travels as a function of the angle of elevation at which is it is launched. The maximum distance is attained for angles of around 45°.
5. Assume that you leave your home walking at a normal pace, realize you have forgotten your homework and run home, and then run even faster to school. You sit for a while in a classroom and then walk leisurely home. Now plot your distance from home as a function of time.

BONUS QUESTION

Assume that water is pouring into each of these containers at a constant rate. The height of water in the container is a function of the volume of liquid. Sketch a graph of this function for each container.

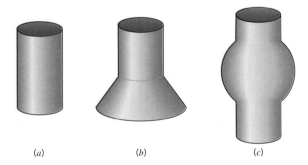

(a)   (b)   (c)

Discussion/Analysis

Are your graphs similar to those generated by the rest of the class? Can you agree as a class as to the basic shape of each of the graphs? Are there instances in which the graphs could look quite different?

EXPLORATION 2.2

Deducing Formulas to Describe Data

Objectives

- find and describe patterns in data
- deduce functional formulas from data tables
- extend patterns using functional formulas

Material/Equipment

None required

Procedure

Class Discussion

1. Examine the following data tables. In each table, look for a pattern in terms of how y changes when x changes. Explain in your own words how to find y in terms of x.

(a)

x	y
0	0.0
1	0.5
2	1.0
3	1.5
4	2.0

(b)

x	y
0	5
1	8
2	11
3	14
4	17

2. Assuming that the pattern continues indefinitely, use the rule you have found to extend the data table to include negative numbers for x.

3. Check your extended data tables. Did you find only one value for y given a particular value for x?

4. Use a formula to describe the pattern that you have found. Do you think this formula describes a function? Explain.

On Your Own

1. For each of the data tables (c)–(h), explain in your own words how to find y in terms of x. Then extend each table using the rules you have found.

(c)

x	y
0	0
1	1
2	4
3	9
4	16

(d)

x	y
0	0
1	1
2	8
3	27
4	64

(e)

x	y
0	0
1	2
2	12
3	36
4	80

Hint: For Table (e) think about some combination of data in Tables (c) and (d).

(f)	x	y
	−2	0
	0	10
	5	35
	10	60
	100	510

(g)	x	y
	0	−1
	1	0
	2	3
	3	8
	4	15

(h)	x	y
	0	3
	10	8
	20	13
	30	18
	100	53

2. For each table, construct a formula to describe the pattern you have found.

Discussion/Analysis

With a Partner

Compare your results. Do the formulas that you have found describe functions? Explain.

Class Discussion

Does the rest of the class agree with your results? Remember that formulas that look different may give the same results.

Exploration-Linked Homework

1. **a.** For each of the following data tables, explain in your own words how to find y in terms of x. Using the rules you have found, extend the data tables to include negative numbers.

(i)	x	y
	−10	10.0
	0	0.0
	3	0.9
	8	6.4
	10	10.0

(j)	x	y
	0	−3
	1	1
	2	5
	3	9
	4	13

b. For each table, find a formula to describe the pattern you have found. Does your formula describe a function? Explain.

2. Make up a functional formula, generate a data table, and bring the data table on a separate piece of paper to class. The class will be asked to find your rule and express it as a formula.

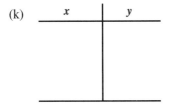

(k)

x	y

Rates of Change

Overview

In Chapter 2 we examined graphs and tables that showed change over time. We saw that the U.S. population has been increasing, the percentage of people living in rural areas has been decreasing, and the number of civil disturbances has cycled up and down. Here we examine different mathematical techniques for measuring such changes.

The *average rate of change* is one of the most useful measures for describing change. The average rate of change of the U.S. population with respect to time is calculated by dividing the change in the population by the change in time. On a graph, the average rate of change represents the slope of the line connecting two points. Change over time can also be described by calculating either the absolute or the percentage change in size.

The first Exploration asks you to identify an issue from a data set and to construct the strongest possible case for a point of view. You can use what you have learned in the first three chapters to argue strenuously one side of an issue. Then you are asked to construct an equally convincing argument for the opposing side. The second Exploration helps you see why rates of change are "averages."

After reading this chapter you should be able to:

- define the average rate of change
- calculate the slope of a line
- describe change over time in several different ways
- use mathematical tools to make convincing arguments

See Excel or graph link file USPOP.

Table 3.1

Population of the United States 1790 to 1995

Year	In Millions
1790	3.9
1800	5.3
1810	7.2
1820	9.6
1830	12.9
1840	17.1
1850	23.2
1860	31.4
1870	39.8
1880	50.2
1890	63.0
1900	76.0
1910	92.0
1920	105.7
1930	122.8
1940	131.7
1950	151.3
1960	179.3
1970	203.3
1980	226.5
1990	248.7
1992	255.4
1994	260.7
1995	263.0

3.1 AVERAGE RATES OF CHANGE

We can think of the U.S. population as a function of time. Table 3.1 and Figure 3.1 are two representations of that function. They show the changes in the size of the U.S. population since 1790, the year the U.S. government conducted its first decennial census. Time, as usual, is the independent variable and population size is the dependent variable.

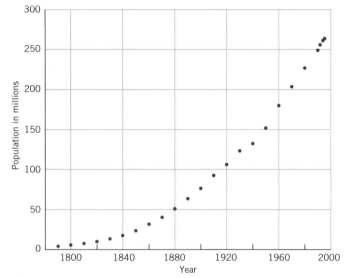

Figure 3.1   Population of the United States.

Source: U.S. Bureau of the Census, Current Population Reports in *The American Almanac 1995–1996 and 1996–1997: Statistical Abstract of the United States,* 1995 and 1996. (Data before 1960 excludes Alaska and Hawaii.)

Figure 3.1 clearly shows that the size of the U.S. population has been growing over the last two centuries, and growing at what looks like an increasingly rapid rate. How can the change in population over time be described quantitatively? Any two points on the graph of the data can be used to find how much the population has changed during the time period between them.

Suppose we look at the change in the population between 1900 and 1992. In 1900 the population was 76.0 million; by 1992 the population had grown to 255.4 million. How much did the population increase? Since it rose from 76.0 million to 255.4 million, it increased by the difference between these two values.

$$\text{change in population size} = 255.4 - 76.0 \text{ million}$$

$$= 179.4 \text{ million}$$

In Figure 3.2, we have drawn two parallel horizontal lines through the points (1900, 76.0) and (1992, 255.4) to the vertical population axis. The 179.4 million population change is represented by the difference in the height of the two points or the change in the variable on the vertical axis.

Knowing that the population changed by 179.4 million tells us nothing about how rapid the change was; this change clearly represents much more dramatic growth if it

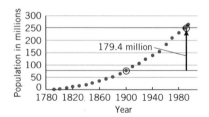

Figure 3.2   Population change: 179.4 million.

happened over 20 years than if it happened over 200 years. In this case, the length of time over which the change in population occurred is

$$\text{change in years} = 1992 - 1900 \text{ years}$$
$$= 92 \text{ years}$$

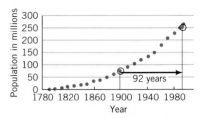

Figure 3.3   Time change: 92 years.

In Figure 3.3 we have drawn two parallel vertical lines from the points (1900, 76.0) and (1992, 255.4) down to the horizontal time axis. The 92-year change is represented by the horizontal difference in the positions of the two points.

If a line segment is drawn connecting the two points, it forms the hypotenuse of the right triangle sketched in Figure 3.4. The length of the horizontal section of the triangle represents a change of 92 years, and the length of the vertical section of the triangle represents a change of 179.4 million in population size.

To find the *average rate of change* in population per year from 1900 to 1992, divide the change in the population by the change in the number of years:

$$\text{average rate of change} = \text{change in population size/change in years}$$
$$= 179.4 \text{ million people/92 years}$$
$$= 1.95 \text{ million people/year}$$

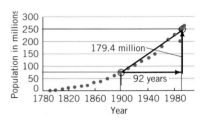

Figure 3.4   Population change over 92 years.

This calculation shows that "on average" the population grew at a rate of 1.95 million people per year from 1900 to 1992. In the phrase "million people/year" the slash sign represents division and is read as "per," so the phrase is read as "million people per year."

Limitations on the Use of Average Rates of Change

Average rates of change have the same limitations as any *average*. Although the average rate of change of the U.S. population from 1900 to 1992 was 1.95 million per year, it is highly unlikely that in each year the population grew by exactly 1.95 million. The number 1.95 million per year is, as the name states, an *average*. If the arithmetic average or *mean* height of students in your class is 67 inches, you don't expect every student to be 67 inches tall.

The average rate of change depends entirely on the end points you use to calculate the rate   If the data points do not all lie on a straight line, the average rate of change varies for different intervals. For instance, if you compare the average rates of change in population for the time intervals 1840 to 1940 and 1880 to 1980 you get two different values (Table 3.2). You can see on the graphs that the line segment is much steeper from 1880 to 1980 than from 1840 to 1940 (Figures 3.5 and 3.6). Different intervals give different impressions of the rate of change in the U.S. population, so it is important to state which end points are used.

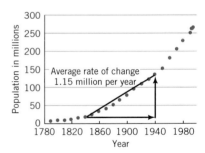

Figure 3.5   Average rate of change: 1840–1940.

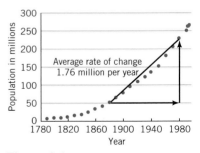

Figure 3.6   Average rate of change: 1880–1980.

Table 3.2

Time Interval	Change in Time	Change in Population	Average Rate of Change
1840 to 1940	100 yrs	$131.7 - 17.1 = 114.6$ million	$\dfrac{114.6 \text{ million}}{100 \text{ yr}} = 1.15$ million/yr
1880 to 1980	100 yrs	$226.5 - 50.2 = 176.3$ million	$\dfrac{176.3 \text{ million}}{100 \text{ yr}} = 1.76$ million/yr

The average rate of change does not reflect all the fluctuations in population size that may occur between the end points. The average rate of change over a large interval gives an overall average estimate; for more specific information, the average rate of change can be calculated for smaller intervals.

Algebra Aerobics 3.1

1. Suppose your weight five years ago was 135 pounds and your weight today is 143 pounds. Find the average rate of change in your weight with respect to time.

2. The United States imported electronic equipment and accessories valued at $18.3 trillion in 1989 and valued at $27.5 trillion in 1993. Find the average rate of change from 1989 to 1993 in the value of the imported electronics.

3. Table 3.3 indicates the number of deaths in motor vehicle accidents in the United States.

Find the average rate of change:

a. From 1970 to 1980

b. From 1980 to 1992

4. If you have access to the course software, open up "C1: U.S. Population" in *Rates of Change*. Clicking and dragging on the mouse will fix one end point and vary a second end point. Track the corresponding average rate of change calculation to develop an intuition for how its values vary.

Table 3.3

Annual Deaths in Motor Vehicle Accidents

1970	1980	1990	1992
114,638	105,718	91,983	86,777

3.2 A CLOSER LOOK AT AVERAGE RATES OF CHANGE

The concept of the average rate of change may give a better sense of what is happening if we compare rate of change calculations over different intervals. One way to do this is to pick a fixed interval size and then calculate the average rate of change for successive intervals. On the U.S. population data, using an interval size of one decade, we can add a third column to the data table in which each entry represents the average population growth *per year* (i.e., the average annual rate of change) during the previous 10 years.[1] If we "connect the dots" on the original graph in Figure 3.1, these numbers would represent the slopes of the small line segments connecting adjacent data points. A few of these calculations are shown in the last column in Table 3.4.

What is happening to the average rate of change over time? Start at the top of the third column and scan down the numbers. Notice that until 1910 the average rate of change increases every year. Not only is the population growing every decade, but it is growing at an increasing rate. It's like a car that is not only moving forward, but accel-

---

[1] We have omitted the information about the years between 1992 and 1995 given in Table 3.1, since the time period was not a full decade.

Table 3.4

Average Annual Rates of Change of U.S. Population from 1790 to 1990

Year	Population (in millions)	Average Annual Rate for Prior Decade (in millions/yr)	Sample Calculations
1790	3.9	Data not available	
1800	5.3	0.14	$0.14 = (5.3 - 3.9)/10$
1810	7.2	0.19	
1820	9.6	0.24	
1830	12.9	0.33	
1840	17.1	0.42	$0.42 = (17.1 - 12.9)/10$
1850	23.2	0.61	
1860	31.4	0.82	
1870	39.8	0.84	
1880	50.2	1.04	
1890	63.0	1.28	
1900	76.0	1.30	
1910	92.0	1.60	
1920	105.7	1.37	
1930	122.8	1.71	
1940	131.7	0.89	$0.89 = (131.7 - 122.8)/10$
1950	151.3	1.96	
1960	179.3	2.80	
1970	203.3	2.40	
1980	226.5	2.32	
1990	248.7	2.22	

erating. A feature that was not so obvious in the original data is now evident: in the decades 1910–1920 and 1930–1940 and in the decades following 1960 we see a decreasing rate of growth. It's like a car decelerating—it is still moving forward but it is slowing down.

The graph (in Figure 3.7) with years on the horizontal axis and average rates of change on the vertical axis shows more clearly how the average rate of change fluctuates over time. The first point, corresponding to the year 1800, shows an average rate of change of 0.14 million people/year for the decade 1790–1800. The rate 1.71, corresponding to the year 1930, means that from 1920 to 1930 the population was increasing at a rate of 1.71 million people per year.

Take the time to interpret a few points; this graph is somewhat more abstract than the previous ones. The pattern of growth was fairly steady up until about 1910. Why did it change? During the decades when the rate of growth was decreasing, did more people die, or were there fewer people born? Did the immigration rate slow down? Any one of these reasons would slow the growth rate. To form some hypotheses you may need to review history. For instance, a possible explanation for the slowdown in the decade prior to 1920 might be World War I and the 1918 flu epidemic that by 1920 had killed nearly 20,000,000 people, including about 500,000 Americans.

In Figure 3.7, the steepest decline in the average rate of change is between 1930 and 1940. One obvious suspect for the big slowdown in population growth in the 1930s is the Great Depression. Look back at Figure 3.1, the graph that shows the overall growth in the U.S. population. The decrease in the average rate of change in the

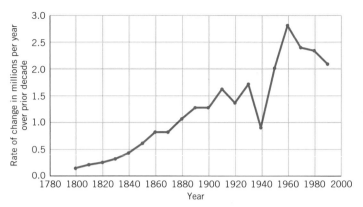

Figure 3.7   Average rates of change in the U.S. population by decade.

Something to think about

What might be some reasons for the slowdown in population growth from the 1960s onward? Where could you turn to find some evidence to test your hypotheses? Questions arising from one set of data often prompt the search for more data.

1930s is large enough to show up in our original graph as a visible slowdown in population growth.[2] Why is there a drastic decline in growth in the 1930s? Perhaps during the Great Depression people just couldn't afford to have children.

Algebra Aerobics 3.2

Table 3.5

World Population

Year	Total Population (in millions)	Average Rate of Change
1800	910	n.a.
1850	1130	
1900	1600	
1950	2510	
1970	3702	
1980	4456	
1990	5293	
1993	5555	
1996	5771	

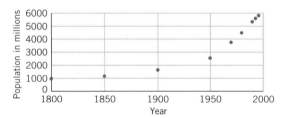

Figure 3.8   World population.

1. Table 3.5 and Figure 3.8 show estimates for world population between 1800 and 1996. (See Excel or graph link file WORLDPOP.)

    a. Fill in the third column on the table by calculating the average annual rate of change.

    b. Graph the average annual rate of change versus time.

    c. During what period was the average annual rate of change the largest?

    d. Describe in general terms what happened to the average rate of change between 1880 and 1996.

Source for table and graph: U.S. Bureau of the Census, Current Population Reports in *The American Almanac 1995–1996 and 1996–1997: Statistical Abstract of the United States,* 1995 and 1996. (Data before 1960 excludes Alaska and Hawaii.)

---

[2] If you zoomed in on the original graph, the other fluctuations in growth rate would show up as well.

3.3 GENERALIZING THE CONCEPT OF THE AVERAGE RATE OF CHANGE

Although our examples so far have described the change in one variable over time, the notion of the *average rate of change* can be used to describe the change in any variable with respect to another. If you have a graph that represents a plot of data points, then the average rate of change between two points is always the change in the vertical variable divided by the change in the horizontal variable.

$$\text{average rate of change} = \frac{\text{change in vertical variable}}{\text{change in horizontal variable}}$$

If the variables represent real world quantities, which have units of measure (e.g., dollars or years), then the average rate of change should be represented in terms of the appropriate units.

$$\text{units of average rate of change} = \frac{\text{units of vertical variable}}{\text{units of horizontal variable}}$$

For example, the units might be dollars/year (read as dollars per year) or pounds/person (read as pounds per person).

On a graph, the average rate of change represents the *slope* of the line connecting two points. The slope is an indicator of the steepness of the line. We often refer to the change in the vertical variable as the *rise* and the change in the horizontal variable as the *run*. The slope is the rise divided by the run.

If (x_1, y_1) and (x_2, y_2) are two points, then $(y_2 - y_1)$ gives the *rise*. This difference is often denoted by Δy (read as "delta y"), where Δ is the Greek letter capital D (think of D as representing difference).

$$\Delta y = y_2 - y_1$$

Similarly, the *run* (delta x) can be represented by:

$$\Delta x = x_2 - x_1$$

$$\text{slope} = \frac{rise}{run} = \frac{y_2 - y_1}{x_2 - x_1} = \frac{\Delta y}{\Delta x} = \frac{\text{change in } y}{\text{change in } x}$$

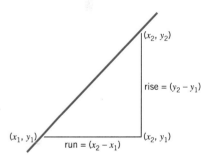

Figure 3.9   Slope = rise/run.

The average rate of change represents a *slope*. Given two points (x_1, y_1) and (x_2, y_2)

$$\text{average rate of change} = \text{slope} = \frac{rise}{run} = \frac{y_2 - y_1}{x_2 - x_1} = \frac{\Delta y}{\Delta x} = \frac{\text{change in } y}{\text{change in } x}$$

A note about calculating slopes: it doesn't matter which point is first   Given two points, (x_1, y_1) and (x_2, y_2), it doesn't matter which one we use as the first point when we calculate the slope; the result is the same either way. We do need to be consistent in the order in which the coordinates appear in the numerator and

The Anthology reading "Slopes" describes many of the practical applications of slopes, from cowboy boots to handicap ramps.

the denominator. If y_1 is the first term in the numerator, then x_1 must be the first term in the denominator.

$$\text{slope} = \frac{y_2 - y_1}{x_2 - x_1} \qquad \text{or} \qquad \text{slope} = \frac{y_1 - y_2}{x_1 - x_2}$$

Example   1

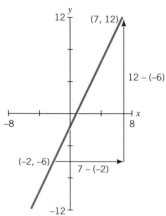

Figure 3.10 $(y_1 - y_2)/(x_1 - x_2) = (y_2 - y_1)/(x_2 - x_1)$.

Given the two points $(-2, -6)$ and $(7, 12)$, we can calculate the slope treating $(-2, -6)$ as (x_1, y_1) and $(7, 12)$ as (x_2, y_2) (Figure 3.10). Then:

$$\text{slope} = \frac{y_2 - y_1}{x_2 - x_1} = \frac{12 - (-6)}{7 - (-2)} = \frac{18}{9} = 2$$

Or, we could have used -6 and -2 as the first terms in the numerator and denominator, respectively. Either way we obtain the same answer.

$$\text{slope} = \frac{y_1 - y_2}{x_1 - x_2} = \frac{-6 - 12}{-2 - 7} = \frac{-18}{-9} = 2$$

• • •

An important aside: proving it doesn't matter which point is first   We saw in the preceding example that the order of points didn't matter when we calculated the slope of the line connecting $(-2, -6)$ and $(7, 12)$. We can show that this is true given any two points (x_1, y_1) and (x_2, y_2). Suppose we use y_2 and x_2 as the first terms in the numerator and denominator, respectively. Then:

$$\text{slope} = \frac{y_2 - y_1}{x_2 - x_1}$$

multiply by $\dfrac{-1}{-1}$

$$= \frac{-1}{-1} \cdot \frac{y_2 - y_1}{x_2 - x_1}$$

simplify

$$= \frac{-y_2 + y_1}{-x_2 + x_1}$$

rearrange terms

$$= \frac{y_1 - y_2}{x_1 - x_2}$$

We end up with an equivalent expression in which y_1 and x_1 are now the first terms in the numerator and denominator, respectively.

Example   2

Construct a graph of the data in Table 3.6. Calculate the average rate of change of the percentage of the U.S. population living in rural areas from 1850 to 1940 and interpret the result.

Table 3.6									
Percentage of U.S. Population Living in Rural Areas									
Year	**%**	**Year**	**%**	**Year**	**%**	**Year**	**%**	**Year**	**%**
1850	84.7	1880	71.8	1910	54.3	1940	43.5	1970	26.4
1860	80.2	1890	64.9	1920	48.8	1950	36.0	1980	26.3
1870	74.3	1900	60.3	1930	43.8	1960	30.1	1990	24.8

Source: U.S. Bureau of the Census, Historical Statistics—Colonial Times to 1970, in *The American Almanac 1993–1994: Statistical Abstract of the United States,* 1993.

SOLUTION

The data are graphed in Figure 3.11.

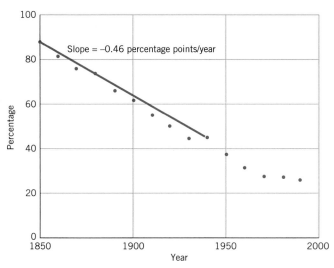

Figure 3.11   Percentage of the U.S. population living in rural areas.

If we connect the two end points (1850, 84.7%) and (1940, 43.5%) we see that the line slopes down, since the percentage of the population that is rural is *declining*. Here we have an example in which the slope or average rate of change is negative.

$$\text{average rate of change} = \frac{\text{change in percent of population that is rural}}{\text{change in years}}$$

$$= \frac{\Delta \text{ percent of population that is rural}}{\Delta \text{ years}}$$

$$= \frac{43.5\% - 84.7\%}{1940 - 1850}$$

$$= \frac{-41.2\%}{90 \text{ yr}}$$

$$= -0.46\%/\text{yr}$$

This means that the percentage of the population that lived in rural areas decreased on the average by 0.46 percentage point (a little under one-half of a percent) each year from 1850 to 1940. That may not seem like very much on an annual basis, but in less than a century the rural population went from being the large majority (84.7%) to less than half of the U.S. population (43.5%). By 1990 the rural population represented less than 25% of the total population.

Something to think about

Were 1850 and 1940 good choices for end points, or would another time period have given an average rate of change (or slope) more typical of all the data? What seems to be happening to the average rate of change between 1970 and 1990? What kind of social and economic implications does a population shift of this magnitude have on society?

The data in Table 3.7 and the scatter plot in Figure 3.12 show median family income during the 1970s and 1980s. (See also Algebra Aerobics 2.1 in Chapter 2.) How could we use the information to make a case that families are better off? Worse off?

Example   3

Table 3.7

Year	Median Family Income in Constant 1990 Dollars
1973	35,474
1974	34,205
1975	33,328
1976	34,359
1977	34,528
1978	35,361
1979	35,262
1980	33,346
1981	32,190
1982	31,738
1983	32,378
1984	33,251
1985	33,689
1986	35,129
1987	35,632
1988	35,565
1989	36,062
1990	35,353

Source: U.S. Bureau of the Census, Historical Statistics—Colonial Times to 1970, in *The American Almanac 1993–1994: Statistical Abstract of the United States,* 1993.

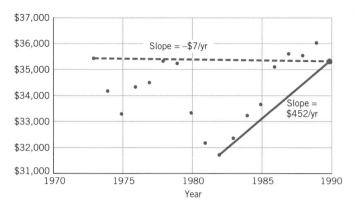

Figure 3.12   Median family income in 1990 dollars.

SOLUTION

To make an optimistic case that families are better off, we could choose as end points (1982, $31738) and (1990, $35353). Then

$$\text{average rate of change} = \frac{\text{change in median income}}{\text{change in years}}$$
$$= \frac{\$35,353 - \$31,738}{1990 - 1982}$$
$$= \frac{\$3615}{8 \text{ yr}}$$
$$= \$452/\text{yr}$$

So between 1982 and 1990 the median family income *increased* by $452 per year. We can see this reflected in Figure 3.12 in the steep positive slope of the line connecting (1982, $31738) and (1990, $35353).

A gloomy interpretation of the same data could be constructed merely by changing the left hand end point to (1973, $35474). Then

$$\text{average rate of change} = \frac{\text{change in median income}}{\text{change in years}}$$
$$= \frac{\$35,353 - \$35,474}{1990 - 1973}$$
$$= \frac{-\$121}{17 \text{ yr}}$$
$$= -\$7/\text{yr}$$

So between 1973 and 1990 the median family income *decreased* by $7 per year! A glimpse at Figure 3.12 shows a slight negative slope of the dotted line connecting (1973, $35474) and (1990, $35353).

It is important to state the period of time represented by such calculations. Both average rates of change are correct, but they certainly give very different impressions of the well-being of American families.

Given Table 3.8 of civil disturbances over time, plot and then connect the points, and (without doing any calculations) indicate on the graph when the average rate of change between adjacent data points is positive ($+$), and negative ($-$), or zero (0).

Example 4

SOLUTION

The data are graphed in Figure 3.13. The $+$, $-$, and 0 indicate when the average rate of change between adjacent points is positive, negative, or zero, respectively. This is, of course, the same thing as indicating when the slope of the line segment is positive, negative, or zero. The largest average rate of change, or steepest slope, seems to be between the Jan.–Mar. and Apr.–June counts in 1968. The largest negative average rate of change appears later in the same year (1968) between the July–Sept. and Oct.–Dec. counts.

Table 3.8		
Civil Disturbances in U.S. Cities		
	Period	Count
1968	Jan.–Mar.	6
	Apr.–June	46
	July–Sept.	25
	Oct.–Dec.	3
1969	Jan.–Mar.	5
	Apr.–June	27
	July–Sept.	19
	Oct.–Dec.	6
1970	Jan.–Mar.	26
	Apr.–June	24
	July–Sept.	20
	Oct.–Dec.	6
1971	Jan.–Mar.	12
	Apr.–June	21
	July–Sept.	5
	Oct.–Dec.	1
1972	Jan.–Mar.	3
	Apr.–June	8
	July–Sept.	5
	Oct.–Dec.	5

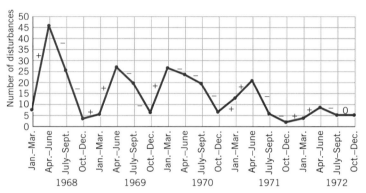

Figure 3.13   Civil disturbances during the years 1968–1972.

Source: Introduction to the Practice of Statistics, by Moore and McCabe. Copyright © 1989 by W.H. Freeman and Company. Used with permission.

Civil disturbances between 1968 and 1972 occurred in cycles: the largest numbers occurred in the summer months and the smallest in the winter months. The peaks decrease over time. What was happening in America that might correlate with the peaks? This was a tumultuous period in our history. Many previously silent factions of society were finding their voices. Recall that in April 1968 Martin Luther King was assassinated and that in January 1973 the last American troops were withdrawn from Vietnam.

Plot the points $(-3, -10)$ and $(3, 2)$ and calculate the slope of the line on which they lie.

Example 5

SOLUTION

The points $(-3, -10)$ and $(3, 2)$ and the line connecting them are plotted in Figure 3.14. The slope of the line is $\Delta y/\Delta x$, which equals

$$\frac{y_2 - y_1}{x_2 - x_1} = \frac{2 - (-10)}{3 - (-3)}$$

$$= \frac{2 + 10}{3 + 3}$$

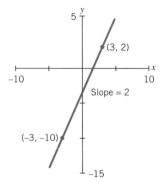

Figure 3.14   Slope = 2.

$$= \frac{12}{6}$$

$$= \frac{2}{1}$$

$$= 2$$

Writing 2 as the ratio 2/1 indicates that if x increases by 1 unit, then y will increase by 2 units. So starting at any point on the line, if you move one unit to the right, you would have to move up 2 units to return to the line.

Example   6   Plot the points $(-6, 3)$ and $(4, -7)$ and calculate the slope of the line on which they lie.

SOLUTION

The points and the lines connecting them are plotted in Figure 3.15. The slope of the line is $\Delta y/\Delta x$, which equals

$$\frac{y_2 - y_1}{x_2 - x_1} = \frac{3 - (-7)}{-6 - 4}$$

$$= \frac{3 + 7}{-10}$$

$$= \frac{10}{-10}$$

$$= \frac{1}{-1}$$

$$= -1$$

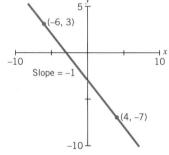

Figure 3.15   Slope = −1.

Writing the number -1 as the ratio $-1/1$ indicates that for an increase of 1 unit in x there is a corresponding decrease of 1 unit in y. Starting at any point on the line, if you move 1 unit to the right, you must move 1 unit down in order to stay on the line.

Algebra Aerobics 3.3

1. Specify the intervals on the graph in Figure 3.16 for which the average rate of change between adjacent data points is approximately zero.

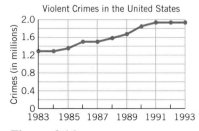

Figure 3.16

2. Specify the intervals on the graph in Figure 3.17 for which the average rate of change between adjacent data points appears positive, negative, or zero.

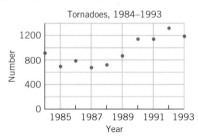

Figure 3.17

3. Find the slope of the line that passes through the given points.

    a. $(-2, 3)$ and $(4, -5)$ **b.** $(3, 500)$ and $(7, 500)$

4. Recalculate the slopes in Examples 5 and 6 reversing the order of the points. Are your results the same as the answers in the text?

5. Average velocity is the change in distance divided by the change in time. Open "C3: Average Velocity & Distance" in *Rates of Change*. What is the relationship between the graphs on the left and on the right? (Hint: Try initially setting all the average velocities to the same value.)

3.4 OTHER WAYS TO DESCRIBE CHANGE OVER TIME

Describing Change over Time with Numbers

In the popular press you may encounter several different ways of describing change over time. They are all legitimate but may give very different impressions. For example, here are three different ways to describe the change in population between 1790 and 1800 (Table 3.9).

1. Absolute change

    The U.S. population increased by 1.4 million between 1790 and 1800.

$$\text{(population in 1800)} - \text{(population in 1790)} = 5.3 - 3.9$$
$$= 1.4 \text{ million}$$

2. Percentage change

    The U.S. population grew by almost 36% between 1790 and 1800.

$$\frac{\text{change in population}}{\text{original 1790 population}} = \frac{1.4 \text{ million}}{3.9 \text{ million}}$$
$$= 0.359 \text{ or } 36\%$$

3. Rate of change

    The U.S. population grew at a rate of 0.14 million per year during the 1790s.

$$\frac{\text{change in population}}{\text{change in years}} = \frac{1.4 \text{ million}}{10 \text{ years}}$$
$$= 0.14 \text{ million per year}$$

Table 3.9	
U.S. Population	
Year	Population (in millions)
1790	3.9
1800	5.3

Adding Words and Graphs

If we include suggestive vocabulary (the italicized words in the following examples) and a graph constructed to support a particular viewpoint, we can influence the interpretation of information. In Washington, D.C., this would be referred to as "putting a spin on the data." Take a close look at the sentences below that are revisions of the sentences in 1, 2, and 3 above. Note that in each case the accompanying graphs display *exactly* the same data, a plot of the two points (1790, 3.9) and (1800, 5.3) with a line connecting them.

1. The U.S. population increased by *only* 1.4 million between 1790 and 1800 (Figure 3.18).

    Stretching the scale of the horizontal axis relative to the vertical axis makes the slope of the line look almost flat and hence minimizes the impression of change.

Exploration 3.1 gives you a chance to put your own "spin" on data.

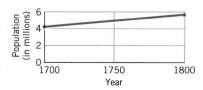

Figure 3.18 U.S. population (in millions).

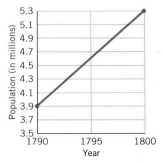

Figure 3.19   U.S. population (in millions).

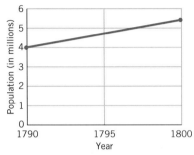

Figure 3.20   U.S. population (in millions).

2. The U.S. population had an *explosive* growth of over 35% between 1790 and 1800 (Figure 3.19).

   Cropping the vertical axis (which now starts at 3.5 instead of 0) and stretching the scale of the vertical axis relative to the horizontal axis makes the slope of the line look steeper and emphasizes the impression of dramatic change.

3. The U.S. population grew at a *reasonable* rate of 0.14 million per year during the 1790s (Figure 3.20).

   Visually the steepness of the line here seems to lie roughly halfway between the previous two graphs. *In fact, the slope of 0.14 million/yr is precisely the same for all three graphs.*

   How could you decide upon a reasonable interpretation of the data? You might try to put the data in context by asking: How does the growth between 1790 and 1800 compare to other decades in the history of the United States? How does it compare to growth in other countries at the same time? Was this rate of growth easily accommodated or did it strain national resources and overload the infrastructure?

   Although the three graphical examples are exaggerated, you need to be aware that a statistical claim is never completely free of bias. For every statistic that is quoted, others have been left out. This does not mean that you should discount all statistics. However, you do need to become an educated consumer, constantly asking common-sense questions and then coming to your own conclusions.

Algebra Aerobics 3.4

1. According to the U.S. Social Security Administration, Social Security total annual payments have increased from $120.472 billion in 1980 to $285.980 billion in 1992.

   a. What is the dollar increase?

   b. What is the percentage increase?

   c. What is the average rate of change?

2. Open "C4: Distortion by Clipping and Scaling" in *Rates of Change*. By choosing different end points and different axis scales, construct 3 very different impressions of the change in median family income.

3. Use the data in (Table 3.10) on immigration into the United States between 1901 and 1990 to answer the following questions:

   a. In which decade did America have the most immigrants?

   b. Between which two consecutive decades was there the largest change (increase or decrease) in the number of immigrants?

   i. Describe the absolute change as a number.

   ii. Describe the change as a percentage.

Table 3.10

Immigration: 1901–1990

Period	Population (in 1000s)
1901 to 1910	8795
1911 to 1920	5736
1921 to 1930	4107
1931 to 1940	528
1941 to 1950	1035
1951 to 1960	2515
1961 to 1970	3322
1971 to 1980	4493
1981 to 1990	7338

CHAPTER SUMMARY

The average rate of change of y with respect to $x = \dfrac{\text{change in } y}{\text{change in } x}$.

If the variables represent real world quantities that have units of measure (e.g., dollars or years), then the average rate of change has units that should be used.

$$\text{units of average rate of change} = \frac{\text{units of vertical variable}}{\text{units of horizontal variable}}$$

For example, the units might be dollars/year (read as dollars per year) or pounds/person (read as pounds per person).

On a graph, the average rate of change represents a slope or the

$$\frac{\text{change in the vertical variable}}{\text{change in the horizontal variable}}$$

Given two points (x_1, y_1) and (x_2, y_2)

$$\text{average rate of change} = slope = \frac{rise}{run} = \frac{y_2 - y_1}{x_2 - x_1} = \frac{\Delta y}{\Delta x}$$

where $\Delta y = y_2 - y_1$ and $\Delta x = x_2 - x_1$.

It doesn't matter which one of the two points (x_1, y_1) or (y_2, y_2) you use as the first point when you take the differences; the result is the same either way. You just have to be consistent.

$$slope = \frac{y_2 - y_1}{x_2 - x_1} \quad \text{or} \quad slope = \frac{y_1 - y_2}{x_1 - x_2}$$

Change over time can be described using absolute numbers, percentages, or average rates of change. By using suggestive vocabulary and graphs, the same data can be presented and perceived in very different ways.

EXERCISES

1. The following table and graph illustrate the average annual salary of professional baseball players from 1987 to 1993.

Year	Salary ($ Millions)	Rate of Change over Prior Year
1987	0.41	n.a.
1988	0.44	
1989	0.50	
1990	0.60	
1991	0.85	
1992	1.03	
1993	1.08	

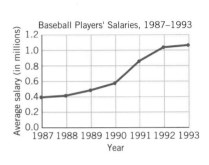

Baseball Players' Salaries, 1987–1993

a. Fill in the third column in the table.

b. During which 1-year interval was the average rate of change the smallest? Compare that part of the graph to the rest of the graph.

c. During which 1-year interval was the rate of change the largest? How does that part of the graph compare to the rest of the graph?

2. **a.** In 1980, aerospace industry profits in the United States were $2.59 billion, whereas in 1990 profits were $4.45 billion. Find the average annual rate of change from 1980 to 1990.

b. In 1992, the aerospace industry showed a loss (negative profit) of $1.84 billion. Find the average annual rate of change from 1990 to 1992.

3. Using the information in the table on college completion:

a. Plot the data, labeling both axes and the coordinates of the points.

b. Calculate the average rate of change (in percentage points per year).

c. Write a topic sentence summarizing what you think is the central idea to be drawn from these data.

Year	Percentage of Persons 25 Years Old and Over Completing 4 or More Years of College
1940	4.6%
1990	21.3%

Source: U.S. Bureau of the Census, Current Population Reports, P-60 Series, in *The American Almanac 1993–1994: Statistical Abstract of the United States,* 1993.

4. Using the information in the table on high school completion:

a. Plot the data, labeling both axes and the coordinates of the points.

b. Calculate the slope of the line connecting the two points and express it in units of percentage points per year.

c. Write a topic sentence summarizing what you think is the central idea to be drawn from these data on high school completion.

Year	Percentage of Persons 25 Years Old and Over Completing 4 or More Years of High School
1940	24.5%
1990	77.6%

Source: U.S. Bureau of the Census, Current Population Reports, in *The American Almanac 1993–1994: Statistical Abstract of the United States,* 1993.

5. Plot each pair of points and calculate the slope of the line that passes through them.

a. (3, 5) and (8, 15)

b. (−1, 4) and (7, 0)

c. (5, 9) and (−5, 9)

6. Read the Anthology article "Slopes" and describe two practical applications of slopes, one of which is from your own experience.

7. Use the accompanying table on life expectancy to answer the following questions. (See also Excel or graph link file LIFEXPEC.)

Average Number of Years of Life Expectancy in the United States by Race and Sex, Since 1990

Life Expectancy at Birth by Year	White Males	White Females	Black Males	Black Females
1900	46.6	48.7	32.5	33.5
1950	66.5	72.2	58.9	62.7
1960	67.4	74.1	60.7	65.9
1970	68.0	75.6	60.0	68.3
1980	70.7	78.1	63.8	72.5
1990	72.7	79.4	64.5	73.6
1994	73.2	79.6	67.5	75.8

Source: U.S. National Center for Health Statistics, in *The American Almanac 1995–1996 and 1997–1998: Statistical Abstract of the United States,* 1995 and 1997.

a. What was the average rate of change in years of life expectancy over time for white females between 1900 and 1950?

b. Which group had the largest average rate of change between 1900 and 1994?

c. Describe the gain in life expectancy for black females in at least two different ways.

8. Choose one of the following and write a statement commenting on the validity of the growth projection you calculate. When you draw the graphs that are called for, plot time on the horizontal axis.

a. Look up the data for the growth of a fetus in the womb (many books on childbirth have this information). Make a chart on graph paper showing length in inches or centimeters versus time in weeks or months up until birth. What is the average rate of change for the 9-month period? If the child were to grow at the same rate for the 24 months after birth as it did for the last month in the womb, how tall would he or she be?

b. Find data for growth of a human from birth to 21 years (if you know it about yourself, that will do; otherwise find the information in medical references on pediatrics). Plot the data on a graph in inches or centimeters versus time in years. What is the average rate of change for the 21-year total? What is the average rate of change for the first year? If humans continued to grow at the same rate as they grew in their first year, how tall would they be at age 21?

c. Find population data for the world or a country that interests you. Plot population versus years in 10-year intervals from 1900. What is the average rate of change for the entire century up to the present time? What is the average rate of change for the last full decade? If population continues to grow at the rate of the last full decade, what will be the total population in the year 2050? If you prefer, you may use rain forest or fossil fuel data instead and calculate their depletion rates.

9. Use the information in the accompanying table to answer the following questions:

**Persons 25 Years Old and Over
Who Have Completed 4 Years of
High School or More**

	1940	1990
White	26.1%	79.1%
Black	7.3%	66.2%
Asian/Pacific Islander	22.6%	80.4%

Data extracted from "Educational Attainment," *Population Profile of the United States 1991*, U.S. Bureau of the Census, Current Population Reports, P-23, No. 173.

a. What has been the average rate of change (of percentage points per year) of completion of high school from 1940 to 1990 for whites? For blacks? For Asian/Pacific Islanders?

b. If these rates continue, what percentage of whites, of blacks, of Asian/Pacific Islanders will have finished high school in the year 2000?

c. Write a 60 second summary describing the key elements in the high-school completion data. Include rates of change and possible projections for the near future.

d. If these rates continue, in what year would 100% of whites have completed 4 years of high school or more? In what year 100% of blacks? 100% of Asian/Pacific Islanders? Do these projections make sense?

10. The following table and graph show the change in median age of the U.S. population from 1850 through the present and projected into the next century. (See also Excel or graph link file USMEDAGE.)

**Median Age of U.S. Population
1850–2050 (Data for 1995–2050
are projected)**

Year	Median Age
1850	18.9
1860	19.4
1870	20.2
1880	20.9
1890	22.0
1900	22.9
1910	24.1
1920	25.3
1930	26.4
1940	29.0
1950	30.2
1960	29.5
1970	28.0
1980	30.0
1990	32.8
1991	33.1
1995	34.0
2000	35.5
2025	38.1
2050	39.0

Source: U.S. Bureau of the Census, Current Population Reports in *The American Almanac 1995–1996: Statistical Abstract of the United States*, 1995.

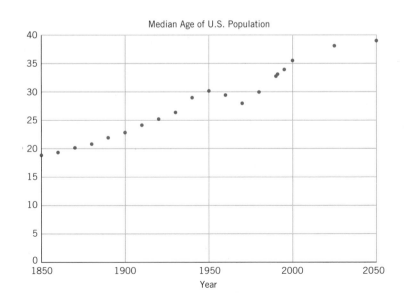

a. Specify the longest time period over which the median age has been increasing.

b. Calculate the average rate of change between the two end points of this time period.

c. What is the projected average rate of change between 2000 and 2050?

d. What changes in society could make this change in median age possible? Does an increase in median age necessarily mean that more people are living longer?

A graphing calculator or computer is helpful for the following parts.

e. Generate a partial third column to the table, starting in 1860 and ending in 1990, that for each listed year contains the average rate of change of median age over the previous decade.

f. Plot average rate of change versus years (columns 3 and 1 in your table).

g. Identify periods with negative average rates of change and suggest reasons for the declining rates.

11. Given the following graph of a function, specify the intervals over which:

a. the *function* is positive, negative, or zero.

b. the *rate of change between any two points in the interval* is positive, is negative, or is zero.

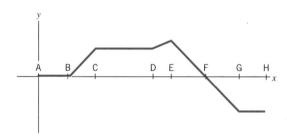

12. Given the following graph of a function, specify the intervals over which:

a. the *function* is increasing, decreasing, or constant.

b. the *rate of change between any two points in that interval* is increasing, decreasing, or constant.

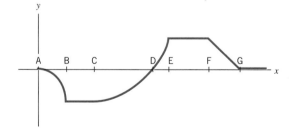

13.   The accompanying data and graphs give a picture of the two major methods of news communication in the United States. (See also Excel or graph link files NEWPRINT and ONAIRTV.)

Year	Newspapers (in thousands of copies printed)	Number of Newspapers Published	Year	Newspapers (in thousands of copies printed)	Number of Newspapers Published
1915	28,777	2580	1955	56,147	1760
1920	27,791	2042	1960	58,882	1763
1925	33,739	2008	1965	60,358	1751
1930	39,589	1942	1970	62,108	1748
1935	38,156	1950	1975	60,655	1756
1940	41,132	1878	1980	62,202	1745
1945	48,384	1749	1985	62,766	1676
1950	53,829	1772	1990	62,324	1611

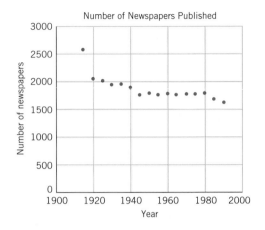

Year	Number of On-Air TV Stations
1950	98
1955	411
1960	515
1965	569
1970	677
1975	706
1980	734
1985	883
1990	1092

Source: Universal Almanac, copyright © 1993 by John W. Wright. Reprinted with permission of Andrews & McNeal. All rights reserved.

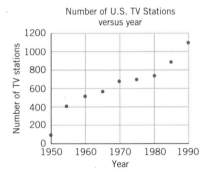

a.   Describe in three different ways the trend in number of newspapers published between 1915 and 1990.

b.   Use the U.S. population numbers from Table 3.1 to calculate and compare the number of copies of newspapers *per person* in 1920 and in 1990.

c.   What is the average annual rate of growth of TV stations for each decade since 1950? What is the average annual rate of decline of newspapers published for the same decades? Graph the results. (Use of a graphing calculator or computer is recommended.)

d.   If TV stations continue to be started at the same rate as in the decade from 1980 to 1990, how many will there be by the year 2000? Do you think this is likely to be a reasonable projection, or is it overly large or small judging from past rates of growth?

e.   What trends do you see in the dissemination of news as reflected in these data? If you were seeking a job in the news reporting industry, what appear to be the greatest employment opportunities? What other data would be helpful in understanding how news is delivered in our society?

14. The following table and graph show the number of new AIDS cases reported in Florida from 1986 to 1994.

Year	Number of New AIDS Cases
1986	1,031
1987	1,633
1988	2,650
1989	3,448
1990	4,018
1991	5,471
1992	5,086
1993	10,958
1994	8,617

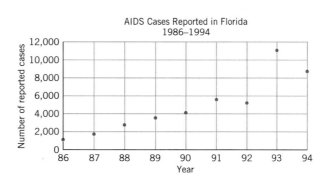

AIDS is a relatively recent phenomenon and its rate of growth in heavily populated states is very important in planning public health programs.

a. In a third column starting in 1987, calculate the average rate of change of AIDS cases over each preceding year. What are the units? Plot the average rate of change versus time.

b. Find something encouraging to say about these data by using numerical facts such as percentages, average rates of change, or absolute change.

c. Find numerical support for something discouraging to say about the data.

d. Write a paragraph describing the status of AIDS in Florida. Give a balanced report on the number of cases. Use data that will be helpful and informative. Mention percentages and average rates of change in your comments.

15. The accompanying graph shows the number of Nobel Prizes awarded in science for various countries between 1901 and 1974. It contains accurate information but gives the impression that the number of prize winners declined drastically in the 1970s, which was not the case. What flaw in the construction of the graph leads to this impression?

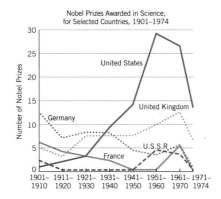

Source: Edward R. Tufte, *The Visual Display of Quantitative Information* (Connecticut: Graphics Press, 1983).

16. The following four graphs show black persons' income as a percentage of white persons' income from 1950 to 1975. Graph (*a*) makes the results look more positive than graph (*b*); graph (*c*) seems to suggest change, whereas graph (*d*) seems to imply stability. Describe how the axes have been altered in each graph to convey these impressions.

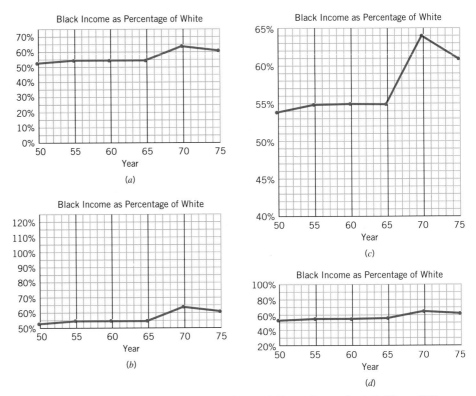

Source: Lou Ferleger and Lucy Horowitz, *Statistics for Social Change,* Boston: South End Press, 1980.

17. The following graphs are an approximation of the costs of doctors' bills and Medicare between 1963 and 1979.

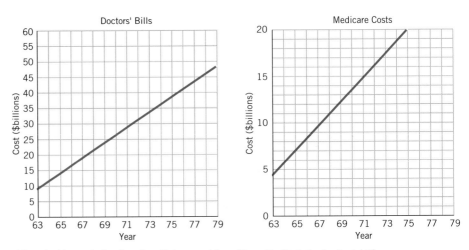

Adapted with permission from Lou Ferleger and Lucy Horowitz, *Statistics for Social Change.*

a. Which *appears* to have been growing at a faster average annual rate of change: doctors' bills or Medicare costs? Why?

b. Which actually grew at a faster average annual rate of change? How can you tell?

18. Describe the decline in air pollutant concentrations between 1985 and 1994 in three different ways.

National Ambient Air Pollutant Concentrations: 1985 to 1994
[Air Quality Standard is 0 Parts per Million (ppm)]

	1985	1987	1988	1989	1990	1991	1992	1993	1994
Carbon monoxide (ppm)	6.97	6.69	6.38	6.34	5.87	5.55	5.18	4.88	5.57

Source: U.S. National Center for Health Statistics, in *The American Almanac 1995–1996 and 1997–1998: Statistical Abstract of the United States,* 1995 and 1997.

19. The following table and graph are both representations of the accumulated debt of the federal government as a function of time. (See also Excel and graph link file FEDDEBT.)

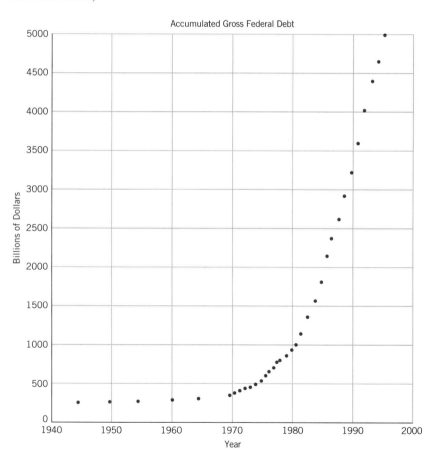

Year	Billions of Dollars
1945	260
1950	257
1955	274
1960	291
1965	322
1970	361
1971	408
1972	436
1973	466
1974	484
1975	541
1976	629
1977	706
1978	777
1979	829
1980	909
1981	994
1982	1137
1983	1371
1984	1564
1985	1817
1986	2120
1987	2346
1988	2601
1989	2868
1990	3206
1991	3599
1992	4003
1993	4410
1994	4644
1995 (est.)	4961

a. Generate a third column for the table containing the average annual rate of change of the debt during the previous interval. Describe what these numbers represent.

b. Graph the average rates of change from your third column versus time. Describe what your graph and the data show.

Source: U.S. Office of Management and Budget, "Budget of the United States Government," annual, in *The American Almanac 1995–1996 and 1996–1997: Statistical Abstract of the United States,* 1995 and 1996.

20. Laws to regulate environmental pollution in America are a very recent phenomenon, with the first federal regulations appearing in the 1950s. The Clean Air Act, passed in 1963 and amended in 1970, established for the first time uniform national air pollution standards. The act placed national limits and a timetable on three classes of automotive pollutants: hydrocarbons (HC), carbon monoxide (CO), and nitrogen oxides (NO). All new cars for each particular year could not exceed these standards. The following tables show the results. (See also Excel or graph link files POLLUTE for emission standards per mile, and EMISSION for total annual automobile emission estimates.)

National Automobile Emission Control Standards

Model Year Applicable	HC Grams/ Mile	CO Grams/ Mile	NO_2* Grams/ Mile
Pre-1968	8.7	87	4.4
1968	6.2	51	n.r.**
1970	4.1	34	n.r.
1972	3.0	28	n.r.
1973	3.0	28	3.1
1975	1.5	15	3.1
1975 (C)***	0.9	9	2.0
1976	1.5	15	3.1
1977	1.5	15	2.0
1978	1.5	15	2.0
1979	1.5	15	2.0
1980	0.41	7	2.0
1981 and beyond	0.41	3.4	1.0

*NO_2, nitrogen dioxide, is a form of nitrogen oxide.

**No requirement.

***California standards.

Source: Paul Portney, ed., *Current Issues in U.S. Environmental Policy,* Baltimore, Md.: Johns Hopkins University Press, 1978, p. 76.

National Automobile Emissions Estimates (million metric tons per year)

Year	HC	CO	NO_2
1970	28.3	102.6	19.9
1971	27.8	103.1	20.6
1972	28.3	104.4	21.6
1973	28.4	103.5	22.4
1974	27.1	99.6	21.8
1975	25.3	97.2	20.9
1976	27.0	102.9	22.5
1977	27.1	102.4	23.4
1978	27.8	102.1	23.3

Source: Environmental Quality—1980: The Eleventh Annual Report of the Council on Environmental Quality, p. 170.

a. Argue that The Clean Air Act was a success.

b. Argue that The Clean Air Act was a failure. (Hint: How is it possible that the emissions per vehicle mile were down but the total amount of emissions did not improve?)

21. Open "L6: Changing Axis Scales" in *Linear Functions*. Generate a line in the upper left-hand box. The same line will appear graphed in the three other boxes, but with the axes scaled differently. Describe how the axes are rescaled in order to create such different impressions.

EXPLORATION 3.1

Having It Your Way

Objective

- construct arguments supporting opposing points of view from the same data

Materials/Equipment

- *Student Statistical Portraits* for the University of Massachusetts/Boston and for the University of Southern Mississippi or for the student body at your institution
- computer with spreadsheet program and printer or graphing calculator with projection system (optional)
- graph paper and/or overhead transparencies

Procedure

Working in Small Groups:

Examine the data and graphs from the *Student Statistical Portrait* for either the University of Massachusetts/Boston or the University of Southern Mississippi. Explore how you would use the data to construct arguments that support at least two different points of view. Decide on the arguments you are going to make and divide up tasks among your team members.

The Rules of the Game:

- Your arguments need only to be a few sentences long but you need to use graphs and numbers to support your position. You may only use legitimate numbers, but you are free to pick and choose the ones that best support your case. If you construct your own graphs, you may use whatever scaling you wish on the axes.
- For any data that represent a time series, as part of your argument, pick two appropriate end points and calculate the associated average rate of change.
- Use "loaded" vocabulary (e.g., "surged ahead," "declined drastically"). This is your chance to be outrageously biased, write absurdly flamboyant prose, and commit egregious sins of omission.
- Decide as a group how to present your results to the class.

Suggested Topics

Your instructor might ask your group to construct one or both sides of the arguments on one topic. If you're using data from your own institution, answer the questions provided by your instructor.

Using data from the *Student Statistical Portrait* from the University of Southern Mississippi:

1. Use the data on enrollment by ethnic group and gender from "10-Year Enrollment Trends" to support each of the following cases:
    a. You are a student activist trying to convince the Board of Trustees that they have not done enough to increase diversity on the USM campus.
    b. You are an administrator arguing that your office has done a good job of increasing the diversity of the student body at USM.
2. Use data on "Retention of Freshmen" and "10-Year Enrollment Trends" to support each of the following cases:
    a. You are a student trustee lobbying the State Legislature for more money for USM.
    b. You are a taxpayer writing an editorial to the local paper arguing that USM does not need more money.
3. Make the case that USM as an institution is growing and then the case that USM is shrinking. Decide on the roles you would assume in making these arguments. Use data from the "10-Year Enrollment Trends" and "Degrees Conferred" tables.

Using data from the *Student Statistical Portrait* from the University of Massachusetts/Boston:

1. You are the Dean of the College of Management. Use the data on "SAT Scores of New Freshmen by College/Program" to make the case that:
    a. The freshmen admitted to the College of Management are not as prepared as the students in the College of Arts and Sciences and therefore you need more resources to support the freshmen in your program.
    b. The freshmen admitted to the College of Management are better prepared than the students in the College of Arts and Sciences and therefore you would like to expand your program.
2. You are an assistant provost lobbying the state legislature. Use the data on "Undergraduate Admissions" to present a convincing argument that:
    a. UMass is becoming less desirable as an institution for undergraduates and so more funds are needed to strengthen the undergraduate program.
    b. UMass is becoming more desirable as an institution for undergraduates and so more funds are needed to support the undergraduate program.
3. Make the case that the university as an institution is growing and then the case that the university is shrinking. Decide on the roles you would assume in making these arguments. Use graphs and data on "Enrollment Trends" and graphs and data on "Degrees Conferred."

Exploration-Linked Homework

With your partner prepare a short class presentation of your arguments, using, if possible, overhead transparencies or a projection panel. Then write individual 60 second summaries to hand in.

EXPLORATION 3.2

Why "Average" Rate of Change?

Objective

- explore two different methods for calculating average rate of change

Materials/Equipment

- computer with spreadsheet program or graphing calculators
- U.S. Population Data 1790 to 1995 (Table 3.1) in Excel or graph link file USPOP

Procedure

Class demonstration using two different methods to calculate the average rate of change

Method A

Enter population data into a spreadsheet or graphing calculator. Calculate the average rate of change between the two end points of 1790 and 1990 using the formula

$$\text{average rate of change} = \frac{\text{change in population}}{\text{change in years}}$$

Method B

Calculate the average rate of change for each 10-year interval from 1790 to 1990. Compare your results with Table 3.4. Sum these rates and then calculate the mean of all the average rates of change from 1790 to 1990. Compare your answer to the answer you found with Method A above. Discuss why your answers are called an *average* rate of change.

Discussion/Analysis

With a Partner:

- Calculate the average rate of change from 1790 to 1990 using Method B and 20-year intervals. Compare the answer with your answers using 10-year intervals.
- Calculate the average rate of change from 1790 to 1990 using Method B and intervals that are not the same size. (For example, include both 10-year and 20-year intervals.) Did you find the same answer as the above methods? What happens when you use intervals that are not all the same size?
- Do you think Method B would work with overlapping or noncontiguous intervals? (An example of overlapping intervals would be intervals from 1900 to 1920 and 1910 to 1930. An example of noncontiguous intervals would be intervals from 1900 to 1920 and 1940 to 1960.) Explain your answer.

Class Discussion:

- Explain in your own words how to summarize Method B. What qualifications do you need to make if you use Method B to calculate the average rate of change over an interval. Explain why.
- What additional data would you need to use Method B to calculate the average rate of change from 1790 to 1994?

4

When Rates of Change Are Constant

Overview

Linear functions describe relationships in which the average rate of change is constant. Linear functions are fundamental in mathematics and its applications to real data. They can be used to model a wide variety of relationships between two variables, for example, between the median weight of female infants and time, or the number of concerts and average audience size per concert, or the percentage of students taking SAT tests and their scores.

The graphs of linear functions are straight lines and we learn to describe them with equations. We look at different ways to construct the equation of a line, examine the special cases of horizontal, vertical, parallel, and perpendicular lines, and then study equivalent ways of describing linear relationships.

The Explorations provide an opportunity to discover patterns in the graphs of linear equations and make generalizations about the properties of linear equations.

After reading this chapter you should be able to:

- recognize when a rate of change is constant
- define and evaluate a linear function
- construct and graph linear functions under different conditions
- understand direct proportionality
- recognize and construct horizontal, vertical, parallel, and perpendicular lines
- describe linear relationships in several different ways

4.1 A FIRST LOOK AT LINEAR RELATIONSHIPS

What If the U.S. Population Had Grown at a Constant Rate? A Hypothetical Example

In Section 3.2, we calculated the average rate of change in the U.S. population between 1790 and 1800 to be 0.14 million people per year. What if the average rate of change had remained constant? What if in every decade after 1790, the U.S. population had grown at exactly the same rate, namely at 0.14 million per year? That would mean that starting with a population estimated at 3,900,000 in 1790, the population would grow by exactly 140,000 people each year, or 1.4 million people per decade. The slopes of all the little line segments connecting adjacent population data points would be identical—namely 0.14 million/year. The graph would be a straight line, indicating a constant rate of change.

Instead, on the original population graph, the slopes of the line segments are increasing, and thus the graph curves upward. When we calculated the actual rate of change in the decades subsequent to 1790, not only was the growth rate positive each year, but the growth rate generally increased from decade to decade. Table 4.1 and Figure 4.1 compare the hypothetical and actual results.

Table 4.1

U.S. Population

Year	Actual Population in Millions	Population in Millions IF the Average Rate of Change Remained Constant at 0.14 Million/Yr
1790	3.9	3.9
1800	5.3	5.3
1810	7.2	6.7
1820	9.6	8.1
1830	12.9	9.5
1840	17.1	10.9
1850	23.2	12.3
1860	31.4	13.7
1870	39.8	15.1
1880	50.2	16.5
1890	63.0	17.9
1900	76.0	19.3
1910	92.0	20.7
1920	105.7	22.1
1930	122.8	23.5
1940	131.7	24.9
1950	151.3	26.3
1960	179.3	27.7
1970	203.3	29.1
1980	226.5	30.5
1990	248.7	31.9
1995	263.0	32.6

Source: U.S. Bureau of the Census, Current Population Reports in *The American Almanac 1995–1996 and 1996–1997: Statistical Abstract of the United States,* 1995 and 1996. (Data before 1960 exclude Alaska and Hawaii.)

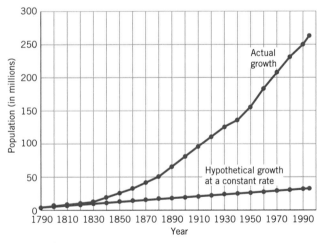

Figure 4.1   U.S. population: a hypothetical example.

A Real Example of Constant Rate of Change

How do children grow? Professionals in many different fields, such as psychology, sociology, and medicine, study questions of human growth and development. Data about height, weight, and psychological maturity are used to measure the growth of young children. An established norm can help parents and doctors determine whether a child's development is progressing satisfactorily.

According to the standardized growth and development charts used by many American pediatricians, the median weight for girls during their first 6 months of life increases at an almost constant rate. Starting at 7.0 lb at birth, for each additional month of life the female median weight increases by 1.5 lb. When the average rate of change is constant, we often drop the word "average" and just use *rate of change*. Thus, 1.5 pounds per month is the rate of change of girls' weight from birth to six months.

Since each age corresponds to a unique median weight, the median weight for baby girls is a function of age and since the rate of change is constant it is a linear function. Table 4.2 and Figure 4.2 show two representations of the function. Note that the domain is between 0 and 6 months.

Table 4.2

Median Weight for Girls

Age (in months)	Weight (in pounds)
0	7.0
1	8.5
2	10.0
3	11.5
4	13.0
5	14.5
6	16.0

Data derived from the Ross Growth and Development Program, Ross Laboratories, Columbus, Ohio.[1]

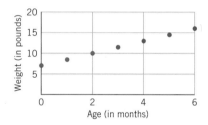

Figure 4.2   Median weight for girls.

The rate of change, 1.5 lb per month, means that as age increases by 1 month, weight increases by 1.5 lb. On the graph, if we move 1 unit (1 month) to the right, then we need to move up 1.5 units (1.5 lb) to find the next data point. Just as in the hypothetical case of the constant U.S. population growth, the graph shows the points lying on a straight line.

Finding a Mathematical Model for the Relationship Between Female Infant Median Weights and Age

Can we find an equation, a *mathematical model*, that gives female infant weight as a function of age? Using Table 4.2, we can think of the weight at birth (when age = 0)

[1] Ross Laboratories adapted data from PVV Hamill, TA Drizd, CL Johnson, RB Reed, AF Roche, WM Moore. "Physical Growth: National Center for Health Statistics Percentiles," *Am J Clin Nutr* 32:607–629, 1979. Data from the Fels Longitudinal Study, Wright State University School of Medicine; Yellow Springs, Ohio.

as our base value. In Figure 4.2, the base value represents the intercept on the vertical axis. In this case the base value is 7.0 lb; that is, the graph crosses the weight axis at 7lb. The median weight for girls starts at 7.0 lb at birth and increases by 1.5 lb every month. Table 4.3 shows the pattern in the monthly weight gain.

Table 4.3
Patterns in Median Weight for Girls

Age (in months)	Weight (in lb)	Pattern	Generalized Expression
0	7.0	7.0 + 1.5(0) =	7.0 + 1.5(age in months)
1	8.5	7.0 + 1.5(1) =	7.0 + 1.5(age in months)
2	10.0	7.0 + 1.5(2) =	7.0 + 1.5(age in months)
3	11.5	7.0 + 1.5(3) =	7.0 + 1.5(age in months)
4	13.0	7.0 + 1.5(4) =	7.0 + 1.5(age in months)
5	14.5	7.0 + 1.5(5) =	7.0 + 1.5(age in months)
6	16.0	7.0 + 1.5(6) =	7.0 + 1.5(age in months)

A general expression for this relationship between weight and age is:

$$\text{median weight for girls} = 7.0 + 1.5 \cdot (\text{age in months})$$

where the median weight is in lb. If we let W represent the median weight in lb and A the age in months, we can write this equation more compactly as:

$$W = 7.0 + 1.5A$$

This equation can be used as a *model* of the relationship between age and median weight for baby girls in their first 6 months of life.

Since our equation represents quantities in the real world, each term in the equation has units attached to it. W (which represents weight) and 7.0 are in lb, 1.5 is in lb/month, and A (which represents age) is in months. The rules for canceling units are the same as the rules for canceling numbers in fractions. The units of the term $1.5A$ are

$$\left(\frac{\text{lb}}{\text{month}}\right)\text{month} = \text{lb}$$

The units of the right-hand side and left-hand side of our equation must match, and they do. That is, in the equation

$$W = 7.0 + 1.5A$$

the units are

$$\text{lb} = \text{lb} + \left(\frac{\text{lb}}{\text{month}}\right)\text{month}$$
$$= \text{lb} + \text{lb}$$
$$\text{lb} = \text{lb}$$

Our equation $W = 7.0 + 1.5A$ defines W as a function of A, since for each value of A (age) the equation determines a unique value of W (weight). If we name the function f, then $f(A) = W$.

Something to think about

If the median birth weight for baby boys is the same as for baby girls, but boys put on weight at a faster rate, which numbers in the model would change and which would stay the same? What would you expect to be different about the graph?

As a model, the function $f(A) = 7.0 + 1.5A$ holds for values of A between 0 and 6 months; so the domain of our function is $0 \leq A \leq 6$. Negative values of age are meaningless and the data for the median weights for girls older than 6 months will deviate from this equation. We would certainly not expect a female to continue to gain 1.5 lb per month for the rest of her life!

We derived the function using only ages that were integer values (whole numbers of months), but the function can be used to *interpolate* to noninteger values of A between 0 and 6. For instance, we can use the function $f(A) = 7.0 + 1.5A$ to predict the median weight for girls who are 2.5 months old.

When $A = 2.5$

$$f(2.5) = 7.0 + 1.5(2.5)$$
$$= 7.0 + 3.75$$
$$= 10.75 \text{ lb}$$

We say the point (2.5, 10.75) is *a solution to* or *satisfies* the equation $W = 7.0 + 1.5A$.

The original data set contained seven *discrete* points, whereas Figure 4.3 shows the continuous function $f(A) = 7.0 + 1.5A$, where A can be *any* real number from 0 to 6.

The Program "L4: Finding 2 Points on a Line" in Linear Functions *will give you practice in finding solutions to equations.*

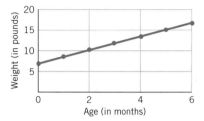

Figure 4.3 Median weight for girls.

Looking at This Function in the Abstract

If we think of the function $f(A) = W$ where $W = 7.0 + 1.5A$ as describing a relationship between two abstract (and unitless) variables W and A, then the natural domain of the function is *all* real numbers, not just those between 0 and 6. For instance, we can let $A = -6.0$ and then use the equation to find the associated value of W, which is $f(-6.0) = 7.0 + (1.5)(-6.0)$ or -2.0. Table 4.4 lists points that satisfy the equation. The graph of *all* the points that satisfy the equation $W = 7.0 + 1.5A$ is a line that extends indefinitely in both directions (Figure 4.4).

Table 4.4	
A	$W = 7.0 + 1.5A$
-6	-2.0
-5	-0.5
-4	1.0
-3	2.5
-2	4.0
-1	5.5
0	7.0
1	8.5
2	10.0
3	11.5
4	13.0
5	14.5
6	16.0
7	17.5
8	19.0

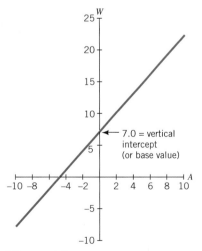

Figure 4.4 Graph of $W = 7.0 + 1.5A$.

From Figure 4.4 and Table 4.4 we can see that the line crosses the vertical axis at the base value of 7.0. If we evaluate $f(A)$ when $A = 0$ we have

$$f(0) = 7.0 + (1.5)(0)$$
$$= 7.0$$

So when $A = 0$, $W = 7.0$. We call 7.0 the *vertical intercept*.

The number 1.5, the coefficient of the independent variable A, is the rate of change or slope of the line. For any two points on the line, the average rate of change between them is always 1.5. We will prove this in the next section.

On Figure 4.4, this means that starting anywhere on the line, if we move (or "run") 1 unit to the right, we must move up (or "rise") 1.5 units to return to the line. In particular, as illustrated in Figure 4.5, if we start at the base value of 7.0 and repeatedly go over 1 and up 1.5 units, we'll keep coming back to the line.

In Figure 4.5, if you move or run 2 units to the right, how many units would you have to rise up to return to the graph of the function? What if you move or run 1 unit *to the left*? (Hint: think of a negative rise as a fall.) What if you move or run 2 units to the left?

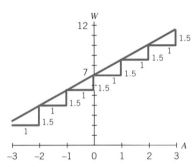

Figure 4.5   For each 1-unit increase in A, W increases by 1.5 units.

Algebra Aerobics 4.1

1. Determine which, if any, of the following points satisfy the equation $W = 7.0 + 1.5A$.
   a. $(2, 10)$ b. $(5, 13.5)$

2. From Figure 4.3, try to estimate the weight W of a baby girl who is 4.5 months old. Then, use the equation $W = 7.0 + 1.5A$ to calculate the corresponding value for W. How close is your estimate?

3. From the same graph, try to estimate the age of a baby girl who weighs 14 lb. Then use the equation to calculate the value for A.

4. Select any two points of the form (A, W) from Table 4.4 that satisfy the equation $W = 7 + 1.5A$. Use these points to verify that the rate of change between them is 1.5.

5. Find two points that satisfy the equation $W = 7 + 1.5A$ that are *not* in Table 4.4. Try using a negative value for A, then find the corresponding value for W. Then try a very large positive value for A and find the corresponding value for W.

6. If you calculate the slope using the two points you generated in Problem 2, what do you think your answer will be? Now do the calculations. Were you right?

4.2 LINEAR FUNCTIONS

The General Linear Form

In the equation

$$W = 7.0 + 1.5A$$

7.0 is the base value or vertical intercept and 1.5 is the rate of change or slope. This equation is in the form

dependent variable = base value + (rate of change) · (independent variable)

Any equation that can be written in this form is called a *linear function*. Its graph is a straight line.

A *linear function* $y = f(x)$ can be represented by an equation of the form

$$y = b + mx$$

Its graph is a straight line where
  m is the *slope* or *rate of change* of y with respect to x
  b is the *vertical intercept* or base value, the value of y when x is zero.

Exploration 4.1(a) and (b) allow you to examine the effects of m and b on the graph of a linear function.

Alternate Ways of Writing a Linear Function

Linear functions can also be written in a form you might have seen before:

$$y = mx + b$$

Social scientists tend to prefer the $y = b + mx$ format, because placing the b term first suggests starting at a base value (when $x = 0$) and then adding on multiples of x. Both forms describe equivalent equations and are used interchangeably.

Note: you do need to be careful about the order of the terms when identifying the slope and vertical intercept. For example,

  in the linear function $y = 15 - 2x$, we have $b = 15$ and $m = -2$

  but in the linear function $y = 15x - 2$, we have $b = -2$ and $m = 15$

When a linear equation is written in function form, the slope is always the coefficient of the independent variable and the vertical intercept is the constant term.

Although x and y are the traditional mathematical choices for the independent and dependent variables, there is nothing sacred about their use, particularly in other disciplines. We could just as well use A and W, as in the example in the previous section, or H_{men} and H_{women} for height of men and women, or I and E for personal income and education, or whatever else is appropriate.

Why Does the Equation $y = b + mx$ Represent a Function?

Given any value for x, the equation $y = b + mx$ determines one and only one corresponding value for y and hence describes y as a function of x. Thus, the use of the function notation $f(x) = b + mx$ is appropriate. For example, if $b = -75$ and $m = 25$, we can find a unique value for y from any x using the equation $y = -75 + 25x$.

Why Is b the y-Intercept?

The program "L1: m & b Sliders" in Linear Functions will help you visualize the effects of changing m and b.

The graph of $y = b + mx$ will cross the y-axis when $x = 0$. When $x = 0$, then

$$y = b + m(0)$$
$$= b$$

When $x = 0$ in the equation $y = b + mx$, the value of y will always equal b.

For example, let $y = 4.3 - 2.7x$. Then $b = 4.3$ and when $x = 0$,

$$y = 4.3 - 2.7(0)$$
$$= 4.3$$

So when $x = 0$, $y = b$.

Why Is m the Slope or Rate of Change?

We will show that if a linear function is in the form $y = b + mx$, then m, the coefficient of the independent variable, is equal to

$$\text{the rate of change} = \frac{\text{change in } y}{\text{change in } x} = \frac{\Delta y}{\Delta x}$$

If (x_1, y_1) and (x_2, y_2) are two distinct points that satisfy the function $y = b + mx$, then the following equations are true:

$$y_1 = b + mx_1 \tag{1}$$
$$y_2 = b + mx_2 \tag{2}$$

Subtracting Equation (1) from Equation (2) gives us:

$$y_2 - y_1 = (b + mx_2) - (b + mx_1)$$

clear the parentheses $= b + mx_2 - b - mx_1$

combine the b's $= mx_2 - mx_1$

factor out m $y_2 - y_1 = m(x_2 - x_1)$

Since y is a function and (x_1, y_1) and (x_2, y_2) are two different points, $x_2 \neq x_1$. So $x_2 - x_1 \neq 0$ and we can divide both sides by $(x_2 - x_1)$ to get:

$$\frac{y_2 - y_1}{x_2 - x_1} = m \tag{3}$$

In Chapter 3 we learned that the left side of this equation represents the rate of change or slope of the line between the two arbitrarily chosen points (x_1, y_1) and (x_2, y_2). Equation (3) shows that this rate of change is always equal to the value of m.

We have also just shown that we can calculate the rate of change or slope from any two points that satisfy a linear equation.

Any two points that lie on a line can be used to calculate the slope. That is, if (x_1, y_1) and (x_2, y_2) are two distinct points that satisfy the function $y = b + mx$, then

$$m = \frac{y_2 - y_1}{x_2 - x_1}$$

"C2: Constant Rates of Change" in Rates of Change *helps you see that the rate of change between any two points on a line remains constant.*

Algebra Aerobics 4.2a

1. Identify the slope, m, and the vertical intercept, b, of the line with the given equation:
    a. $y = 5x + 3$
    b. $y = 5 + 3x$
    c. $y = 5x$
    d. $y = 3$
    e. $f(x) = 7.0 - x$
    f. $h(x) = -11x + 10$

2. Calculate the slope of the line that passes through the given points.
    a. $(4, 1)$ and $(8, 11)$
    b. $(-3, 6)$ and $(2, 6)$
    c. $(0, -3)$ and $(-5, -1)$

3. If $f(x) = 50 - 25x$, evaluate
    a. $f(0)$
    b. $f(-2)$
    c. $f(2)$
    d. Use (a) and (b) to verify that the slope is -25.

Another Example of a Linear Function

For children between the ages of 2 and 12 years, the relationship between median height and age can be approximated by the linear function $H = g(A)$ where

$$H = 30.0 + 2.5A$$

and A = age (in years) and H = median height (in inches).[2] We can consider the function g both as an abstract expression with a domain of all real numbers and as a model for the relationship between age and median height with a domain of $2 \le A \le 12$ (denoted by the dotted lines on Figure 4.6).

The number 2.5 is the slope or the rate of change of H with respect to A. In our model, that means that the median height increases 2.5 inches per year for children between the ages of 2 and 12.

The number 30.0 is the vertical or H-intercept, the value of H when $A = 0$. Note that though the point $(0, 30.0)$ satisfies the equation, it is outside the domain of this model.

In the model, H and 30.0 are in inches, 2.5 is in inches per year, and A is in years. If we think of $H = 30.0 + 2.5A$ as describing a relationship between two abstract variables A and H, then all the terms are unitless.

[2] Information adapted from R.E. Behrman, V.C. Vaughan, III, eds. *Nelson Textbook of Pediatrics*, sr. ed. Waldo E. Nelson, 12th ed. (Philadelphia: W.B. Saunders, 1983) p. 19.

Table 4.5 and Figure 4.6 show a set of points that satisfy the equation.

Table 4.5	
A	H
-3	22.5
-2	25.0
-1	27.5
0	30.0
1	32.5
2	35.0
3	37.5
4	40.0
5	42.5
6	45.0
7	47.5
8	50.0
9	52.5
10	55.0
11	57.5
12	60.0
13	62.5
14	65.0
15	67.5

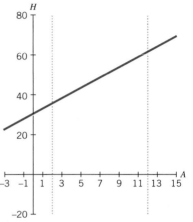

Figure 4.6   Graph of $H = 30.0 + 2.5A$.

Algebra Aerobics 4.2b

1. If $S(x) = 20{,}000 + 1000x$ describes the annual salary in dollars for a person who has worked for x years for the Acme Corporation, then what is the unit of measure for 20,000? For 1000?

2. In the equation $W = 7 + 1.5A$, the equation in units of measure only is

$$\text{lb} = \text{lb} + \left(\frac{\text{lb}}{\text{month}}\right)\text{months}$$

Note how month "cancels out" and you are left with:

$$\text{lb} = \text{lb} + \text{lb}$$

Rewrite $S(x) = 20{,}000 + 1000x$ as an equation relating only the units of measure.

$$\text{dollars} = ? + (?)(?)$$

3. If $C = 15P + 10$ describes the relation between the number of persons (P) in a dining party and the total cost in dollars (C) of their meals, what is the unit of measure for 15? For 10? Now rewrite this as an equation relating only units of measure.

4. What is the range for the function $H = 30.0 + 2.5A$ when the domain is $2 \le A \le 12$?

4.3 GENERATING LINEAR EQUATIONS (THREE CASES)

We can recognize that certain equations such as $y = 2000 + 1200x$ or $F = 1.5W - 2.7$ are linear functions. We now examine how to generate a linear equation given different types of information about the function. Here we show three distinct ways to find an expression of the form

$$y = b + mx$$

where m is the slope (or rate of change of y with respect to x) and b is the y-intercept. For each case, we generate a small table of points that satisfy the function and then graph the function.

Case 1: The Slope (m) and Vertical Intercept (b) Are Known

This is the easiest case, since all we need to do is substitute the values for m and b directly into the general linear equation.

Find the equation of the line with a slope of 3/5 and a vertical intercept of -2. Generate a small table of points that satisfy the equation and then graph the equation. Interpret the slope.   **Example   1**

SOLUTION

The equation $y = -2 + (3/5)x$ matches the general form of a linear equation $y = b + mx$ (m is the slope and b is the y-intercept). The domain is all the real numbers. Table 4.6 and Figure 4.7 show points that satisfy the equation.

Table 4.6	
x	y
-3.0	-3.8
-2.0	-3.2
-1.0	-2.6
0.0	-2.0
1.0	-1.4
2.0	-0.8
3.0	-0.2

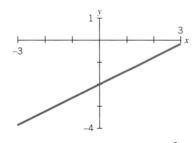

Figure 4.7   Graph of $y = -2 + \dfrac{3}{5}x$.

Since the slope 3/5 equals the change in y divided by the change in x, we can start at any point on the line, move 5 units to the right, and then move up 3 units to return to the line. We can also write 3/5 as 0.6 or 0.6/1. So moving 1 unit to the right, the line rises 0.6 units. This is more easily seen in Table 4.6, where y increases by 0.6 for every increase in x of 1 unit.

Example   2

Harvard Pilgrim Healthcare, an HMO in Massachusetts, distributes the following recommended weight formula for men: "Give yourself 106 lb for the first 5 ft, plus 6 lb for every inch over 5 ft tall." Construct a mathematical model for the relationship. Use your equation to determine the recommended weight for a male 5 ft 8 in. and for a male 6 ft 2 in. tall.

SOLUTION

We could let h represent height in inches, but since we are concerned only with heights over 5 ft, we instead let h = height in inches in *excess* of 5 ft and w = weight in pounds. Then according to the recommendations, when $h = 0$ in. (for a 5-ft male), $w = 106$ lb. So the w-intercept is 106 lb. For each additional inch in height, the recommendations permit an additional 6 lb. The rate of change of weight with respect to height is a constant 6 lb/in. The linear function $f(h) = w$ where

$$w = 106 + 6h$$

gives us a mathematical model for the relationship. h = height in inches above 5 ft, and w = weight in lb. We can think of 106 lb as our base value and 6 lb/in. as our constant rate of change. Note that w and 106 are in pounds, 6 is in pounds per inch, and h is in inches.

In our model, as in all models, we need to think about realistic values for the domain. The recommendations are for men 5 ft (or 60 in.) and taller. Our estimate for a reasonable maximum male height is 7 ft (or 84 in.) Hence the constraints on h, the possible values for the number of inches above 5 ft, would be $0 \leq h \leq 24$ in.

For a 5 ft 8 in. male, $h = 8$. The recommended weight would be:

$$f(8) = 106 + 6(8)$$
$$= 106 + 48$$
$$= 154 \text{ lb}$$

A 6 ft 2 in. male is 74 in. tall. Since 5 ft = 60 in., then this male would be 14 in. taller than 5 ft. So $h = 14$ in. His recommended weight would be:

$$f(14) = 106 + 6(14)$$
$$= 106 + 84$$
$$= 190 \text{ lb}$$

Table 4.7 and Figure 4.8 are both representations of this function.

Something to think about

According to the HMO recommendations, what region of Figure 4.8 represents people who are overweight? What region represents people who are underweight?

Table 4.7

h, height in inches in excess of 5 ft	w, weight in lb
0	106
4	130
8	154
12	178
16	202
20	226
24	250

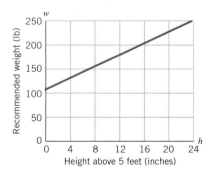

Figure 4.8   Recommended weight for males between 5 ft and 7 ft tall.

Find the equation of the linear function graph in Figure 4.9. Describe the slope of the line.

Example   3

SOLUTION
The y-intercept is -1, so the function is of the form $f(x) = -1 + mx$. We can use any two points on the graph to calculate m, the slope. If, for example, we take $(-3, 0)$ and $(3, -2)$, then

$$\text{slope} = \frac{\text{(change in } y)}{\text{(change in } x)} = \frac{-2 - 0}{3 - (-3)} = \frac{-2}{6} = \frac{-1}{3}$$

Hence the equation is:

$$y = -1 - \frac{1}{3}x.$$

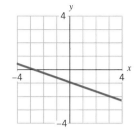

Figure 4.9   Graph of a linear function.

One way to describe the slope is to think of the slope as the rise/run $= -1/3$. If we start at any point on the line and run 3 units to the right and then *drop down* 1 unit (because of the minus sign), we will return to the line. Thinking of the rise/run in its equivalent form $1/(-3)$, starting at any point on the line, if we run 3 units *to the left* (because of the minus sign), we would have to rise 1 unit in order to return to the line.

Algebra Aerobics 4.3a

For Problems 1 and 2, find an equation, make an appropriate table, and sketch the graph of:
1. A line with slope 1.2 and vertical intercept -4.
2. A line with slope -400 and vertical intercept 300 (be sure to think about axes scales).

For Problems 3 and 4, find an equation to represent:
3. Current salary after x years if the starting salary is $12,000 with annual increases of $3000.
4. **a.** Total cost to produce x items if the startup cost is $200,000 and the production cost per item is $15.
    b. Why is the total cost per item less if the item is produced in large quantities?

Case 2: The Slope (m) and a Point Are Known

If the slope, m, and the coordinates of any point on the line are given, we only need to find the value of the vertical intercept, b, in order to construct the equation.

Find the linear function with a slope of 3 and a graph that passes through the point $(-1, -2)$.

Example   4

SOLUTION

Choose variables: Let x be the independent variable and y be the dependent variable.
Find m: We are told the slope is 3, so $m = 3$. The function has the form:

$$y = b + 3x$$

Find b: The point $(-1, -2)$ satisfies the equation. So we can substitute the values -1 for x and -2 for y and solve the equation for b.

substitute known x and y	$-2 = b + 3(-1)$
solve for the unknown b	$-2 = b - 3$
add 3 to both sides	$-2 + 3 = b - 3 + 3$
	$1 = b$

The equation: Now we can substitute the values for m and b directly to get

$$y = 1 + 3x$$

which describes y as a function of x.

The domain: Since there are no constraints on x, the domain is all real numbers.

Table and graph: Table 4.8 and Figure 4.10 are both representations of the function $y = 1 + 3x$.

Table 4.8

x	$y = 1 + 3x$
-3	-8
-2	-5
-1	-2
0	1
1	4
2	7
3	10

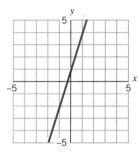

Figure 4.10   Graph of $y = 1 + 3x$.

Example 5

In 1996 a TV announcer reporting on health care said that since 1965, when the country had 1.7 million hospital beds, the number of beds had been declining by 20,000 (or 0.02 million) a year. What equation represents this statement? In what year does the equation predict we will have only 1 million beds?

SOLUTION

Choose variables: We are looking for an equation that describes the number of hospital beds, N, in terms of time, T, the independent variable. If we assume that the number of beds has been declining at a constant rate, then we can use a linear function in the form

$$N = b + mT$$

as a model. Since we are only interested in years from 1965 on, the computations will be simpler if we let $T =$ number of years since 1965.

Find m: The statement that the number of beds is declining by 20,000, or equivalently 0.02 million beds a year, tells us that the average rate of change of N with respect to T is -0.02 million beds/year. So $m = -0.02$ million beds/year. The function has the form:

$$N = b - 0.02T$$

Now we have to stop and think about the units for each term. T is in years, -0.02 is in millions of beds/yr. Years/years cancel out so $0.02T$ will be in millions of beds. The units for b and N will also be millions of beds.

Find b:

We know that in 1965 the nation had 1.7 million beds. T is years since 1965, so at 1965, $T = 0$. N is millions of beds and in 1965 $N = 1.7$. So the point $(0, 1.7)$ is a solution to our equation. Substituting these values gives us:

$$1.7 = b - (0.02)(0)$$

or

$$1.7 = b$$

We might have also recognized $(0, 1.7)$ as the vertical intercept and substituted 1.7 for b right away.

The equation:

Our equation is

$$N = 1.7 - 0.02T$$

where T = years since 1965, and N = number of hospital beds in millions. N is a linear function of T.

The domain:

The values for T start at 0 years (for the year 1965). Since the announcement was made in 1996, the domain for T would be $0 \leq T \leq 31$.

Table and graph:

Table 4.9 and Figure 4.11 are representations of the function $N = 1.7 - 0.02\,T$.

Table 4.9	
Number of Hospital Beds	
T (Years Since 1965)	N (Number of Beds in Millions)
0	1.7
5	1.6
10	1.5
15	1.4
20	1.3
25	1.2
30	1.1

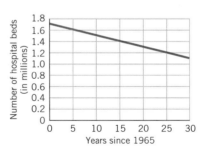

Figure 4.11   Number of hospital beds since 1965, $N = 1.7 - 0.20T$.

We can use the equation to predict when the number of hospital beds would reach 1 million. Since N is measured in millions of beds, we are asking when N would equal 1 million. Setting $N = 1$ we get

$$1 = 1.7 - 0.02T$$

subtract 1.7 from both sides

$$-0.7 = -0.02T$$

divide both sides by -0.02

$$35 = T$$

If the rate of decline stays constant at 20,000 hospital beds a year, 35 years after 1965, in the year 2000 we will have 1 million hospital beds.

The program "L3: Finding a Line Through 2 Points" in Linear Functions gives you practice in this skill.

Case 3: Two Points (x_1, y_1) and (x_2, y_2) Are Known

In this case, both the slope (m) and the y-intercept (b) have to be determined. Given two distinct points (x_1, y_1) and (x_2, y_2) on the line, we can always calculate the slope. Once we have found the slope, we are back to Case 2 where the slope and one of those points are used to find b.

Example   6

Find the equation of the line passing through $(-3, -5)$ and $(-1, 3)$.

SOLUTION

Choose variables: Let x be the independent variable and y be the dependent variable.

Find m: Since both $(-3, -5)$ and $(-1, 3)$ are on the line,

$$m = \frac{3 - (-5)}{-1 - (-3)} = \frac{8}{2} = 4$$

The slope, m, is 4, and substituting that value for m, the equation is:

$$y = b + 4x$$

Find b: Since the line passes through $(-1, 3)$, we can substitute -1 for x and 3 for y in the equation to find b:

$$3 = b + 4(-1)$$
$$3 = b - 4$$
$$b = 7$$

The equation: We can now substitute these values for m and b directly into the equation to get

$$y = 7 + 4x$$

This equation represents y as a linear function of x.

The domain: Since x has no constraints, the domain is all real numbers.

Table and graph: Table 4.10 and Figure 4.12 show solutions to the equation $y = 7 + 4x$.

Table 4.10	
x	y
-3	-5
-1	3
0	7
1	11
3	19

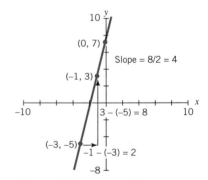

Figure 4.12   Graph of $y = 7 + 4x$.

This strategy of using two points on the line to find the equation will prove useful in the next section, where we examine fitting linear models to data sets.

Algebra Aerobics 4.3b

For each situation described in Problems 1 and 2, write an equation, make an appropriate table with two or three points, and sketch the graph.

1. A line has a slope of -4 and passes through the point $(-3, 1)$.

2. Salary increases by $1.60 per hour for every year of education, and with 10 years of education, the hourly salary is $8.50.

3. We could have used either point in Example 6 to solve for b. Calculate again using the other point $(-3, -5)$ to verify that $b = 7$.

4. Find an equation for the line through the points $(-2, 4)$ and $(3, 1)$.

4.4 ESTIMATING LINEAR MODELS FOR DATA

We've seen linear models for the median weight of baby girls between birth and 6 months and for the median height of children between the ages of 2 and 12 years. These models come from observational studies of children. In the following study, children's heights were measured monthly over several years as a part of a study of nutrition in developing countries. Table 4.11 and Figure 4.13 show the mean heights of a group of 161 children in Kalama, Egypt.

Table 4.11	
Mean Height of Kalama Children	
Age (months)	**Height (cm)**
18	76.1
19	77.0
20	78.1
21	78.2
22	78.8
23	79.7
24	79.9
25	81.1
26	81.2
27	81.8
28	82.8
29	83.5

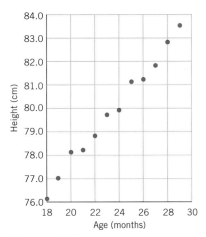

Figure 4.13 Mean height of children in Kalama, Egypt.

Source: Introduction to the Practice of Statistics by David S. Moore and George P. McCabe. Copyright © 1989 by W.H. Freeman and Company. Used with permission.

See Excel or graph link file KALAMA and the program "R5: Age-Height Data: Kalama, Egypt" in Linear Regression. *You can use "R1: Regression Line: m & b Sliders" also in* Linear Regression *to estimate a best fit line for a random set of data.*

Although the data points do not lie exactly on a straight line, the overall pattern seems clearly linear. Rather than generating a line through two of the data points, try eyeballing a line that approximates all the points. A ruler or a piece of black thread laid down through the dots will give you a pretty accurate fit. Figure 4.14 shows a line that approximates the data points. This line does not necessarily pass through any of the original data points. In Chapter 5, we will see that there are ways to use technology to find a "best fit" line.

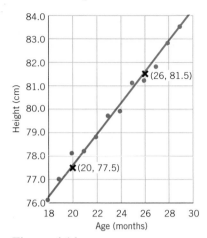

Figure 4.14   Estimated coordinates of two points on the line.

What is an equation for our linear model? We can use the strategy from Case 3, where we found the equation for a line from two known points. The estimated coordinates of two points *on the line* are labeled on the graph. Note that they are *not* taken from the original data in Table 4.11. We can use these points to find the average rate of change or slope of that line:

$$m = \frac{(81.5 - 77.5)\text{ cm}}{(26 - 20)\text{ months}}$$

$$= \frac{4.0\text{ cm}}{6\text{ months}}$$

$$= 0.67\text{ cm/month}$$

The rate of change is the most important number in our linear model. It is a compact way of summarizing how height changes with age. *Our model predicts that for each additional month between 18 and 29 months, an "average" child will grow about 0.67 centimeter.*

Now we know our linear equation is of the form:

$$y = b + 0.67x \tag{1}$$

where x = age in months and y = mean height in centimeters.

How can we find b? We have to resist the temptation to read b directly from the graph, using the coordinates where our line crosses the height axis. Why? Isn't b always the vertical intercept? As is frequently the case in social science graphs, both the horizontal and the vertical axes are cropped. At a quick glance, what looks like the origin, the point (0, 0), actually has coordinates (18, 76). Because the horizontal axis is cropped, we can't read the vertical intercept off the graph. We'll have to calculate it using the methods introduced in Case 3.

Since the line passes through (20, 77.5) we have:

substitute (20, 77.5) in Equation (1) $77.5 = b + (0.67)(20)$

simplify $77.5 = b + 13.4$

solve for b $b = 64.1$

Our linear equation modeling the Kalama data is:

$$y = 64.1 + 0.67x$$

where x = age in months and y = height in centimeters. It is a linear function that offers a compact summary of the Kalama data.

What is the domain of this model? The data were collected on children aged 18 to 29 months. We don't know its predictive value outside these ages, so

$$\text{domain} = \text{values of } x \text{ where } 18 \le x \le 29$$

Note that although the y-intercept is necessary to write the equation for the line, it lies outside of the set of values for which our model applies. We have no reason to expect that growth patterns between 18 and 29 months can be projected back to give the height of a newborn baby as 64.1 cm.

Compare Figure 4.15 to Figure 4.14. They both show graphs of the same equation $y = 64.1 + 0.67x$. In Figure 4.14 both axes are cropped, while Figure 4.15 includes the origin (0, 0). On Figure 4.15 the vertical intercept is now visible and the dotted lines indicate the region that applies to our model. So, a word of warning when reading graphs: look carefully to see if the axes have been cropped.

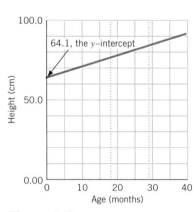

Figure 4.15   Graph of $y = 64.1 + 0.67x$ that includes the origin (0, 0). Dotted lines show region that models Kalama data.

A messier example   The Kalama data set was fairly clean, partially because we didn't have all the raw data. Table 4.11 contains one height value for each month that represents the mean of the children's heights, and does not give all of the individual heights. For many data sets there may be many *y* values associated with one *x* value. In these cases the scatter plots look more like clouds of data.

Figure 4.16 shows mean SAT verbal scores versus percentage of seniors taking the SAT in 1989 for each state in the United States. Each point is labeled with an abbreviation of the state's name. The point labeled WA near the middle of the graph with approximate coordinates (40, 450) means that in 1989, 40% of seniors in the state of Washington took the SAT test and their mean verbal score was 450. There appears to be a general downward trend. As the percentage of students taking the test increases, the mean SAT verbal score decreases. We can sketch a best fit line through the data (as shown on the scatter plot), estimate its equation, and interpret its slope.

The article "Verbal SAT Scores" from which this data was taken appears in The Anthology of Readings

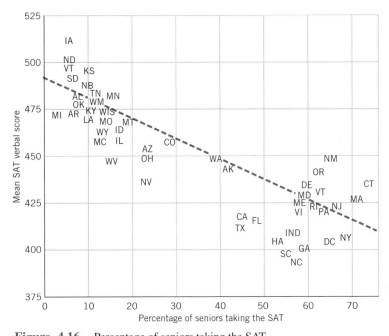

Figure 4.16   Percentage of seniors taking the SAT.

Source: "Verbal SAT Scores," *Quantitative Examples,* edited by John Truxal, NLA Monograph Series, Suny Research Foundation, 1991.

Choose variables: Let *P* be the independent variable representing the percentage of seniors. Let *S* be the dependent variable representing the mean SAT verbal score. The equation of the line is of the form:

$$S = b + mP$$

Find *m*:   Do a rough hand-drawn sketch of the line (for example, the dotted line in the scatter plot). In order to calculate the slope, we must estimate the coordinates of two points *on the line*, for instance (10, 480) and (60, 430). Then we have

$$m = \frac{(480 - 430) \text{ SAT points}}{(10 - 60)\%} = \frac{50 \text{ SAT points}}{-50\%} = -1.0 \text{ SAT points}/\%$$

On average for each additional 1% of seniors taking the SAT verbal test, our model predicts that the mean verbal score was roughly 1 point lower. Equivalently, when the percentage of seniors taking the test increased by 10, the mean SAT verbal score dropped on average by 10 points.

Substituting -1 for m, the equation for the line is

$$S = b - P$$

Find b: We estimated that the line passes through (60, 430), so we can substitute the values 60 and 430 for P and S in the equation of the line to find b.

$$430 = b - 60$$

$$490 = b$$

The equation: Substituting this value for b, we get

$$S = 490 - P$$

which is a linear function where P = the percentage of seniors taking the SAT verbal test, and S = the mean SAT verbal score.

The domain: Given the plotted data, a reasonable domain for P might be $5\% \leq P \leq 75\%$.

Table and graph: The SAT scores predicted by our linear model (*not* the actual scores) are included in the Table 4.12 and Figure 4.17. Note that in Figure 4.17 as in Figure 4.16 the vertical axis is cropped.

Something to think about

1. How well do the predicted SAT scores match the real scores?

2. Look at the Anthology article "Verbal SAT Scores." What very different conclusions might you draw looking at the SAT scores as a single variable as opposed to looking at their relationship to the percentage taking the SAT? What are some questions this model suggests that you should ask when comparisons are drawn between test scores of American students and students from other countries?

Table 4.12	
P (percentage)	S (SAT score)
10	480
20	470
30	460
40	450
50	440
60	430
70	420
80	410

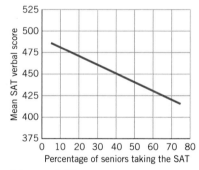

Figure 4.17 Mathematical model for percentage of seniors taking SAT as a function of mean verbal score.

"The Paramount Convenience of Straight Lines"[3]

When analyzing relationships between two variables, linear relationships are of special importance, not because most relationships are linear but because straight lines are easily drawn and analyzed. By stretching a thread across a scatter plot, we can fit a straight line by eye almost as well as a computer does. Since linear equations are easy to manipulate and interpret, lines are often used as first approximations to patterns in data.

[3] This title and the ideas in this subsection are from Douglas Riggs, *The Mathematical Approach to Physiological Problems*, (Cambridge, MA: MIT Press, 1976) pp. 57–58.

Algebra Aerobics 4.4

1. **a.** Figure 4.18 shows the number of students in the United States who were 35 or older between the years of 1980 and 1995. Sketch a line through the data.

   b. Estimate the coordinates of two points on the line and use them to find the slope.

   c. Let x = number of years since 1980 and y = the number of U.S. students 35 or older. Find the equation for your linear model.

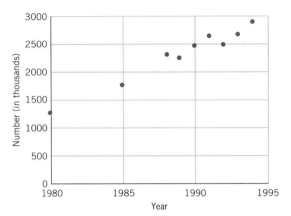

Figure 4.18 U.S. population 35 years old and above enrolled in school.

Source: U.S. National Center for Education Statistics, *Digest of Education Statistics,* annual.

4.5 SPECIAL CASES

Case 1: Direct Proportionality

The simplest possible relationship between two variables is when one is equal to a constant times the other. For instance, in the Illinois sales tax example $T = 0.08P$; the sales tax T equals a constant, 0.08, times P, the purchase price. We say that T is *directly proportional to P* or that *T varies directly with P*.

How can we recognize direct proportionality? Linear functions in the form

$$y = mx$$

describe a relationship where y is directly proportional to x. Notice that we could write the equation $y = mx$ as

$$y = 0 + mx$$

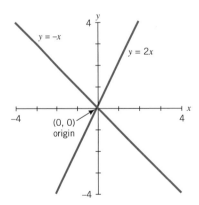

Figure 4.19 Graphs of two relationships in which y is directly proportional to x. Note that both graphs are lines that go through the origin.

which would be in the general linear form $y = b + mx$ where $b = 0$. So, direct proportionality is a special case of the linear function where b, the y-intercept, is 0. Hence the graph of two variables which are directly proportional to each other will pass through the origin, the point $(0, 0)$. Figure 4.19 shows the graphs of two relationships in which y is directly proportional to x: namely $y = 2x$ and $y = -x$. Note that in both cases the graphs are straight lines that pass through the origin.

If y is directly proportional to x, is x directly proportional to y?

If $y = mx$, where $m \neq 0$
the relationship between x and y is described by saying that
y *is directly proportional to* x
or that
y *varies directly with* x.

Example   1

Suppose you go on a road trip, driving at a constant speed of 60 miles per hour. After 1 hour, you will be 60 miles away from where you started; after 2 hours, you will be 120 miles away; and so on. The distance d (in miles) you have traveled varies directly with the time t (in hours) according to the formula: $d = 60t$.

Example   2

Suppose you are paid by the hour for performing a certain job. If your pay rate is $7/hr, then E, the amount of money earned (in dollars), after H hours can be represented by $E = 7H$; that is, E is directly proportional to H.

Example   3

On July 7, 1994, you could buy 1.345 Canadian dollars for 1 U.S. dollar. Find a linear function that converts American dollars into Canadian dollars at this rate.

SOLUTION

Choose variables: Let US$ = number of American dollars (the independent variable) and CD$ = number of Canadian dollars (the dependent variable)

Find b: When US$ = 0, the value of CD$ = 0, so $b = 0$.

Find m: Each additional American dollar corresponds to 1.345 Canadian dollars; thus, $m = 1.345/1 = 1.345$.

Construct function: Substituting, the equation of the linear function is:

$$CD\$ = 1.345 \cdot (US\$)$$

The number of Canadian dollars you receive is directly proportional to the number of American dollars you convert.

The domain: US$ $\geq$ 0.

Table and graph: Table 4.13 and Figure 4.20 convert American dollars into Canadian dollars.

Table 4.13

US$	CD$
0	13.45(0) = 0
1	1.345(1) = 1.345
10	1.345(10) = 13.45
20	1.345(20) = 26.90
30	1.345(30) = 40.35

Figure 4.20
Exchange rate for American to Canadian dollars.

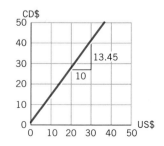

If y is directly proportional to x, and $y = -10$ when $x = 5$, find the value of x when $y = -8$.

Example   4

SOLUTION

Since y is directly proportional to x, the equation relating x and y is of the form $y = mx$. We know $(0, 0)$ is a solution and we are given that $(5, -10)$ is also a solution. We can use these two points to calculate m.

$$m = \frac{\text{change in } y}{\text{change in } x} = \frac{-10 - 0}{5 - 0} = -2$$

Hence the equation is $y = -2x$. Substituting -8 for y gives

$$-8 = -2x$$

So

$$x = 4$$

• • •

Sometimes it is more convenient to think of a linear equation of the form $y = mx$ in its equivalent form:

$$\frac{y}{x} = m \qquad (\text{where we assume that } x \text{ is not equal to } 0)$$

We call the quotient y/x the *ratio* of y to x. When x and y are *directly proportional*, the ratio of y to x remains constant, and is equal to m, the slope of the line.

> When two variables x and y are *directly proportional*
> then the *ratio y/x* is constant.

If the student/faculty ratio is 12/1, and an institution has 3000 students, how many faculty does it have?

Constant proportions are very common in cooking. A good oil and vinegar dressing is 3 parts olive oil to 2 parts balsamic vinegar, or a great shandy is 12 parts beer to 32 parts ginger ale. These proportions are rates of change that can be used to make any desired quantity, for one person or a large party.

a. How much oil is needed for this salad dressing recipe when 7 tablespoons of vinegar are used?

b. How much ginger ale is needed to make a shandy when 21 pints of beer are used?

SOLUTION

a. The ratio of
$$\text{oil/vinegar} = 3/2$$
$$\text{oil} = (3/2) \cdot (\text{vinegar})$$
$$\text{oil} = 1.5 \cdot (\text{vinegar})$$

Substituting 7 tbs for vinegar we get:
$$\text{oil} = 1.5(7 \text{ tbs})$$
$$\text{oil} = 10.5 \text{ tbs}$$

Example   5

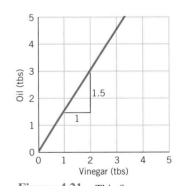

Figure 4.21   This figure shows a graph of the amount of oil needed for salad dressing as a function of the amount of vinegar. The graph is a straight line through the origin with a slope of 1.5.

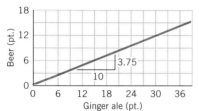

Figure 4.22   This figure shows a graph of the amount of beer needed for a shandy as a function of the amount of ginger ale. The graph is a straight line through the origin with a slope of $\dfrac{3}{8}$ or 0.375.

So 7 tablespoons of vinegar corresponds to 10.5 tablespoons of oil in the oil and vinegar dressing.

b.   The ratio of

$$\text{beer/ginger ale} = 12/32$$
$$= 3/8$$

So

$$\text{beer} = (3/8) \cdot (\text{ginger ale})$$
$$\text{beer} = 0.375 \cdot (\text{ginger ale})$$

If we substitute 21 pts for beer then

$$21 \text{ pts} = 0.375 \cdot (\text{ginger ale})$$
$$56 \text{ pts} = \text{ginger ale}$$

So 21 pts of beer requires 56 pts of ginger ale in our shandy recipe.

Algebra Aerobics 4.5a

1. Find a function that represents the relationship between distance, d, and time, t, of a moving object using the data in Table 4.14. Is d directly proportional to t? Which is a more likely choice for the object, a person walking or a moving car?

Table 4.14

t (hr)	d (miles)
0	0
1	5
2	10
3	15
4	20

2. For each of Tables 4.15–4.17, determine whether the variables vary directly with each other. Represent each relationship with an equation.

Table 4.15

x	y
−2	6
−1	3
0	0
1	−3
2	−6

Table 4.16

x	y
0	5
1	8
2	11
3	14
4	17

Table 4.17

G (gasoline bought, number of gallons)	C (cost of gasoline, dollars)
1	1.50
2	3.00
3	4.50
4	6.00
5	7.50

3. On the scale of a map 1 inch represents a distance of 35 miles. What is the distance between two points that are 4.5 inches apart on the map?

Special Case 2: Horizontal Lines

If a line is horizontal (Figure 4.23), then as the values for the independent variable increase, the values for the dependent variable stay the same. No matter what the change in x, the corresponding change in y is 0, so the slope, m, is 0. The equation of the line has the form $y = (0)x + b$, or simply

$$y = b$$

This means that y does not depend on x at all. No matter what value x assumes, the value for y remains the same. Hence y always has the same constant value, b.

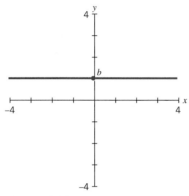

Figure 4.23 A horizontal line has a slope of 0 and an equation $y = b$.

Find the equation of a horizontal line through the point (2, 3).

Example 6

SOLUTION

Let x be the independent variable and y be the dependent variable. Plot the point (2, 3) and sketch a horizontal line through it (see Figure 4.24). Note that each point on the line is 3 units above the x-axis. Since y is always equal to 3, then the equation should be $y = 3$. Table 4.18 shows a set of points that lie on the line.

Table 4.18	
x	$y = 3$
-4	3
-3	3
-2	3
-1	3
0	3
1	3
2	3
3	3
4	3

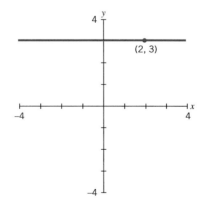

Figure 4.24 Graph of $y = 3$.

The equation $y = 3$ does represent y as a function of x, since for each value of x there is one and only one value of y, namely 3.

Example   7

Table 4.19 and Figure 4.25 show the past shipments of two kinds of computer disk drives: the older 5.25 inch disk drives and the newer, smaller 3.5-inch or less disk drives. A business person would call this a "changing product mix." Describe the shipment history of the 5.25 inch disk drives and construct an equation to represent shipments.

Table 4.19

Worldwide Flexible Disk Drive Shipments—All Manufacturers (in thousands)

	1983	1984	1985	1986	1987	1988	1989	1990	1991	1992	1993
5.25-inch drives	10,490	15,616	12,995	16,287	16,232	16,574	14,887	14,946	15,020	17,291	14,711
3.5-inch or less drives	438	1,972	3,268	6,199	12,337	18,115	23,216	29,446	32,819	41,790	51,152

Source: Disk/Trend, Nov. 1994 report. Reprinted with permission.

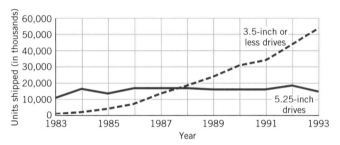

Figure 4.25   Worldwide flexible disk drive shipments.

SOLUTION

The shipments for 5.25-inch drives remained fairly flat between 1983 and 1993 at about 15,000 thousand drives per year. Mathematically, we could describe the horizontal line approximating 5.25-inch disk drive shipments by:

$$S_{5.25} = 15,000$$

where $S_{5.25}$ = shipments for 5.25-inch drives measured in thousands (see Figure 4.26).

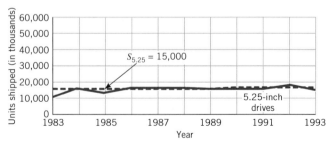

Figure 4.26   Worldwide flexible disk drive shipments. The horizontal dotted line approximates shipments for 5.25-inch disk drives.

Special Case 3: Vertical Lines

Every point on a vertical line (as in Figure 4.27) has the same x-coordinate. So the x value is constant and does not depend upon y. The equation is simply:

$$x = c$$

Where c is some constant. This equation does not represent y as a function of x, since for each x there is not associated one and only one value for y. For example, the equation $x = 1$, plane (see Figure 4.27) is not a function. When x is 1, there are infinitely many associated y values, so the graph fails the vertical line test.

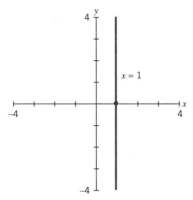

Figure 4.27   Graph of the vertical line $x = 1$.

If we try to calculate the slope between two points on the line, the change in x is 0.

$$\text{slope} = \frac{\Delta y}{\Delta x} = \frac{\Delta y}{0} \text{ which is undefined.}$$

For vertical lines the slope is undefined and we can't use the standard form of the linear equation.

Find an equation for the vertical line that passes through the point $(-2, 1)$ (Figure 4.28).

Example 8

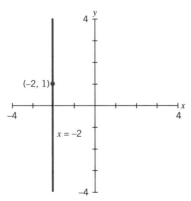

Figure 4.28   Graph of the vertical line through $(-2, 1)$.

SOLUTION

Let x be the independent variable and y the dependent variable. Plot the point $(-2, 1)$ and sketch a vertical line through it. Each point on the line must have an x value of -2, no matter what the y value. So the equation is:

$$x = -2$$

Example 9

Vertical lines are relatively rare in the social sciences, but one theoretical example stems from economics. Economists talk about demand curves that depict the relationship between the market price of a commodity and the consumers' demand for that commodity. (We take a closer look at demand curves in Chapter 6.) Generally higher prices are associated with decreased consumer demand. So the demand curve normally has a negative slope when quantity is graphed on the horizontal axis and price on the vertical axis. If the demand for a commodity (computer scientists or Mighty Morphin Power Rangers®) is so great that an increase in price has little effect on the quantity demanded, then the demand curve becomes close to a vertical line, and demand is said to be *inelastic* (Figure 4.29).

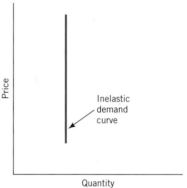

Figure 4.29   Graph of inelastic demand.

Algebra Aerobics 4.5b

1. In each case, find an equation for a horizontal line that passes through the given point.

   a. $(3, -5)$ **c.** $(-3, 5)$

   b. $(5, -3)$ **d.** $(-5, 3)$

2. An employee for an aeronautical corporation has a starting salary of $25,000 per year. After working there for 10 years and not receiving any raises, he decides to seek employment elsewhere. Write an equation and graph the employee's salary as a function of the time he was employed with this corporation. What is the domain for this function? What is the range?

3. In each case, find an equation for the vertical line that passes through the given point.

   a. $(3, -5)$ **c.** $(-3, 5)$

   b. $(5, -3)$ **d.** $(-5, 3)$

Special Case 4: Parallel Lines

If two linear relationships between the same variables have equal rates of change, their graphs are two lines with equal slopes, so the lines are parallel. Geometrically this means that the lines always maintain the same vertical distance apart and never cross. For example, the two lines $y = 2.0 - 0.5x$ and $y = -1.0 - 0.5x$ each have a slope of -0.5 and thus are parallel (see Figure 4.30).

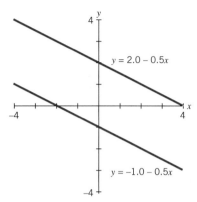

Figure 4.30   Parallel lines have equal slopes.

Between 1975 and 1985 the lines representing growth in the percentage of medical degrees awarded to women doctors and women dentists in the United States were roughly parallel. Table 4.20 and Figure 4.31 show the percentage of medical and dental degrees awarded to women during this time period.

Example   10

Table 4.20

Percentage of Medical Degrees Conferred on Women

	1975	1980	1985
Doctors	13.1	23.4	30.4
Dentists	3.1	13.3	20.7

Source: U.S. National Center for Education Statistics, "Digest of Education Statistics," annual, in *The American Almanac 1995–1996: Statistical Abstract of the United States,* 1995.

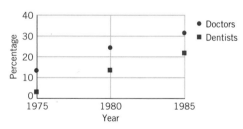

Figure 4.31   Percentage of medical and dental degrees conferred on women.

The data points for female doctors' and dentists' degrees lie roughly on parallel lines, each with a slope of about 1.7%/yr. That means that between 1975 and 1985 the percentage of degrees awarded to female doctors and female dentists both increased by about 1.7 percentage points per year. If both continue to grow at the same rate, the percentage of dentists cannot catch up with the percentage of doctors (until all doctors are women, an unlikely possibility). Similar processes may be at work causing these increases, a topic for further study by social scientists.

Example 11

Something to think about

Clearly, other factors could be considered in making predictions on sales of a new product. What else might play a role in the sales of new optical disk drives? How might your graph change if you considered these factors?

In the business world, each industry has its own set of commercial "market forecasters," and each of them has developed proprietary models for forecasting industry trends.[4] A typical forecast would track 10 years of historical data and then predict 2–5 years into the future. If a new product is being introduced, then there are obviously no historical data available. One strategy is to look at another product previously introduced into the same market and assume that the new product sales will follow a parallel pattern.

For example, a company in Massachusetts made the following prediction based on past sales. The company was thinking of introducing a new optical disk drive. Past disk drive sales were noted (see Example 7) and predictions were made based on that history. Figure 4.32 shows the simplified graph of worldwide shipments of 3.5-inch or less disk drives and the predicted shipments for the new optical disk drives. The graphs show the classic "hockey stick" pattern of growth: Initial sales are fairly flat, but then as the product catches on, sales take a dramatic upward turn.

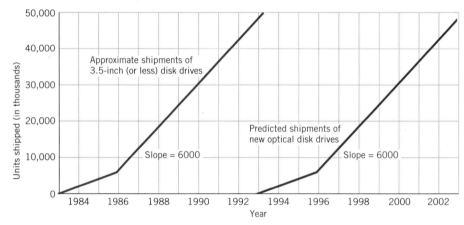

Figure 4.32   Parallel growth for 3.5-inch disk drive and new optical disk drives.

The rate of change during the rapid growth period of 3.5-inch (and less) disk drive shipments, and hence the prediction for the equivalent period for the new optical disk drive, is an increase of approximately 6000 thousand (or 6 million total) units per year. The graphs of past and predicted shipments are two parallel line segments each with a slope of 6000 thousand units per year.

Special Case 5: Perpendicular Lines

Consider a line whose slope is given by v/h. Now imagine rotating the line 90 degrees clockwise to generate a second line perpendicular to the first (Figure 4.33).

What would the slope of this new line be?

The positive vertical change, v, becomes a positive horizontal change. The horizontal change, h, becomes a negative vertical change.

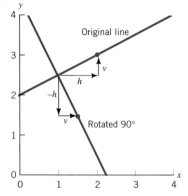

Figure 4.33   Perpendicular lines $m_2 = -1/m_1$.

---

[4] Dataquest and The Yankee Group are two such market forecasters in the computer industry.

The slope of the original line is $\dfrac{v}{h}$, and the slope of the line rotated 90 degrees clockwise is $\dfrac{-h}{v}$. Note that $\dfrac{-h}{v} = \dfrac{-1}{(v/h)}$, which is the original slope inverted and multiplied by -1.

In general, the slope of a perpendicular line is the negative reciprocal of the slope of the original line. If the slope of a line is m_1, then the slope, m_2, of a perpendicular line is $-1/m_1$.

This is true for any pair of perpendicular lines for which slopes exist. It does not work for horizontal and vertical lines since vertical lines have undefined slopes.

Something to think about

1. If two lines are perpendicular to each other, what do you know about the units for the two variables?
2. Describe the equation for any line perpendicular to the horizontal line $y = b$.

Find the equation of a line that is perpendicular to the line $y = 3 - 2x$ and that passes through the point $(-2, -3)$.

Example 12

SOLUTION

The slope of the given line, $y = 3 - 2x$, is -2, so the slope of any line perpendicular to it will be the negative reciprocal of -2. Hence, the new slope is $-(1/-2) = 1/2$. The equation of the new line is of the form

$$y = b + \frac{1}{2}x.$$

Since the point $(-2, -3)$ lies on the new line, we can substitute -2 for x and -3 for y in order to find b.

$$-3 = b + \left(\frac{1}{2}\right)(-2)$$
$$-3 = b - 1$$
$$-2 = b$$

The equation for the new line is:

$$y = -2 + \frac{1}{2}x$$

Figure 4.34 shows its graph.

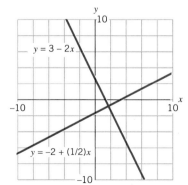

Figure 4.34   Graph of $y = 3 - 2x$ and $y = -2 + \left(\dfrac{1}{2}\right)x.$

Algebra Aerobics 4.5c

1. Find the equation of a line that is parallel to $y = 4 - x$ and that passes through the origin.

2. Find the equation of a line that is parallel to $W = 358.9C + 2500$ and passes through $(4, 1000)$.

3. Find the slope of a line perpendicular to each of the following.

    a. $y = 4 - 3x$ **b.** $y = x$ **c.** $y = 3.1x - 5.8$

4. **a.** Find an equation for the line which is perpendicular to $y = 2x - 4$ and passes through $(3, -5)$. Graph both lines on the same coordinate system.

    b. Find the equations of two other lines that are perpendicular to $y = 2x - 4$ but do not pass through the point $(3, -5)$.

    c. How do the three lines from (a) and (b) that are perpendicular to $y = 2x - 4$ relate to each other?

4.6 EQUIVALENT WAYS OF DESCRIBING LINEAR RELATIONSHIPS

Most of the linear equations we've worked with so far have been in the form $y = b + mx$ or $y = mx + b$. By rearranging terms and multiplying by a real number, we could rewrite these equations in the form $Ax + By = C$ (where A, B, and C are constants). Sometimes this equation is the more natural format. For example, if you have $100 to buy food for a party and liter bottles of Coke cost $1.50 and bags of pretzels cost $1.00, then:

    (the amount you spend on Coke) + (amount you spend on pretzels) = $100.00

If x = number of bottles of Coke and y = number of bags of pretzels, then

    the amount you spend on Coke = (cost per bottle)(number of bottles) = $1.50x$
    the amount you spend on pretzels = (cost per bag)(number of bags) = $1.00y$

So the linear equation

$$1.50x + 1.00y = 100.00 \qquad (1)$$

describes the relationship between x, the number of Coke bottles, and y, the number of pretzel bags. This is in the form $Ax + By = C$ where $A = 1.50$, $B = 1.00$, and $C = 100.00$.

If we want to know how much Coke we could buy, we could solve for x:

given	$1.50x + 1.00y = 100.00$
subtract $1.00y$ from both sides	$1.50x = 100.00 - 1.00y$
divide both sides by 1.50	$x = \dfrac{(100.00 - 1.00y)}{1.50}$
divide terms by 1.50	$x = \left(\dfrac{100.00}{1.50}\right) - \left(\dfrac{1.00}{1.50}\right)y$
simplify fractions	$x = \dfrac{200}{3} - \dfrac{2}{3}y \qquad (2)$

Similarly if we want to know how many bags of pretzels we could buy, we solve for y in terms of x to get:

$$y = 100.00 - 1.50x \qquad\qquad (3)$$

So Equations (1), (2), and (3) are all equivalent.

A Summary of Different Ways to Describe a Linear Relationship

We've now seen a number of different ways to represent a linear relationship between two variables. These are summarized in the following table.[5]

Two variables x and y are said to be *linearly related* if any *one* of the following equivalent conditions is shown to be true:

1. All of the pairs (x, y) lie on a straight line.
2. For a fixed amount of change in x there is a fixed amount of change in y.
3. The average rate of change of y with respect to x is constant.
4. There exist two constants, m (the slope) and b (the y-intercept) such that the equation $y = b + mx$ is true for all pairs (x, y).
5. There exist three constants, A, B, and C, such that the equation $Ax + By = C$ holds true for all pairs (x, y). (B is not equal to 0.)

Algebra Aerobics 4.6

1. Determine if each of the following describes y as a linear function of x.

   a. $y = 2x$

   b. $y = 3$

   c. $41x + 34y = -11$

   d. $x = 4$

   e. $y = x^2$

   f.

   Table 4.21

x	y
1	2
2	5
3	10

   g.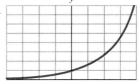

   Figure 4.35

2. For the examples in Problem 1 that are linear functions, find m and b.

3. Adding minerals or organic compounds to water lowers its freezing point. Antifreeze for car radiators contains glycol (an organic compound) for this purpose. Table 4.22 shows the effect of salinity (dissolved salts) on the

---

[5] Note: this summary ignores the special case of vertical lines.

freezing point of water. Salinity is measured in the number of grams of salts dissolved in 1000 grams of water. So our units for salinity are in parts per thousand, abbreviated as *ppt*.

Is the relationship between the freezing point and salinity linear? If so, give at least three equivalent ways of describing the relationship.

Table 4.22	
Relationship between Salinity and Freezing Point	
Salinity (ppt)	**Freezing point (°C)**
0	0.00
5	−0.27
10	−0.54
15	−0.81
20	−1.08
25	−1.35

Source: Data adapted from *Oceanography: An Introduction to the Planet Oceanus* by Paul R. Pinet (St. Paul, Minnesota: West Publishing Company, 1992) p. 522.

CHAPTER SUMMARY

A *linear function* has the form $y = f(x)$ where

$$y = b + mx$$

Its graph is a straight line where b is the *vertical intercept*, or value of y when x is 0 and m is the *slope*, or *rate of change* of y with respect to x. We have

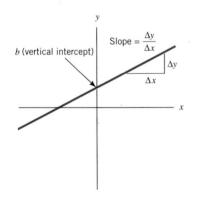

$$m = \frac{\text{change in } y}{\text{change in } x} = \frac{\Delta y}{\Delta x}$$

Linear functions are often written in the form $y = mx + b$.

Any two points that lie on a line can be used to calculate the slope. That is, if (x_1, y_1) and (x_2, y_2) are two distinct points that satisfy the function $y = b + mx$, then

$$m = \frac{y_2 - y_1}{x_2 - x_1}$$

We can generate the equation of a line if we know:

m and b, or

m and any point on the line, or

two points on the line.

For a data set whose graph exhibits a linear pattern, we can visually position a line to fit the data. The equation of this line offers an approximate but compact description of the data.

Lines are of special importance because they are easily drawn and manipulated. Lines offer the simplest choice for a rough description of patterns in data.

Special cases of lines:

1. *Direct proportionality*

    y is said to be *directly proportional to x* or to *vary directly with x* if there is a *real number m*, such that:

    $$y = mx$$

    This equation is a linear function in which the y-intercept is 0, and the graph goes through the origin.

2. *Horizontal lines*

    When the slope, m, is 0, the general linear equation becomes $y = b$, which is the equation of a horizontal line through the point $(0, b)$.

3. *Vertical lines*

    A vertical line has an equation of the form $x = c$. Its slope is undefined.

4. *Parallel lines*

    Two lines are parallel if they have the same slope. For example, $y = 5 + 2x$ and $y = -7 + 2x$ are parallel since they both have a slope of 2.

5. *Perpendicular lines*

    Two lines are perpendicular if their slopes are negative reciprocals. For example, the lines $y = 5 + 2x$ and $y = 8 - (1/2)x$ are perpendicular since $-1/2$ is the negative reciprocal of 2.

Two variables x and y are said to be linearly related if any *one* of the following equivalent conditions is shown to be true:

1. All of the pairs (x, y) lie on a straight line.
2. For a fixed amount of change in x there is a fixed amount of change in y.
3. The average rate of change of y with respect to x is constant.
4. There exist two *constants, m* (the slope), and b (the y-intercept) such that the equation $y = b + mx$ is true for all pairs (x, y).
5. There exist three *constants*, A, B, and C such that the equation $Ax + By = C$ holds true for all pairs (x, y). (B is not equal to 0.)

This list excludes the special case of vertical lines.

EXERCISES

In Exercises 1–3, find an equation, generate a small table of solutions, sketch the graph, and interpret the slope.

1. A line that has a vertical intercept of -2 and a slope of 3.

2. A line that crosses the vertical axis at 3.0 and has a rate of change of -2.5.

3. A line that has a vertical intercept of 1.5 and a slope of 0.

4. The equation $K = 4F - 160$ models the relationship between F, the temperature in degrees Fahrenheit, and K, the number of cricket chirps per minute for the snow tree cricket.

    a. Assuming F is the independent variable and K is the dependent variable, identify the slope and vertical intercept in the above equation.

    b. Identify the units for K, 4, F, and -160.

    c. Generate a small table of points that satisfy the equation. Be sure to choose realistic values for F.

    d. Calculate the slope directly from two data points. Is this value what you expected? Why?

    e. Graph the equation with F on the horizontal axis and K on the vertical axis.

    f. What is a reasonable domain for this model? Indicate this domain on the graph.

5. The equation $F = 32 + (9/5)C$ gives the relationship between degrees Fahrenheit, F, and degrees Celsius, C.

    a. Assuming C is the independent variable and F is the dependent variable, what are the values of the slope and the vertical intercept?

    b. Identify the units for F, 32, (9/5), and C.

    c. Generate a small table of points that satisfy the equation.

    d. Calculate the slope directly from two data points. Is this value what you expected? Why?

    e. Graph the equation with C on the horizontal axis and F on the vertical axis.

    f. At what temperature are the Celsius and Fahrenheit temperatures equal? Show your work.

6. Using the equations in the previous two exercises, find an equation that describes a relationship between K, the number of cricket chirps per minute, and C, the temperature in degrees Celsius.

7. The women's recommended weight formula from Harvard Pilgrim Healthcare says: "Give yourself 100 lb for the first 5 ft plus 5 lb for every inch over 5 ft tall." Find a mathematical model for this relationship. Specify a reasonable domain for the function and then graph it.

8. Construct a linear equation that satisfies each of the following properties.

    a. A negative slope and a positive y-intercept.

    b. A positive slope and a vertical intercept of -10.3.

    c. A constant rate of change of $1300/yr.

9. Estimate *b* (the *y*-intercept) and *m* (the slope) for each of the following graphs. Then, for each graph, write the corresponding linear function. Be sure to note the scales on the axes.

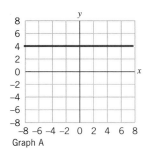
Graph A

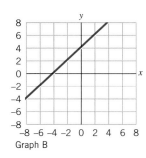
Graph B

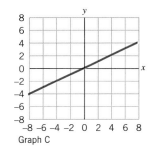
Graph C

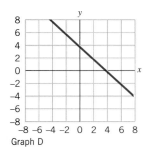
Graph D

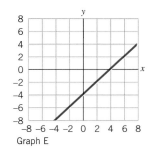
Graph E

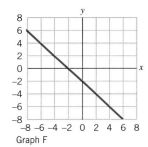
Graph F

10. (Requires course software.) Open the program *Linear Functions.*
    a. Click on the button for "L2: Finding *m* & *b*" The program will randomly generate the graph of a line. Try predicting the values for *m* and *b*. Set the sliders to your predicted values and click on "Graph the line" to check your predictions. Keep generating new lines (with both integer and real coefficients) until you are confident in your ability to predict *m* and *b*.
    b. Close L2 and open "L3: Finding a Line Through 2 Points." Click on two points on the graph and then click on "Draw." Now calculate the values for *m* and *b*, and write down the equation for the line. Click on "*m* & *b*" and "Equation" to check your answers. Now click on the yellow box and repeat the process until you can find *m* and *b* and the equation accurately each time.
    c. Close L3 and open "L4: Finding 2 Points on a Line." The computer will have randomly generated the equation of a line that is listed at the top right of the screen. You are asked to click on two points that lie on the line. Clicking on "Graph the line" will show you the graph of the original line. If the two graphs do not coincide, click on "Clear & Retry" and try again. Repeat this process with new equations until you are 100% accurate in your guesses.

11. The following scatter plot shows the relationship between literacy rate (the percentage of the population that can read and write) and infant mortality rate (infant deaths per 1000 live births) for 91 countries. The raw data are described in Part III "The Nations Data Set" on the CD-ROM and are contained in the

Excel or graph link file NATIONS. (You might wish to identify the outlier, the country with about a 20% literacy rate and a low infant mortality rate of about 40 per 1000 live births.) Construct a linear model. Show all your work and clearly identify the variables and units. Interpret your results.

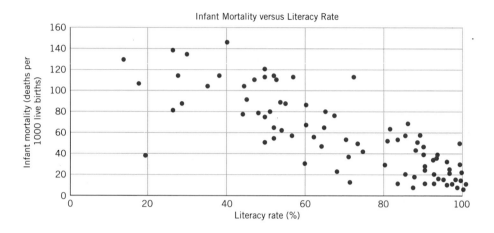

Infant Mortality versus Literacy Rate

Olympic Shot Put	
Year	**Feet Thrown**
1900	46
1904	49
1908	48
1912	50
1920	49
1924	49
1928	52
1932	53
1936	53
1948	56
1952	57
1956	60
1960	65
1964	67
1968	67
1972	70
1976	70
1980	70
1984	70
1988	74
1992	71
1996	71

12.   The following table and graph show data for the men's Olympic 16-lb shot put.

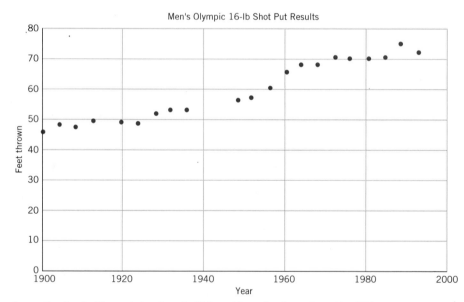

Men's Olympic 16-lb Shot Put Results

Source: Reprinted with permission from the Universal Press Syndicate. From *The 1996 Universal Almanac.* (The data points are missing for the early 1940s since the Olympics were not held during World War II.)

a.   Draw a line approximating the data and find its equation. Show your work and interpret your results. For ease of calculation you may wish to think of 1900 as year 0 and let your *x*-coordinate measure the number of years since 1900. So the point (1900, 46) would become (0, 46); (1904, 48) would become (4, 48), etc.

b.   If the shot put results continued to change at the same rate, in what year would you predict that the winner will put the shot a distance of 80 feet? Does this seem like a realistic estimate? Why or why not?

13. Construct two linear equations that
    a. are parallel to each other
    b. intersect at the same point on the y-axis
    c. both go through the origin
    d. are perpendicular to each other

14. For (a) and (b), find the value of *t* if *m* is the slope of the line that passes through the given points.
    a. $(3, t)$ and $(-2, 1)$ $m = -4$
    b. $(5, 6)$ and $(t, 9)$ $m = 2/3$

15. For each of the following graphs compare the *m* values and then the *b* values. You don't need to do any calculations or determine the actual equations. Do the lines have the same slope? Are the slopes both positive or both negative, or is one negative and one positive? Do they have the same y-intercept?

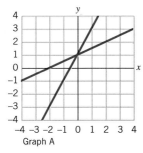
Graph A

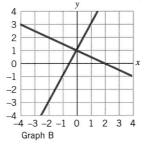
Graph B

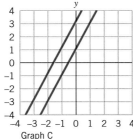
Graph C

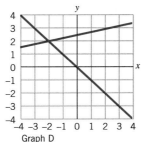
Graph D

16. Find the equation of the line for each of the following set of conditions:
    a. slope $= -4$ and y-intercept $= 5.4$
    b. slope $= 2$ and line passes through the point $(5, -4)$
    c. line passes through the points $(3, -4)$ and $(6, 11)$
    d. line passes through the points $(1, -1)$ and $(0, -5)$
    e. line is parallel to $2y - 7x = y + 4$ and passes through the point $(-1, 2)$

17. Find the equation of the line in the form $y = mx + b$ for each of the following sets of conditions. Show your work.
    a. Slope is -1.62 and y-intercept is 4.20.
    b. Slope is \$1400/yr and line passes through the point (10 yr, \$12,000).

   c. Line passes through (1940, 1.3%) and (1990, 11.3%).

   d. Line passes through (1971, $460) and (1979, $410).

   e. Equation is $1.48x - 2.00y + 4.36 = 0$.

   f. Line is parallel to a line with the equation $y = 0.44 + 0.83x$ and passes through the point (0, 3.89).

   g. Line is horizontal and passes through (1.0, 7.2).

   h. Line is vertical and passes through (275, 1029).

   i. Line is perpendicular to $y = -2x + 7$ and passes through (5, 2).

18. Rewrite each of these equations in the form $y = b + mx$. In each case identify the slope and the y-intercept and then do a quick sketch:

   a. $4 = y - 6x$ **d.** $(x/2) + (y/3) = 1$

   b. $2x + 3y = -6$ **e.** $1/x = 1/(y + 2)$

   c. $0.2y - 0.04x = 1.3$ **f.** $x + y = 0$

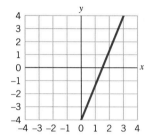

19. Find the equation of the line shown on the graph at left. Use this equation to create two new graphs, taking care to label the scales on your new axes. For one of your graphs, choose scales that make the line appear steeper than in the original graph. For your second graph, choose scales that make the line appear less steep than in the original graph.

20. Suppose that:

     for 8 years of education, the mean salary for women is approximately $6800,

     for 12 years of education, the mean salary for women is approximately $11,600, and

     for 16 years of education, the mean salary for women is approximately $16,400.

   a. Plot this information on a graph.

   b. What sort of relationship does this information suggest between mean salary for women and education? Justify your answer.

   c. Generate an equation that could be used to model the data from the limited information given (letting E = years of education and W = personal wages). Show your work.

21. The percentage of dentistry degrees awarded to women in the United States between 1970 and 1990 is shown in the table and graph below.

Percentage of Dentistry Degrees Awarded to Women

	1970	1975	1980	1985	1990
%	0.9	3.1	13.3	20.7	30.9

Source: U.S. National Center for Education Statistics, "Digest of Education Statistics," annual, in *The American Almanac 1993–1994: Statistical Abstract of the United States,* 1993.

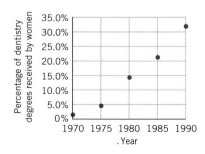

The data show that since 1970 the percentage of dentists who are women has been rising.

a. Sketch a line to represent the data points. What is the rate of change of percentage of dentistry degrees awarded to women according to your line?

b. According to your estimate of the rate of increase, when will 100% of dentistry degrees be awarded to women?

c. Since it seems extremely unlikely that 100% of dentistry degrees will ever be granted to women, comment on what is likely to happen to the rate of growth of women's degrees in dentistry; sketch a likely graph for the continuation of the data into the next century.

d. What other data might be helpful in making an estimate for the future of women in dentistry?

22. The following table and graph show the mortality rate (in deaths per 1000) for male and female infants in the United States from 1980 to 1990.

Infant Mortality Rates (per 1000)

Year	Male	Female
1980	13.9	11.2
1981	13.1	10.7
1982	12.8	10.2
1983	12.3	10.0
1984	11.9	9.6
1985	11.9	9.3
1986	11.5	9.1
1987	11.2	8.9
1988	11.0	8.9
1989	10.8	8.8
1990	10.3	8.1

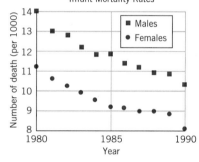

a. Sketch a line through the graph of the data for the female infant mortality rates. Does the line seem to be a reasonable model for the data? What is the approximate slope of the line through these points? (Show your work.)

b. Now sketch a line through the male mortality rates. Is it a reasonable approximation? Estimate its slope. (Show your work.)

c. List at least two important conclusions from the data set.

23. The y-axis, the x-axis, the line $x = 6$, and the line $y = 12$ determine the four sides of a 6 by 12 rectangle in the first quadrant of the xy plane. Imagine that this rectangle is a pool table. There are pockets at the four corners and at the points $(0, 6)$ and $(6, 6)$ in the middle of each of the longer sides. When a ball bounces off one of the sides of the table, it obeys the "pool rule," the slope of the path after the bounce is the negative of the slope before the bounce. (Hint: It helps to sketch the pool table on a piece of graph paper first.)

a. Your pool ball is at $(3, 8)$. You hit it toward the y-axis, along the line with slope 2.

    i. Where does it hit the y-axis?

    ii. If the ball is hit hard enough, where does it hit the side of the table next? And after that? And after that?

    iii. Does it ultimately return to (3, 8)? Would it do this if the slope had been different from 2? What is special about the slope 2 for this table?

b. A ball at (3, 8) is hit toward the y-axis, and bounces off it at (0, 16/3). Does it end up in one of the pockets? If so, what are the coordinates of that pocket?

c. Your pool ball is at (2, 9). You want to shoot it into the pocket at (6, 0). Unfortunately, there is another ball at (4, 4.5) which may be in the way.

    i. Can you shoot directly into the pocket at (6, 0)?

    ii. You want to get around the other ball by bouncing yours off the y-axis. If you hit the y-axis at (0, 7), do you end up in the pocket? Where do you hit the line $x = 6$?

    iii. Bouncing off the y-axis at (0, 7) didn't work. However perhaps there is some point (0, b) on the y-axis from which the ball would bounce into the pocket at (6, 0). Try to find that point.

    iv. Where, exactly, should you bounce off the y-axis to land in the pocket at (6, 0)?

EXPLORATION 4.1.A

Looking at Lines

Objective

- find patterns in the graphs of linear equations in the form $y = mx + b$

Materials/Equipment

- computer with course software "L1: m & b Sliders." and "L2: Finding m & b." both in *Linear Functions*.

Procedure

Working in pairs, try the following explorations and compare your findings with each other. After comparing your findings write down your observations.

Open the program *Linear Functions* and click on the button "L1: m & b Sliders." This program will allow you to construct your own lines and to explore the effects of m and b on the graph of the equation.

1. What is the effect of "m" on the graph of the equation?

   Fix a value for b. Construct four graphs with the same value for b but with different values for m. Continue to explore the effects of m on the graph of the equation. Jot down your observations about the effect on the line when m is positive, negative, or equal to zero, and when $|m|$ (the absolute value or distance of m from zero) increases. Do you think your conclusions work for values of m that are not on the slider?

   Choose a new value for b and repeat your experiment, jotting down notes. Are your observations still valid? Compare your observations with your partner and summarize in your notebook the effect of m on the graph of the equation.

2. What is the effect of "b" on the graph of the equation?

   Fix a value for m. Construct four graphs with the same value for m but with different values for b. What is the effect on the graph of changing b? Jot down your observations. Would your conclusions still hold for values of b that are not on the slider?

   Choose a new value for m and repeat your experiment, again keeping jottings. Are your observations still valid? Compare your observations with your partner. Write a short summary which describes the effect of b on the graph of the equation.

3. Constructing lines under certain constraints

    Construct the following sets of lines using "L1: *m* & *b* Sliders." Be sure to write down the equations for the lines you construct. What generalizations can you make about the lines in each case? Are the slopes or the *y*-intercepts of the lines related in some way?

    Construct any line. Then construct another line that has a steeper slope, and then construct one that has a shallower slope.
    Construct three *parallel lines*.
    Construct three lines with the *same y-intercept*, the point where the line crosses the *y*-axis.
    Construct a pair of lines that are *horizontal*.
    Construct a pair of lines that go *through the origin*.
    Construct a pair of lines that are *perpendicular* to each other.

    In your notebook, write a 60 second summary of what you have learned about the equations of lines.

4. Predicting the equation of a given line.

    Close the "L1: *m* & *b* Sliders." program by clicking on the square exit button in the upper left corner of the screen. Open the program "L2: Finding *m* & *b*." This program randomly generates a line that you will see on your screen. Your task is to find the equation that represents that line. Predict values for *m* and *b* and use your sliders to check your predictions.

EXPLORATION 4.1.B

Looking at Lines with a Graphing Calculator

Objective

- find patterns in the graphs of linear equations in the form $y = mx + b$

Materials/Equipment

- graphing calculator (instructions for the TI-82 and the TI-83 are available in the Graphing Calculator Workbook)

Procedure

Getting Started

Set your calculator to the integer window setting. For the TI-82 or TI-83 do the following:

```
WINDOW FORMAT
Xmin=-47
Xmax=47
Xscl=10
Ymin=-31
Ymax=31
Yscl=10
```

1. Press ZOOM , select [6:ZStandard].
2. Press ZOOM , select [8:ZInteger], ENTER .
3. Press WINDOW to see whether the settings are the same as the duplicated screen image.

Working in pairs

Try the following explorations and compare your findings with each other.

1. What is the effect of "m" on the graph of the equation $y = mx$?

   a. Case 1: $m > 0$

   Enter the following functions into your calculator and then sketch the graphs by hand. To get you started, try $m = 1$, 2, and 5. Try a few other values of "m" where $m > 0$.

$$Y1 = x$$
$$Y2 = 2x$$
$$Y3 = 5x$$
$$Y4 = \cdots$$
$$Y5 =$$
$$Y6 =$$

   Compare your observations with your partner. In your notebook describe the effect of multiplying x by a positive value for m in the equation $y = mx$.

b.   Case 2: $m < 0$

Begin by comparing the graphs of the lines when $m = 1$ and $m = -1$. Then experiment with other negative values for m and compare the graphs of the equations.

$$Y1 = x$$
$$Y2 = -x$$
$$Y3 = \cdots$$
$$Y4 =$$
$$Y5 =$$
$$Y6 =$$

Alter your description in Procedure 1a to describe the effect of multiplying x by any real number "m", for $y = mx$ (remember to also explore what happens when $m = 0$).

2.   What is the effect of "b" on the graph of an equation, $y = mx + b$.

a.   Enter the following into your calculator and then sketch the graphs by hand. To get started try, $m = 1$ and $b = 0$, 20, and -20. Try other values for b as well.

$$Y1 = x$$
$$Y2 = x + 20$$
$$Y3 = x - 20$$
$$Y4 = \cdots$$
$$Y5 =$$
$$Y6 =$$

b.   Discuss with your partner the effect of adding any number "b" to x for $y = x + b$. (Hint: Use "trace" to find where the graph crosses the y-axis.) Record your comments in your notebook.

c.   Choose another value for m and repeat the exercise. Are your observations still valid?

3.   Construct a summary describing the effects of both m and b on the linear equation $y = mx + b$.

4.   Constructing lines under certain constraints

Construct the following sets of lines using your graphing calculator. Be sure to write down the equations for the lines you construct. What generalizations can you make about the lines in each case? Are the slopes or the y-intercepts of the lines related in any way?

Construct any line. Then construct another line that has a steeper slope, then one that has a shallower slope.

Construct three *parallel lines*.

Construct three lines with the *same y-intercept*.

Construct a pair of lines that are *horizontal*.

Construct a pair of lines that go *through the origin*.

Construct a pair of lines that are *perpendicular* to each other.

In your notebook, write a 60 second summary of what you have learned about the equations of lines.

5.   On graph paper plot the points (2, 1) and (4, 5). Then, draw a line through the points so that the line intersects the x- and y-axes.

a.   Determine the rate of change or the slope of this line.

b.   Estimate the y-intercept.

c.   Find the equation of the line $y = mx + b$ that goes through the points (2, 1) and (4, 5). What are the values of m and b? Check to see whether the points satisfy your equation. If not, make the necessary adjustments to your equation.

Looking for Links: A Case Study in Education and Income

Overview

Robert Reich, former U.S. Secretary of Labor for the Clinton administration, says that "Learning is the key to earning." Is this true? Is there a relationship between education and income? We examine how a social scientist might answer this question.

A case study approach is used to explore possible ways in which education and income may be related. We start by exploring a large data set extracted from the U.S. Census and examine how certain relationships can be summarized by fitting lines to data. The resulting linear models are called *regression lines.* The next steps often involve trying to answer questions raised by the analysis. How good is the model? What other variables should be taken into consideration? Why are the variables correlated?

"Fitting lines to relationships is the major tool of data analysis." (Edward Tufte in *Data Analysis of Politics and Policy.*) Yet we need to be cautious in how we interpret our findings. A classic case, showing a relationship between the number of radios and the proportion of insane people, illustrates some of the pitfalls.

The Explorations provide a hands-on opportunity to use U.S. Census data to conduct your own case study in education and income in the United States.

After reading this chapter you should be able to:
- analyze a large data set based on U.S. Census data
- use regression lines to summarize data
- understand the distinction between correlation and causation
- conduct your own case study on the relationship between education and income

The articles "U.S. Government Data Collection" and "U.S. Government Definition of Census Terms" give a closer look at federal data collection and, in particular, the Current Population Survey.

Read the article "Who Collects Data and Why" for an overview of the types of data collected by institutions other than the federal government and the purposes to which the information is put.

5.1 A CASE STUDY USING U.S. CENSUS DATA

Does more education mean more income? The answer may seem obvious. We may reasonably expect that having more education gives access to higher paying jobs. Is this indeed the case? We explore how a social scientist might start to answer these questions, using a random sample from U.S. Census data. Our data set, called FAM 1000, provides information on 1000 individuals and their families.

The Bureau of the Census, as mandated by the Constitution, conducts a nationwide census every 10 years. In order to collect up-to-date information, the Census Bureau also conducts a monthly survey of American households called the Current Population Survey, or CPS. The CPS is the largest survey taken between census years. The CPS is based on data collected each month from approximately 60,000 households. Questions are asked about race, education, housing, number of people in the household, income, and employment status.[1] The March survey is the most extensive. Our sample of census data (FAM 1000) has been extracted from the March 1996 Current Population Survey. It contains information about 1000 individuals randomly chosen from those who worked at least one week in 1995.

Table 5.2 shows all of the information for 52 of the 1000 individuals in FAM 1000 and Table 5.1 is a *data dictionary* with short definitions for each data category. Think of the data as a large array of rows and columns of facts. Each row represents all the information obtained from one particular respondent about his or her family. Each column contains the coded answers of all the respondents to one particular question. Try deciphering the information below that comes from the first row of our data array in Table 5.2.

region	cencity	famsize	faminc	marstat	sex	race	reorgn	age	occup	educ	wkswork	hrswork	pwages	ptotinc	yrft
2	2	2	$51,832	1	2	1	8	54	5	12	52	40	$34,500	$35,219	1

Referring to the data dictionary, we learn that the respondent lives in the Midwest, in the suburbs, in a family with two people that had an annual family income of $51,832 in 1995. (The income figures are for the prior year, in this case 1995.) The respondent is presently not married, is female, white, non-Hispanic and 54 years old. She works in administrative support, has a 12th grade education, worked 52 weeks in 1995, usually 40 hours a week, earned $34,500 in personal wages and salary, had a personal total income of $35,219, and considered herself a year-round, full-time worker.

Table 5.1

Data Dictionary for March 1996 Current Population Survey*

Variable	Definition	Unit of Measurement (Code and Allowable Range)	Variable	Definition	Unit of Measurement (Code and Allowable Range)
REGION	Census region	1 = Northeast 2 = Midwest 3 = South 4 = West	CENCITY	Residence location	1 = Central City 2 = Suburban 3 = Rural 4 = Not identified

[1] The results of the survey are used to estimate numerous economic and demographic variables, such as the size of the labor force, the employment rate, and income and education levels. The results are widely quoted in the popular press and are published monthly in *The Monthly Labor Review* and *Employment and Earnings,* irregularly in the *Current Population Reports* and *Special Labor Force Reports,* and yearly in the *Statistical Abstract of the United States* and *The Economic Report of the President.*

Table 5.1 (*Continued*)

Data Dictionary for March 1996 Current Population Survey*

Variable	Definition	Unit of Measurement (Code and Allowable Range)	Variable	Definition	Unit of Measurement (Code and Allowable Range)
FAMSIZE	Family size	Range is 1 to 39	EDUC	Years of education	1 = Less than first grade
FAMINC	Family income	Range is −$400,000 to $24,000,000			4 = First, second, third, or fourth grade
MARSTAT	Marital status	0 = Presently married			6 = Fifth or sixth grade
		1 = Presently not married			8 = Seventh or eighth grade
SEX	Sex	1 = Male			9 = Ninth grade
		2 = Female			10 = Tenth grade
RACE	Race of respondent	1 = White			11 = Eleventh grade
		2 = Black			12 = Twelfth grade
		3 = American Indian, Aleut Eskimo			HG = High school graduate (diploma or equivalent)
		4 = Asian or Pacific Islander			SC = college, but no degree
		5 = Other			AO = Associate's degree in occupation/vocation program
REORGN	Ethnic origin of respondent (Hispanic or other)	1 = Mexican American			AP = Associate's degree in academic program
		2 = Chicano			
		3 = Mexican (Mexicano)			BD = Bachelor's degree
		4 = Puerto Rican			MD = Master's degree (M.A., M.S., M.S.W., M.B.A.)
		5 = Cuban			
		6 = Central or South American			
		7 = Other Spanish			PD = Professional school degree (M.D., D.D.S., D.V.M., L.L.B., J.D.)
		8 = All other			
		9 = Don't know			
		10 = Not available			PH = Doctoral degree (Ph.D., Ed.D.)
AGE	Age	Range is 18 to 90			
OCCUP	Occupation group of respondent	0 = Not in universe	Note: In the FAM 1000 data, substitutions of numbers for symbols were made in order to plot points. For example, we replaced HG (high school graduate) with 12, representing 12 years of school. SC (some college but no degree), we estimate as 14, or 2 years more than high school. Similar estimates were made for the higher education categories.		
		1 = Executive, administrative, managerial			
		2 = Professional specialty			
		3 = Technicians and related support			
		4 = Sales			
		5 = Administrative support, including clerical	WKSWORK	Weeks worked in 1995 (even for a few hours)	Range 1 to 52
		6 = Private household service			
		7 = Protective service			
		8 = Other service occupations	HRSWORK	Usual hours worked	Range is 1 to 99 per week in 1995
		9 = Precision production, craft and repair			
		10 = Machine operators, assemblers and inspectors	PWAGES	Total personal wages	Range is $0 to $400,000
		11 = Transportation and materials moving equipment	PTOTINC	Personal total income	Range is −$10,000 to $600,000
		12 = Handlers, equipment cleaners, helpers, etc.	YRFT	Employed full-time year-round	1 = full-time all year-round
		13 = Farming, forestry, and fishing			2 = part-time all year-round
					3 = full-time part of the year
		14 = Armed forces			4 = part-time part of the year

* Adapted from the U.S. Census Bureau, *Current Population Survey,* March 1996, by Jie Chen, Computing Services, University of Massachusetts, Boston.

Table 5.2
Information from 52 Respondents in the FAM 1000 Data Set*

region	cencity	famsize	faminc	marstat	sex	race	reorgn	age	occup	educ	wkswork	hrswork	pwages	ptotinc	yrft
2	2	2	$51,832	1	2	1	8	54	5	12	52	40	$34,500	$35,219	1
2	2	8	$29,464	1	2	2	8	53	4	12	52	40	$25,000	$29,464	1
2	2	3	$139,330	0	1	1	8	49	1	16	52	40	$103,000	$106,262	1
1	4	3	$78,092	0	2	1	8	56	2	16	42	50	$41,000	$41,596	3
4	4	1	$4,600	1	2	1	8	21	2	14	52	19	$4,600	$4,600	2
4	4	3	$106,530	1	2	1	8	20	8	14	52	20	$5,020	$5,030	2
3	1	2	$39,047	0	2	1	8	57	8	9	52	3	$300	$900	2
4	2	4	$82,322	1	2	1	8	18	0	11	2	35	$306	$306	3
2	4	5	$34,932	0	2	4	8	46	10	8	52	38	$11,269	$11,269	1
3	3	2	$64,555	0	2	1	8	33	2	18	52	50	$31,000	$31,100	1
3	2	1	$26,700	1	2	1	8	47	0	14	22	40	$7,500	$26,700	3
2	1	3	$30,875	1	2	2	8	36	8	11	52	40	$30,000	$30,000	1
3	2	2	$186,116	0	1	1	8	60	4	11	52	40	$50,000	$120,558	1
4	2	3	$79,426	1	2	4	8	30	2	18	52	10	$300	$300	2
4	2	4	$124,959	1	1	1	8	29	8	14	52	40	$25,000	$25,000	1
3	4	2	$13,512	0	1	2	8	57	9	9	52	40	$13,500	$13,512	1
1	4	1	$33,277	1	1	1	8	37	10	12	52	48	$33,000	$33,277	1
4	3	3	$13,700	1	1	1	8	46	0	11	1	40	$500	$500	3
1	1	2	$43,616	0	2	1	8	70	4	14	24	30	$0	$15,478	4
4	2	3	$69,040	0	1	2	8	57	9	14	52	40	$40,000	$39,020	1
1	1	4	$56,622	0	2	1	8	50	2	18	52	40	$36,040	$39,111	1
2	4	3	$27,805	1	2	1	8	38	2	16	52	40	$14,600	$27,805	1
2	3	2	$21,388	0	2	1	8	20	8	14	52	25	$7,236	$7,295	2
3	4	4	$16,197	0	2	1	8	32	4	14	52	35	$7,696	$15,072	1
4	2	1	$45,209	1	1	2	8	28	9	12	52	70	$50,000	$45,209	1
2	2	1	$22,477	1	2	1	8	35	2	16	52	40	$20,456	$22,477	1
3	2	4	$127,846	0	1	1	8	45	2	20	52	36	$120,000	$123,923	1
3	3	2	$42,000	0	2	1	8	41	2	14	26	20	$10,000	$10,000	4
1	4	6	$50,140	1	2	1	8	18	13	8	52	30	$0	$0	2
1	2	3	$31,000	0	2	2	8	47	2	12	52	39	$14,000	$14,000	1
4	2	4	$129,500	0	1	1	8	46	5	16	52	40	$45,000	$70,500	1
3	4	6	$53,658	0	1	1	8	46	2	14	52	40	$43,000	$43,008	1
3	3	2	$13,085	0	1	1	8	59	0	12	21	40	$5,200	$5,242	3
4	1	2	$86,068	0	2	1	8	43	2	14	50	36	$51,000	$51,034	1
3	3	5	$164,255	0	2	1	4	42	2	14	52	60	$66,000	$66,000	1
2	3	3	$36,604	1	1	1	8	22	12	16	52	30	$3,000	$3,036	2
3	2	4	$55,724	0	1	1	1	31	9	14	52	40	$29,000	$29,012	1
4	1	4	$974	1	2	1	8	52	0	12	2	20	$724	$974	4
2	2	1	$29,900	1	2	1	8	58	5	14	52	40	$20,200	$29,900	1
3	1	1	$10,000	1	2	2	8	37	5	16	52	40	$10,000	$10,000	1
1	1	6	$66,700	1	2	1	6	26	8	14	52	12	$5,000	$7,600	2
1	2	3	$102,076	0	2	4	8	51	2	14	52	32	$32,000	$41,038	2
1	1	2	$26,000	1	2	1	8	49	3	14	52	40	$26,000	$26,000	1
3	3	2	$48,226	0	1	1	8	65	1	12	52	40	$30,000	$30,113	1
4	3	2	$25,727	0	2	1	8	57	0	16	20	25	$4,000	($5,982)	4
1	2	2	$35,299	0	1	1	8	49	7	12	52	40	$15,600	$15,600	1
4	2	4	$28,400	0	2	1	6	24	6	4	52	30	$10,400	$10,400	2
4	2	5	$28,164	0	2	1	8	35	4	16	46	40	$11,619	$13,666	3
1	2	6	$117,329	1	2	1	8	22	2	16	25	40	$6,500	$6,513	3
2	3	4	$52,765	1	1	1	8	21	12	14	39	20	$3,500	$3,500	4
4	2	1	$3,000	1	1	1	8	19	8	9	26	20	$3,000	$3,000	4
2	1	5	$23,597	0	1	1	6	41	8	14	4	40	$1,600	$1,621	3

* From the U.S. Census Bureau, *Current Population Survey,* March 1996. Data can be accessed through the Census Bureau website, at http://www.bls.census.gov/cps/cpsmain.htm

5.2 SUMMARIZING THE DATA: REGRESSION LINES

What Is the Relationship Between Education and Income?

In the physical sciences, the relationship among variables is often quite direct; if you hang a weight on a spring, it is clear, even if the exact relationship is not known, that the amount the spring stretches is definitely dependent on the heaviness of the weight. Further, it is reasonably clear that the weight is the *only* important variable; the temperature, the phase of the moon, etc., can safely be neglected.

In the social and life sciences it is usually difficult to tell whether one variable truly depends on another. For example, it is certainly plausible that a person's income depends in part on how much formal education he or she has had, since we may suspect that having more education gives access to higher paying jobs, but many other factors certainly also play a role. Some of these factors, such as the person's age or type of work, are measured in the FAM 1000 data set; others, such as family background or good luck, may not have been measured, or even be measurable. Despite this complexity, we attempt to determine as much as we can by first looking at the relationship between income and education alone.

We begin to explore this question with a *scatter plot* of education and personal wages from the FAM 1000 data set. If we hypothesize that income depends on education, then the convention is to graph education on the horizontal axis. Each ordered pair of data values gives a point with coordinates of the form

(education, personal wages)

Take a moment to examine the scatter plot in Figure 5.1. Each point refers to a particular respondent and has two coordinates. One coordinate gives the years of education and the other gives the personal wages. For example, the point near the top represents an individual with 16 years of education who makes about $150,000 in personal wages. The other information that we have about the person is not shown.

You can use the FAM 1000 data to follow the discussion in the text and/or to conduct your own case study. (See Explorations 5.1 and 5.2.) The CD-ROM contains: FAM 1000 Census Graphs, which provides easy to use interactive software for analyzing the FAM 1000 data; the full FAM 1000 data stored as the Excel file FAM1000; and condensed versions of the data for use with a graphing calculator. Descriptions of the graph link files are in Sec. 5 of the Graphing Calculator Workbook.

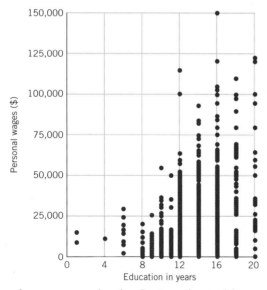

Figure 5.1 Personal wages versus education. In attempting to pick a reasonable scale to display personal wages on this graph, the vertical axis was cropped at $150,000, which meant excluding from the display the points for five outliers: five individuals who each earned wages between $250,000 and $300,000.

By referring back to the original data set, we can find out more information about the outlier. This person is a 39-year-old, married, white male, who works full-time in administrative support, holds a bachelor's degree, has two people in his family, and lives in the Northeast outside a city. Some or all of this additional information may well be relevant, but it is clearly too much to display on one graph.

There appears to be an overall trend up and to the right. Everyone with fewer than 10 years of school earns less than $30,000, whereas many of those with 10 or more years of school earn more than $30,000. This suggests that there is some relationship between education and income.

How might we think of the relationship between these two variables? Clearly, personal income is not a function of education in the mathematical sense since people who have the same amount of education earn widely different amounts. The scatter plot obviously fails the vertical line test.

However, we may ask the following question: Suppose that, in order to form a simple description of these data, we were to insist on finding a simple functional description. And suppose we insisted that this simple relationship be a linear function. In Chapter 4, we fit linear functions to data informally. A formal mathematical procedure called *regression analysis* lets us determine what linear function is the "best" approximation to the data; the resulting "best fit" line is called a *regression line* and is similar in spirit to reporting only the mean of a set of data, rather than the entire data set. "It is often a useful and powerful method of summarizing a set of data."[2]

If you are interested in a standard technique for generating regression lines, a summary of the method used in the course software is provided in the reading "Linear Regression Summary."

Technically, for any candidate linear function, we may determine the error in how this function represents the data set by adding up the squares of the vertical distances between the line and the data points; the regression line is the line that makes this error as small as possible. The calculations necessary to compute this line are tedious, although not difficult, and are easily carried out by computer software and graphing calculators. The *FAM 1000 Census Graphs* program can be used to find regression lines. The various techniques for determining regression lines are beyond the scope of this course.

In Figure 5.2, we show the FAM 1000 data set, along with a regression line determined from the data points. The equation of the line is

$$\text{prs wages}_{\text{all data}} = -15{,}840 + 2930 \cdot \text{years of education}$$

Since personal wages are in dollars, the units for the term 15,840 must be in dollars, and the units for 2930 must be in dollars per year.

This line is certainly more concise than the original set of 1000 data points. Looking at the graph, you may judge with your eyes to what extent the line is a good description of the original data set. Note from the equation that the vertical intercept, $-15{,}840$, of this line is *negative;* so according to this regression line, people with five or fewer years of education earn *negative* wages, even though all wages in the original data set are positive. The linear model is clearly inaccurate near the left edge of the graph, so we should exclude these points from the domain of our model. The number 2930 represents the slope of the regression line, or the average rate of change of personal wages with respect to years of education. Thus, this model predicts that for each additional year of education, individual personal wages increased by $2930.

The data points are widely scattered about the line, for reasons that are clearly not captured by the linear model. Some of this scatter is simply a result of randomness and the fact that each data point is a different individual; however, some of the scatter may result from other variables such as age, which a more sophisticated analysis could include.

[2] Edward Tufte, *Data Analysis of Politics and Policy,* (Upper Saddle River, NJ: Prentice Hall, 1974) p. 65.

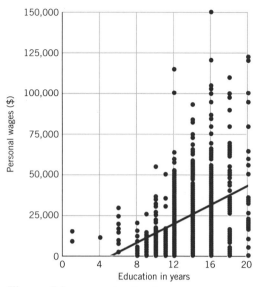

Figure 5.2   Regression line for personal wages versus education.

We emphasize that, although we can construct an approximate linear model for any data set, this does not mean that we really believe that the data truly represent a linear relationship. In the same way, we may report the mean to summarize a set of data, without believing that the data values are all the same number. In both cases, there are features of the original data set that may or may not be important and that we do not report.

Another way to look for structure in this data set is to eliminate clutter by grouping together all people with the same years of education, and to plot the *mean* personal wages of each group. The resulting graph (Figure 5.3) is sometimes called the *graph of averages*. Note that the vertical scale now only goes up to $60,000.

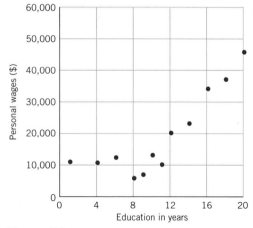

Figure 5.3   Mean personal wages versus education.

This plot is less cluttered. For each year of education, only the single mean income point has been graphed. For instance, the point corresponding to 12 years of education has a vertical value of approximately $20,000; hence the mean of the personal wages of everyone in the FAM 1000 data with 12 years of education is about $20,000. The pattern is clearer: The upward trend to the right is more obvious in this graph.

Every time we construct a simplified representation of an original data set, we should ask ourselves what information has been suppressed. For the graph of averages, we have suppressed the spread of data in the vertical direction. We also do not see how many different people are represented by a single data point at each year of education; for example, there are only *two* people with 1 year of education, but several hundred with 16 years, and each of these sets is represented by a single point in the graph of averages.

In the graph of averages, we can see some additional features of the data that have not been evident in either the full scatter plot or in the equation of the straight-line model. For example, there are small jumps upward at 16 years and at 20 years, representing completing college and graduate school, respectively. The graph is nearly flat below 12 years; perhaps if you do not graduate from high school, it does not matter very much how many years you went to school. Starting at 12 years it slopes upward, perhaps representing the payoff from college and graduate education. All of these observations are completely suppressed by the simple linear model.

We can fit a line to the graph of averages by the same method of linear regression. The equation of this straight line is

$$\text{prs wages}_{\text{mean data}} = -2300 + 1980 \cdot \text{years of education}$$

-2300 is the vertical intercept and 1980 is the slope or rate of change of personal wages with respect to education. Again, since the vertical intercept is negative, we would ignore the left end segment of the line. This model predicts that for each additional year of education the *mean* personal wage increases by $1980. Note that this linear model predicts *mean* personal wages for the *group,* not personal wages for an individual.

Figure 5.4 shows the average data, and the straight line representing the fit to the averages. We have also superimposed the original linear fit (the thinner line) to the full

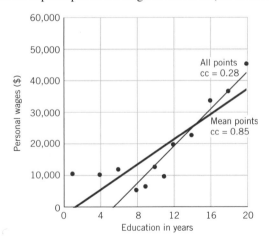

Figure 5.4   Regression lines for wages versus education.

data set. Both of these straight lines are reasonable answers to the question, "What straight line best describes the relationship between education and income?" and the difference between them indicates the uncertainty in answering such a question. We may argue that the benefit in income of each year of education is $1980, or $2930, or somewhere between these values.

Regression Line: How Good a Fit?

Once we have determined a line that approximates our data, we must ask, "How good a fit is our regression line?" To help answer this, statisticians calculate a quantity called the *correlation coefficient*. This number is computed by most regression software and graphing calculators, and we have included it on our graphs, labeled "cc."[3] The correlation coefficient is always between -1 (negative slope and no scatter; the data points fit exactly on the line) and 1 (positive slope and no scatter; the data points fit exactly on the line). The closer the absolute value of the correlation coefficient is to 1, the better the fit and the stronger the linear association between the variables.

The programs R1-R4 and R7 in Linear Regression *can help you visualize the links among scatter plots, best fit lines, and correlation coefficients.*

Remember that the absolute value of a number, N, is expressed in symbols as $|N|$. It means the distance from zero without taking into account the sign. If $N \geq 0$, then $|N| = N$. If $N < 0$, then $|N| = -N$. Thus $|0.89| = 0.89$ and $|-0.89| = 0.89$.

A small correlation coefficient (with absolute value close to zero) indicates that the variables do not depend linearly on each other. This may be because there is no relationship between them, or because there is a relationship that is something more complicated than linear. In future chapters we discuss many possible non-linear functional relationships.

The reading "The Correlation Coefficient" will tell you how to calculate the correlation coefficient and give you an intuitive sense of how the size of the correlation coefficient helps to answer the question, "How good a fit to the data is the regression line?"

There is no absolute answer to the question of when a correlation coefficient is "good enough" to say that the linear regression line is a good fit to the data. A fit to the graph of averages always gives a higher correlation coefficient than a fit to the original data set because the scatter has been smoothed out. When in doubt, plot all the data along with the linear model, and use your best judgment. The correlation coefficient is only a tool that may help you decide among competing models or interpretations.

Algebra Aerobics 5.2

1. a. Evaluate each of the following:

$$|0.65| \qquad |-0.68| \qquad |-0.07| \qquad |0.70|$$

b. List the absolute values in part (a) in ascending order from the smallest to the largest.

2. The following equation represents the best fit regression line for mean personal wages versus years of education for the 296 people in the FAM 1000 data who live in the southern region of the U.S.

$$W = -3460 + 2120E$$

where W = mean personal wages and E = years of education; cc = 0.76.

a. Identify the slope of the regression line, the vertical intercept, and the correlation coefficient.

b. What does the slope mean in this context?

c. By what amount does this regression line predict that mean personal wages for those who live in the South change for 1 additional year of education? For 10 additional years of education?

---

[3] We use the label cc for the correlation coefficient in the text and software in order to minimize confusion. In a statistics course the correlation coefficient is usually referred to as *Pearson's r* or just *r*.

3. Figure 5.5 shows regression lines and corresponding correlation coefficients for four different scatter plots. Which of the lines describes the strongest linear relationship between the variables? Which of the lines describes the weakest linear relationship?

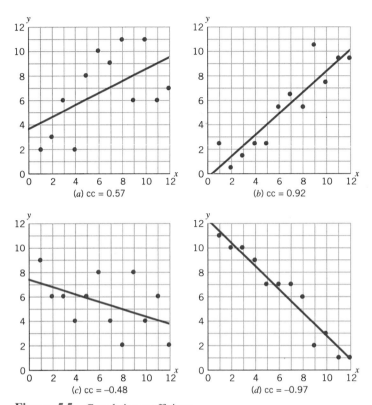

Figure 5.5   Correlation coefficients.

5.3 CORRELATION VERSUS CAUSATION

One is tempted to conclude that increased education *causes* increased income. This may be true, but the model we have used does not offer conclusive proof. This model can show how strong or weak a relationship exists between variables but does not answer the question, "Why are the variables related?" We need to be cautious in how we interpret our findings.

Regression lines show *correlation,* not *causation.* We say that two events are correlated when there is a statistical link. If we find a regression line with a correlation coefficient that is close to 1 in absolute value, a strong relationship is suggested. In our previous example, education is positively correlated with personal wages. If education increases, personal wages increase. Yet this does not prove that education causes an increase in personal wages. The reverse might be true; that is, an increase in personal

In "North Dakota, Math Country," New York Senator Daniel Patrick Moynihan offers a humorous strategy for improving student math scores in each state.

wages might cause an increase in education. The correlation may be due to other factors altogether. It might occur purely by chance or be jointly caused by yet another variable. Perhaps both educational opportunities and income levels are strongly affected by parental education or a history of family wealth. Thus a third variable, such as parental socioeconomic status, may better account for both more education and higher income. We call this type of variable that may be affecting the results a *hidden variable*.

Figure 5.6 shows a clear correlation between the number of radios and the proportion of insane people in England between 1924 and 1937. (People were required to have a license to own a radio.)

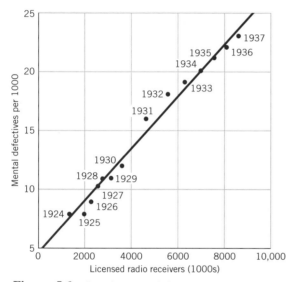

Figure 5.6   A curious correlation?

From E. A. Tufte, *Data Analysis for Politics and Policy.* ©1974, p. 90. Reprinted by permission of Prentice Hall, Inc., Upper Saddle River, N.J.

Are you convinced that radios cause insanity? Or are both variables just increasing with the years? We tend to accept as reasonable the argument that an increase in education causes an increase in personal wages, because the results seem intuitively possible and they match our preconceptions. But we balk when asked to believe that an increase in radios causes an increase in insanity. Yet the arguments are based on the same sort of statistical reasoning. The flaw in the reasoning is that statistics can only show that events occur together or are correlated, but *statistics can never prove that one event causes another.* Any time you are tempted to jump to the conclusion that one event causes another because they are correlated, think about the radios in England!

Something to think about

How good are the data? Return to the questions raised about data in Chapter 1. What questions would you raise about the data collected on education and income by the U.S. Bureau of the Census?

5.4 NEXT STEPS: RAISING MORE QUESTIONS

When a strong link is found between variables, often the next step is to raise questions whose answers may provide more insight into the nature of the relationship. How can the evidence be strengthened? If we use other income measures, such as total personal income or total family income, will the relationship between education and income still hold? Are there other variables that affect the relationship?

Does Income Depend on Age?

We started our exploration by looking at how income depended on education, because it seemed natural that more education might lead to more income. But it is equally plausible that a person's income might depend on his or her age for several reasons: People generally earn more as they advance through their working careers; they gain on-the-job experience in addition to formal education; and their incomes drop when they eventually retire. We can examine the FAM 1000 data to look for evidence to support this hypothesis.

It's hard to see much when we plot all the data points. But the plot of mean personal wages versus age (Figure 5.7) seems to suggest that up until about age 55, as age increases, mean personal wages increase. After age 55, as people move into middle age and retirement, mean personal wages tend to decrease. So age does seem to affect personal wages, in a way that is roughly consistent with our intuition. But the relationship appears to be nonlinear, and so linear regression may not be a very effective tool to explore this dependence.

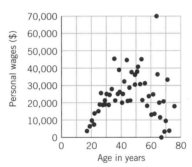

Figure 5.7   Mean personal wages versus age.

Something to think about

What other variables do you think affect income? Which measure of income would you use to test your hypothesis? For example, the FAM 1000 census data include the following measures of income: personal wages, personal total income, family income, and personal wages per hour.

We see that the FAM 1000 data set contains internal relationships that are not obvious on a first analysis. We might also investigate the relationship between education and age. Age may be acting as a hidden variable influencing the relationship between education and income.

There are a few simple ways to attempt to minimize the effect of age. For example, we can restrict our analysis to individuals who are all roughly the same age. We could construct a scatter plot showing income versus education only for people aged between 30 and 35. This sample still would include a very diverse collection of people. More sophisticated strategies involve statistical techniques such as *multivariable analysis,* a topic beyond the scope of this course.

Does Income Depend on Gender?

We can continue to look for simple relationships in the FAM 1000 data set by using some of the other variables to sort the data in different ways. For example, we can look at whether the relationship between income and education is different for men and women. To do this, we compute the mean personal wages for each year of education for men and women separately. By plotting the results, we begin to see some patterns.

Figures 5.8 and 5.9 indicate that in general, personal wages for both men and women increase as education increases.

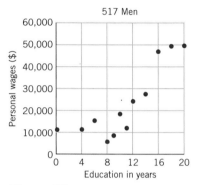

Figure 5.8 Mean personal wages versus education for men.

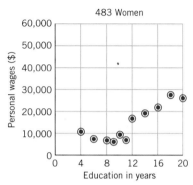

Figure 5.9 Mean personal wages versus education for women.

Something to think about

The graphs in Fig. 5.8 and 5.9 raise interesting questions. For example, in this particular data set, the mean personal wage for women is approximately the same for both 18 years and 20 years of education. The same is true for men. Will we find the same pattern if we look at a larger sample of data? Do you think this observation remains true if we examine specific professions?

If we put the data for men and women on the same graph (Figure 5.10), it is easier to make comparisons. We can see in Figure 5.10 that the mean personal wages of men are consistently higher than the mean personal wages of women, except at 8 years of education. We can also examine the best fit lines for mean personal wages versus education for men and for women shown in Figure 5.11.

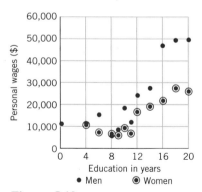

Figure 5.10 Mean personal wages versus education for women and for men.

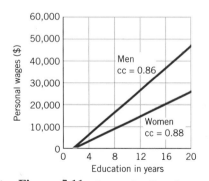

Figure 5.11 Regression lines for mean personal wages versus education for women and for men.

The linear model for mean personal wages for men is given by:

$$W_{\text{men}} = -3990 + 2510E$$

where E = years of education, W_{men} = mean personal wages for men. The correlation coefficient is 0.86. The rate of change of mean personal wages with respect to education is approximately $2510 per year. For males in this set, the mean personal wage increases by roughly $2510 for each additional year of education.

For women the comparable linear model is:

$$W_{\text{women}} = -2600 + 1420E$$

where E = years of education, W_{women} = mean personal wages for women. The correlation coefficient is 0.88. As you might predict from the relative status of men and women in the United States work force, the rate of change for women is much lower. For women in this sample, the model predicts that the mean personal wages increase by only $1420 per year for each additional year of education. In Figure 5.11, we can see that the regression line for men is steeper than the one for women when plotted on the same grid. In addition, as we can see in Figure 5.10, the mean personal wage for any particular number of years of education (except 8 years) is consistently lower for women than for men. The disparity seems to increase with the level of education.

The regression lines graphed in Figure 5.11 estimate mean personal wages for *all* the women and men in FAM 1000. What other variables have we ignored that we may want to consider in a more refined analysis of the impact of gender on personal income? We could consider type of job or amount of work or other things we think may be important.

- Type of job
  Do women and men make the same salaries when they hold the same types of jobs? We could compare only people within the same profession, and ask whether the same level of education corresponds to the same level of personal wages for women as for men. There are many more questions, such as: Are there more men than women in higher paying professions? Do men and women have the same access to higher paying jobs, given the same level of education?

- Amount of work
  Typically part-time jobs pay less than full-time jobs and more women hold part-time jobs than men. In addition there are usually more women than men who are unemployed. We can examine the personal wages of only those who are working full-time, and determine whether it is still true that given equal amounts of education, women get paid less on average than men.

Something to think about

Next steps often involve hypothesizing, "Why?" Why do you think gender affects the relationship between education and income?

Conclusion

The real world is a messy place and when we try to understand it, our best mathematics can be defeated by the deficiencies of actual data.

Econometric theory is like an exquisitely balanced French recipe, spelling out precisely with how many turns to mix the sauce, how many carats of spice to add, and for how many milliseconds to bake the mixture at exactly 474 degrees of temperature. But when the statistical cook turns to raw materials, he finds that hearts of cactus fruit are unavailable, so he substitutes chunks of cantaloupe; where the recipe calls for vermicelli he uses shredded wheat; and he substitutes green garment dye for curry, ping-pong balls for turtle's eggs and, for Chalifougnac vintage 1883, a can of turpentine.[4]

[4] Stefan Valavanis, *Econometrics: An Introduction to Maximum Likelihood Methods.* (New York: McGraw-Hill, 1959) p. 83.

In this chapter, we have illustrated some of the methods by which economists and statisticians try to make sense out of the society in which all of us live. We have also indicated some of the tremendous difficulties involved in drawing simple conclusions from data about real people in a complex society.

Questions like the ones here are actively discussed and debated every day, in Washington, D.C., in local government, in business and industry. As an educated member of society, you need to evaluate evidence that is used to form conclusions and . make policy. You need to be able to understand and evaluate the strengths and weaknesses of arguments presented to you, and to be able to form and defend your own judgments. It is hoped that mathematics is one of the tools you use to understand and participate in the world around you.

CHAPTER SUMMARY

Using a case study approach, this chapter examines analytical tools and the difficulties in applying them to draw conclusions about real people in a complex society. In the physical sciences, the relationship of variables is often quite direct. In the social and life sciences it is usually difficult to tell whether one variable truly depends upon another. One of the most important tools of social scientists for studying relationships is fitting lines to data. Computers or graphing calculators can be used to find a best fit line, called a *regression line,* for a scatter plot. A number called the *correlation coefficient* indicates the strength of the linear correlation between the variables. Its value is always between -1 and 1; the closer its absolute value is to 1, the better the fit and the stronger the linear association between the variables.

Regression lines show *correlation,* not *causation.* In this chapter we found that income is positively correlated with education. But that does not imply that increased education *causes* increased income. Regression lines can be used to describe the relationship between education and income, but they do not explain why the relationship exists. We examined some of the many factors that need to be taken into account to explain this relationship.

Exercises

In Exercises 1–4, the data analyzed are from the FAM 1000 data files and the income measure is personal total income. Personal total income includes personal wages and other sources of income, such as interest and dividends on investments. The numbers in the equations are rounded to the nearest 10. You can generate your own regression lines using the FAM1000 Excel file or the software programs in *FAM 1000 Census Graphs* (if you have access to a computer), or using the graph link files FAM1000A, FAM1000B, FAM1000C, or FAM1000D (if you have access to a graphing calculator). A description of the graph link files is contained in Section 5 of the Graphing Calculator Workbook.

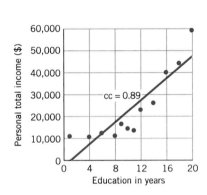

Personal Total Income = –3690 + 2530 · educ yrs

1. The graph and regression line at left show mean personal total income versus years of education.
    a. What is the slope of the regression line?
    b. Interpret the slope in this context.
    c. By what amount does this regression line predict that mean personal total income changes for 1 additional year of education? For 10 additional years of education?
    d. What features of the data are not well described by the regression line?
    e. What are some of the limitations of the data and the model in making predictions about income?
    f. Describe how your findings on mean personal total income relate to the findings in the text for personal wages.

2. The following equation represents a best fit regression line for mean personal total income of white males versus years of education.

    Personal total income = −4980 + 3240 · years of education

    The correlation coefficient is 0.91 and the sample size is 469 white males.
    a. What is the rate of change of mean personal total income with respect to years of education?
    b. Generate a small table with three points that lie on this regression line. Use two of these points to calculate the slope of the regression line.
    c. How does this slope relate to your answer to part (a)?
    d. Sketch the graph.

3. From the FAM 1000 data, the best fit regression line for mean personal total income of white females versus years of education is:

    Personal total income = −4780 + 1770 · years of education

    The correlation coefficient is 0.84 and the sample size is 406 white females.
    a. Interpret the number 1770 in this equation.
    b. Generate a small table with three points that lie on this regression line. Use two of these points to calculate the slope of the regression line.
    c. How does this slope relate to your answer to part (a)?
    d. Sketch the graph.

e. Describe some differences between mean personal total income versus education for white females and for white males (see Exercise 2). What are some of the limitations of the model in making this comparison?

4. From the FAM 1000 data, the best fit regression line for mean personal total income for the census category of nonwhite females versus years of education is:

$$\text{Personal total income} = -2480 + 1560 \cdot \text{years of education}$$

The correlation coefficient is 0.86 and the sample size is 77 nonwhite females.

Exercise 3 gives the best fit regression line and cc for mean personal total income for white females versus years of education. The following figures show the graph of averages for the mean personal total income for white and nonwhite females.

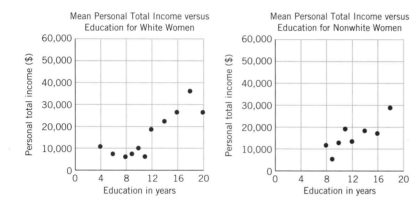

Write a 60 second summary comparing the relationship between education and personal total income for white and nonwhite females.

5. The term "linear regression" was first coined in 1903 by Karl Pearson as part of his efforts to understand the way physical characteristics are passed from generation to generation. He assembled and graphed measurements of the heights of fathers and their sons, from more than a thousand families. The independent variable, F, was the height of the fathers. The dependent variable, S, was the mean height of the sons who all had fathers with the same height. The best fit line for the data points had a slope of 0.516, which is much less than 1. If on average, the sons grew to the same height as their fathers, the slope would equal 1. Tall fathers would have tall sons and short fathers would have equally short sons. Instead, the graph shows that whereas the sons of tall fathers are still tall, they are not (on average) as tall as their fathers. Similarly the sons of short fathers are not as short as their fathers. Pearson termed this *regression*; the heights of sons *regress* back toward the height that is the mean for that population.

The equation of the regression line is $S = 33.73 + 0.516F$, where $F = $ heights of fathers in inches and $S = $ mean heights of sons in inches.

a. Interpret the number 0.516 in this context.

b. Use the regression line to predict the mean height of sons whose fathers are 64 inches tall and of those whose fathers are 73 inches tall.

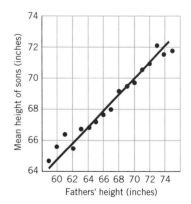

From Snedecor and Cochran, *Statistical Methods,* 8th ed; by permission of the Iowa State University Press, Copyright © 1967.

c. If there were over 1000 families, why are there only 17 data points on this graph?

6. Read the excerpt in the Anthology from *Performing Arts—The Economic Dilemma.* Describe what the regression line tells you about the relationship between attendance per concert and the number of concerts for a major orchestra.

See Excel or graph link file EDUCOSTS.

7. (Optional use of graphing calculator or computer.)

The data and the graph below give the mean annual cost for tuition and fees at public and private four-year colleges in the United States since 1985.

Year	Public Ed ($)	Private Ed ($)
1985	1386	6843
1987	1651	8118
1988	1726	8771
1989	1846	9451
1990	2035	10348
1991	2159	11379
1992	2410	12192
1993	2604	13055
1994	2820	13874
1995	2982	14510

Source: U.S. Bureau of the Census, Current Population Reports in *The American Almanac 1996–1997: Statistical Abstract of the United States,* 1996.

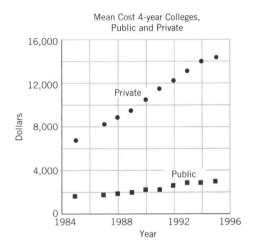

It is clear that the cost of education is going up, but public education is still less expensive than private. The graph suggests that costs of both public and private education versus time can be roughly represented as straight lines.

a. Either using technology or sketching by hand, construct two lines to represent the data.

b. Using your two linear models, calculate the rate of change of education cost per year for public and for private education. (If you sketched the lines by hand, estimate the coordinates of two points that lie on the line, and then estimate the slope.)

c. If the costs continue to rise at the same rates for both sorts of schools, what would be the respective costs for public and private education in the year 2000? Does this seem plausible to you? Why or why not?

8. How did a Princeton professor's statistical analysis influence a decision by a federal judge in Philadelphia to give a Pennsylvania senate seat to a losing Republican candidate? See the reading "His Stats Can Oust a Senator or Price a Bordeaux."

9. (Requires graphing calculator or computer.)

See Excel or graph link file KALAMA.

In Chapter 4, we generated by hand a linear model for the mean heights of a group of 161 children in Kalama, Egypt.

| Mean Height of Kalama Children | | | |
Age (months)	Height (cm)	Age (months)	Height (cm)
18	76.1	24	79.9
19	77.0	25	81.1
20	78.1	26	81.2
21	78.2	27	81.8
22	78.8	28	82.8
23	79.7	29	83.5

Source: Introduction to the Practice of Statistics by Moore and McCabe. Copyright © 1989 by W.H. Freeman and Company. Used with permission.

a. Use a graphing calculator to find the best fit regression line.

b. Identify the variables for your model, specify the domain, and interpret the slope and vertical intercept in this context. How good a fit is your line?

c. Use your line to predict the mean height for children 26.5 months old.

10. The following table shows the calories per minute burned by a 154-lb person moving at speeds from 2.5 to 12 mi/hr. (Note: a fast walk is about 5 mi/hr; faster than that is considered jogging or slow running.) Marathons, about 26 miles long, are now run in slightly over 2 hours, so that top distance runners are approaching a speed of 13 mi/hr.)

See Excel or graph link file CALORIES.

Speed (mi/hr)	Calories per Minute
2.5	3.0
3.0	3.7
3.5	4.2
3.8	4.9
4.0	5.5
4.5	7.0
5.0	8.3
5.5	10.1
6.0	12.0
7.0	14.0
8.0	15.6
9.0	17.5
10.0	19.6
11.0	21.7
12.0	24.5

a. Plot the data.

b. Does it look as if the relationship between speed and calories per minute is linear? If so, generate a linear model. Identify the variables and a reasonable domain for the model, and interpret the slope and vertical intercept. How well does your line fit the data?

c. Describe in your own words what the model tells you about the relationship between speed and calories per minute.

See Excel or graph link file
SMOKERS.

11. (Optional use of graphing calculator or computer.)

The following table shows (for years 1965 to 1993 and for people 18 and over) the total percentage of cigarette smokers, the percentage of males that are smokers, and the percentage of females that are smokers.

	Smokers		
Year	**% of Total Population 18 and Older**	**% of All Males**	**% of All Females**
1965	42.4	51.9	33.9
1974	37.1	43.1	32.1
1979	33.5	37.5	29.9
1983	32.1	35.1	29.5
1985	30.1	32.6	27.9
1987	28.8	31.2	26.5
1988	28.1	30.8	25.7
1990	25.5	28.4	22.8
1991	25.6	28.1	23.5
1992	26.5	28.6	24.6
1993	24.2	26.2	22.3

Source: U.S. National Center for Health Statistics, in *The American Almanac 1995–1996 and 1997–1998: Statistical Abstract of the United States,* 1995 and 1996.

a. By hand, draw a scatter plot of the percentage of all current smokers 18 and older versus time.

   i. Calculate the average rate of change from 1965 to 1993.

   ii. Calculate the average rate of change from 1990 to 1993.

   Be sure to specify the units in each case.

b. On your graph, sketch an approximate regression line. By estimating coordinates of points on your regression line, calculate the average rate of change of the percentage of total smokers with respect to time.

c. Using a calculator or computer, generate a regression line for the percentage of all smokers 18 and older as a function of time. (You may wish to set 1965 as year 0 and let the independent variable represent the number of years since 1965.) Record the equation and the correlation coefficient. How good a fit is this regression line to the data? Compare the rate of change for your hand-generated regression line to the rate of change for the technology-generated regression line.

d. Generate and record regression lines (and their associated correlation coefficients if you are using technology) for the percentages of both male and female that are smokers as functions of time.

e. Write a summary paragraph using the results from your graphs and calculations to describe the trends in smoking from 1965 to 1993. Would you expect this overall trend to continue? Why?

See Excel or graph link file
SKIJUMP.

12. (Optional use of graphing calculator or computer.)

The following table is a list of U.S. ski jumping records set on the 90-meter jumping hill at Howelsen Hill, Steamboat Springs, Colorado. Note that the record was broken more than once in the years 1950 and 1963.

Year	Distance (ft)	Ski Jumper
1916	192	Ragnar Omtvedt
1917	203	Henry Hall
1950	301	Gordon Wten
1950	305	Merrill Barber
1950	307	Art Devlin
1951	316	Ansten Samuelstuen
1963	318	Gene Kotlarek
1963	322	Gene Kotlarek
1978	354	Jim Denny

Source: Tread of Pioneers Museum, Steamboat Springs, Colorado.

a. Plot the data on a scatter plot. (You may want to define 1916 to be year 0 and let your independent variable be the number of years since 1916.)

b. Do the data seem to be linear? Explain your answer.

c. Assuming the data are linear, construct a best fit line (by hand or using technology). Carefully identify your variables, and discuss the meaning of the slope and the vertical intercept within the context of this problem. How good a fit is your line?

d. Use your linear model to predict the record ski jump in the year 2000. Does your estimate seem reasonable?

13. (Requires computer or TI-83 graphing calculator.)

Examine the section entitled "Nations Data Set" in *Part III: Exploring on Your Own* on the CD-ROM. It contains a printout of the NATIONS data set and a data dictionary. What pairs of variables would you suspect to be linearly related? Explore the data to find out which variables you want to study. Choose one pair and generate a best fit regression line. Interpret your results in a 60 second summary. What are your hypotheses as to why these variables may be related? What hidden factors might be influencing the relationship?

See Excel or graph link file NATIONS.

14. (Requires graphing calculator or computer.)

In the exercises in Chapter 4, we saw the following table and graph for the men's Olympic 16-pound shot put.

See Excel or graph link file SHOTPUT.

Olympic Shot Put

Year	Feet Thrown	Year	Feet Thrown
1900	46	1956	60
1904	49	1960	65
1908	48	1964	67
1912	50	1968	67
1920	49	1972	70
1924	49	1976	70
1928	52	1980	70
1932	53	1984	70
1936	53	1988	74
1948	56	1992	71
1952	57	1996	71

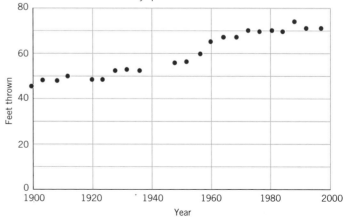

Source: Reprinted with permission from the Universal Press Syndicate. From *The 1996 Universal Almanac.* (The data points are missing for the early 1940s since the Olympics were not held during World War II.) 1996 data from http://sports.yahoo.com/oly/lgns/960726/tfmshotmed.html.

a. Using a graphing calculator or computer, find the equation of the best fit line and, if possible, its correlation coefficient. When you enter the data, you may want to set 1900 as year 0 and let the independent variable be the number of years since 1900. Is the line a good fit?

b. Describe in your own words your linear model for the data, interpreting each of the terms in your equation. Be sure to specify the units.

See Excel or graph link file
FARMPOP.

15. (Requires graphing calculator or computer.)

The accompanying table contains data on the farm population for each decade from 1880 to 1990. Farm population consists of all persons living on farms in rural areas.

Farm Population: 1880–1990

Year	Population (thousands)	Percentage of Total U.S. Pop.
1880	21,973	43.8
1890	24,771	42.3
1900	29,875	41.9
1910	32,077	34.9
1920	31,974	30.1
1930	30,529	24.9
1940	30,547	23.2
1950	23,048	15.3
1960	15,635	8.7
1970	9,712	4.8
1980	6,051	2.7
1990	4,591	1.9

Source: For 1880–1970, U.S. Bureau of Census, *Historical Statistics of the United States: Colonial Times to 1970.* For 1970 on, U.S. Bureau of Census, *Statistical Abstract of the United States,* (Washington, D.C., 1996).

a. Make scatter plots of the farm population in absolute numbers over time and in percentage of total population over time.

b. Is a linear model appropriate for describing either of the two scatter plots? If so, generate a regression line and interpret your equation.

c. If you selected a different time period, would you find a stronger association between farm population and time? Confirm your prediction using a graphing calculator or function graphing program.

16. (Requires computer and access to the Internet.)

United States census data can be downloaded from the Internet. Here are a few projects that have "homepages" that describe how to use the Internet to access census data.

• http://www.bls.census.gov/ gives access to census data and projects that have homepages. In particular, http://www.census.gov/cps/cpsmain.htm is the homepage for the Current Population Survey, a subset of whose data is used throughout this chapter.

• http://www.hist.umn.edu/~ipums/ will bring you to the Integrated Public Use Microdata Series (IPUMS), Minnesota Historical Census Projects at the University of Minnesota. IPUMS claims to be the "largest publicly accessible computerized database on a human population." It includes U.S. population census data from 1850 to the present.

- http://www.psc.lsa.umich.edu/SSDAN/ is the address of the homepage for the Social Science Data Analysis Network (SSDAN) at the Population Studies Center, University of Michigan. SSDAN provides "tailor-made" data sets from U.S. census data from 1950 to the present.

Think of a question that you would like to answer about the population of your city or state. Use census data available from one of these projects to find data that can help you answer your question.

17. (Optional use of graphing calculator or computer.)

The following table and graph show the winning running time in minutes for women in the Boston marathon.

See Excel or graph link file MARATHON.

Reprinted with permission from the Universal Press Syndicate. From *The 1993 Universal Almanac*. 1996 data from http://www.bostonmarathon.org/

Women's Boston Marathon Winning Times

Year	Time (minutes)
1972	190
1973	186
1974	167
1975	162
1976	167
1977	166
1978	165
1979	155
1980	154
1981	147
1982	150
1983	143
1984	149
1985	154
1986	145
1987	145
1988	145
1989	145
1990	145
1991	144
1992	144
1993	145
1994	142
1995	145
1996	147

a. Generate a line that approximates the data (by hand or with a graphing calculator or function graphing program). You may wish to set 1972 as year 0. If you are using technology, specify the correlation coefficient. Interpret the slope of your line in this context.

b. If the marathon times continued to change at the rate in your linear model, predict the winning running time for the women's marathon in 2010. Does that seem reasonable? If not, why not?

c. The graph seems to flatten out after about 1981. Construct a second regression line for the data from 1981 on. If you're using technology, what is the correlation coefficient for this line? What would this line predict for the winning running time for the women's marathon in 2010? Does this estimate seem more realistic than your previous estimate?

d. Write a short paragraph summarizing the trends in the Boston Marathon winning times for women.

See Excel or graph link file
MENSMILE.

Record Times for Men's Mile

Year	Time (minutes)
1913	4.24
1915	4.21
1923	4.17
1931	4.15
1933	4.13
1934	4.11
1937	4.11
1942	4.10
1942	4.10
1942	4.08
1943	4.04
1944	4.03
1945	4.02
1954	3.99
1954	3.97
1957	3.95
1958	3.91
1962	3.91
1964	3.90
1965	3.89
1966	3.86
1967	3.85
1975	3.85
1975	3.82
1979	3.82
1980	3.81
1981	3.81
1981	3.81
1981	3.79
1985	3.77
1993	3.74

18. (Optional use of graphing calculator or computer.)

The following table and graph show the world record times for the men's mile. Note that several times the standing world record was broken more than once during a year.

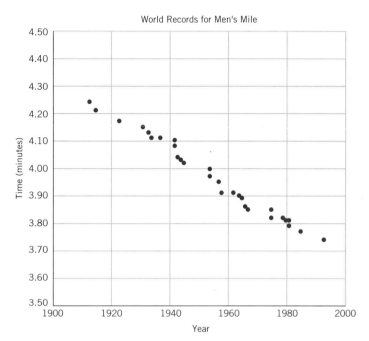

Data extracted from the website at http://www.uta.fi/~csmipe/sports/eng/mwr.html.

a. Generate a line that approximates the data (by hand, with a graphing calculator or a function graphing program). You may wish to set 1910 or 1913 as year 0. If you are using technology, specify the correlation coefficient. Interpret the slope of your line in this context.

b. If the world record times continued to change at the rate specified in your linear model, predict the record time for the men's mile in 2010. Does your prediction seem reasonable? If not, why not?

c. In what year would your linear model predict the world record to be 0 minutes? Since this is impossible, what do you think is a reasonable domain for your function? Describe what you think would happen in the years after those contained in your domain.

d. Write a short paragraph summarizing the trends in the world record times for the men's mile.

Boiling Temperature of Water

Altitude (ft above sea level)	Degrees F	Degrees C
0	212	100
2,000	208	98
5,000	203	95
7,500	198	92
10,000	194	90
15,000	185	85
30,000	158	70

19. The temperature at which water boils is affected by the difference in atmospheric pressure at different altitudes above sea level. The classic cookbook *The Joy of Cooking* by Irma S. Rombauer and Marion Rombauer Becker gives the data you see at left (rounded to the nearest degree) on the boiling temperature of water at different altitudes above sea level.

a. Plot boiling temperature in degrees Centigrade, °C, versus the altitude. Find a formula to describe the boiling temperature of water, in °C, as a function of altitude.

b. Plot boiling temperature in degrees Fahrenheit, °F, versus the altitude. Find a formula to describe the boiling temperature of water, in °F, as a function of altitude.

c. The highest point in the United States is Mount McKinley in Alaska, at 20,320 feet above sea level; the lowest point is Death Valley in California, at 285 feet below sea level. You can think of distances below sea level as negative altitudes from sea level. At what temperature in degrees Fahrenheit will water boil in each of these locations according to your formula?

d. A healthy human has a normal temperature of around 98.6 degrees Fahrenheit. A certain amount of water in the human body is necessary for life; at what altitude does your formula predict that the water in the body will boil?

e. A scientist who develops a formula for observed phenomena is interested in testing the limits of its accuracy. It is interesting to ask, and then test, whether there is an altitude at which water can be made to boil at 0° Centigrade, the freezing point of water at sea level. (Note that airplane cabins are pressurized to near sea level atmospheric pressure conditions to avoid unhealthy conditions resulting from high altitude.)

20. (Optional use of graphing calculator or computer.)

The following table and graph show the world distance records for the women's long jump. Several times a new long jump record was set more than once during a given year.

See Excel or graph link file LONGJUMP.

a. Generate a line that approximates the data (by hand or with a graphing calculator or function graphing program). You may wish to set 1950 or 1954 as year 0. If you are using technology, specify the correlation coefficient. Interpret the slope of your line in this context.

Women's Long Jump Records

Year	Distance (meters)	Year	Distance (meters)
1954	6.28	1976	6.99
1955	6.28	1978	7.07
1955	6.31	1978	7.09
1956	6.35	1982	7.15
1956	6.35	1982	7.20
1960	6.40	1983	7.21
1961	6.42	1983	7.27
1961	6.48	1983	7.43
1962	6.53	1985	7.44
1964	6.70	1986	7.45
1964	6.76	1986	7.45
1968	6.82	1987	7.45
1970	6.84	1988	7.45
1976	6.92	1988	7.52

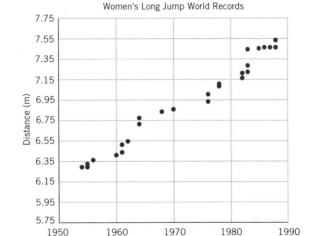

Data extracted from the web site at http://www.uta.fi/~csmipe/sports/eng/mwr.html.

b. If the world record distances continued to change at the rate described in your linear model, predict the world record distance for the women's long jump in the year 2000.

c. What would your model predict for the record in 1943? How does this compare with the actual 1943 record of 6.25 meters? What do you think would be a reasonable domain for your model? What do you think the data would look like for years outside your specified domain?

EXPLORATION 5.1

A Case Study on Education and Income in the United States: Part I

Objective

- find possible relationships between education and income using the FAM 1000 data

Materials / Equipment

If using a computer:

- course software: "F2: Education / Wage Regression" and "F3: Regression with Multiple Subsets" in the *FAM 1000 Census Graphs* program.
- printer and overhead transparencies (optional)

If using a graphing calculator:

- graphing calculators with graph link capabilities
- graph link files FAM1000A, FAM1000B, FAM1000C, FAM1000D.
- printout of the FAM 1000 graph link files (See Sec. 5 in Graphing Calculator Workbook.)

Related Readings

"Money Income" and "Educational Attainment," *Population Profile of the United States, 1995,* Bureau of the Census, Current Population Reports, Special Studies Series P23-189.

"Wealth, Income, and Poverty," from *The Data Game: Controversies in Social Science Statistics* by Mark Maier.

"Education and Income" chart from *The New York Times,* January 28, 1993.

Class Discussion

We will examine the relationship between education and income in the United States. We start our case study by discussing the Current Population Survey and the information provided in the FAM 1000 data set. If a demonstration computer is available, view the program "F2: Education / Wage Regression" in the *FAM 1000 Census Graphs* program, and talk about the results. Discuss the generation of regression lines and correlation coefficients, and their interpretation.

Procedure

Working with Partners

Work in pairs using two computers or two calculators, if possible, so you can compare two regression lines more easily. You will study the nature of the relationship between education and different income variables for various groups of people. You'll be asked to write a summary of your findings and to present your conclusions to the class.

1. Finding regression lines

   If using a computer:

   Open "F3: Regression with Multiple Subsets" in *FAM 1000 Census Graphs*. This program allows you to find regression lines for education versus income for different income variables and for different groups of people. Select (by clicking on the appropriate box) one of the four income variables: personal wages, personal wages per hour, personal total income, or family income. Then select at least two regression lines that it would make sense to compare (e.g., men versus women, white versus nonwhite, two or more regions of the country). You should do some browsing through the various regression line options to pick those that are the most interesting. Print out your regression lines (on overhead transparencies if possible).

   If using graphing calculators and graph link files:

   Discuss the nature of summary data sets in the graph link files. Make a decision on the type of comparison you want to make. Use data from the appropriate graph link files and find regression lines. Hard copy of the data in the various graph link files and instructions for transferring and using TI graph link files are included in the Graphing Calculator Workbook.

2. For each of the regression lines you choose, work together with your partner to record the following information in your notebooks.

   The equation of the regression line:
   The variables represent:
   The subset of the data the line represents (i.e., men? nonwhites?):
   cc =
   Whether or not the line is a good fit and why:
   The slope =
   Interpretation of the slope (e.g., for each additional year of education, average total personal wages rise by such and such an amount):

Discussion/Analysis

With your partner, explore ways of comparing the two regression lines. How do the two slopes compare? Can you say that one group is better off? Is that group better off no matter how many years of education? What factors are hidden or not taken into account?

Exploration-Linked Homework

1. Prepare a 60 second summary of your results. Discuss with your partner how to present your findings. What are the limitations of the data? What are the strengths and weaknesses of your analysis? What factors are hidden or not taken into account? What questions are raised? You may wish to consult the related readings or find additional sources to add to your analysis.

2. If you have access to a computer, use "F3: Regression with Multiple Subsets" in *FAM 1000 Census Graphs* to look for possible relationships between income and another variable in the FAM 1000 data (e.g., age, family size, or number of work weeks). Describe your results.

EXPLORATION 5.2

A Case Study on Education and Income in the United States: Part II

Objectives

- study education and income separately using data from the *FAM 1000* data set
- explore how to make comparisons using graphs and numerical descriptors

Materials/Equipment

- computer with course software "F1: Histograms" in *FAM 1000 Census Graphs* program

Related Readings

"Money Income" and "Educational Attainment," *Population Profile of the United States, 1995,* Bureau of the Census, Current Population Reports, Special Studies Series P23-189.

Procedure

We continue our case study by looking at income and education as separate variables. For your analysis, you can use a spreadsheet or the program "F1: Histograms" in *FAM 1000 Census Graphs* on the CD-ROM, and the related readings.

Working with Partners

Working with a classmate compare the income distribution of two groups and explore whether there are differences between the groups. For example, how does the distribution of income differ for men and women? How do income levels compare in various regions of the country or among various racial groups?

1. Open "F1: Histograms" in the *FAM 1000 Census Graphs* program. It can create the histograms of your choice from FAM 1000 data. You can change the size of the interval by clicking on the arrow on the bottom of the screen and then clicking "Histogram." Experiment by creating histograms for different measures of income, such as personal wages, total personal income, and wages per hour for different groups of people. You can choose different interval sizes to see what patterns emerge. Note that each bar of the histogram includes the left end point but excludes the right end point.

2. Choose two sets of people you wish to compare, for instance men versus women, or people living in two different regions of the country, or members of two different racial groups. See whether there are differences between the two groups in one of the income measures. (Click on the income measure, then click on the subset you want to examine and then click "Histogram.") One partner can work on one group and the other partner can do the second group. Discuss your observations. Pick a comparison that you think is interesting and print out your histograms.

3. Make jottings on the differences and similarities between the two groups for the income variable you have chosen. Include numerical descriptors such as mean and median.

Discussion/Analysis

Discuss with your partner what type of arguments you can make, the limitations of the data, and the questions that are raised by your analysis of the data.

Exploration-Linked Homework

1. Examine the related readings to find data and arguments that relate to the variables you have been analyzing in the class exploration. The U.S. Census Current Population Reports profile American income and education levels using the data from the Current Population Surveys from which our FAM 1000 data are extracted.

2. Write a 60 second summary describing your findings. Start with a topic sentence and use numerical descriptors and information from your graphs to support your topic sentence. Conclude your paragraph by summarizing your findings or suggesting a plausible explanation for your findings. Attach the histograms you used.

3. Write a paragraph on the limitations of the data you used or on the limitations of your analysis.

4. Conduct a similar exploration to study whether there are differences in the educational attainment of two different groups of people.

When Lines Meet

Overview

A collection of two or more equations relating the same variables is called a *system of equations*. A system that contains equations relating income and taxes can be used to compare different tax schemes. Two equations relating quantity and price can represent consumers' demands versus producers' willingness to supply a product. The place where the graphs of two equations meet often has particular significance. The point of intersection can indicate when supply equals demand, or the income at which taxes are the same under two different tax plans. A point where the graphs of all the equations in a system meet is a *solution* to the system.

This chapter explores how to construct systems of linear equations and how to solve them. We examine how to interpret the solutions in situations where each line is used to represent a social or physical reality.

The Exploration involves analyzing two different state income tax plans presented to the Massachusetts voters in 1994.

After reading this chapter you should be able to:

- construct and graph systems of linear equations
- find a solution for a system of two linear equations
- interpret intersection points
- use systems of linear equations to model social or physical situations
- construct and graph piecewise linear functions

6.1 AN ECONOMIC COMPARISON OF SOLAR VERSUS CONVENTIONAL HEATING SYSTEMS

On a planet with limited fuel resources, heating decisions involve both monetary and ecological considerations. Solar heating is appealing because the fuel, solar rays, is free and is easily delivered without pipes or trucks. Why then is solar heat rarely used?

Solar heating can work well in places that receive a fair amount of sunlight during the heating season. However, the installation cost is so high that it takes many years of low-cost operation to make up for the initial investment.[1]

Typical costs for three different kinds of heating systems for a three-bedroom housing unit are given in Table 6.1.

Table 6.1

Typical Costs for Three Heating Systems

Type of System	Installation Cost ($)	Operation Cost/Year ($/yr)
Electric	5,000	1,100
Gas	12,000	700
Solar	30,000	150

Solar heating is clearly the most costly to install and the least expensive to run. Electric heating, conversely, is the cheapest to install and the most expensive to run. By converting this information to equation form, we can find out when the solar system begins to pay back the initially higher cost. If no allowance is made for inflation or changes in fuel price, the general equation for the total cost, C, is:

$$C = \text{installation cost} + (\text{annual operating cost})(\text{years of operation}).$$

If we let n equal the number of years of operation and use the data from Table 6.1, we can construct the following linear equations:

$$C_{\text{electric}} = 5000 + 1100n$$
$$C_{\text{gas}} = 12,000 + 700n$$
$$C_{\text{solar}} = 30,000 + 150n$$

Together they form a *system of linear equations*. Table 6.2 gives the cost data at 5-year intervals, and Figure 6.1 shows the costs over a 40-year period for the three heating systems.

Let's compare the costs for gas and electric heat graphed against years of operation in Figure 6.2. The point of intersection shows where the lines predict the same total cost for both gas and electricity, given a certain number of years of operation. From the graph, the total cost of operation is about $24,000 for each type of heating system at about 17 years.

---

[1] There are several reasons for the high installation cost. A backup heating system, such as a wood-burning stove or an electric heating system, is needed for any long periods of cloudy weather. The capture and storage of the sun's heat requires a lot of space in water drums, rocks, or special thick floors and walls. The building must have a particular orientation in relation to the sun and a series of solar collection panels and piping maintained on the roof.

	Table 6.2		
	Heating System Total Costs		
Year	Electric ($)	Gas ($)	Solar ($)
0	5,000	12,000	30,000
5	10,500	15,500	30,750
10	16,000	19,000	31,500
15	21,500	22,500	32,250
20	27,000	26,000	33,000
25	32,500	29,500	33,750
30	38,000	33,000	34,500
35	43,500	36,500	35,250
40	49,000	40,000	36,000

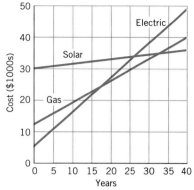

Figure 6.1 Comparison of home heating costs.

Compare the relative costs to the left and right of the point of intersection. How many years of operation does it take for the total cost of gas to become less than the total cost of electricity? When would the total cost of gas be higher? Figure 6.2 shows that gas is less expensive than electricity to the right of the intersection point, after approximately 17 years of operation, and that electricity is less expensive than gas to the left.

To find the breakeven point exactly, we find the coordinates of the intersection point. When the gas and electric equations intersect, the coordinates satisfy both equations. At the point of intersection, the total cost of electric heat, $C_{electric}$, equals the total cost of gas heat, C_{gas}. Thus, the two expressions for the total cost can be set equal to each other to find the coordinates of the intersection point.

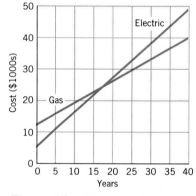

Figure 6.2 Gas versus electric.

$$C_{electric} = 5000 + 1100n \quad (1)$$
$$C_{gas} = 12,000 + 700n \quad (2)$$

set Equation (1) equal to Equation (2) $\qquad C_{electric} = C_{gas}$

substitute $\qquad 5000 + 1100n = 12,000 + 700n$

subtract 5000 from each side $\qquad 1100n = 7000 + 700n$

subtract 700n from each side $\qquad 400n = 7000$

divide each side by 400 $\qquad n = 17.5$ years

When $n = 17.5$ years, the total cost for electric or gas heating is the same. The total cost can be found by substituting this value for n in Equation (1) or (2).

substitute 17.5 for n in Equation (1) $\qquad C_{electric} = 5000 + 1100(17.5) \qquad (1)$

$$= 5000 + 19,250$$

$$C_{electric} = \$24,250$$

Since we claim that the pair of values (17.5, 24250) satisfies both equations, we need to check, when n is equal to 17.5 years, that C_{gas} is also $24,250.

substitute 17.5 for n in Equation (2) $\qquad C_{gas} = 12,000 + 700(17.5)$

$$= 12,000 + 12,250$$

$$C_{gas} = \$24,250$$

The coordinates (17.5, 24250) satisfy both equations. When $n = 17.5$ years, then

$C_{electric} = C_{gas} = \$24{,}250$. The point (17.5, 24250) is called a *solution* to the system of these two equations. We were able to find accurate values for the coordinates of the point of intersection by using the equations. We found that 17.5 years after installation, a total of \$24,250 has been spent on heat for either an electric or a gas heating system. Electricity is a more economical fuel supply for 17.5 years, then gas becomes cheaper.

The intersection points for electric versus solar heating and gas versus solar heating can be estimated using the graphs in Figures 6.3 and 6.4. The actual solutions to these systems will be found later in the text and in the homework exercises.

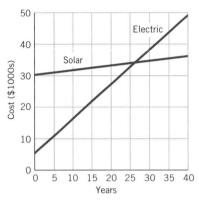

Figure 6.3   Electric versus solar heating.

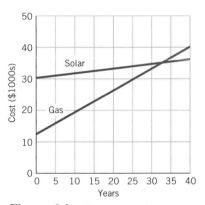

Figure 6.4   Gas versus solar heating.

We made many simplifying assumptions when constructing this model. We ignored inflation and rising fuel prices, and we did not consider inevitable repair costs. We did not take into account what economists call *opportunity costs*. Opportunity costs include, for instance, the 40-years worth of interest that we could have collected if we had invested the dollars (in stocks or bonds or savings accounts) that we used to pay for startup costs. More sophisticated mathematical models might also consider the cost of depleting fuel resources and the risks of generating nuclear power.

Algebra Aerobics 6.1

1. Our model tells us that electric systems are cheaper than gas for 17.5 years of operation. Using Figure 6.1, estimate the interval over which gas is the cheapest of the three heating systems. When does solar heating become the cheapest system compared to the other two heating systems?

2. For the following system of equations:

$$4x + 3y = 9$$
$$5x + 2y = 13$$

    a. Determine whether (3, −1) is a solution.

    b. Show why (−1, 3) is not a solution for this system.

3. **a.** Show that the following equations are equivalent.

$$4x = 6 + 3y$$
$$12x - 9y = 18$$

    b. How many solutions are there for the system of equations in part (a)?

6.2 FINDING SOLUTIONS TO SYSTEMS OF LINEAR EQUATIONS

In the heating example, we found a solution for a system of two equations by finding the point where the graphs of the two equations intersected. At the point of intersection, the equations share the same value for the independent variable and the same value for the dependent variable.

A collection of linear equations that involve the same variables is called a *system of linear equations* or a *set of simultaneous linear equations*.

A pair of real numbers is a *solution* to a system of linear equations in two variables if and only if the pair of numbers is a solution to each equation in the system. On the graph of a system of linear equations, a solution is a point at which all the lines intersect.

We begin analyzing a system of equations by finding out if there is a solution, that is, if the graphs of the equations intersect.

Visualizing Solutions to Systems of Linear Equations

For a single linear equation, the graph of its solutions is a line and every point on the line is a solution. This equation has an unlimited number of solutions. In a system of two linear equations, a solution must satisfy both equations. We can easily visualize what might happen. If we graph two different straight lines and the lines intersect (Figure 6.5), there is only one solution—at the intersection point. If the lines are parallel, they never intersect and there are no solutions (Figure 6.6).

If the lines are not parallel, the coordinates of the point of intersection can be estimated by inspecting the graph. For example, in Figure 6.5, the two lines appear to cross near the point where the x-coordinate is -1 and the y-coordinate is approximately 1.5. For better estimates, we can use a zoom function on a graphing calculator

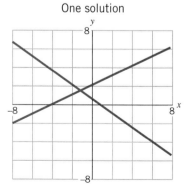

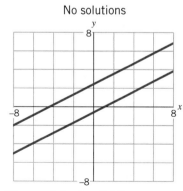

Figure 6.5 Lines intersect at a single point.

Figure 6.6 Parallel lines never intersect.

Infinitely many solutions

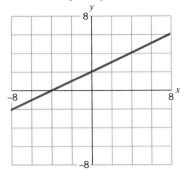

Figure 6.7   The two equations represent the same line. Every point on the line is a solution for both equations.

or function graphing program. Most function graphing programs can generate values accurate to several decimal places.

How else might the graphs of two lines be related? We may find that the two equations represent the same line. Consider, for example, the following two equations:

$$y = 0.5x + 2 \tag{1}$$
$$3y = 1.5x + 6 \tag{2}$$

If we multiply each side of Equation (1) by 3, we obtain Equation (2). The two equations are *equivalent*, since any solution of one equation is also a solution of the other equation. Both equations represent the same line. There are an infinite number of points on that line and they are all solutions for both equations (Figure 6.7).

Using Equations to Find Solutions

A system of two linear equations can be solved in several ways. The form of the equations usually determines the most efficient method. All of the methods we examine here give the same final answer. In each case, a solution consists of a pair of numbers (or a set of pairs of numbers), which we often write in the form (x, y).

Case 1: Both Equations Are in Function Form, $y = mx + b$
Strategy: Set the Two Expressions for y Equal to Each Other

Example   1

Find the solution for this system of equations

$$y = 2x + 8 \tag{1}$$
$$y = -3x - 7 \tag{2}$$

We are looking for a pair of values (x, y) that satisfies *both* equations. At the intersection point, both equations have the same value for x and the same value for y. Thus, to find the solution, we can set the expressions for y equal to each other.

The steps for this strategy are:

1.   Set the two expressions for y equal to each other. Then, solve the resulting equation to find the value for x.
2.   Substitute this value for x in one of the two original equations to find the corresponding value for y.
3.   Check your results in the other original equation.

It is a good idea to graph the equations to verify that the solution you have found gives the coordinates of the intersection point.

SOLUTION
In both equations y is written in terms of x. We can set the expressions for y equal to each other.

$$2x + 8 = -3x - 7$$

then solve for x $\qquad 5x = -15$

$$x = -3$$

The two lines cross when $x = -3$. In order to find the y value at the point of intersection, we can substitute -3 for x in either of the two original equations. Using Equation (1):

$$y = 2x + 8 \qquad\qquad (1)$$

substitute -3 for x $\qquad y = 2(-3) + 8$

multiply $\qquad\qquad = -6 + 8$

$$y = 2$$

We can check that the pair of values $(-3, 2)$ does indeed work in *both* equations. Using Equation (2), we verify that when $x = -3$, then $y = 2$.

$$y = -3x - 7 \qquad\qquad (2)$$

substitute -3 for x $\qquad y = -3(-3) - 7$

simplify $\qquad\qquad = 9 - 7$

$$y = 2$$

So $x = -3$ and $y = 2$ is the solution to the system. In Figure 6.8 we get visual confirmation that $(-3, 2)$ represents the intersection point of the two lines and thus is a solution to the system.

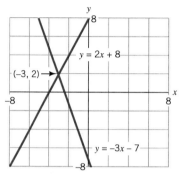

Figure 6.8   Graphs of $y = 2x + 8$ and $y = -3x - 7$.

Algebra Aerobics 6.2a

1. Solve the following systems of equations.

   a. $y = x + 4$ **b.** $y = -1700 + 2100x$ **c.** $F = C$

       $y = -2x + 7$ $y = 4700 + 1300x$ $F = 32 + (9/5)C$

Case 2: Only One Equation Is in Function Form
Strategy: Substitution

Find the point (if any) where the graphs of the following two linear equations intersect.

$$6x + 7y = 25 \qquad\qquad (1)$$

$$y = 15 + 2x \qquad\qquad (2)$$

Example 2

An intersection is a solution to the system. At the intersection the value for y in both equations is the same. Thus we can substitute the expression for y from Equation (2) into Equation (1). The steps for this strategy are as follows.

1. If one equation gives y in terms of x (or if y can easily be represented in terms of x), substitute that expression for y in the other equation.

2. Solve the resulting equation for the value for x.
3. Substitute this value for x into either original equation to find the corresponding y value.
4. Check your answer in the other original equation.

SOLUTION

Here Equation (2) is in function form, but Equation (1) isn't. We could put Equation (1) into function form and apply the strategy in Case 1 by setting the two expressions for y equal to each other. However, if we solve Equation (1) for y, the resulting expression for y is not easy to use; thus, *substitution* involves easier computation.

given $\qquad\qquad\qquad\qquad\qquad\qquad\qquad 6x + 7y = 25 \qquad\qquad$ (1)

and $\qquad\qquad\qquad\qquad\qquad\qquad\qquad\quad y = 15 + 2x \qquad\qquad$ (2)

In Equation (1) substitute expression for y from Equation (2).

$$6x + 7(15 + 2x) = 25$$

simplify $\qquad\qquad\qquad\qquad\quad 6x + 105 + 14x = 25$

$$20x = -80$$

$$x = -4$$

We can use one of the original equations to find the value for y when $x = -4$. Using Equation (2):

$$y = 15 + 2x \qquad\qquad (2)$$

substitute -4 for x $\qquad\qquad\quad y = 15 + 2(-4)$

multiply $\qquad\qquad\qquad\qquad\quad y = 15 - 8$

$$y = 7$$

We can check that $(-4, 7)$ satisfies Equation (1) by substituting -4 for x and 7 for y and verifying that the result is a true statement.

$$6x + 7y = 25 \qquad\qquad (1)$$

$$6(-4) + 7(7) = 25$$

$$-24 + 49 = 25$$

true statement $\qquad\qquad\qquad\qquad 25 = 25$

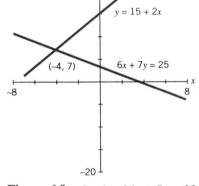

Figure 6.9   Graphs of $6x + 7y = 25$ and $y = 15 + 2x$.

Figure 6.9 shows a graph of the two equations and the intersection point at $(-4, 7)$.

Algebra Aerobics 6.2b

1. Solve the following systems.

    a. $y = x + 3$ **b.** $z = 3w + 1$ **c.** $x = 2y - 5$ **d.** $r - 2s = 5$

            $5y - 2x = 21$ $9w + 4z = 11$ $4y - 3x = 9$ $3r - 10s = 13$

Case 3: Neither Equation Is in Function Form
Strategy: Elimination

Another method called *elimination* is convenient when neither equation can be represented in a function form that is easy to use. The strategy is to eliminate one variable by adding the equations together. Elimination occurs if the coefficients of one variable are opposites of each other (for example, $2x$ and $-2x$, or $-4y$ and $4y$).

Example 3

The following system can be solved by adding the equations and thus eliminating the x terms.

$$2y + 3x = -6 \qquad (1)$$
$$5y - 3x = 27 \qquad (2)$$

SOLUTION

combine Equation (1) with
Equation (2) to
eliminate the x terms

$$\begin{aligned} 2y + 3x &= -6 \\ 5y - 3x &= 27 \\ \hline 7y + 0x &= 21 \\ 7y &= 21 \end{aligned}$$

divide by 7 $\qquad\qquad\qquad y = 3$

Now we can solve for x by substituting $y = 3$ in one of the original equations. Using Equation (1):

let $y = 3$ $\qquad\qquad 2(3) + 3x = -6$

multiply $\qquad\qquad\qquad 6 + 3x = -6$

add -6 to both sides $\qquad\quad 3x = -12$

divide by 3 $\qquad\qquad\qquad x = -4$

We can check this solution in the other equation. Using Equation (2)

$$5y - 3x = 27$$

let $x = -4$ and $y = 3$ $\qquad 5(3) - 3(-4) = 27$

multiply $\qquad\qquad\qquad 15 + 12 = 27$

true statement $\qquad\qquad\qquad 27 = 27$

The solution to the system is $x = -4$ and $y = 3$, or $(-4, 3)$.

● ● ●

In this case our work was simplified since the coefficients of x (namely 3 and -3) were opposites of each other, so the sum of the x terms was 0. When neither variable has coefficients in different equations that are opposites of each other, the strategy is to

1. Find equivalent equations for one or both equations such that when the equations are added together, one variable is eliminated or cancels out.
2. Solve the new equation for the remaining variable.
3. Substitute the value found in step 2 into one of the original equations to determine a value for the eliminated variable.
4. Check your answer in the other original equation.

Example 4

Solve this system of equations:

$$4x - 3y = 13 \tag{1}$$

$$3x - 5y = -4 \tag{2}$$

SOLUTION

In this system we can find equivalent equations such that the coefficients of one of the variables are opposites of each other. There are many ways to accomplish this. First choose which variable is to be eliminated. Sometimes the computation required to eliminate one variable is easier than the computation involved to eliminate the other. As in this case, it usually doesn't matter. If we choose to eliminate y, then we need to find out how to make the coefficients of y opposites of each other. An easy way is to multiply both sides of Equation (1) by 5 and both sides of Equation (2) by -3. The result is a system of equations (Equations (3) and (4)) that is equivalent to our original system, but now the coefficients of y are 15 and -15.

multiply each side of
Equation (1) by 5 $5(4x - 3y) = 5(13)$

simplify $20x - 15y = 65 \tag{3}$

multiply each side of
Equation (2) by -3 $-3(3x - 5y) = -3(-4)$

simplify $-9x + 15y = 12 \tag{4}$

Add Equation (3) and Equation (4) $20x - 15y = 65$
to eliminate the y terms $\underline{-\ 9x + 15y = 12}$

                                          $11x\qquad\quad = 77$

divide by 11 $x = 7$

Solve Equation (1) for y when $x = 7$.

$$4x - 3y = 13 \tag{1}$$

let $x = 7$ $4(7) - 3y = 13$

multiply $28 - 3y = 13$

add -28 to each side $-3y = -15$

divide each side by -3 $y = 5$

Use Equation (2) to check whether $x = 7$ and $y = 5$ is a solution.

$$3x - 5y = -4 \tag{2}$$

let $x = 7$ and $y = 5$ $3(7) - 5(5) = -4$

multiply $21 - 25 = -4$

true statement $-4 = -4$

Figure 6.10 shows the graphs of the two equations and the intersection point at $(7, 5)$.

Something to think about

Why is the sum of the left side of two equations equal to the sum of their right sides?

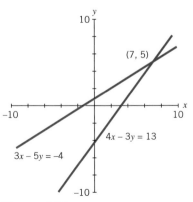

Figure 6.10   Graphs of $4x - 3y = 13$ and $3x - 5y = -4$.

Algebra Aerobics 6.2c

1. Solve each system of equations using the method you think is most efficient.

    a. $2y - 5x = -1$
    $3y + 5x = 11$

    b. $3x + 2y = 16$
    $2x - 3y = -11$

    c. $t = 3r - 4$
    $4t + 6 = 7r$

    d. $z = 2000 + 0.4(x - 10{,}000)$
    $z = 800 + 0.2x$

Special Cases: How Can You Tell if There Is No Unique Intersection Point?

A system with no solution

Solve the following system of two linear equations.

Example 5

$$y = 10{,}000 + 1500x \qquad (1)$$
$$2y - 3000x = 50{,}000 \qquad (2)$$

If we assume the two lines intersect, then the y values for Equations (1) and (2) are the same at some point. We can try to find this value for y by using substitution as in Case 2. We substitute the expression for y from Equation (1) into Equation (2).

$$2(10{,}000 + 1500x) - 3000x = 50{,}000$$

simplify $\qquad\qquad 20{,}000 + 3000x - 3000x = 50{,}000$

false statement $\qquad\qquad\qquad\quad 20{,}000 = 50{,}000 \ (\text{???})$

What could this possibly mean? Where did we go wrong? If we return to the original set of equations and solve Equation (2) for y in terms of x, we can see why there is no solution for this system of equations.

$$2y - 3000x = 50{,}000 \qquad (2)$$

add $3000x$ to both sides $\qquad\qquad 2y = 50{,}000 + 3000x$

divide by 2 $\qquad\qquad\qquad y = 25{,}000 + 1500x \qquad (3)$

Now examine Equations (1) and (3). Equation (3) is a rewritten form of the original Equation (2).

$$y = 10{,}000 + 1500x \qquad (1)$$
$$y = 25{,}000 + 1500x \qquad (3)$$

We can see that we have two parallel lines, both with the same slope of 1500 but different y-intercepts (10,000 and 25,000, respectively). So there is no intersection point for these lines (see Figure 6.11). There is no value of x that yields the same value of y in both equations, because the value of y in Equation (3) must always be 15,000 greater than the corresponding y value in Equation (1). Our initial premise, that the two y values were equal at some point, allowed us to substitute one y expression for another. However, analyzing these equations allows us to discover that our premise was incorrect.

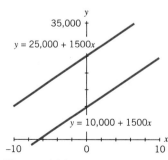

Figure 6.11 Graphs of $y = 25{,}000 + 1500x$ and $y = 10{,}000 + 1500x$.

Example 6

A system with infinitely many solutions

Solve the following system:

$$45x = -y + 33 \qquad (1)$$

$$2y + 90x = 66 \qquad (2)$$

As always, there are multiple ways of solving the system. Let's put both functions in function form.

solve Equation (1) for y	$45x = -y + 33$	(1)
add y to both sides	$y + 45x = 33$	
add $-45x$ to both sides	$y = -45x + 33$	

solve Equation (2) for y	$2y + 90x = 66$	(2)
add $-90x$ to both sides	$2y = -90x + 66$	
divide by 2	$y = -45x + 33$	

We can now see that the original equations are both equivalent to $y = -45x + 33$. Thus, the two equations represent the same line. Every point on the line determined by Equation (1) is also on the line determined by Equation (2), so there are infinitely many solutions. In Figure 6.12 the graphs of the two equations are identical, and every point on the line is a solution to the system.

Figure 6.12   Graph of $y = -45x + 33$.

Algebra Aerobics 6.2d

1. Solve each system of equations and check your answers.

   a. $5y + 30x = 20$
   $y = -6x + 4$

   b. $y = 1500 + 350x$
   $2y = 700x + 3500$

   c. $10u + 7v = 11$
   $14v = 22 - 20u$

2. Construct a system of two linear equations in which there is no solution for the system.

6.3 INTERSECTION POINTS REPRESENTING EQUILIBRIUM

When systems of equations are used to model phenomena, an intersection point often represents a state of equilibrium. In this section we examine how economists use and interpret intersection points.

Supply and Demand Curves

In economics, a graph of two lines or curves is often used to model the relationship between supply and demand.[2] The horizontal axis represents the quantity of a product

---

[2] Note that economists use the term "supply and demand *curves*" even though the graphs are frequently drawn as straight lines.

(measured perhaps in thousands of units per month) and the vertical axis represents the market price of the product (perhaps in dollars per unit).

The demand curve represents the point of view of the consumer. It shows the relationship between the market price of a product and the quantity demanded of that product. It typically has a negative slope. When the price of a product is high, consumers are reluctant to buy it, so the quantity demanded is low. As the price of the product decreases, consumers are more willing to buy, so the quantity demanded increases. Economists assume that the quantity demanded depends on the price; that is, price is considered the independent variable that determines demand.[3] But you should note they usually graph price on the vertical axis.

The supply curve represents the producer's point of view. It shows the relationship between market prices and the amounts that producers are willing to supply. It typically has a positive slope. When the market price of a product is low, the profit is smaller, and producers will supply a smaller quantity of the product. If the price rises, a better profit can be made and producers will make larger quantities of the product to sell.

The intersection of the two curves represents the point at which supply equals demand (Figure 6.13). At the price indicated by the intersection point, the quantity demanded by consumers equals the quantity supplied by producers. The intersection point is called the *equilibrium point*, and the price at that point is called the *equilibrium price*. At any lower price, demand would exceed supply; at any higher price, supply would exceed demand.

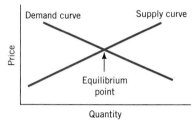

Figure 6.13   Supply and demand curves.

Questions   What if the overall demand increases? What if consumers want more of the product and are willing to pay a higher price to get it? What happens to the equilibrium point?

Discussion   The graph in Figure 6.14 shows what happens when overall demand increases. A *shift* of the whole demand curve upward to the right means that more will be bought at each price. Pick a particular price, P, and trace that price horizontally to the original demand curve to locate the quantity consumers have previously demanded, Q_{old}, at that price. If overall demand has increased, then that same price P is now associated with a higher quantity, Q_{new}. Since this reasoning holds for *any* price P, all points on the original demand curve shift to the right, forming a new demand curve.

The supply curve has not changed. But the shift in the demand curve moves the equilibrium point up and to the right (Figure 6.15). The new equilibrium point represents a larger quantity and a higher price. In other words, an increase in overall consumer demand causes producers to make a larger quantity that sells at a higher price.

[3] For this analysis, economists also assume "other things being equal." See Paul Samuelson. *Economics: An Introductory Analysis* (New York: McGraw Hill, 1964) p. 58ff.

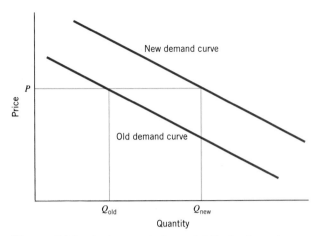

Figure 6.14   An increase in demand shifts the demand curve to the right.

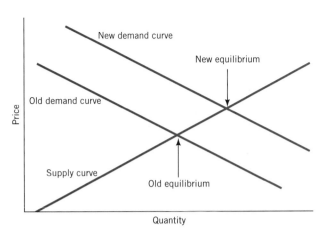

Figure 6.15   An increase in demand shifts the equilibrium point to the right and up.

6.4 GRADUATED VERSUS FLAT INCOME TAX: USING PIECEWISE LINEAR FUNCTIONS

Simple Tax Models

The New York Times *article "How a Flat Tax Would Work for You and for Them" discusses the trade-offs in using a flat tax.*

Income taxes may be based on either a flat or a graduated tax rate. With a flat tax rate, no matter what the income level, everyone is taxed at the same percentage. Flat taxes are often said to be unfair to those with lower incomes, because paying a fixed percentage of income is considered more burdensome to someone with a low income than to someone with a high income. A graduated tax rate means that people with higher incomes are taxed at a higher rate. Such a tax is called *progressive* and is generally less popular with those who have high incomes. Whenever the issue appears on the ballot, the pros and cons of the graduated versus the flat tax rate are hotly debated in the news media and paid political broadcasting. Of the 42 states with an income tax, 35 had a graduated income tax in 1994.

For the taxpayer there are two primary questions in comparing the effect of flat and graduated tax schemes. For what income level will the taxes be the same under both plans? And, given a certain income level, how will taxes differ under the two plans?

Taxes are influenced by many factors. Exemptions and deductions may be subtracted from income. Married people may file as couples, singles, or heads of household. For our comparisons of flat and graduated income tax plans, we examine one filing status and assume that exemptions and deductions have already been subtracted from income.

A flat tax model   Flat taxes are uniquely determined by income. So we can use income, i, as the independent variable, flat taxes, T_F, as the dependent variable, and write T_F as a function, f, of i. If the flat tax rate is 15% (or 0.15 in decimal form), then $T_F = f(i)$ where

$$T_F = 0.15i$$

This flat tax plan is represented in Table 6.3 and Figure 6.16.

Table 6.3

Taxes Under Flat Tax Plan

Income After Deductions ($)	Taxes at Flat Tax Rate of 15% ($)
0	0
10,000	1500
20,000	3000
30,000	4500
40,000	6000
50,000	7500

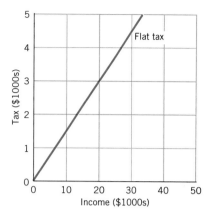

Figure 6.16   Flat tax at a rate of 15%.

A graduated tax model: a piecewise linear function   Under a graduated income tax, the tax rate changes for different portions of the income after deductions. Let's consider a graduated tax where the first $10,000 of income is taxed at a 10% rate and any income in excess of $10,000 is taxed at a 20% rate. For example, if your income after deductions is $30,000, then your taxes under this plan are:

$$\text{graduated tax} = (10\% \text{ of } \$10,000) + (20\% \text{ of income over } \$10,000)$$
$$= 0.10(\$10,000) + 0.20(\$30,000 - \$10,000)$$
$$= 0.10(\$10,000) + 0.20(\$20,000)$$
$$= \$1000 + \$4000$$
$$= \$5000$$

In general, if an income is over $10,000, then under this plan

$$\text{graduated tax} = 0.10(\$10,000) + 0.20(i - \$10,000)$$
$$= \$1000 + 0.20(i - \$10,000)$$

This graduated tax plan is represented in Table 6.4 and Figure 6.17.

Table 6.4

Taxes Under Graduated Tax Plan

Income After Deductions ($)	Taxes ($)
0	0
5,000	500
10,000	1000
20,000	1000 + 2000 = 3000
30,000	1000 + 4000 = 5000
40,000	1000 + 6000 = 7000
50,000	1000 + 8000 = 9000

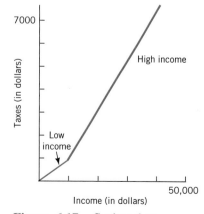

Figure 6.17   Graduated tax.

The graph of the graduated tax is the result of piecing together two different line segments that represent the two different formulas used to find taxes. The short segment represents taxes for low incomes between $0 and $10,000 and the longer, steeper segment represents taxes for higher incomes that are greater than $10,000.

To find the algebraic expression for the graduated tax, we need to consider that the formula used for finding taxes is different for different levels of income. Functions that use different formulas for different intervals of the domain are said to be *piecewise defined*. Since each income determines a unique tax, we can define the graduated tax T_G as a piecewise function, g, of income i. Graduated taxes $T_G = g(i)$ can be represented by the piecewise function

$$T_G = \begin{cases} 0.10i & \text{for } i \leq \$10,000 \\ 1000 + 0.20(i - 10,000) & \text{for } i > \$10,000 \end{cases}$$

The value of the independent variable i for income determines which formula to use to evaluate the function. This function is called a *piecewise linear function*, since each piece consists of a different linear formula. The formula for incomes equal to or below $10,000 is different from the formula for incomes above $10,000.

To find $g(\$8000)$, the value of T_G when $i = \$8000$, we use the upper formula in the definition since income, i, in this case is less than $10,000.

For $i \leq \$10,000$ $\qquad\qquad\qquad\qquad$ $g(i) = 0.10i$

substituting $8,000 for i $\qquad\qquad$ $g(\$8000) = 0.10(\$8000)$

$\qquad\qquad\qquad\qquad\qquad\qquad\qquad$ $= \$800$

To find $g(\$40,000)$, we use the lower formula in the definition, since in this case income is greater than $10,000.

For $i > \$10,000$ $\qquad\qquad\qquad\qquad$ $g(i) = \$1000 + 0.20(i - \$10,000)$

substituting $40,000 for i $\qquad\quad$ $g(\$40,000) = \$1000 + 0.20(\$40,000 - \$10,000)$

$\qquad\qquad\qquad\qquad\qquad\qquad\qquad$ $= \$1000 + 0.20(\$30,000)$

$\qquad\qquad\qquad\qquad\qquad\qquad\qquad$ $= \$7000$

Comparing the two tax models   In Figure 6.18 we compare the different tax plans by plotting the flat tax and graduated tax equations on the same graph.

The intersection points indicate the incomes at which the amount of tax is the same under both plans. From the graph, we can estimate the coordinates of the two points as (0, 0) and (20000, 3000). That is, under both plans with $0 income you pay $0 taxes and with $20,000 in income you would pay approximately $3000 in taxes.

To find the actual intersection we need to set $T_F = T_G$. We know that $T_F = 0.15i$. Which of the two expressions do we use for T_G? The answer depends upon what value of income, i, we consider. For $i \leq \$10,000$, we have $T_G = 0.10i$.

If $\qquad\qquad\qquad\qquad\qquad\qquad\qquad\qquad$ $T_F = T_G$

and $i \leq \$10,000$, then $\qquad\qquad\qquad$ $0.10i = 0.15i$

this can only happen when $\qquad\qquad\quad$ $i = 0$

If $i = 0$ both T_F and T_G equal 0, therefore one intersection point is indeed (0, 0).

For $i > \$10,000$, we use $T_G = \$1000 + 0.20(i - \$10,000)$ and again set $T_F = T_G$.

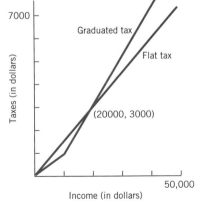

7000

Taxes (in dollars)

Graduated tax

Flat tax

(20000, 3000)

50,000

Income (in dollars)

Figure 6.18   Flat tax versus graduated tax.

If	$T_F = T_G$
then	$0.15i = \$1000 + 0.20(i - \$10,000)$
apply distributive property	$0.15i = \$1000 + 0.20i - (0.20)(\$10,000)$
multiply	$0.15i = \$1000 + 0.20i - \$2000$
combine terms	$0.15i = -\$1000 + 0.20i$
add $-0.20i$ to each side	$-0.05i = -\$1000$
divide by -0.05	$i = -\$1000/(-0.05)$
	$i = \$20,000$

So each plan requires the same tax for an income of $20,000. How much tax is required? We can substitute $20,000 for i into either the flat tax formula or the graduated tax formula for incomes over $10,000 and solve for the tax. Using the flat tax function,

given	$T_F = 0.15i$
if $i = \$20,000$, then	$T_F = (0.15)(\$20,000)$
	$= \$3000$

We can check to make sure that when $i = \$20,000$, the graduated tax, T_G, will also be $3000.

given	$T_G = \$1000 + 0.20(i - \$10,000)$
if $i = \$20,000$, then	$T_G = \$1000 + 0.20(\$20,000 - \$10,000)$
perform operations	$= \$3000$

The other intersection point is, as we estimated, (20000, 3000).

Individual voters want to know what impact these different plans will have on their taxes. From the graph in Figure 6.18, we can see that to the left of the intersection point at (20000, 3000), the flat tax is *greater* than the graduated tax for the same income. To the right of this intersection point, the flat tax is *less* than the graduated tax for the same income. So for incomes *less than* $20,000, taxes are *greater* under the flat tax plan and for incomes *greater than* $20,000, taxes will be *less* under the flat tax plan.

Something to think about

Some people argue that flat taxes are *regressive* or place an unfair burden on those people with low incomes. We have assumed that deductions were the same under both plans. If we drop this assumption, how could we make a flat tax plan less burdensome for people with low incomes?

Algebra Aerobics 6.4a

1. In our simple tax models, the amount of flat tax is a function f of income i where $f(i) = 0.15i$ and the amount of graduated tax is described by the piecewise linear function

$$g(i) = \begin{cases} 0.10i & \text{for } i \leq \$10,000 \\ 1000 + 0.20(i - 10,000) & \text{for } i > \$10,000 \end{cases}$$

   a. Use Figure 6.18 to predict for each of the following pairs which would be larger and by how much.
      i. $f(\$5000)$ $g(\$5000)$
      ii. $f(\$15,000)$ $g(\$15,000)$
      iii. $f(\$40,000)$ $g(\$40,000)$

   b. Now use the equations to evaluate each term in part (a). Calculate the actual differences, and compare to your predictions.

2. Construct a new graduated tax function, if
   a. the tax is 5% on the first $50,000 of income, and 8% on any income in excess of $50,000.
   b. the tax is 6% on the first $30,000 of income and 9% on any income in excess of $30,000.

The Case of Massachusetts

The state of Massachusetts currently has a flat tax of 5.95% on earned income, but there has been an ongoing debate about whether or not to change to a graduated income tax. We can apply the same sort of analysis used in the previous model to decide who would benefit under the proposed new tax.

The Massachusetts flat tax, T_F, can be represented by a function, f, in which the independent variable, i, represents earned income after deductions and exemptions. Then $T_F = f(i)$

where $$T_F = 0.0595i$$

In 1994 Massachusetts voters considered a proposal called Proposition 7. Proposition 7 would have replaced the flat tax rate with graduated income tax rates (called marginal rates) as shown in Table 6.5.

Table 6.5
Massachusetts Graduated Income Tax Proposal

Filing Status	Marginal Rates		
	5.5%	8.8%	9.8%
Married/Joint	<$81,000	$81,000–$150,000	$150,000+
Married/Separate	<$40,500	$40,500– $75,000	$75,000+
Single	<$50,200	$50,200– $90,000	$90,000+
Head of Household	<$60,100	$60,100–$120,000	$120,000+

Source: The Office of the Secretary of State, Michael J. Connolly, Boston, Massachusetts.

In Exploration 6.1 you can analyze the impact of Proposition 7 on people using different filing statuses.

Questions   For what income and filing status would the taxes be equal under both plans? Who will pay less tax and who will pay more under the graduated income tax plan? We analyze the tax for a single person and leave the analyses of the other filing categories for you to do in the Exploration.

Discussion   The proposed graduated income tax is designed to tax at higher rates only that portion of the individual's income which exceeds a certain threshold. For example, for those who file as single people, the graduated tax rate means that earned income under $50,200 would be taxed at a rate of 5.5%. Any income between $50,200 and $90,000 would be taxed at 8.8%, and any income over $90,000 would be taxed at 9.8%.

The taxes for a single person earning $100,000 would be the sum of three different dollar amounts:[4]

$$5.5\% \text{ on the first } \$50,200 = (0.055)(\$50,200)$$
$$= \$2761$$

$$8.8\% \text{ on the next } \$39,800 \text{ (the portion of income between } \$50,200 \text{ and } \$90,000)$$
$$= (0.088)(\$39,800)$$
$$\approx \$3502$$

$$9.8\% \text{ on the remaining } \$10,000 = (0.098)(\$10,000)$$
$$= \$980$$

---

[4] Technically, we should be calculating 5.5% on the first $50,199, but 5.5% of $50,199 rounds to $2761.

The total graduated tax would be $2761 + $3502 + $980 = $7243. Under the flat tax rate the same individual pays 5.95% of $100,000 = 0.0595($100,000) or $5950.

Table 6.6 shows the differences between the flat tax and the proposed graduated tax plan for single people at several different income levels.

Table 6.6		
Massachusetts Taxes		
Flat Rate versus Graduated Rate for Single People		
Income After Exemptions and Deductions ($)	**Current Mass. Flat Tax at 5.95% ($)**	**Graduated Tax Under Prop. 7 ($)**
0	0	0
25,000	1488	1375
50,000	2975	2750
75,000	4463	4943
100,000	5950	7243

We can write T_F, flat taxes for single people, as a linear function, f, of income, i.

$$T_F = f(i)$$

where

$$T_F = 0.0595i$$

We can write T_G, graduated taxes for single people, as a piecewise linear function, g, of income i.

$$T_G = g(i)$$

where

$$T_G = \begin{cases} 0.055i & \text{for } 0 \leq i < \$50,200 \\ \$2761 + 0.088(i - \$50,200) & \text{for } \$50,200 \leq i \leq \$90,000 \\ \$6263 + 0.098(i - \$90,000) & \text{for } i > \$90,000 \end{cases}$$

Note that $2761 in the second formula of the definition is the tax on the first $50,200 of income (5.5% of $50,200), and $6263 in the third formula of the definition is the sum of the taxes on the first $50,200 and the next $39,800 of income (5.5% of $50,200) + (8.8% of $39,800). The flat tax and the graduated income tax for single people are compared in Figure 6.19.

For what income would single people pay the same tax under both plans? An intersection point on the graph indicates when taxes are equal. One intersection point occurs at (0, 0). That makes sense since under either plan if you have zero income, you pay zero taxes. The second intersection point is at approximately (58000, 3500). That means for an income of approximately $58,000, the taxes are the same and are approximately $3500. We can use our equations to find an exact value for the intersection.

To find the intersection point we set $T_F = T_G$. From the graph in Figure 6.19, we estimated that this happens when income is approximately $58,000 (i.e., at an income, i, such that $50,200 < i \leq $90,000). Thus, we set the following functions equal:

flat tax $\qquad\qquad\qquad\qquad\qquad T_F = 0.0595i$

graduated tax for $50,200 < i < $90,000 $\qquad T_G = \$2761 + 0.088(i - \$50,200)$

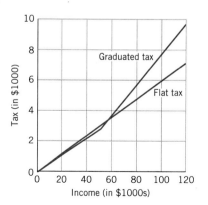

Figure 6.19 Massachusetts graduated tax versus flat tax.

Set $T_F = T_G$ $0.0595i = \$2761 + 0.088(i - \$50,200)$

$0.0595i = \$2761 + 0.088i - \$4418$

$0.0595i = -\$1657 + 0.088i$

$0.0285i = \$1657$

$i \approx \$58,140$

Note that $\$50,200 < \$58,140 \leq \$90,000$, so the value for i falls within the constraints set on i.

Having found a value for i, we now need to find the corresponding value for taxes when $i = \$58,140$. We can use either of the original functions and substitute the value found for i.

use flat tax $T_F = 0.0595i$

let $i = \$58,140$ $T_F = (0.0595)(\$58,140)$

$= \$3460$

We can check to see whether the coordinates (58140, 3460) also satisfy the function T_G.

use graduated tax for
$\$50,200 < i \leq \$90,000$ $T_G = \$2761 + 0.088(i - \$50,200)$

let $i = \$58,140$ $T_G = \$2761 + 0.088(\$58,140 - \$50,200)$

$= \$2761 + 0.088(\$7940)$

$\approx \$2761 + \$699$

$\approx \$3460$

The coordinates (58140, 3460) satisfy both equations.

Single people in Massachusetts with incomes below about \$58,140 would pay less in taxes under a graduated tax plan whereas those with incomes above \$58,140 would pay less in taxes under the flat tax plan.

Algebra Aerobics 6.4b

1. The function $f(i)$ represents the current state tax in Massachusetts and the function $g(i)$ represents the proposed graduated tax, in which the independent variable, i, represents earned income after deductions and exemptions.

$f(i) = 0.0595i$

$$g(i) = \begin{cases} 0.055i & \text{for } 0 \leq i < \$50,200 \\ \$2761 + 0.088(i - \$50,200) & \text{for } \$50,200 \leq i \leq \$90,000 \\ \$6263 + 0.098(i - \$90,000) & \text{for } i > \$90,000 \end{cases}$$

Use the graph in Figure 6.19 to predict for each of the following pairs which would be the larger and by approximately how much. Then evaluate each term and compare your estimates to the actual differences.

a. $f(\$30,000)$ $g(\$30,000)$
b. $f(\$60,000)$ $g(\$60,000)$
c. $f(\$120,000)$ $g(\$120,000)$

CHAPTER SUMMARY

A collection of linear equations that relates the same variables is called a *system of linear equations* or a set of *simultaneous linear equations*. A pair of real numbers is a *solution* to a system of linear equations in two variables if and only if the pair of numbers is a solution to each equation.

On the graph of a system of linear equations, a solution appears as an intersection point. In a system of two linear equations, if the lines intersect once, there is one solution. If the lines are parallel, they never intersect, and there are no solutions. If the two equations are equivalent, they represent the same line and there are an infinite number of solutions.

One solution

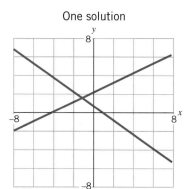

Lines intersect at a single point.

No solutions

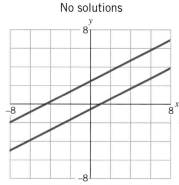

Parallel lines never intersect.

Infinitely many solutions

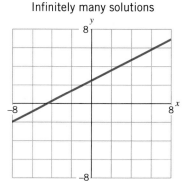

The two equations represent the same line. Every point on the line solves both equations.

There are three basic methods for finding precise solutions to systems of two linear equations in two unknowns:

1.   Setting the expressions for the dependent variable equal to each other
2.   Substitution
3.   Elimination

Functions that use different formulas for different intervals of the domain are said to be *piecewise defined*. Some functions, like the graduated income tax, can be constructed out of pieces of several different linear functions. They are called *piecewise linear* functions.

EXERCISES

1. Predict the *number* of solutions to each of the following systems. Give reasons for your answer. You don't need to find any actual solutions.

   a. $y = 20{,}000 + 700x$
           $y = 15{,}000 + 800x$

   b. $y = 20{,}000 + 700x$
           $y = 15{,}000 + 700x$

   c. $y = 20{,}000 + 700x$
           $y = 20{,}000 + 800x$

2. Calculate the solution(s), if any, to each of the following systems of equations. Use any method you like. Then graph each system and indicate on the graph any solutions.

   a. $y = -11 - 2x$
           $y = 13 - 2x$

   b. $t = -3 + 4w$
           $-12w + 3t + 9 = 0$

   c. $y = 2200x - 700$
           $y = 1300x + 4700$

   d. $3x = 5y$
           $4y - 3x = -3$

   In some of the following examples you may wish to round off your answers.

   e. $y = 2200x - 1800$
           $y = 1300x + 4700$

   f. $y = 4.2 - 1.62x$
           $1.48x - 2y + 4.36 = 0$

   g. $4r + 5s = 10$
           $2r - 4s = -3$

   h. $2x + 3y = 13$
           $3x + 5y = 21$

   i. $xy = 1$
           $x^2y + 3x = 2$
           (A nonlinear system! Hint: solve $xy = 1$ for y and use substitution.)

3. In the text the following cost equations were given for gas and solar heating,

   $$C_{gas} = 12{,}000 + 700n$$
   $$C_{solar} = 30{,}000 + 150n$$

   where n represents the number of years since installation and the cost represents the total accumulated costs up to and including year n.

   a. Sketch the graph of this system of equations.

   b. What do the coefficients 700 and 150 represent on the graph, and what do they represent in terms of heating costs?

   c. What do the constant terms 12,000 and 30,000 represent on the graph? What does the difference between 12,000 and 30,000 say about the costs of gas versus solar heating?

   d. Label the point on the graph where gas and solar heating costs are equal. Make a visual estimate of the coordinates, and interpret what the coordinates mean in terms of heating costs.

   e. Use the equations to find a better estimate for the intersection point. To simplify the computations, you may want to round values to whole numbers. Show your work.

   f. When is the total cost of solar heating more expensive? When is gas heating more expensive?

4. Answer the questions in Exercise 3 (with suitable changes in wording) for the cost equations for electric and solar heating.

   $$C_{electric} = 5000 + 1100n$$
   $$C_{solar} = 30{,}000 + 150n$$

5. Assume you have $2000 to invest for 1 year. You can make a safe investment that yields 4% interest a year or a risky investment that yields 8% a year. If you want to combine safe and risky investments to make $100 a year, how much of the $2000 should you invest at the 4% interest? How much at the 8% interest?

(Hint: Set up a system of two equations in two variables, where one equation represents the total amount of money you have to invest and the other equation represents the total amount of money you want to make on your investments.)

6. If $y = b + mx$, solve for values for m and b by constructing two linear equations in m and b for part (a) and two linear equations in m and b for part (b).

   a. When $x = 2$, $y = -2$ and when $x = -3$, $y = 13$

   b. When $x = 10$, $y = 39.5$ and when $x = 1.5$, $y = -3.5$

7. The following are formulas predicting future raises for four different groups of union employees. N represents the number of years from the date of contract. Each salary represents the salary that will be earned during the given year.

group A:	salary $= 30{,}000 + 1500\,N$
group B:	salary $= 30{,}000 + 1800\,N$
group C:	salary $= 27{,}000 + 1500\,N$
group D:	salary $= 21{,}000 + 2100\,N$

   a. Will group A ever earn more per year than group B? Explain.

   b. Will group C ever catch up to group A? Explain.

   c. Which group will be making the highest yearly salary in 5 years? How much will that be?

   d. Will group D ever catch up to group C? If so, after how many years and at what salary?

   e. How much total salary would an individual in each group have earned 3 years after the contract?

8. In 1997 AT&T offered two long-distance calling plans. The "One Rate" plan charged a flat rate of $0.15 per minute. The "One Rate Plus" plan charged a service fee of $4.95 a month plus $0.10 per minute.

   a. Consider the monthly telephone cost as a function of the number of minutes of phone use that month. Construct a cost function (covering a one-month period) for each calling plan.

   b. Sketch both functions on the same grid.

   c. Use your graph to estimate when the costs will be the same under both plans.

   d. Now calculate when the costs will be equal, label the corresponding point on the graph, and interpret your answer.

9. Solve the following system of three equations in three variables:

   $$2x + 3y - z = 11 \qquad (1)$$
   $$5x - 2y + 3z = 35 \qquad (2)$$
   $$x - 5y + 4z = 18 \qquad (3)$$

   We can apply the same basic strategy of elimination as we did in solving two equations in two variables. We systematically eliminate variables until we end up with one equation in one variable.

a. Use Equations (1) and (2) to eliminate one variable, creating a new equation (4) in two variables.

b. Use Equations (1) and (3) to eliminate the same variable as in part (a). You should end up with a new equation (5) that has the same variables as Equation (4).

c. Equations (4) and (5) represent a system of two equations in two variables. Solve the system.

d. Find the corresponding value for the variable eliminated in part (a).

e. Check your work by making sure your solution works in all three original equations.

10. a. Using the strategy described in Exercise 9, solve the following system:

$$2a - 3b + c = 4.5 \tag{1}$$
$$a - 2b + 2c = 0 \tag{2}$$
$$3a - b + 2c = 0.5 \tag{3}$$

b. How many distinct equations would you need in order to solve a system with five variables? With n variables?

11. Assume that you are given the graph of a standard model for supply and demand as shown in Section 6.3 of the text.

a. What happens if the overall demand decreases, that is, at each price level consumers purchase a smaller quantity? Sketch a graph showing the supply curve and both the original and the changed demand curve. Label the graph. How has the demand curve changed? Describe the shift in the equilibrium point in terms of price and quantity.

b. What happens if the supply decreases, that is, at each price level manufacturers produce lower quantities? Sketch a graph showing the demand curve and both the original and the changed supply curve. Label the graph. How has the supply curve changed? Describe the shift in the equilibrium point in terms of price and quantity.

12. When studying populations (human or otherwise), the two primary factors affecting population size are the birth rate and the death rate. There is abundant evidence that, other things being equal, as the population density increases, the birth rate tends to decrease and the death rate tends to increase. (See Edward O. Wilson and William H. Bossert's *A Primer of Population Biology*, Sunderland, Mass.: Sinauer Associates, Inc., 1971, p. 104.)

a. Generate a rough sketch showing birth rate as a function of population density. Note the units for population density on the horizontal axis are the number of individuals for a given area. The units on the vertical axis represent a rate, such as the number of individuals per 1,000. Now add to your graph a rough sketch of the relationship between death rate and population density. In both cases you may assume the relationship is linear.

b. At the intersection point of the two lines the growth of the population is zero. Why? *Note:* We are ignoring all other factors such as immigration.

The intersection point is called the *equilibrium point*. At this point the population is said to have stabilized, and the size of the population that corresponds to this point is called the *equilibrium number*.

c. What happens to the equilibrium point if the overall death rate decreases, that is at each value for population density the death rate is lower? Sketch a graph showing the birth rate and both the original and the changed death rate. (Label the graph carefully.) Describe the shift in the equilibrium point.

d. What happens to the equilibrium point if the overall death rate increases? Analyze as in part (c).

13. Use the information in Exercise 12 to answer the following questions:

a. What if the overall birth rate increases? (That is, if at each population density level the birth rate is higher.) Sketch a graph showing the death rate and both the original and the changed birth rate. Be sure to label the graph carefully. Describe the shift in the equilibrium point.

b. What happens if the overall birth rate decreases? Analyze as in part (a).

14. The accompanying table, taken from a pediatric text, provides a set of formulas for the approximate "average" height and weight of normal infants and children.

Age	Weight in Kilograms	Weight in Pounds
At birth	3.25	7
3–12 months	(age in months + 9)/2	(age in months) + 11
1–5 years	2 • (age in years) + 8	5 • (age in years) + 17
6–12 years	3.5 • (age in years) − 2.5	7 • (age in years) + 5
Age	Height in Centimeters	Height in Inches
At birth	50	20
At 1 year	75	30
2–12 years	6 • (age in years) + 77	2.5 • (age in years) + 30

Source: R.E. Behrman and V.C. Vaughan (eds.) *Nelson Textbook of Pediatrics,* 12th edition (Philadelphia: W.B. Saunders, 1983) p. 19.

For children from birth to 12 years of age, construct and graph a piecewise linear function for each of the following (assuming age is always the independent variable):

a. weight in kilograms

b. weight in pounds (How does this model compare to the model for female infants in Section 4.1?)

c. height in centimeters

d. height in inches

Note that at 1 year or 12 months, you have two formulas for weight to choose from in the table and they may give you conflicting values. In this case, you need to decide which formula you wish to use for weight at 1 year. Be sure that your function notation reflects your choice.

15. Construct a piecewise linear function for each of the following graphs.

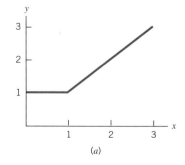

(a)

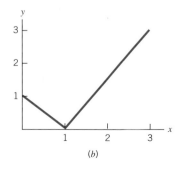

(b)

16. Suppose a flat tax was 10% of income. Suppose a graduated tax was a fixed $1,000 for any income $\leq\$20,000$, plus 20% of any income in excess of $20,000.

a. Construct functions for the flat tax and the graduated tax.

b. Construct a small table of values:

Income	Flat Tax	Graduated Tax
$0		
$10,000		
$20,000		
$30,000		
$40,000		

c. Graph both tax plans and estimate any point(s) at which the two plans would be equal.

d. Use the function definitions to calculate any point(s) at which the two plans would be equal.

e. For what levels of income would the flat tax be more than the graduated tax? For what levels of income would the graduated tax be more than the flat tax?

f. Are there any conditions in which an individual might have negative income under either of the above plans, that is, the amount of taxes would exceed an individual's income? This is not as strange as it sounds. For example, many states impose a minimum corporate tax on a company, even if it is a small one-person operation with no income in that year.

17. Construct a small table of values and graph the following piecewise linear functions. In each case specify the domain.

a. $f(x) = \begin{cases} 5 & \text{for } x < 10 \\ -15 + 2x & \text{for } x \geq 10 \end{cases}$

b. $g(t) = \begin{cases} 1 - t & \text{for } -10 \leq t \leq 1 \\ t & \text{for } 1 < t < 10 \end{cases}$

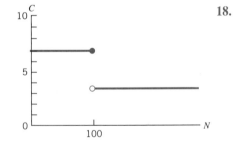

18. The pieces of a piecewise linear function may not always be "connected." The following is a graph of a *step function*, called this since its graph looks like a series of steps.

Note: when there might be confusion about whether an endpoint is included in a line, we draw a solid dot ● to indicate when the point is included, and a hollow dot ○ to indicate when it is not.

a. What does C equal when $N < 100$? When $N = 100$? When $N > 100$?

b. Write a piecewise linear function that represents this graph.

c. If we let $N =$ number of xerox copies of 1 page, and $C =$ cost per copy, then this function tells you what a national copy chain charges per copy. Describe its pricing policy.

d. If the page you are copying is part of a group of pages that are to be copied and collated (bound together), the price structure changes. The cost per copy for a page that will be collated is 7¢ per copy for the first 5,000 copies, and 3.5¢ per copy thereafter. Draw a new graph and construct a new step function to reflect this.

Challenge questions:

e. Using the information in part (c), construct and graph a piecewise linear function that describes not the individual cost, but the *total* cost, *T*, of making *N* copies of a page.

f. Modify the function you constructed in part (e), to reflect the total cost of making *N* copies of a page that is part of a group that will be collated.

19. Find out what kind of income tax (if any) your state has. Is it flat or graduated? Construct and graph a function that describes your state's income tax for one filing status. Identify the income tax for various income levels.

20. **a.** Mississippi state tax rates for single people are shown in the following table. Construct and graph a piecewise linear function for Mississippi's graduated income tax for a single person.

Mississippi Tax Rates for a Single Person	
Income after Deductions ($)	Marginal Tax Rate (%)
≤5000	3.00
5001 to 10,000	4.00
>10,000	5.00

b. Compare Missouri's graduated income tax for a single person as shown in the accompanying table with that of Mississippi shown in part (a). For what net income levels would a single person pay less tax in Mississippi? For what net income levels would a single person pay less tax in Missouri? (Net income refers to income after deductions.)

Missouri State Tax for a Single Person	
Income after Deductions ($)	Marginal Tax Rate (%)
≤1000	1.50
1001–2000	2.00
2001–3000	2.50
3001–4000	3.00
4001–5000	3.50
5001–6000	4.00
6001–7000	4.50
7001–8000	5.00
8001–9000	5.50
>9000	6.00

c. What other financial considerations should be taken into account in comparing taxes in Mississippi and Missouri?

21. A graduated income tax is proposed in Borduria to replace an existing flat rate of 8% on all income. The new proposal is that persons will pay no tax on their first $20,000 of income, 5% on income over $20,000 and less than or equal to $100,000, and 10% on their income over $100,000.

(Note: Borduria is a fictional totalitarian state in the Balkans that figures in the adventures of Tintin.)

a. Construct a table of values that shows how much tax persons will pay under both the existing 8% flat tax and the proposed new tax for each of the following incomes: $0; $20,000; $50,000; $100,000; $150,000; $200,000.

b. Construct a graph of tax dollars versus income for the 8% flat tax.

c. On the same graph plot tax dollars versus income for the proposed new graduated tax.

d. Construct a function that describes tax dollars under the existing 8% tax as a function of income.

e. Construct a piecewise function that describes tax dollars under the proposed new graduated tax rates as a function of income.

f. Use your graph to estimate income level for which the taxes are the same under both plans. What plan would benefit people with incomes below your estimate? What plan would benefit people with incomes above your estimate?

g. Use your equations to find the coordinates that represent the point at which the taxes are the same for both plans. Label this point on your graph.

h. If the median income in the state is $27,000 and the mean income is $35,000, do you think the new graduated tax would be voted in by the people? Explain your answer.

See "How a Flat Tax Would Work, for You and for Them" for a description of proposed flat tax plans.

22. Steve Forbes, a candidate for President of the United States in the 1996 election, advocated a flat tax plan. A variety of plans were debated in this election. One plan recommended a flat tax rate of 19% on income after deductions. The table below gives the 1996 federal tax rates on income after deductions for single people.

1996 Tax Rate Schedules

Schedule X—Use if your filing status is **Single**

If the amount on Form 1040, line 37, is: Over—	But not over—	Enter on Form 1040, line 38	of the amount over—
$0	$24,000	- - - - - - 15%	$0
24,000	58,150	**$3,600.00 +** 28%	24,000
58,150	121,300	**13,162.00 +** 31%	58,150
121,300	263,750	**32,738.50 +** 36%	121,300
263,750	- - - - - -	**84,020.50 +** 39.6%	263,750

a. Construct a function representing flat taxes of 19% where income, i, represents income after deductions.

b. Construct a piecewise linear function representing the graduated federal tax for single people in 1996, where income, i, represents income after deductions.

c. Graph your two functions on the same grid. Estimate from your graph any intersection points for the two functions.

d. Use your equations to calculate more accurate values for the points of intersection.

e. Interpret your findings. Assume deductions are treated the same under both the flat tax plan and the graduated tax plan. What do the intersection points tell you about the differences between the tax plans?

23. Heart health is a prime concern, because heart disease is the leading cause of death in the United States. Aerobic activities such as walking, jogging, and running are recommended for cardiovascular fitness, because they increase the heart's strength and stamina.

a. A typical training recommendation for a beginner is to walk at a moderate pace of about 3.5 miles per hour (or approximately 0.0583 miles per minute) for 20 minutes. Construct a function that describes the distance traveled *in miles, $D_{beginner}$*, as a function of time, T, *in minutes*, for someone maintaining this pace. Construct a small table of solutions and graph the function, using a reasonable domain.

b. A more advanced training routine is to walk at a pace of 3.75 miles per hour (or 0.0625 miles per minute) for 10 minutes, and then jog at 5.25 miles per hour (or 0.0875 miles per minute) for 10 minutes. Construct a piecewise linear function that gives the total distance, $D_{advanced}$, as a function of time, T, *in minutes*. Generate a small table of solutions and plot the graph of this function on your graph in part (a).

c. Do these two graphs intersect? If so, what do the intersection point(s) represent?

EXPLORATION 6.1

Flat versus Graduated Income Tax: Who Benefits?

Objectives

- compare the effect of different tax plans on individuals in different income brackets
- interpret intersection points

Materials/Equipment

- spreadsheet or graphing calculator (optional). If using a graphing calculator, see examples on graphing piecewise functions in the Graphing Calculator Workbook.
- graph paper
- enclosed background reading and worksheet on tax rate plans for Massachusetts

Procedure

If you need practice using the graphing calculator to graph systems of equations and find intersection points, see the Graphing Calculator Workbook.

The questions we explore are:

For what income will taxes be equal under both plans?

Who will benefit under the graduated income tax plan as compared to the flat tax plan?

Who will pay more taxes under the graduated plan as compared to the flat tax plan?

In a small group or with a partner

1. Review the accompanying background information extracted from Section 6.4, *The Case of Massachusetts,* regarding the flat versus graduated income tax plans in Massachusetts. Choose a category other than single people to study.

2. Using the attached worksheet, generate a data table for taxes under the flat rate plan and the graduated plan for the group you have chosen to study.

3. Model each of these plans using taxes as a function of income. Note that the graduated income tax function will be a piecewise linear function.

4. Graph the functions representing the two tax plans on the same grid. You can do this by hand and check your results using a spreadsheet or graphing calculator. Verify that the points in the data table lie on the graph.

5. Estimate from your graph any intersection points for the two functions. Use your equations to calculate exact values for the points of intersection.

6. (Extra credit) Use your results to make changes in the tax plans. Decide on a different income for which taxes will be equal under both plans. You can use what you know

about the distribution of income in the United States from the FAM 1000 data to make your decision. Alter one or both of the original functions such that both tax plans will generate the same tax given the income you have chosen.

Analysis

- Interpret your findings. What do the intersection points tell you about the differences between the tax plans?
- What information is useful in deciding on the merits of each of the plans?

Exploration-Linked Homework

Reporting your results

Take a stance for or against a graduated income tax in Massachusetts. Using supportive quantitative evidence, write a 60 second summary for a voters' pamphlet advocating your position. Present your arguments to the class.

Background Information (from Section 6.4 in the text)

Flat Rate versus Graduated Rate: *The Case of Massachusetts*

The state of Massachusetts currently has a flat tax rate of 5.95%. There has been an ongoing debate as to whether or not to change to a graduated income tax plan.

In Massachusetts a graduated income tax proposal, called Proposition 7, was on the ballot in 1994. Proposition 7 proposed that the state's flat tax on earned income be replaced with a graduated income tax as follows:

Graduated Tax Plan for Various Income Levels			
Filing Status	**5.5%**	**8.8%**	**9.8%**
Married/joint	<$81,000	$81,000–150,000	$150,000+
Married/separate	<$40,500	$40,500– 75,000	$75,000+
Single	<$50,200	$50,200– 90,000	$90,000+
Head of household	<$60,100	$60,100–120,000	$120,000+

This tax rate structure is designed to tax at higher rates only that portion of the individual's income that exceeds a threshold income. For example, for those who file as single people, the graduated tax rate means that earned income under $50,200 would be taxed at a rate of 5.5% whereas any income over $50,200 would be taxed at higher rates. If a single person earned $100,000, then he or she would be taxed at 5.5% on the first $50,200, 8.8% on the next $39,800 (the portion of income between $50,200 and $90,000), and 9.8% on the final $10,000.

Worksheet for Exploration 6.1. Flat Versus Graduated Income Tax: Who Benefits?

Selected filing status: _____

<table>
<tr><td colspan="3" align="center">Data Table for Massachusetts Taxes
Flat versus Graduated Rate</td></tr>
<tr><th>Income after
Exemptions and
Deductions ($)</th><th>Current Flat Tax ($)</th><th>Graduated Tax
under Proposition 7 ($)</th></tr>
<tr><td>0</td><td></td><td></td></tr>
<tr><td>25,000</td><td></td><td></td></tr>
<tr><td>50,000</td><td></td><td></td></tr>
<tr><td>75,000</td><td></td><td></td></tr>
<tr><td>100,000</td><td></td><td></td></tr>
<tr><td>150,000</td><td></td><td></td></tr>
<tr><td>200,000</td><td></td><td></td></tr>
</table>

Function for current flat tax plan

$$T_F =$$

Function for Proposition 7 tax plan

$$T_G = \left\{ \right.$$

Include a graph of the two functions.

Estimation of intersection point(s) from graph _____

Use your equations to find the intersection point(s). Show your work.

PART

Algebra in the Physical and Life Sciences

OVERVIEW

Part II introduces additional algebraic tools used by scientists to model physical phenomena. Here we examine several families of functions—exponential, power, polynomial, and logarithmic—and show their many uses in the physical and life sciences and in other disciplines.

Astronomy must deal with numbers from the incredibly large to the extremely small. It provides a meaningful context for the development of a number notation and graphing system that can help us describe the amazing range of sizes found in the universe. Scientific notation is the international method for compactly expressing numbers of any size. Logarithmic scales provide a convenient way to display numbers of widely differing sizes.

Biology provides the setting for a discussion of modeling growth and decay. We analyze the exponential growth of bacteria and see how the same model can be applied in nuclear physics, music, epidemiology, and finance.

Power functions help explain the relationship in animals between size and shape, the properties of gases, and why scuba divers don't hold their breath.

We turn next to physics for an understanding of how scientists can uncover the "rules of the universe," such as the basic physical laws that govern gravity. We study the motion of freely falling bodies and construct quadratic models to describe their behavior.

Logarithmic functions allow us to solve exponential equations. They are used to describe the brightness of stars and the intensity of sounds. By plotting data on graphs with logarithmic scales on one or both axes, we learn simple strategies for recognizing when an exponential or a power function is an appropriate mathematical model.

Deep Time and Deep Space

Overview

We inhabit a minuscule portion of a vast universe that has been evolving for billions of years. Understanding where we fit in the history of the known universe and where we are relative to other planets and galaxies has been traditionally the province of philosophers and religious leaders. Scientists now tackle these questions, and we have come to some understanding of the age and size of our universe.

We now know that the scale of the universe is truly "astronomical." *Scientific notation* allows us easily to specify and compare sizes from the subatomic to the cosmic. Scientific notation, like our number system in general, is based on *powers of 10*. The laws of exponents govern the manipulation of expressions written in base 10 and other bases. *Logarithms base 10* are used to scale graphs and to construct measurement systems that deal with numbers of widely varying size.

The first Exploration helps you develop an understanding of the relative size and age of various objects in the universe, while offering practice in the laws of exponents and unit conversions. The second Exploration is based on Kepler's discoveries about the laws governing planetary motion.

After reading this chapter you should be able to:

- write expressions in scientific notation
- convert between English and metric units
- perform operations on expressions of the form a^n and $a^{m/n}$
- compare numbers of widely differing sizes
- calculate logarithms base 10 and plot numbers on a logarithmic scale

7.1 POWERS OF TEN

Measuring Time and Space

On a daily basis we encounter quantities measured in tenths, tens, hundreds, or perhaps thousands. Finance or politics may bring us news of "1.2 billion people living in China" or "a federal debt of $6 trillion." In the physical sciences the range of numbers encountered is much larger. *Scientific notation* was developed not only to write but to help understand and compare the sizes found in our universe, from the largest object we know, the observable universe, to the tiniest, the minuscule quarks oscillating inside the nucleus of an atom. We use examples from deep space and deep time to demonstrate the power of scientific notation.

The Big Bang

The poem "Imagine" offers a creative look at the Big Bang.

In 1929 the American astronomer Edwin Hubble published an astounding paper that established the basic framework for measuring time and space in our universe. He claimed that the universe was expanding. From his observations of galaxies Hubble predicted that, depending on its total mass, the universe would either expand forever or collapse back upon itself.[1] Since then most astronomers and cosmologists have become convinced that Hubble's controversial theory is correct, and they have started to investigate its consequences. If the universe is expanding, they asked, when did it start to expand, and how big was it at that time? The commonly accepted answer to this question is that somewhere between 8 and 15 billion years ago the universe began an explosive expansion from an infinitesimally small point. This event is referred to as the "Big Bang" and the universe has been expanding ever since. If this is correct, then everything that we have ever seen or measured has occurred within the last 15 billion or so years.[2]

Something to think about

Scientists don't know what existed before the Big Bang, or what will happen to the universe if it does eventually collapse back upon itself. One hypothesis is that our universe oscillates—endlessly cycling through a Big Bang, an outward expansion, followed by an eventual collapse back into an infinitely small point, only to be reborn with another Big Bang. Scientists also speculate about what may exist outside our universe. Some believe there are multiple other universes, which we currently cannot detect. What do you think?

Deep Space

Measuring the universe: the metric system   The international scientific community and most of the rest of the world uses the *metric system*, a system of measurements based on the meter (which is about 39.37 inches, a little over 3 feet). In daily life Americans have resisted converting to the metric system and still use the *English system* of inches, feet, and yards. So in order to understand scientific measurements we need to know how to convert between metric and English units.[3] Table 7.1

---

[1] Cosmologists are unable to estimate the total mass of the universe, since they are in the embarrassing position of not being able to find about 90% of it. Scientists call this missing mass *dark matter*, which describes not only its invisibility but scientists' own mystification. Some recent experiments at Los Alamos National Laboratory in New Mexico have produced evidence that some particles called *neutrinos*, previously thought to be weightless, may actually have mass and may constitute a major component of dark matter.

[2] The actual age of the universe is still a hotly debated topic. Observations made by the repaired Hubble space telescope give estimates for the age of the universe somewhere between 8 and 12 billion years. Yet cosmologists are fairly certain that the oldest stars in the Milky Way are at least 14 billion years old, a troubling contradiction to resolve. Keep reading the newspaper for the latest figures.

[3] Eventual conversion to a metric system seems inevitable. There is once again a renewed push for a "metric America."

shows the conversions for three standard units of length: the meter, the kilometer, and the centimeter. Most of the measurements in the examples in this chapter and throughout Part II will be in metric units.

For a real appreciation of the size of things in the universe, we highly recommend the video and related book by Philip and Phylis Morrison entitled Powers of Ten: About the Relative Size of Things. *A room in the Smithsonian Museum in Washington, D.C., is dedicated just to running this remarkable video.*

Table 7.1
Conversions from Metric to English for Some Standard Units

Metric Unit	Abbreviation	In Meters	Equivalence in English Units	Informal Conversion
meter	m	1 meter	3.28 feet	The width of a twin bed, a little more than a yard
kilometer	km	1000 meters	0.62 mile	A casual 12-minute walk, a little over a half a mile
centimeter	cm	0.01 meter	0.39 inch	The width of a pencil, a little under a half an inch

The observable universe   Current measurements with the most advanced scientific instruments generate a best guess for the radius of the observable universe at about 100,000,000,000,000,000,000,000,000 meters, or "one hundred trillion trillion meters." Obviously we need a more convenient way to read, write, and say this number.

100,000,000,000,000,000,000,000,000 is a 1 with 26 zeros after it. You could also think of it as 26 tens multiplied together. In order to avoid writing a large number of zeros, exponents can be used as a shorthand.

- 10^{26} means: a 1 with 26 zeros after it
- 10^{26} means: $10 \cdot 10 \cdot 10 \cdot \ldots \cdot 10$, the product of twenty-six 10s.
- 10^{26} is read as "ten to the twenty-sixth" or "ten to the twenty-sixth power."

So the estimated size of the radius of the observable universe is 10^{26} meters.

> When n is a positive integer, 10^n is defined as the product of n 10s, or equivalently a 1 followed by n zeros.

The Milky Way   Through a telescope we can see some galaxies and clusters of galaxies, most of them elliptical in shape. With the naked eye, we can see about four galaxy clusters, members of what is called the Local Group. Most of the objects we see in the night sky are the luminous balls of gas we call stars that are in our own galaxy, the Milky Way. There are approximately 100 billion (100,000,000,000 or 10^{11}) stars in the Milky Way. Most of these are thought to have planets circling them, though the first confirmation of a planet outside our solar system came only in 1993. The Milky Way has a relatively unusual spiral shape, and Earth is near the edge of one of the long spiral arms. The radius of the Milky Way is approximately 1,000,000,000,000,000,000,000 or 10^{21} meters.

Our solar system   The few other sky objects we can see regularly with the naked eye (outside of the moon, airplanes, and human-manufactured satellites) are other planets in our own solar system. Unlike stars, planets do not glow; like the moon, we see them by the light they reflect from the sun.

The approximate radius of our solar system is 1,000,000,000,000 or 10^{12} meters. In the metric system the prefix *tera-* means 10^{12}, so this distance is one *terameter*.

Our sun   The radius of our own sun, a modest star about half-way through its life cycle, is approximately 1,000,000,000 or 10^9 meters (1 billion meters). In the metric system, *giga-* is the prefix for 10^9, so the distance is one *gigameter*.

Earth   Our home, the third planet orbiting our sun, has a radius of somewhat less than 10,000,000 or 10^7 meters.

Us   Human beings are roughly in the middle of the scale of measurable objects in the universe. Human heights, including children's, vary from about one-third of a meter to two meters. In the wide scale of objects in the universe, a rough estimate for human height is 1 meter.

In order to continue the system of writing all sizes using powers of 10, we need a way to express 1 as a power of 10. Since $10^3 = 1000$, $10^2 = 100$, and $10^1 = 10$, a logical way to continue would be to say that $10^0 = 1$. Since reducing a power of 10 by 1 is equivalent to dividing by 10, the following calculations give justification for defining 10^0 as equal to 1.

$$10^2 = \frac{10^3}{10} = \frac{1000}{10} = 100$$

$$10^1 = \frac{10^2}{10} = \frac{100}{10} = 10$$

$$10^0 = \frac{10^1}{10} = \frac{10}{10} = 1$$

Hence the convention:

> 10^0 is defined to be 1.

Thus a rough estimate for human height is 1 or 10^0 meter.

DNA molecules   A DNA strand provides genetic information for a human being. It is made up of a chain of building blocks called nucleotides. The chain is tightly coiled into a double helix, but stretched out it would measure about 0.01 meter in length. How does this DNA length translate into a power of 10? 0.01 or one hundredth equals $1/10^2$. We define $1/10^2$ to be 10^{-2}. So a DNA strand, uncoiled and measured lengthwise, is approximately 10^{-2} meters or one *centimeter*.

By using negative exponents, we can continue to use powers of 10 to represent numbers less than 1. We want a system in which reducing the power by 1 remains equivalent to dividing by 10. So we define

$$10^2 = 100, \quad 10^1 = 10, \quad 10^0 = 1, \quad 10^{-1} = 0.1, \quad 10^{-2} = 0.01, \quad 10^{-3} = 0.001,$$

and so on, since

$$10^1 = \frac{10^2}{10} = \frac{100}{10} = 10$$

$$10^0 = \frac{10^1}{10} = \frac{10}{10} = 1$$

$$10^{-1} = \frac{10^0}{10} = \frac{1}{10} = 0.1$$

$$10^{-2} = \frac{10^{-1}}{10} = \frac{0.1}{10} = 0.01$$

When n is a positive integer, 10^{-n} is defined to be $1/10^n$, or equivalently, $(n - 1)$ zeros to the right of a decimal point, followed by a 1.

Living cells   The approximate size of the radius of a living cell is

$$0.000\,01 = \frac{1}{100,000} = \frac{1}{10^5} = 10^{-5} \text{ meter.}$$

Note that when we write 10^{-5} in standard decimal notation there are 4 or $5 - 1$ zeros between the decimal point and the 1.

Atoms   Atoms have an approximate average radius of

$$0.000\,000\,000\,1 = \frac{1}{10,000,000,000} = \frac{1}{10^{10}} = 10^{-10} \text{ meter,}$$

a unit commonly called an *Angstrom*.

Hydrogen atoms   The hydrogen atom, the smallest of the atoms, has an approximate radius of

$$0.000\,000\,000\,01 = \frac{1}{100,000,000,000} = \frac{1}{10^{11}} = 10^{-11} \text{ meter.}$$

Protons   A proton, a positively charged particle usually located in the nucleus of the atom, has an approximate radius of

$$0.000\,000\,000\,000\,001 = \frac{1}{1,000,000,000,000,000} = \frac{1}{10^{15}} = 10^{-15} \text{ meter,}$$

or one *femtometer*.

Quarks   Quarks are nature's strange elementary building blocks. Three particular quarks form a proton, three other quarks form a neutron. Different quark combinations make more transitory particles. A quark's size is approximately

$$0.000\,000\,000\,000\,000\,1 = \frac{1}{10,000,000,000,000,000} = \frac{1}{10^{16}} = 10^{-16} \text{ meter.}$$

Summary of Powers of Ten

The following tabulation summarizes the powers of 10 notation.[4]

Positive powers of 10

$$10^9 = 1,000,000,000$$
$$10^6 = 1,000,000$$
$$10^3 = 1,000$$
$$10^2 = 100$$
$$10^1 = 10$$

Special case: zero power of 10 $10^0 = 1$

Negative powers of 10

$$10^{-1} = 0.1$$
$$10^{-2} = 0.01$$
$$10^{-3} = 0.001$$
$$10^{-6} = 0.000\ 001$$
$$10^{-9} = 0.000\ 000\ 001$$

When n is a positive integer:

$10^n = 10 \cdot 10 \cdot 10 \cdot \ \ldots \ \cdot 10$ the product of n 10s or 1 followed by n zeros

$10^{-n} = \dfrac{1}{10^n}$ 1 divided by 10 n times or a decimal point followed by $(n - 1)$ zeros and a 1

$10^0 = 1$

Multiplying by 10 is equivalent to moving the decimal point one place to the right. Dividing by 10 is equivalent to moving the decimal point one place to the left.

Algebra Aerobics 7.1

1. Express as a power of 10.
   a. 10,000,000,000   b. 0.000 000 000 000 01

2. Express in standard notation (without exponents).
   a. 10^{-8} b. 10^{13}

---

[4] The metric prefixes for multiples or subdivisions by powers of 10 are:

atto-	a	10^{-18}	centi	c	10^{-2}	kilo-	k	10^3
femto-	f	10^{-15}	deci-		10^{-1}	mega-	M	10^6
pico-	p	10^{-12}	(unit)		10^0	giga-	G	10^9
nano-	n	10^{-9}	deka-		10^1	tera-	T	10^{12}
micro-	μ	10^{-6}	hecto-		10^2	peta-	P	10^{15}
milli-	m	10^{-3}				exa-	E	10^{18}

Thus millimeter is abbreviated mm; kilometer km; and megameter Mm. Abbreviations are usually not used for deci, deka, and hecto.

3. Express as a power of 10 and then in standard notation.
   a. a nanosecond   **c.** a gigabyte
   b. a decimeter

4. Rewrite each measurement in meters, first using a power of 10 and then standard notation.
   a. 7 cm   **b.** 9 mm   **c.** 5 km

7.2 SCIENTIFIC NOTATION

In the previous examples we estimated the sizes of objects to the nearest power of 10 without worrying about the exact numbers. Getting the power of 10 correct is the first consideration when dealing with physical objects. You have to be in the ballpark before you can find your exact seat. The first step is to know whether something is closer to a thousand or to 10 billion meters in size. Once the power of 10 is known the number can be refined to a greater level of accuracy.

For example, a more accurate measure of the radius of a hydrogen atom is 0.000 000 000 052 9 meter across. This can be written more compactly by using *scientific notation*; that is, we can rewrite it as a number times a power of 10.

In order to convert a number into scientific notation:

- Write down the first nonzero digit and all the digits following it. Put a decimal point right after the first digit. This new number is called the *coefficient*. In this case the coefficient is 5.29.

- Figure out what power of 10 is needed to convert the coefficient back to the original number. In this case the decimal place has to move 11 places to the left to get 0.<u>000 000 000 052</u> 9. This is equivalent to dividing 5.29 by 10 eleven times, or dividing it by 10^{11}.

$$0.000\ 000\ 000\ 052\ 9 = \frac{5.29}{10^{11}}$$
$$= 5.29\left(\frac{1}{10^{11}}\right)$$

$1/10^{11} = 10^{-11}$ so we get:
$$= 5.29 \cdot 10^{-11}$$

This number is now said to be in scientific notation.[5]

When a number is negative, then its coefficient is negative. Thus, the number $-0.000\ 000\ 000\ 052\ 9$ in scientific notation is $-5.29 \cdot 10^{-11}$. The coefficient is -5.29.

The coefficient, N, can be positive or negative. But we insist that the coefficient

---

[5] Most calculators or computers automatically translate a number into scientific notation when it is too large or small to fit into the display. The notation is often slightly modified by using the letter E (short for "exponent") to replace the "times 10 to some power" part, so $3.0 \cdot 10^{26}$ may appear as $3.0\,\text{E}+26$. The number after the E tells how many places and the sign ($+$ or $-$) indicates in which direction to move the decimal point of the coefficient.

stripped of its sign, called its *absolute value* and written $|N|$, be such that $1 \leq |N| < 10$.

Remember when calculating the absolute value of N, if $N \geq 0$, then $|N| = N$. For example: $|2| = 2$, $|11.57| = 11.57$, and $|0| = 0$. But if $N < 0$, then $|N| = -N$. When N is negative, $-N$ is actually positive. For example, $|-2| = -(-2) = 2$ and $|-11.57| = -(-11.57) = 11.57$. The important fact is that no matter what the value of N, the absolute value of N, or $|N|$, will always be greater than or equal to 0.

The expression $|N|$ is read "the absolute value of N."

> If $N \geq 0$, then $|N| = N$.
> If $N < 0$, then $|N| = -N$.

So whether N is positive, negative, or zero, $|N| \geq 0$.

Any number, positive or negative, can be written in scientific notation, that is, written as the product of a coefficient N multiplied by 10 to some power, where $1 \leq |N| < 10$.

A number is in *scientific notation* if it is in the form

$$N \cdot 10^n$$

where $1 \leq |N| < 10$ and n is an integer (positive, negative, or zero).

Thus 2,000,000 or 2 million or $2 \cdot 10^6$ are all correct representations of the same number. The one you choose to use depends on the context.

Deep Time

Powers of 10 and scientific notation can be used to record the progress of the universe through time.

Age of the universe   One current estimate places the age of our universe at about 15,000,000,000 years or about 15 billion years. Using scientific notation we would say the universe is $1.5 \cdot 10^{10}$ years old.

Age of Earth   Earth is believed to have been formed during the last third of the universe's existence—about 4,600,000,000 or 4.6 billion years ago. In scientific notation the age is $4.6 \cdot 10^9$ years.

Age of Pangaea   About 200,000,000 or 200 million years ago, all Earth's continents collided to form one giant land mass now referred to as Pangaea. Pangaea existed $2.0 \cdot 10^8$ years ago.

Age of human life   *Homo sapiens* first walked on Earth about 100,000 or $1.0 \cdot 10^5$ years ago. In the life of the universe, this is almost nothing. If all of time, from the Big Bang to today, were scaled down into a single year, with the Big Bang on January 1, our early human ancestors would not appear until about 10:30 p.m. on December 31, New Year's Eve.

This metaphor is played out in Carl Sagan's video Cosmos *and in his book* The Dragons of Eden, *which is excerpted in the Anthology of Readings.*

Algebra Aerobics 7.2

1. Avogadro's number is $6.02 \cdot 10^{23}$. A mole of any substance is defined to be Avogadro's number of molecules of that substance. Express this number in standard notation.

2. The distance between Earth and its moon is 384,000,000 meters. Express this in scientific notation.

3. An *Angstrom* (denoted by Å), a unit commonly used to measure atoms, is 0.000 000 01 cm. Express its size using scientific notation.

4. The width of a DNA double helix is approximately 2 nanometers or $2 \cdot 10^{-9}$ meters. Express the width in standard notation.

5. Express in standard notation.
    a. $-7.05 \cdot 10^8$ b. $-4.03 \cdot 10^{-5}$

6. Express in scientific notation.
    a. $-43,000,000$ b. $-0.000\ 008\ 3$

7.3 EXPRESSIONS OF THE FORM a^n

Humans have a particular fondness for the number 10, perhaps because we have 10 fingers and 10 toes. Whether we count money or stars, people or planets, we represent the answer in a number built up out of powers of 10. Apart from the fact that we are accustomed to a number system based on 10, there is nothing particularly special about 10. The same rules for writing and manipulating expressions of the form 10^n apply to general expressions of the form a^n, such as 7^4, $(-5.1)^3$, or x^{-2}.

To answer questions about how far away a star is, how fast light travels, or how much bigger the sun is than Earth, we need to be able to manipulate expressions of the form a^n.

> In the expression a^n, where a is any real number, the number a is called the *base* and n is called the *exponent* or *power*.
>
> If a is any nonzero real number and n is a positive integer, then
> $a^n = a \cdot a \cdot a \cdot \ \ldots \ \cdot a$ (the product of n number of a's)
> $a^{-n} = 1/a^n$ (1 divided by a, n times)
> $a^0 = 1$

Something to think about

What can we say about the value of $(-1)^n$ when n is an even integer? An odd integer? For any real number a, what can we say about $(-a)^n$ when n is an even integer? An odd integer? Does it matter if a is positive or negative?

Since light takes time to travel, everything we see is from the past. When you look in the mirror, you see yourself not as you are, but as you were nanoseconds ago. Suppose you look up tonight at the bright star Deneb. Since Deneb is 1,600 light years away, how old is the image you are seeing? If Deneb had gone supernova 1000 years ago, how long would it take for us to realize?

Even more disconcertingly, what we see as simultaneous events do not necessarily occur simultaneously. Consider the two stars Betelgeuse and Rigel in the constellation Orion. Betelgeuse is 300 and Rigel 500 light years away. The images we see simultaneously were actually generated how many years apart?

Multiplication

When multiplying very large or very small numbers, there is an advantage to working in scientific notation. Calculations are simplified and answers are more compact. For example, suppose we want to calculate the distance between Earth and the star Deneb, one of the brightest stars in our sky. The distance that light travels in 1 year, called a *light year*, is approximately 5.88 trillion miles.[6] Deneb is 1600 light years from Earth.

Since   1 light year = 5,880,000,000,000 miles

then the distance from Earth to Deneb is:

$$1600 \text{ light years} = (1600) \cdot (5,880,000,000,000 \text{ miles})$$
$$= (1.6 \cdot 10^3) \cdot (5.88 \cdot 10^{12} \text{ miles})$$
$$= (1.6 \cdot 5.88) \cdot (10^3 \cdot 10^{12}) \text{ miles}$$
$$\approx 9.4 \cdot (10^3 \cdot 10^{12}) \text{ miles}$$
$$\approx 9.4 \cdot 10^{15} \text{ miles}$$

So Deneb is approximately $9.4 \cdot 10^{15}$ or 9.4 quadrillion miles away.

The product $10^3 \cdot 10^{12}$ means

$$(10 \cdot 10 \cdot 10) \cdot (10 \cdot 10 \cdot 10 \cdot 10 \cdot 10 \cdot 10 \cdot 10 \cdot 10 \cdot 10 \cdot 10 \cdot 10 \cdot 10).$$

The number of times 10 is multiplied together is $3 + 12 = 15$

The answer is written more compactly as $10^3 \cdot 10^{12} = 10^{15}$

The exponent 15 is the sum of the exponents 3 and 12. This example illustrates a general rule based on the definition of exponents: To *multiply* two terms with exponents and the *same* base (in this case 10), *add* the exponents. If you forget the rule and need to perform calculations, think about what the exponent does to the base.

The general rule for multiplying two terms with exponents and the same base is:

If a is any nonzero real number and m and n are integers,

$$a^m \cdot a^n = a^{(m+n)}$$

Example   1

The rule works no matter what the base.

$$7^3 \cdot 7^2 = (7 \cdot 7 \cdot 7) \cdot (7 \cdot 7) = 7^{3+2} = 7^5$$
$$w^3 \cdot w^5 = (w \cdot w \cdot w) \cdot (w \cdot w \cdot w \cdot w \cdot w) = w^{3+5} = w^8$$

The rule still holds when one or more power is negative. We can see why by expressing x^{-n} as $1/x^n$. If $x \neq 0$, then

$$x^2 \cdot x^{-5} = x^2 \cdot \frac{1}{x^5} = \frac{x^2}{x^5} = \frac{1}{x^3} = x^{-3}$$

---

[6] Note that a light year is a *distance* measurement, not a *time* measurement.

Note that $-3 = 2 + (-5)$. The answer could have been calculated directly as

$$x^2 \cdot x^{-5} = x^{2+(-5)} = x^{-3}$$

The rule for applying negative powers is the same whether a is an integer or a fraction:

$$a^{-n} = 1/a^n \qquad \textit{where } a \neq 0$$

For example, $\qquad \left(\dfrac{1}{2}\right)^{-1} = \dfrac{1}{(1/2)^1} = 1 \div \left(\dfrac{1}{2}\right) = 1 \cdot \left(\dfrac{2}{1}\right) = 2$

In general, $\qquad \left(\dfrac{a}{b}\right)^{-n} = \dfrac{1}{(a/b)^n} = 1 \div \left(\dfrac{a}{b}\right)^n = 1 \cdot \left(\dfrac{b}{a}\right)^n = \left(\dfrac{b}{a}\right)^n$

Using the rule for multiplying numbers with the same base and the rule for negative powers we have:

$$\left(\dfrac{1}{2}\right)^{-11}\left(\dfrac{1}{2}\right)^{-2} = \left(\dfrac{1}{2}\right)^{-13} = \left(\dfrac{2}{1}\right)^{13} = 2^{13} = 8192 \approx 8.2 \cdot 10^3$$

Common Errors

1. Don't confuse a term such as $-x^4$ with $(-x)^4$. For example, $-2^4 = -(2^4) = -16$. But $(-2)^4 = (-2)(-2)(-2)(-2) = +16$. You have to remember what is being raised to the power. In the expression -2^4, order of operations says to compute the power first, before applying the negation sign. In the expression $(-2)^4$, everything inside the parentheses is raised to the fourth power. So 2 is negated first, and then the result (-2) is raised to the fourth power.

   For example, $(-3a)^2$ means that everything in the parentheses is squared.

   $$(-3a)^2 = (-3a)(-3a) = (-3)(-3)a \cdot a = 9a^2$$

   However in $-3a^2$, the exponent applies only to the base a.

2. It is important to remember that in order to combine the exponents of two multiplied terms, the terms *must have the same base*. For example,

   $$(9300)^2 \cdot (9300)^6 = (9300)^8$$

   But an expression like $81^2 \cdot 47^6$, which means $(81 \cdot 81) \cdot (47 \cdot 47 \cdot 47 \cdot 47 \cdot 47 \cdot 47)$, cannot be simplified by adding exponents since *the bases are not equal*.

3. The sum of terms with the same base, such as $10^2 + 10^3$, *cannot* be simplified by adding exponents. Be careful to avoid this common error. The sum $10^2 + 10^3 \neq 10^5$. We can verify this by calculating the values of both sides.

   $$10^2 + 10^3 = 100 + 1000 \text{ or } 1100$$

   but

   $$10^5 = 100,000$$

   A sum of terms with the *same exponents and bases* can be simplified, although not by adding powers. For example $10^3 + 10^3 = 2 \cdot 10^3$. Why can these terms be combined?

   use the distributive property
   to rewrite $10^3 + 10^3$ $\qquad\qquad 10^3 + 10^3 = 10^3(1 + 1)$

   and then add $\qquad\qquad\qquad\qquad\qquad\qquad = 10^3 \cdot 2$

   and use the commutative property $\qquad\qquad\quad = 2 \cdot 10^3$

Something to think about

Using the distributive property, rewrite the expression $10^3 + 10^4$ by pulling out a common factor.

Examples are:

$7^3 \cdot 6^5$ and $7^3 + 6^5$ cannot be simplified since the bases 7 and 6 are different

$6^5 \cdot 6^5 \cdot 6^5 = 6^{15}$, but $6^5 + 6^5 + 6^5 = 3 \cdot 6^5$

Algebra Aerobics 7.3a

1. Simplify where possible, leaving the answer in a form with exponents.

   a. $10^5 \cdot 10^7$ d. $5^5 \cdot 6^7$

   b. $8^6 \cdot 8^{-4}$ e. $7^3 + 7^3$

   c. $z^{-5} \cdot z^{-4}$ f. $(-5)^2$ and -5^2

2. A typical TV signal, traveling at the speed of light, takes $3.3 \cdot 10^{-6}$ second to travel 1 kilometer. Estimate how long it would take the signal to travel across the United States (a distance of approximately 4300 km).

3. Multiply through and simplify $x^{-2}(x^5 + x^{-6})$.

Division

In comparing two objects of about the same size it is common to subtract one size from the other and say, for instance, that one person is 6 inches taller than another. This method of comparison is not effective for objects that have vastly different sizes. To say that the difference between the estimated radius of our solar system (1 terameter or 1,000,000,000,000 meters) and the average size of a human (about 10^0 or 1 meter) is $1,000,000,000,000 - 1 = 999,999,999,999$ meters is not particularly useful. In fact, since our measurement of the solar system certainly isn't accurate to within 1 meter, this difference is meaningless.

So instead of subtracting one size from another, a more useful method for comparing objects of wildly different size is to calculate the ratio of the two sizes. For example, the radius of the sun is approximately 10^9 meters and the radius of Earth is about 10^7 meters. One way to answer the question "How many *times* larger is the sun than Earth?" is to form the ratio of the two radii:

$$\frac{\text{radius of the sun}}{\text{radius of Earth}} = \frac{10^9 \text{ meters}}{10^7 \text{ meters}}$$

We can use the rule for multiplying terms of the form a^n to simplify this division problem.

$$\frac{10^9 \text{ meters}}{10^7 \text{ meters}} = 10^9 \cdot \left(\frac{1}{10^7}\right)$$

use negative powers $\quad = 10^9 \cdot 10^{-7}$

add exponents $\quad = 10^{9+(-7)}$

simplify $\quad = 10^2$

The units cancel, so 10^2 is unitless. The radius of the sun is approximately 10^2 or 100 times larger than the radius of Earth.

The exponent 2 is the difference of the original exponents, 9 and 7, that is,

$9 - 7 = 2$. To *divide* two exponential terms with the same base (in this case 10), *subtract* the exponent in the denominator from the exponent in the numerator.

If a is any nonzero real number and m and n are integers, $\dfrac{a^m}{a^n} = a^{(m-n)}$.

The following examples illustrate the division rule.

$$10^8/10^3 = 10^{8-3} = 10^5$$
$$10^2/10^6 = 10^{2-6} = 10^{-4}$$
$$6^2/6^{-7} = 6^{2-(-7)} = 6^9$$
$$(-5)^2/(-5)^6 = (-5)^{2-6} = (-5)^{-4}$$
$$z^8/z^3 = z^{(8-3)} = z^5 \qquad (z \neq 0)$$

According to the U.S. Bureau of the Census, in 1995 the estimated gross federal debt was \$4.96 trillion and the estimated U.S. population was 264 million. What was the approximate federal debt *per person*?

Example 2

SOLUTION

$$\frac{\text{federal debt}}{\text{U.S. population}} = \frac{4.96 \cdot 10^{12} \text{ dollars}}{2.64 \cdot 10^8 \text{ people}}$$

$$= \left(\frac{4.96}{2.64}\right) \cdot \left(\frac{10^{12}}{10^8}\right) \frac{\text{dollars}}{\text{people}} \approx 1.88 \cdot 10^4 \frac{\text{dollars}}{\text{people}}$$

So the federal debt amounted to about \$1.88 \cdot 10^4$ or \$18,800 per person.

Algebra Aerobics 7.3b

1. Simplify (if possible), leaving the answer in exponent form.

   a. $\dfrac{10^5}{10^7}$ **c.** $\dfrac{3^{-5}}{3^{-4}}$ **e.** $\dfrac{7^3}{7^3}$

   b. $\dfrac{8^6}{8^{-4}}$ **d.** $\dfrac{5^5}{6^7}$

2. **a.** In 1995 Japan had a population of approximately 125.5 million people and a size of about 152.5 thousand square miles. What was the population density, the number of people per square mile?

   b. In 1995 the United States had a population of ap-

   proximately 263.8 million people and a total land area of about 3620 thousand square miles. What was the population density of the United States?

   c. Compare the population densities of Japan and the United States.

3. A high-density diskette has a user capacity of about 1.44 megabytes ($1.44 \cdot 10^6$ bytes). If a hard drive has a capacity of 2 gigabytes ($2 \cdot 10^9$ bytes), how many diskettes would it take to equal the storage capacity of the hard drive?

Raising a Power to a Power

We previously compared the radii of the sun and Earth in answer to the question: "How many times larger is the sun than Earth?" Another way to answer the question is to compare the volumes of each object. The sun and Earth are both roughly spherical. Given a radius, r,

$$\text{the volume of a sphere} = (4/3)\pi r^3$$

The radius of the sun is approximately 10^9 meters, so

$$\text{the volume of the sun} \approx (4/3)\pi(10^9)^3 \text{ m}^3$$

In words, the expression $(10^9)^3$ tells us to take the ninth power of 10 and cube it. There are several ways of performing this calculation. We could evaluate the expression inside of the parentheses and then cube it. Thus

$$(10^9)^3 = (1{,}000{,}000{,}000)^3$$
$$= 1{,}000{,}000{,}000{,}000{,}000{,}000{,}000{,}000{,}000$$

rewrite as a power of 10 $= 10^{27}$

We could also rewrite $(10^9)^3$ using our knowledge about exponents:

definition of exponents $(10^9)^3 = (10^9) \cdot (10^9) \cdot (10^9)$

multiplication rule $= 10^{(9+9+9)}$

simplify $= 10^{27}$

Note that 27, the power of the base 10, can be obtained by multiplying the exponents 9 and 3 in the original expression $(10^9)^3$. This makes sense if you think about the meaning of multiplying expressions with exponents and what the exponent tells us to do. So

$$\text{the volume of the sun} \approx (4/3)\pi(10^9)^3 \text{ m}^3$$
$$= (4/3)\pi 10^{27} \text{ m}^3$$

Since the radius of Earth is about 10^7 meters, the ratio of the two volumes is

$$\frac{\text{volume of the sun}}{\text{volume of Earth}} = \frac{(4/3)\pi(10^9)^3\text{m}^3}{(4/3)\pi(10^7)^3\text{m}^3}$$
$$= \frac{(10^9)^3}{(10^7)^3} \qquad \text{(Note: } (4/3)\pi \text{ and m}^3 \text{ cancel.)}$$
$$= \frac{10^{27}}{10^{21}}$$
$$= 10^6$$

So while the radius of the sun is 100 times larger than the radius of Earth, the *volume* of the sun is approximately $10^6 = 1{,}000{,}000$ or 1 million times larger than the volume of Earth!

In general, to raise a power of a base, a, to a power, multiply the exponents.

If a is any nonzero real number and m and n are integers, $(a^m)^n = a^{(m \cdot n)}$.

Some illustrations of applying the power law for exponents are:

$$(10^2)^6 = 10^{2 \cdot 6} = 10^{12}$$
$$(13^{-8})^3 = 13^{(-8) \cdot 3} = 13^{-24}$$
$$(w^2)^{-7} = w^{2 \cdot (-7)} = w^{-14}$$

Show why $(11^2)^4 = 11^8$.

Example 3

SOLUTION

$$(11^2)^4 = 11^2 \cdot 11^2 \cdot 11^2 \cdot 11^2$$
$$= 11^{2+2+2+2}$$
$$= 11^8$$

Simplify $\dfrac{v^{-2}(w^5)^2}{(v^{-1})^4 w^{-3}}$

Example 4

SOLUTION

apply power law twice $\quad \dfrac{v^{-2}(w^5)^2}{(v^{-1})^4 w^{-3}} = \dfrac{v^{-2} w^{10}}{v^{-4} w^{-3}}$

apply division law twice $\quad = v^{-2-(-4)} w^{10-(-3)}$

simplify $\quad = v^2 w^{13}$

$\bullet \quad \bullet \quad \bullet$

We can summarize the rules for exponents.

If a is any nonzero real number and m and n are integers,

$$a^m \cdot a^n = a^{(m+n)}$$
$$\frac{a^m}{a^n} = a^{(m-n)}$$
$$(a^m)^n = a^{(m \cdot n)}$$

Algebra Aerobics 7.3c

1. Simplify, leaving the answer in exponent form.
   a. $(10^4)^5$ c. $(7^{-2})^{-3}$
   b. $(10^4)^{-5}$ d. $(x^4)^5$

2. The radius of Jupiter, the largest of the planets in our solar system, is approximately $7.14 \cdot 10^4$ km. Estimate the surface area and the volume of Jupiter. (If r is the radius of a sphere, the sphere's surface area equals $4\pi r^2$, and its volume equals $(4/3)\,\pi r^3$.)

3. Simplify:
   a. $\dfrac{t^{-3}t^0}{(t^{-4})^3}$ b. $\dfrac{v^{-3}w^7}{(v^{-2})^3 w^{-10}}$

Estimating Answers

By rounding off numbers and using scientific notation and the rules for exponents, we can often make quick estimates of answers to complicated calculations. In this age of calculators and computers, it is easy to trust an answer that looks correct but, because of an error in entering the data, is off by several powers of 10. Being able to roughly estimate the size of an answer in advance is a critical skill.

Example 5   Estimate the value of $\dfrac{(382{,}152) \cdot (490{,}572{,}261)}{(32{,}091) \cdot (1942)}$. Express your answer in both scientific and standard notation.

SOLUTION

Round off each number
$$\frac{(382{,}152) \cdot (490{,}572{,}261)}{(32{,}091) \cdot (1942)} \approx \frac{(400{,}000) \cdot (500{,}000{,}000)}{(30{,}000) \cdot (2000)}$$

rewrite in scientific notation
$$\approx \frac{(4 \cdot 10^5) \cdot (5 \cdot 10^8)}{(3 \cdot 10^4) \cdot (2 \cdot 10^3)}$$

group the coefficients and the powers of 10
$$\approx \left(\frac{4 \cdot 5}{3 \cdot 2}\right) \cdot \left(\frac{10^5 \cdot 10^8}{10^4 \cdot 10^3}\right)$$

simplify each expression
$$\approx \frac{20}{6} \cdot \frac{10^{13}}{10^7}$$

we get
$$\approx 3.33 \cdot 10^6$$

or in standard notation
$$\approx 3{,}330{,}000$$

Using a calculator on the original problem we get a more precise answer of 3,008,199.595

In 1995 the world population was approximately 5.734 billion people. There are roughly 57.9 million square miles of land on Earth, of which about 22% are favorable for agriculture. Estimate how many people per square mile of farmable land there were in 1995.

Example   6

SOLUTION

$$\frac{\text{size of world population}}{\text{amount of farmable land}} = \frac{5.734 \text{ billion people}}{22\% \text{ of } 57.9 \text{ million square miles}}$$

$$= \frac{5.734 \cdot 10^9 \text{ people}}{(0.22) \cdot (57.9) \cdot 10^6 \text{ sq. mile}}$$

$$\approx \frac{6 \cdot 10^9 \text{ people}}{(0.2) \cdot 60 \cdot 10^6 \text{ sq. mile}}$$

$$\approx \frac{6 \cdot 10^9 \text{ people}}{12 \cdot 10^6 \text{ sq. mile}}$$

$$\approx 0.5 \cdot 10^3 \text{ people/sq. mile}$$

$$\approx 500 \text{ people/sq. mile}$$

So there are roughly 500 people/sq. mile of farmable land in the world. A calculator produces the answer 450 people/sq. mile of farmable land.

Algebra   Aerobics   7.3d

1. Estimate the answer to:
   a. $(0.000\ 297\ 6) \cdot (43{,}990{,}000)$
   b. $\dfrac{453{,}897 \cdot 2{,}390{,}702}{0.004\ 38}$

2. Only about 3/7 of the land favorable for agriculture is actually being farmed. Using the facts in Example 6, estimate the number of people/sq. mile of farmable land that is being used. Should your estimate be larger or smaller than 500 people/sq. mile?

7.4 CONVERTING UNITS

Problems in science constantly require converting back and forth between different units of measure. In order to do so, we need to be comfortable with the laws of exponents and the basic metric and English units (see Table 7.1). The following unit conversion examples describe a strategy based upon *conversion factors*.

Exploration 7.1 will help you understand the relative ages and sizes of objects in our universe and give you practice in scientific notation and unit conversion.

Converting Units within the Metric System

Example   1 — Light travels at a speed of approximately $3.00 \cdot 10^5$ kilometers per second (km/sec). Describe the speed of light in meters per second (m/sec).

SOLUTION

The prefix *kilo-* means thousand. 1 kilometer (km) is equal to 1000 or 10^3 meters (m).

$$1 \text{ km} = 10^3 \text{ m} \qquad (1)$$

Dividing both sides of Equation (1) by 1 km, we can rewrite it as:

$$1 = \frac{10^3 \text{ m}}{1 \text{ km}}$$

If instead we divide both sides of Equation (1) by 10^3 m, we get

$$\frac{1 \text{ km}}{10^3 \text{ m}} = 1$$

The ratios $(10^3 \text{ m})/(1 \text{ km})$ and $(1 \text{ km})/(10^3 \text{ m})$ are called *conversion factors,* because we can use them to convert between kilometers and meters.

When solving unit conversion problems, the crucial question is always; "What is the right conversion factor?" If units in km/sec are multiplied by units in m/km we have

$$\frac{\cancel{\text{km}}}{\text{sec}} \cdot \frac{\text{m}}{\cancel{\text{km}}}$$

and the result is in m/sec. So multiplying the speed of light in km/sec by a conversion factor in m/km will give us the correct units of m/sec. Since a conversion factor always equals 1, we will not change the value of the original quantity by multiplying it by a conversion factor. In this case, we use the conversion factor of $(10^3 \text{ m})/(1 \text{ km})$.

$$3.00 \cdot 10^5 \text{ km/sec} = 3.00 \cdot 10^5 \frac{\cancel{\text{km}}}{\text{sec}} \cdot \frac{10^3 \text{ m}}{1 \cancel{\text{km}}}$$

$$= 3.00 \cdot 10^5 \cdot 10^3 \frac{\text{m}}{\text{sec}}$$

$$= 3.00 \cdot 10^8 \frac{\text{m}}{\text{sec}}$$

Hence light travels at approximately $3.00 \cdot 10^8$ meters per second.

Example   2 — Check your answer in Example 1 by converting $3.00 \cdot 10^8$ m/sec back to km/sec.

SOLUTION

Here we use the same strategy, but now we need to use the other conversion factor.

Multiplying $3.00 \cdot 10^8$ m/sec by $(1 \text{ km})(10^3 \text{ m})$ gives us

$$3.00 \cdot 10^8 \, \frac{\text{m}}{\text{sec}} \cdot \frac{1 \text{ km}}{10^3 \text{ m}} = 3.00 \cdot \frac{10^8 \text{ km}}{10^3 \text{ sec}}$$

$$= 3.00 \cdot 10^5 \, \frac{\text{km}}{\text{sec}}$$

which was the original value given for the speed of light.

Converting between the Metric and English Systems

If light travels $3.00 \cdot 10^5$ km/sec, how many *miles* does light travel in a second?

Example 3

SOLUTION

The crucial question is: "What conversion factor should be used?" From Table 7.1 we know that

$$1 \text{ km} = 0.62 \text{ mile}$$

This equation can be rewritten in two ways:

$$1 = \frac{0.62 \text{ mile}}{1 \text{ km}} \qquad \text{or} \qquad 1 = \frac{1 \text{ km}}{0.62 \text{ mile}}$$

It produces two possible conversion factors:

$$\frac{0.62 \text{ mile}}{1 \text{ km}} \qquad \text{and} \qquad \frac{1 \text{ km}}{0.62 \text{ mile}}$$

Which one will convert kilometers to miles? We need one with kilometers in the denominator and miles in the numerator, namely (0.62 mile)/(1 km).

Multiplying $3.00 \cdot 10^5$ km/sec by the conversion factor (0.62 mile)/(1 km) gives us:

$$3.00 \cdot 10^5 \, \frac{\text{km}}{\text{sec}} \cdot \frac{0.62 \text{ mile}}{1 \text{ km}} = 1.86 \cdot 10^5 \, \frac{\text{mile}}{\text{sec}}$$

So light travels $1.86 \cdot 10^5$ or 186,000 miles/sec.

Something to think about

1. Estimate the number of heartbeats in a lifetime.
2. A nanosecond is 10^{-9} seconds. Modern computers can perform on the order of one operation every nanosecond. Approximately how many feet does light (and hence an electrical signal in a computer) travel in 1 nanosecond?

Using Multiple Conversion Factors

Light travels at $3.00 \cdot 10^5$ km/sec. How many kilometers does light travel in one *year*?

Example 4

SOLUTION

Here our strategy is to use more than one conversion factor to convert from seconds to years. Use your calculator to perform the following calculations.

$$3.00 \cdot 10^5 \, \frac{\text{km}}{\text{sec}} \cdot \frac{60 \text{ sec}}{1 \text{ min}} \cdot \frac{60 \text{ min}}{1 \text{ hr}} \cdot \frac{24 \text{ hr}}{1 \text{ day}} \cdot \frac{365 \text{ days}}{1 \text{ year}} = 94,608,000 \cdot 10^5 \, \frac{\text{km}}{\text{year}}$$

$$\approx 9.46 \cdot 10^7 \cdot 10^5 \frac{\text{km}}{\text{year}}$$

$$= 9.46 \cdot 10^{12} \frac{\text{km}}{\text{year}}$$

So a light year, the distance light travels in one year, is approximately equal to $9.46 \cdot 10^{12}$ kilometers.

Algebra Aerobics 7.4

1. In Section 7.3 we said that a light year was about $5.88 \cdot 10^{12}$ miles. Verify that $9.46 \cdot 10^{12}$ kilometers $\approx$ $5.88 \cdot 10^{12}$ miles.

2. 1 Angstrom $= 10^{-8}$ cm. Express 1 Angstrom in meters.

3. If a road sign says the distance to Quebec is 135 miles, how many kilometers away is it?

4. The mean distance from our sun to Jupiter is $7.8 \cdot 10^8$ kilometers. Express this distance in meters.

5. The distance from Earth to the sun is about 93,000,000 miles. There are 5280 feet in a mile, and a dollar bill is approximately 6 inches long. Estimate how many dollar bills would have to be placed end to end to reach from Earth to the sun.

7.5 EXPRESSIONS OF THE FORM $a^{m/n}$

So far we have derived rules for operating with expressions of the form a^n, where n is any integer. These rules can be extended to expressions of the form $a^{m/n}$ where the exponent is a fraction. We need first to consider what an expression such as $a^{m/n}$ means.

The expression m/n can also be written as $m \cdot (1/n)$ or $(1/n) \cdot m$. If the laws of exponents are consistent, then

$$a^{m/n} = (a^m)^{(1/n)} = (a^{1/n})^{(m)}$$

What does $a^{(1/n)}$ mean? Raising a to the $(1/n)$th power is the opposite of raising a to the nth power; it is an operation called "taking a root."

Square Roots: Expressions of the Form $a^{(1/2)}$

The expression $a^{(1/2)}$ is called the *principal square root* (or just the *square root*) of a and is often written as $\sqrt{a}$. The symbol $\sqrt{}$ is called a radical. The principal square root of a is the *nonnegative* number b such that $b^2 = a$. Both the square of -2 and the square of 2 are equal to 4, but the notation $\sqrt{4}$ is defined as *only the positive root*. If both -2 and 2 are to be considered, we write $\pm\sqrt{4}$, which means "plus or minus the square root of 4."

In the real numbers, $\sqrt{a}$ is not defined when a is negative. For example, in the real numbers the term $(-4)^{(1/2)} = \sqrt{-4}$ is undefined, since there is no real number b such that $b^2 = -4$.

If a is any nonnegative real number,

$$a^{(1/2)} = \sqrt{a}$$

where $\sqrt{a}$ is that nonnegative number b such that $b^2 = a$.

Many calculators and spreadsheet programs have a square root function often labeled $\sqrt{}$ or perhaps "SQRT." You can also calculate square roots by raising a number to the 1/2 or 0.5 power using the $\wedge$ key, as in $4 \wedge .5$. Try using a calculator to find $\sqrt{4}$ and $\sqrt{9}$. What happens on your calculator when you evaluate $\sqrt{-4}$?

In any but the simplest cases where the square root is immediately obvious, you will probably use the calculator. For example, use your calculator to find

$$8^{(1/2)} = \sqrt{8} \approx 2.8284$$

Double check the answer by verifying that $(2.8284)^2 \approx 8$.

Square roots appear quite frequently in real-life computations. The following rules apply to computations with radicals.

$$\sqrt{ab} = \sqrt{a} \cdot \sqrt{b} \quad \text{and} \quad \sqrt{a/b} = \sqrt{a}/\sqrt{b}$$

but $\quad \sqrt{a+b} \neq \sqrt{a} + \sqrt{b} \quad \text{and} \quad \sqrt{a-b} \neq \sqrt{a} - \sqrt{b}.$

For example, $\sqrt{64} = \sqrt{16 \cdot 4} = \sqrt{16} \cdot \sqrt{4} = 4 \cdot 2 = 8$

but $\quad \sqrt{25 + 16} = \sqrt{41} \approx 6.40$ while $\sqrt{25} + \sqrt{16} = 5 + 4 = 9.$

Example 1

The function $S = \sqrt{30d}$ describes the relationship between S, the speed of a car in miles per hour, and d, the distance in feet a car skids after applying the brakes on a dry tar road. Estimate the speed of a car that leaves 40-foot long skid marks on a dry tar road, and the speed that leaves 150-foot long skid marks.

SOLUTION
If $d = 40$ feet, then $S = \sqrt{30 \cdot 40} = \sqrt{1200} \approx 35$, so the car was traveling at about 35 miles per hour.

If $d = 150$ feet, then $S = \sqrt{30 \cdot 150} = \sqrt{4500} \approx 67$, so the car was traveling at almost 70 miles per hour.

Example 2

If we solve for the radius, r, in the formula for the surface of a sphere, $S = 4\pi r^2$, we get $r = \sqrt{\dfrac{S}{4\pi}} = \dfrac{1}{2}\sqrt{\dfrac{S}{\pi}}$. If we assume that Earth has a spherical shape, and we know that its surface area is approximately 200,000,000 square miles, we can use this formula to estimate Earth's radius. Substituting for S we get

$$r = \frac{1}{2}\sqrt{\frac{200{,}000{,}000 \text{ sq. miles}}{\pi}}$$

$$\approx \frac{1}{2}\sqrt{63{,}694{,}268 \text{ sq. miles}}$$

$$\approx \frac{1}{2}(63{,}694{,}268 \text{ sq. miles})^{1/2} \approx \frac{1}{2} \cdot 7980 \text{ miles} = 3990 \text{ miles}$$

So Earth has a radius of about 4000 miles.

$\bullet \quad \bullet \quad \bullet$

Estimating square roots   A number is called a *perfect square* if its square root is an integer. For example, 25 and 36 are both perfect squares since $25 = 5^2$ and $36 = 6^2$, so $\sqrt{25} = 5$ and $\sqrt{36} = 6$. If we don't know the square root of some number x and don't have a calculator handy, we can estimate the square root by bracketing it between perfect squares, a and b, for which we do know the square roots. If $x \geq 1$ and $a < x < b$, then $\sqrt{a} < \sqrt{x} < \sqrt{b}$. For example, to estimate $\sqrt{10}$,

We know $9 < 10 < 16$ where 9 and 16 are perfect squares

so $\sqrt{9} < \sqrt{10} < \sqrt{16}$

and $3 < \sqrt{10} < 4$

Therefore $\sqrt{10}$ lies somewhere between 3 and 4, probably closer to 3. According to a calculator $\sqrt{10} \approx 3.16$.

Example   3

Estimate $\sqrt{27}$.

SOLUTION

we know $25 < 27 < 36$

therefore $\sqrt{25} < \sqrt{27} < \sqrt{36}$

and $5 < \sqrt{27} < 6$

So $\sqrt{27}$ lies somewhere between 5 and 6. Would you expect $\sqrt{27}$ to be closer to 5 or to 6? Check your answer with a calculator.

Algebra Aerobics 7.5a

1. Evaluate each of the following without a calculator.
   a. $81^{1/2}$ b. $144^{1/2}$

2. Simplify and rewrite without radical signs.
   a. $\sqrt{9x}$ b. $\sqrt{\dfrac{x^2}{25}}$ $(x \geq 0)$

3. Use the formula in Example 1 to estimate the speed of a car that leaves 60-foot long skid marks on a dry tar road and the speed that leaves 200-foot long skid marks.

4. Without a calculator, find two consecutive integers between which the given number lies.
   a. $\sqrt{29}$ b. $\sqrt{92}$

*n*th Roots: Expressions of the Form $a^{1/n}$

$a^{(1/n)}$ denotes the *n*th root of *a*, often written as $\sqrt[n]{a}$. The *n*th root of *a* is that number *b* such that $b^n = a$. That is, if $b = a^{(1/n)}$, then $b^n = a$. Note that when taking square roots we usually omit the "2" and write $\sqrt{a}$ instead of $\sqrt[2]{a}$.

For example,

$$8^{(1/3)} = \sqrt[3]{8} = 2 \quad \text{since } 2^3 = 8. \text{ We call 2 the third or cube root of 8.}$$
$$16^{(1/4)} = \sqrt[4]{16} = 2 \quad \text{since } 2^4 = 16. \text{ We call 2 the fourth root of 16.}$$

> If *a* is any nonnegative real number and *n* is a positive integer, then $a^{(1/n)}$ or $\sqrt[n]{a}$ is that number *b* such that $b^n = a$.

If the *n*th root exists, you can find its value on a calculator. For example, to estimate a fifth root, raise the number to the (1/5) or the 0.2 power. So

$$3125^{(1/5)} = \sqrt[5]{3125} = 5$$

Double check your answer by verifying that $5^5 = 3125$.

The volume of a sphere is given by the equation $V = (4/3)\pi r^3$. We can rewrite the formula, solving for the radius as a function of the volume.

Example 4

Given

$$V = (4/3)\pi r^3$$

multiply both sides by 3

$$3V = 4\pi r^3$$

divide by 4π

$$\frac{3V}{4\pi} = r^3$$

take the cube root and switch sides

$$r = \sqrt[3]{\frac{3V}{4\pi}}$$

Estimate the radius of a spherical balloon that has a volume of 4 cubic feet.

Example 5

SOLUTION

Substituting 4 feet^3 for *V* we have

$$r = \sqrt[3]{\frac{3 \cdot 4 \text{ feet}^3}{4\pi}} = \sqrt[3]{\frac{3 \text{ feet}^3}{\pi}} \approx \sqrt[3]{0.955 \text{ feet}^3} \approx 0.98 \text{ feet}$$

So the balloon has a radius of about 1 foot.

Algebra Aerobics 7.5b

1. Evaluate each of the following without a calculator.
    a. $27^{1/3}$ **b.** $16^{1/4}$ **c.** $8^{-1/3}$
2. Use a calculator or spreadsheet to estimate
    a. $\sqrt[4]{1295}$ **b.** $\sqrt[3]{372{,}783}$

3. Estimate the radius of a spherical balloon with a volume of 2 cubic feet.

You may want to do Exploration 7.2 on Kepler's laws of planetary motion after reading this section.

Something to think about

Show why $\sqrt{a \cdot b} = \sqrt{a} \cdot \sqrt{b}$.

Fractional Powers: Expressions of the Form $a^{m/n}$

What about $a^{m/n}$? We can write $a^{m/n}$ either as $(a^m)^{(1/n)}$ or $(a^{1/n})^m$. Writing it as $(a^m)^{(1/n)}$ means that we would first raise the base, a, to the mth power and then take the nth root of that. Writing it as $(a^{(1/n)})^m$ implies first finding the nth root of a and then raising that to the mth power. For example:

$$2^{(3/2)} = (2^3)^{(1/2)}$$
$$= (8)^{(1/2)}$$

using a calculator ≈ 2.8284

Equivalently, $2^{(3/2)} = (2^{(1/2)})^3$
$$\approx (1.414)^3$$
$$\approx 2.8284$$

We could, of course, use a calculator to compute $2^{(3/2)}$ (or $2^{1.5}$) directly by raising 2 to the 3/2 or 1.5 power.

Exponents expressed as ratios, of the form m/n, are called *rational exponents*. The set of laws for simplifying exponential expressions also holds for rational exponents.

If a is any nonnegative number, and p and q are rational numbers, then

$$a^p \cdot a^q = a^{(p+q)}$$
$$a^p/a^q = a^{(p-q)} \qquad (a \neq 0)$$
$$(a^p)^q = a^{(p \cdot q)}$$

We have dealt with integer exponents and rational (fractional) exponents in expressions such as $8^{(2/3)}$. Terms with irrational exponents, such as 2^{π}, are well defined but are beyond the scope of this course. You may recall that irrational real numbers are those that cannot be expressed as a ratio of integers. However the same *rules* for exponents apply.

Find the product of $(\sqrt{5}) \cdot (\sqrt[3]{5})$ leaving the answer in exponent form.

Example   6

SOLUTION

$(\sqrt{5}) \cdot (\sqrt[3]{5}) = 5^{1/2} \cdot 5^{1/3} = 5^{(1/2)+(1/3)} = 5^{5/6}$

According to McMahon and Bonner in *On Size and Life*,[7] common nails range from 1 to 6 inches in length. The weight varies even more, from 11 to 647 nails per pound. Longer nails are relatively thinner than shorter ones. A good approximation of the relationship between length and diameter is given by the equation

Example   7

$$d = 0.07 \, L^{2/3}$$

where d = diameter and L = length, both in inches. Estimate the diameters of nails that are 1, 3, and 6 inches long.

SOLUTION

When $L = 1$ inch, $d = 0.07 \cdot (1)^{2/3} = 0.07 \cdot 1 = 0.07$ inches.
When $L = 3$ inches, then the diameter equals $0.07 \cdot (3)^{2/3} \approx 0.07 \cdot 2.08 = 0.15$ inches.
When $L = 6$ inches, then $d = 0.07 \cdot (6)^{2/3} \approx 0.07 \cdot 3.30 = 0.23$ inches.

Note that L and d are both in inches. Since $d = 0.07L^{2/3}$, then $0.07L^{2/3}$ must also be in inches. The units for $L^{2/3}$ are (in.)$^{2/3}$! So the coefficient, 0.07, must have units of (in.)$^{1/3}$, since (in.)$^{1/3} \cdot$ (in.)$^{2/3} =$ (in.)$^1 =$ in. The coefficients of variables with fractional power are often in strange units that are hard to figure out. Fortunately, knowing the actual units in such cases is less interesting and less important than knowing the power.

Algebra Aerobics 7.5c

1. Find the product expressed in exponent form.
    a. $\sqrt{2} \sqrt[3]{2}$ **b.** $\sqrt{3} \sqrt[3]{9}$ **c.** $\sqrt{5} \sqrt[4]{5}$
2. Find the quotient by first expressing in exponent form. Leave the answer in exponent form.

    a. $\dfrac{\sqrt{2}}{\sqrt[3]{2}}$ **b.** $\dfrac{2}{\sqrt[4]{2}}$ **c.** $\dfrac{\sqrt[4]{5}}{\sqrt[3]{5}}$

3. McMahon and Bonner give the relationship between chest circumference and body weight of adult primates as

    $$c = 17.1m^{3/8}$$

    where m = weight in kilograms and c = chest circumference in centimeters. Estimate the chest circumference of a
    a. 0.25-kilogram tamarin   **b.** 25-kilogram baboon

---

[7] Thomas A. McMahon and John Tyler Bonner. *On Size and Life* (New York: Scientific American Library, Scientific American Books, Inc, 1983).

7.6 ORDERS OF MAGNITUDE

Comparing Numbers of Widely Differing Sizes

We have seen that a useful method of comparing two objects of wildly different sizes is to calculate the ratio rather than the difference of the sizes. The ratio can be estimated by computing *orders of magnitude*, the number of times we would have to multiply or divide by 10 to convert one size into the other. Each power of 10 represents one order of magnitude.

For example, the radius of the observable universe is approximately 10^{26} meters and the radius of our solar system is approximately 10^{12} meters. To compare the radius of the observable universe to the radius of our solar system, calculate the ratio:

$$\frac{\text{radius of the universe}}{\text{radius of our solar system}} \approx \frac{10^{26} \text{ meters}}{10^{12} \text{ meters}}$$

$$\approx 10^{26-12}$$

$$\approx 10^{14}$$

The radius of the universe is roughly 10^{14} times larger than the radius of the solar system; that is, we would have to multiply the radius of our solar system by 10 fourteen times in order to obtain the radius of the universe. Since each factor of 10 is counted as a single order of magnitude, the radius of the universe is *14 orders of magnitude larger* than the radius of our solar system. Equivalently, we could say that the radius of our solar system is *14 orders of magnitude smaller* than the radius of the universe.

When something is one order of magnitude larger than a *reference object*, it is 10 times larger. You *multiply* the *reference size* by 10 to get the other size. If the object is two orders of magnitude larger, it is 100 or 10^2 times larger, so you would multiply the reference size by 100. If it is one order of magnitude smaller, it is 10 times smaller, so you would *divide* the reference size by 10. Two orders of magnitude smaller means the reference size is divided by 100 or 10^2.

The following examples demonstrate order of magnitude comparisons.

- We previously found that the radius of the sun is 100 or 10^2 times larger than the radius of Earth. This is equivalent to saying that the radius of the sun is 2 orders of magnitude larger than the radius of Earth. We also found that the volume of the sun is 10^6 times or 6 orders of magnitude larger than the volume of Earth.

- The radius of the sun, at 10^9 meters, is 20 orders of magnitude larger than the radius of a hydrogen atom, at 10^{-11} meter, since

$$\frac{\text{radius of sun}}{\text{radius of the hydrogen atom}} \approx \frac{10^9 \text{ meters}}{10^{-11} \text{ meter}}$$

$$\approx 10^{9-(-11)}$$

$$\approx 10^{20}$$

So the radius of the sun is 10^{20} times larger than the radius of the hydrogen atom.

- Surprisingly enough, the radius of a living cell is approximately three orders of magnitude *smaller* than one of the single strands of DNA it contains, if the DNA is uncoiled and measured lengthwise.

$$\frac{\text{length of DNA strand}}{\text{radius of the living cell}} = \frac{10^{-2} \text{ meter}}{10^{-5} \text{ meter}}$$
$$= 10^{-2-(-5)}$$
$$= 10^{-2+5}$$
$$= 10^{3}$$

Algebra Aerobics 7.6a

1. If my salary is \$100,000 and you make an order of magnitude more, what is your salary? If Henry makes two orders of magnitude less money than I do, what is his salary?

2. For each of the following pairs, determine the order of magnitude difference:

a. the radius of the sun (10^{9} meters) and the radius of the Milky Way (10^{21} meters).

b. the radius of a hydrogen atom (10^{-11} meter) and the radius of a proton (10^{-15} meter).

A Measurement Scale Based on Orders of Magnitude: The Richter Scale

The *Richter scale*, designed by the American Charles Richter in 1935, allows us to compare the magnitude of earthquakes throughout the world. The Richter scale measures the maximum vertical ground movement (tremors) as recorded on an instrument called a seismograph. The sizes of earthquakes vary widely, so Richter designed the scale to measure order of magnitude differences. The scale ranges from less than 1 to over 8. Each increase of 1 unit on the Richter scale represents an increase of 10 times or one order of magnitude in the maximum tremor size of the earthquake. So an increase from 2.5 to 3.5 indicates a 10-fold increase in maximum tremor size. An increase of 2 units from 2.5 to 4.5 indicates an increase in maximum tremor size by a factor of 10^{2} or 100.

Table 7.2 contains some typical values on the Richter scale along with a description of how humans near the center (called the *epicenter*) of an earthquake perceive its effects. There is no theoretical upper limit on the Richter scale, but the biggest earthquakes measured so far registered as 8.6 on the Richter scale (in Japan and Chile).[8]

The reading "Earthquake Magnitude Determination" describes how earthquake tremors are measured.

[8] The following table gives values on the Richter scale and the approximate number per year (worldwide).

Richter Scale Magnitude	Number per Year (Worldwide)
2	More than 100,000
4.5	A few thousand
7	16 to 18
8	1 or 2

Source: C.C. Plummer and D. McGeary, *Physical Geology,* 5th ed. (Dubuque, Iowa: Wm. C. Brown, 1991) p. 352.

Table 7.2	
Description of the Richter Scale	
Richter Scale Magnitude	Description
2.5	Generally not felt, but recorded on seismographs
3.5	Felt by many people
4.5	Felt by all; some slight local damage may occur
6	Considerable damage in ordinary buildings; a destructive earthquake
7	"Major" earthquake; most masonry and frame structures destroyed; ground badly cracked
8 and above	"Great" earthquake; a few per decade worldwide; almost total or total destruction; bridges collapse, major openings in ground, and tremors visible

Algebra Aerobics 7.6b

1. In 1987 Los Angeles had an earthquake that measured 5.9 on the Richter scale. In 1988 Armenia had an earthquake that measured 6.9 on the Richter scale. Compare the sizes of the two earthquakes using orders of magnitude.

2. In 1983 Hawaii had an earthquake that measured 6.6 on the Richter scale. Compare the size of this earthquake to the largest ever recorded, 8.6 on the Richter scale.

Graphing Numbers of Widely Differing Sizes

Exploration 7.1 asks you to construct a graph using logarithmic scales.

If the sizes of various objects in our solar system are plotted on a standard linear axis, we get the uninformative picture shown in Figure 7.1. The largest value stands alone, and all the others are so small when measured in terameters that they all appear to be zero. When objects of widely different orders of magnitude are compared on a linear scale, the effect is similar to pointing out an ant in a picture of a baseball stadium.

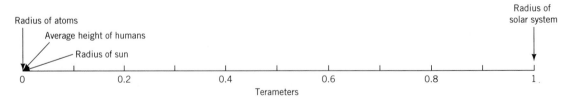

Figure 7.1   Sizes of various objects in the universe on a linear scale. (Note: One terameter = 10^{12} meters.)

A more effective way of plotting sizes with different orders of magnitude is to use an axis that has orders of magnitude (powers of 10) evenly spaced along it. This is called a *logarithmic* or *log scale*. The plot of the previous data graphed on a logarithmic scale is much more informative (Figure 7.2).

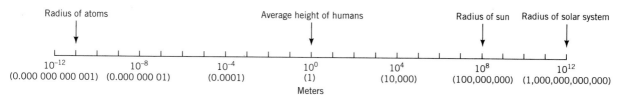

Figure 7.2 Sizes of various objects in the universe on an order of magnitude (or logarithmic) scale.

Graphing sizes on a log scale is extremely useful, but we have to read the scales very carefully. When we use a linear scale, each move of one unit to the right is equivalent to *adding* 1 unit to the number, and each move of k units to the right is equivalent to *adding* k units to the number (Figure 7.3).

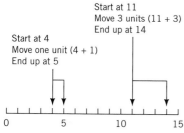

Figure 7.3 Linear scale.

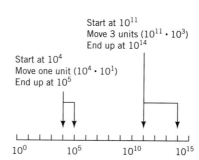

Figure 7.4 Order of magnitude (or logarithmic) scale.

When we use a log scale (see Figure 7.4), we need to remember that now one unit of length represents a change of one order of magnitude. Moving one unit to the right is equivalent to *multiplying by 10*. So moving from 10^4 to 10^5 is equivalent to multiplying 10^4 by 10. Moving three units to the right is equivalent to *multiplying* the starting number by 10^3 or 1000. In effect, a linear scale is an "additive" scale and a logarithmic scale is a "multiplicative" scale.

The ingenuity behind the design of logarithmic scales inspired the following poetic comment by D'Arcy Thompson in his classic book *On Growth and Form*.[9]

It is a remarkable thing, worth pausing to reflect on, that we can pass so easily and in a dozen lines from molecular magnitudes to the dimensions of a Sequoia or a whale. Addition and subtraction, the old arithmetic of the Egyptians, are not powerful enough for such an operation; but Archimedes and Napier elaborated the arithmetic of multiplication. So passing up and down by easy steps, as Archimedes did when he numbered the sands of the sea, we compare the magnitudes of the great beasts and the small, of the atoms of which they are made, and of the world in which they dwell.

[9] D'Arcy Wentworth Thompson, *On Growth and Form* (New York: Dover Publications, Inc., 1992).

Algebra Aerobics 7.6c

1. By rounding the number to the nearest power of 10, find the approximate location of each of the following on the logarithmic scale in Figure 7.2.
   a. The radius of Earth: 6,370,000 meters
   b. The radius of a virus: 0.000 000 7 meter

2. Now plot on Figure 7.2 an object whose radius is:
   a. 2 orders of magnitude smaller than that of Earth
   b. 6 orders of magnitude larger than that of a virus

7.7 LOGARITHMS BASE 10

Logarithms were originally used to help carry out calculations for astronomical problems. The logarithm and the ideas that underlie it have had many other important scientific and practical applications. A slide rule, for example, multiplies numbers by mechanically adding distances that represent their logarithms. We use logarithmic scales to measure physical quantities from sound intensities to earthquake strengths.

Finding the Logarithm of Any Positive Number

Finding the logarithm of powers of 10   When handling very large or very small numbers, we have seen that is it often easier to write the number using powers of 10. For example,

$$100,000 = 10^5.$$

We refer to 10 as the base and 5 as the power (or exponent). We say that:

> "100,000 equals the base 10 to the 5th power."

But we could also rephrase this as

> "5 is the power of the base 10 that is needed to produce 100,000."

The more technical way to say this is

> "5 is the *logarithm* base 10 of 100,000."

In symbols we write:

$$5 = \log_{10} 100,000$$

So the expressions

$$100,000 = 10^5 \qquad \text{and} \qquad 5 = \log_{10} 100,000$$

are two ways of saying exactly the same thing.

The key point to remember is that a logarithm is an exponent.

The *logarithm base 10 of x* is the power of 10 you need to produce x.

$$\log_{10} x = c \qquad \text{means that} \qquad 10^c = x$$

Since

$$1,000,000,000 = 10^9$$

Example   1

then

$$\log_{10} 1,000,000,000 = 9$$

and we say that the logarithm base 10 of 1,000,000,000 is 9. The logarithm of a number tells us the exponent, in this case of the base 10. Since here the logarithm is 9, that means that when we write 1,000,000,000 as a power of 10, the exponent is 9.

Since

$$1 = 10^0$$

Example   2

then

$$\log_{10} 1 = 0$$

and we say that the logarithm base 10 of 1 is 0. Since logarithms represent exponents, this says that when we write 1 as a power of 10, that exponent is 0.

How do we calculate the logarithm base 10 of decimals such as 0.000 01?

Example   3

SOLUTION
Since

$$0.000\ 01 = 10^{-5}$$

then

$$\log_{10} 0.000\ 01 = -5$$

and we say that the logarithm base 10 of 0.000 01 is -5.

•   •   •

In the previous example, we found that the log of a number can be a negative. This makes sense if we think of logarithms as exponents, since exponents can be any real number. But we cannot take the log of a negative number or zero: that is, $\log_{10} x$ is not defined when $x \leq 0$. Why? If $\log_{10} x = c$, where $x \leq 0$, then $10^c = x$ (a number ≤ 0). But 10 to any power will never produce a number that is negative or zero, so $\log_{10} x$ is not defined if $x \leq 0$.

$\log_{10} x$ is not defined when $x \leq 0$.

Table 7.3 gives a sample set of values for x and their associated logarithms base 10. For each value of x, in order to find the logarithm base 10, we write the value as a power of 10, and the logarithm is just the exponent.

Table 7.3
Logarithms of Powers of 10

x	Exponential Notation	$\text{Log}_{10} x$
0.0001	10^{-4}	-4
0.001	10^{-3}	-3
0.01	10^{-2}	-2
0.1	10^{-1}	-1
1	10^{0}	0
10	10^{1}	1
100	10^{2}	2
1,000	10^{3}	3
10,000	10^{4}	4

Most scientific calculators and spreadsheet programs have a LOG function that calculates common logarithms. Try using technology to double check some of the numbers in Table 7.3.

Logarithms base 10 are used frequently in our base 10 number system and are called *common logarithms*. We usually drop the 10 in the notation $\log_{10} x$ and just write $\log x$.

Logarithms base 10 are called *common logarithms*.
$\log_{10} x$ is frequently abbreviated as $\log x$.

Algebra Aerobics 7.7a

1. Without using a calculator, find the logarithm base 10 of:

    a. 10,000,000   **c.** 10,000

    b. 0.000 000 1   **d.** 0.0001

    Now use a calculator to double check your answers.

2. Rewrite the following expressions in an equivalent form using powers of 10.

    a. $\log 100{,}000 = 5$ **c.** $\log 10 = 1$

    b. $\log 0.000\ 000\ 01 = -8$

Finding the logarithm of numbers between 1 and 10   When a number such as 100 or 0.001 can be written as an integer power of 10, it's easy to find its logarithm base 10. Express the number as a power of 10, and the log is just the exponent. But what about the logarithms of other numbers, such as 2?

Estimate log 2.

Example   4

SOLUTION

$$\log 2 = c \qquad \text{is true if and only if} \qquad 10^c = 2$$

So we need to find a number c such that 10 raised to that power will give us 2. Since

$$1 < 2 < 10$$

and

$$1 = 10^0, \ 2 = 10^c, \ 10 = 10^1$$

then

$$10^0 < 10^c < 10^1$$

So we might suspect that

$$0 < c < 1$$

Using a calculator, we can estimate values of 10^c where c is between 0 and 1. Remember that $10^{0.1}$ (or $10^{1/10}$) means the 10th root of 10.

$$10^{0.1} \approx 1.258\ 925$$
$$10^{0.2} \approx 1.584\ 893$$
$$10^{0.3} \approx 1.995\ 263$$
$$10^{0.4} \approx 2.511\ 886$$

So $10^{0.3}$ is very close to 2, whereas $10^{0.4}$ is larger than 2. Hence a good estimate is

$$\log 2 \approx 0.3$$

By trial and error we could come even closer to the actual value.

$$10^{0.31} \approx 2.041\ 738$$
$$10^{0.301} \approx 1.999\ 862$$
$$10^{0.3015} \approx 2.002\ 166$$

Since 1.999 862 is closer to 2 than 2.002 166, we can improve our original estimate to

$$\log 2 \approx 0.301$$

One calculator gave

$$\log 2 = 0.301\ 029\ 996$$

so our estimates were pretty close.

Putting it all together   How can we find the logarithms of numbers that are not between 1 and 10? Our knowledge of scientific notation can help. Recall that a positive number, c, written in scientific notation is of the form

$$c = N \cdot 10^n$$

where $1 \leq N < 10$ and n is an integer. We can put together what we know about finding $\log N$ and $\log 10^n$ to find $\log c$.

Example 5

What is log 2000?

SOLUTION

Write 2000 in scientific notation	$2000 = 2 \cdot 10^3$
substitute $10^{0.301}$ for 2	$\approx 10^{0.301} \cdot 10^3$
(see Example 4 in this section)	
combine powers	$\approx 10^{0.301+3}$
	$2000 \approx 10^{3.301}$
rewrite as a logarithm	$\log 2000 \approx 3.301$

A calculator gives $\log 2000 = 3.301\ 029\ 996$.

Example 6

Find log 0.000 02.

SOLUTION

Write 0.000 02 in scientific notation	$0.000\ 02 = 2 \cdot 10^{-5}$
substitute $10^{0.301}$ for 2	$\approx 10^{0.301} \cdot 10^{-5}$
(see Example 4)	
combine powers	$\approx 10^{0.301-5}$
	$0.00002 \approx 10^{-4.699}$
rewrite as a logarithm	$\log 0.000\ 02 \approx -4.699$

A calculator gives $-4.698\ 970\ 004$.

• • •

The logarithm of any positive number written in scientific notation equals the log of the coefficient plus the exponent of the power of 10. That is, if c written in scientific notation is

$$c = N \cdot 10^n$$

(where $c > 0$ and hence $N > 0$), then

$$\log c = (\log N) + n$$

Once you understand what logarithms represent, you can use a calculator to do the actual computation.

Using logarithms to write any number as a power of 10   We will
see later that in many cases it is useful to write a number as a power of 10.

Write 6,370,000 (the radius of Earth in meters) as a power of 10.

Example 7

SOLUTION

By definition log 6,370,000 is the number c such that $10^c = 6,370,000$. A calculator
gives

$$\log_{10} 6,370,000 \approx 6.804$$

From the definition of logarithms we know that

$$6,370,000 \approx 10^{6.804}$$

• • •

In most practical applications, we are only interested in writing a positive
number as a power of 10. However, we can also write any negative number as a power
of 10.

Write $-0.000\ 35$ as a power of 10.

Example 8

SOLUTION

Recall that we can only compute logarithms of positive numbers. But we can find the
log of the absolute value of a negative number, and then negate the result. For example, log 0.000 35 ≈ -3.456, so

$$0.000\ 35 \approx 10^{-3.456}$$

Hence

$$-0.000\ 35 \approx -10^{-3.456}$$

Algebra Aerobics 7.7b

1. Estimate each of the following, and then check with a
   calculator.
   a. log 3 **b.** log 6 **c.** log 6.37
2. Use the answers from Problem 1 to estimate values for:
   a. log 3,000,000 **b.** log 0.006
   Then use a calculator to check your answers.

3. Write each of the following as a power of 10.
   a. 0.000 000 7 m (the radius of a virus)
   b. 780,000,000 km (the mean distance from our sun
      to Jupiter)
   c. -0.0042
   d. $-5,400,000,000$

CHAPTER SUMMARY

Scientists believe that somewhere between 8 and 15 billion years ago the universe began expanding rapidly from an infinitesimally small point. This event is referred to as the "Big Bang," and the universe has been expanding ever since.

In describing the size of objects in the universe, the scientific community and most countries other than the United States use the metric system, a measurement system based on the meter. The following table shows the conversion into English units for three of the most common metric length measurements.

Metric Unit	Abbreviation	In Meters	Equivalence in English Units	Informal Conversion
meter	m	1 meter	3.28 feet	The width of a twin bed; a little more than a yard
kilometer	km	1000 meters	0.62 mile	A casual 12-minute walk; a little over a half a mile
centimeter	cm	0.01 meter	0.39 inches	The width of a pencil; a little under a half an inch

Powers of 10 are useful in describing the scale of objects. When n is a positive integer we define:

$$10^n = 10 \cdot 10 \cdot 10 \cdot \ \ldots \ \cdot 10 \qquad \text{the product of } n \text{ tens or } 1 \text{ followed by } n \text{ zeros}$$

$$10^{-n} = \frac{1}{10^n} \qquad \text{1 divided by 10 } n \text{ times or a decimal point followed by } (n-1) \text{ zeros and a 1}$$

$$10^0 = 1$$

To write very small or large numbers compactly, we use scientific notation. A number is in *scientific notation* if it is in the form

$$N \cdot 10^n$$

where $1 \le |N| < 10$ and n is an integer (positive, negative, or zero).

The expression $|N|$ denotes "the absolute value of N." No matter what the value of N, the absolute value of N will always be greater than or equal to 0. If $N \ge 0$, then $|N| = N$. If $N < 0$, then $|N| = -N$.

The Laws of Exponents

If a is any nonzero real number and m and n are real numbers then,

$$a^m \cdot a^n = a^{(m+n)}$$

$$\frac{a^m}{a^n} = a^{(m-n)}$$

$$(a^m)^n = a^{(m \cdot n)}$$

Fractional Powers

If a is any nonnegative real number,

$$a^{1/2} = \sqrt{a} \qquad \text{where } \sqrt{a} \text{ is that nonnegative } b \text{ such that } b^2 = a.$$

For example, $\sqrt{16} = 4$ since $4^2 = 16$.

If a is any nonnegative real number and n is a positive integer, then

$$a^{1/n} = \sqrt[n]{a} \quad \text{is that number } b \text{ such that } b^n = a.$$

For example, $8^{1/3} = 2$ since $2^3 = 8$.

If a is any nonnegative number and m and n are integers ($n \neq 0$), then

$$(a^{m/n}) = (a^m)^{1/n} = (a^{1/n})^m$$

For example, $8^{2/3} = (8^{1/3})^2 = 2^2 = 4$ and also $(8^{2/3}) = (8^2)^{1/3} = (64)^{1/3} = 4$.

Orders of Magnitude

When comparing two objects of widely different sizes, it is common to use *orders of magnitude*. For example, to compare the radius of the universe to the radius of the solar system, calculate the ratio of the two sizes.

$$\frac{\text{radius of the universe}}{\text{radius of the solar system}} = \frac{10^{26} \text{ meters}}{10^{12} \text{ meters}}$$
$$= 10^{26-12}$$
$$= 10^{14}$$

The radius of the universe is 10^{14} times larger than the radius of the solar system, so it is 14 orders of magnitude larger than the solar system.

Each factor of 10 is counted as a single order of magnitude. When something is one order of magnitude larger than a *reference object*, it is 10 times larger. If it is one order of magnitude smaller, it is 10 times smaller or one-tenth the size.

Orders of magnitude or logarithmic scales are used to graph objects of widely differing sizes.

Logarithms

The *logarithm base 10 of x* is the power of 10 you need to produce x. So

$$\log_{10} x = c \qquad \text{means that} \qquad 10^c = x$$

We say c is the logarithm base 10 of x. For example, $\log_{10} 6{,}370{,}000 = 6.804$ means that $10^{6.804} = 6{,}370{,}000$.

Logarithms base 10 are called *common logarithms* and $\log_{10} x$ is frequently abbreviated as $\log x$.

Log x is not defined when $x \leq 0$.

EXERCISES

1. Write each expression as a power of 10.

   a. $10 \cdot 10 \cdot 10 \cdot 10 \cdot 10 \cdot 10$ **c.** one billion

   b. $\dfrac{1}{10 \cdot 10 \cdot 10 \cdot 10 \cdot 10}$ **d.** one thousandth

2. Express in standard decimal notation (without exponents).

   a. 10^{-7} **b.** 10^{7}

3. Express each in meters, using powers of 10.

   a. 10 cm   **c.** 3 terameters

   b. 4 km   **d.** 6 nanometers

4. Hubble's Law states that galaxies are receding from one another at velocities directly proportional to the distances separating them. The accompanying graph illustrates that Hubble's Law holds true across the known universe. The plot includes 10 major clusters of galaxies. The boxed area at the lower left represents the galaxies observed by Hubble when he discovered

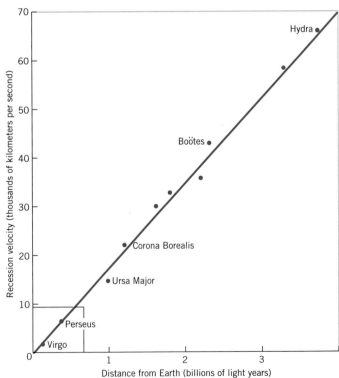

Source: Coming of Age in the Milky Way, by Timothy Ferris; copyright © by Timothy Ferris; by permission of William Morrow & Company, Inc.

the law. The easiest way to understand this graph is to think of Earth as being at the center of the universe (i.e., at 0 distance) and not moving (i.e., at 0 velocity). In other words, imagine Earth at the origin of the graph. (A favorite fantasy of humans.) Think of the horizontal axis as measuring the distance of a galaxy cluster from Earth, and the vertical axis as measuring the velocity at which a galaxy cluster is moving away from Earth. Then answer the following questions.

a. Identify the coordinates of two data points that lie on the regression line that is drawn in on the graph. (Hint: for the horizontal coordinate it is easier to use such numbers as 1 or 2 with the units being billions of light years, and for the vertical coordinate use such numbers as 10 or 20 with units in thousands of kilometers per second.)

b. Use the points in part (a) to calculate the slope of the line. That slope is called the Hubble constant.

c. What does the slope mean in terms of distance from Earth and recession velocity?

d. Construct an equation for your line in the form $y = mx + b$. Show your work.

5. Write each of the following in scientific notation.

a. 0.000 29   **c.** 720,000

b. 654.456   **d.** 0.000 000 000 01

6. Why are the following expressions *not* in scientific notation? Rewrite each in scientific notation.

a. $25 \cdot 10^4$ **b.** $0.56 \cdot 10^{-3}$

7. Write each of the following in standard decimal form.

a. $7.23 \cdot 10^5$ **c.** $1.0 \cdot 10^{-3}$

b. $5.26 \cdot 10^{-4}$ **d.** $1.5 \cdot 10^6$

8. a. Generate a small table of values and plot the function $y = |x|$.

   b. On the same graph, plot the function $y = |x - 2|$.

9. The following amusing graph shows a roughly linear relationship between the "scientifically" calculated age of Earth and the year the calculation was published. For instance in about 1935 Ellsworth calculated that Earth was about 2 billion years old. The age is plotted on the vertical axis and the year the calculation was published on the horizontal axis. The triangle on the horizontal coordinate represents the presently accepted age of Earth.

a. Who calculated that Earth was less than 1 billion years old? Give the coordinates of the points that give this information.

b. In about what year did scientists start putting the age of Earth at over a billion years old? Give the coordinates that represent this point.

c. On your graph sketch an approximation of a best fit line for these points. Use two points on the line to calculate the slope of the line.

d. Interpret the slope of that line in terms of the year of calculation and estimated age of Earth.

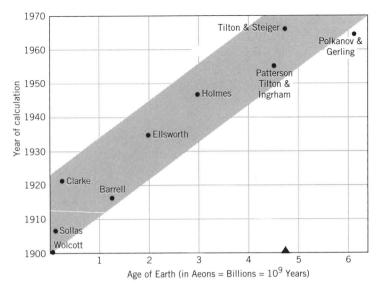

A graph of calculations for the age of Earth.

From *American Scientist,* © 1980, Research Triangle Park, North Carolina.

10.   Evaluate and write the results using scientific notation.

a. $(2.3 \cdot 10^4)(2.0 \cdot 10^6)$ **d.** $\dfrac{3.25 \cdot 10^8}{6.29 \cdot 10^{15}}$

b. $(3.7 \cdot 10^{-5})(1.1 \cdot 10^8)$ **e.** $(6.2 \cdot 10^{52})^3$

c. $\dfrac{8.19 \cdot 10^{23}}{5.37 \cdot 10^{12}}$ **f.** $(5.1 \cdot 10^{-11})^2$

11.   Write each of the following in scientific notation.

a. $725 \cdot 10^{23}$ **d.** $-725 \cdot 10^{23}$

b. $725 \cdot 10^{-23}$ **e.** $-725 \cdot 10^{-23}$

c. $\dfrac{1}{725 \cdot 10^{23}}$

12.   The Republic of China was estimated in 1995 to have about 1,260,000,000 people, and Monaco about 32,000. Monaco has an area of 0.75 sq. miles, and China has an area of 3,704,000 sq. miles.

a.   Express the populations and geographic areas in scientific notation.

b.   By calculating a ratio, determine how much larger China's population is than Monaco's.

c.   What is the population density (people per square mile) for each country?

d.   Write a paragraph comparing and contrasting the population size and density for these two nations.

13.   Simplify when possible, writing your answer as an expression with exponents.

a. $10^4 \cdot 10^3$

f. $4^7 + 5^2$

b. $10^4 + 10^3$

g. $\dfrac{z^7}{z^2}$

c. $10^3 + 10^3$

h. 256^0

d. $x^5 \cdot x^{10}$

i. $\dfrac{3^5 \cdot 3^2}{3^8}$

e. $(x^5)^{10}$

j. $4^5 \cdot (4^2)^3$

14. Estimate the values of the following expressions without a calculator. Show your work, writing your answers in scientific notation. Use a calculator to verify your answers.

a. $(0.000\ 359) \cdot (0.000\ 002\ 43)$

b. $(2{,}968{,}001{,}000) \cdot (189{,}000)$

c. $(0.000\ 079) \cdot (31{,}140{,}284{,}788)$

d. $\dfrac{4{,}083{,}693 \cdot 49{,}312}{213 \cdot 1945}$

15. The distance that light travels in 1 year (a light year) is $5.88 \cdot 10^{12}$ miles. If a star is $2.4 \cdot 10^8$ light years from Earth, what is this distance in miles?

16. A TV signal traveling at the speed of light takes about $8 \cdot 10^{-5}$ second to travel 15 miles. How long would it take the signal to travel a distance of 3000 miles?

17. Convert the following to feet and express your answers in scientific notation.

a. The radius of the solar system is approximately 10^{12} meters.

b. The radius of a proton is approximately 10^{-15} meters.

18. The speed of light is approximately $1.86 \cdot 10^5$ miles/sec.

a. Write this number in decimal form and express your answer in words.

b. Convert the speed of light into meters/year. Show your work.

19. The average distance from Earth to the sun is about 150,000,000 km and the average distance from the planet Venus to the sun is about 108,000,000 km.

a. Express these distances in scientific notation.

b. Divide the distance from Venus to the sun by the distance from Earth to the sun and express your answer in scientific notation.

c. The distance from Earth to the sun is called 1 astronomical unit (1 A.U.) How many A.U. is Venus from the sun?

d. Pluto is 5,900,000,000 km from the sun. How many A.U. is it from the sun?

20. The distance from Earth to the sun is approximately 150 million kilometers. If the speed of light is $3.00 \cdot 10^5$ km/sec, how long does it take light from the sun to reach Earth? If a solar flare occurs right now, how long would it take for us to see it?

21. The National Institutes of Health guidelines suggest that adults over 20 should have a body mass index, or BMI, of under 25. This index is created according to the formula:

$$\text{BMI} = \frac{\text{weight in kilograms}}{(\text{height in meters})^2}$$

a. Given that 1 kilogram = 2.2 pounds and 1 meter = 39.37 inches, calculate the body mass index of President Clinton who is 6 feet 2 inches tall and in 1997 weighed 216 pounds. According to the guidelines, was he overweight or underweight?

b. Most Americans don't use the metric system. So in order to make the BMI easier to use, convert the formula into an equivalent one using weight in pounds and height in inches. Check your new formula by using Clinton's weight and height and confirm that you get the same BMI.

c. The following excerpt from the article "America Fattens Up," (*The New York Times*, October 20, 1996) describes a very complicated process for determining your BMI.

> To estimate your body mass index you first need to convert your weight into kilograms by multiplying your weight in pounds by 0.45. Next, find your height in inches. Multiply this number by .0254 to get meters. Multiply that number by itself and then divide the result into your weight in kilograms. Too complicated? Internet users can get an exact calculation at http://141.106.68.17/bsa.acgl.

Can you do a better job of describing the process?

d. A letter to the editor from Brent Kigner, of Oneonta, N.Y., in response to *The New York Times* article says:

> Math intimidates partly because it is often made unnecessarily daunting. Your article "American Fattens Up" convolutes the procedure for calculating the Body Mass Index so much that you suggest readers retreat to the Internet. In fact, the formula is simple: Multiply your weight in pounds by 703, then divide by the square of your height in inches. If the result is above 25, you weigh too much.

Is Brent Kigner right?

22. Calculate the following:
    a. $4^{1/2}$ **e.** $8^{2/3}$
    b. $-4^{1/2}$ **f.** $-8^{2/3}$
    c. $27^{1/3}$ **g.** $16^{1/4}$
    d. $-27^{1/3}$ **h.** $16^{3/4}$

23. Evaluate when $x = 2$.
    a. $(-x)^2$ **d.** $(-x)^{1/2}$
    b. $-x^2$ **e.** $x^{3/2}$
    c. $x^{1/2}$ **f.** x^0

24. Constellation I

Reduce each of the following expressions to the form: $r^a \cdot n^b$; then plot the exponents as points with coordinates (a, b) on graph paper. Do you recognize the constellation?

a. $\dfrac{(r^5)^2}{n^6 \cdot r^{-1}}$ **d.** $\dfrac{n}{r^{0.8}}$ **g.** $\dfrac{1}{r^{-0.3} \cdot n^{0.6}}$

b. $\dfrac{r^{10}}{r^{21} \cdot (n^{-2})^3}$ **e.** $\dfrac{r^{-6} \cdot r^{-13} \cdot r^7}{(r \cdot n^2)^3}$

c. $\dfrac{(r \cdot n^5)^0 \cdot r \cdot n^3}{r^{-0.4} \cdot n^5}$ **f.** $(r^3 \cdot n^2)^3$

Constellation II

Reduce each of the following expressions to the form $u^a \cdot m^b$; then plot the exponents as points with coordinates (a, b) on graph paper. Do you recognize the constellation?

a. $\dfrac{(u^2)^2 \cdot m}{u^2 \cdot m^{-4}}$ **d.** $\dfrac{(um^2)^3 \cdot u^2}{(um)^4}$ **g.** $\dfrac{(mu)^0 \cdot (u^{10})^{-1} \cdot m^{1/4}}{(m^{-3} \cdot u^{-1/3})^3}$

b. $\dfrac{u^{-9/5} \cdot m^3}{(umu^2)^1 \cdot m^{-1}}$ **e.** $\dfrac{u^{-3/2} \cdot u^{-7/2} \cdot m^1 \cdot (m^3)^3}{(um)^2}$

c. $\dfrac{u^2 \cdot u^{-4}}{u^3 \cdot (m^{-2})^3}$ **f.** $\dfrac{1}{u^{12} \cdot m^{-9}}$

Constellation III

Reduce each of the following expressions to the form $p^a \cdot c^b$; then plot the exponents as points with coordinates (a, b) on graph paper. What constellation appears?

a. $\dfrac{p^7 \cdot p^{-3} \cdot (c^7)^0}{p^4}$ **c.** $\dfrac{p^{-5}}{p^{2.5} \cdot p^2 \cdot c^{-2}}$ **e.** $\dfrac{1}{(p^{0.5})^3 \cdot (c^{-13/9})^3}$

b. $\dfrac{(p^{-2})^{-3} \cdot c^4}{p^{-5} \cdot c^{7.5}}$ **d.** $\dfrac{(p^2 \cdot c^3)^2}{c^5}$

25. Without using a calculator, find two *consecutive* integers such that one is smaller and one is larger than each of the following (for example $3 < \sqrt{11} < 4$). Show your reasoning.

a. $\sqrt{13}$ **b.** $\sqrt{22}$ **c.** $\sqrt{40}$

Now verify your answers using a calculator.

26. The breaking strength S (in pounds) of a three-strand manila rope is a function of its diameter, D (in inches). The relationship can be described by the equation: $S = 1700\, D^{1.9}$. Calculate the breaking strength when D equals:

a. 1.5 in.   **b.** 2.0 in.

27. If a rope is wound around a wooden pole, the pounds of frictional force, F, between the pole and the rope is a function of the number of turns, N, according to the equation $F = 14 \cdot 10^{0.70N}$. What is the frictional force when the number of turns is

a. 0.5   **b.** 1   **c.** 3

28. Water boils (changes from a liquid to a gas) at 100°C (degrees Centigrade). The temperature of the core of the sun is 20 million degrees Centigrade. By how many orders of magnitude is the sun's core hotter than the boiling temperature of water?

29. Determine the order of magnitude difference in the sizes of the radii for
    a. the solar system (10^{12} meters) compared to Earth (10^7 meters)
    b. protons (10^{-15} meter) compared to the Milky Way (10^{21} meters)
    c. atoms (10^{-10} meter) compared to quarks (10^{-16} meter)

30. In order to compare the sizes of different objects, we need to use the same unit of measure.
    a. Convert each of these to meters.
        The radius of the moon is approximately 1,922,400 yards.
        The radius of Earth is approximately 6300 km.
        The radius of the sun is approximately 432,000 miles.
    b. Determine the order of magnitude difference between:
        i. the surface areas of the moon and Earth.
        ii. the volumes of the sun and the moon.

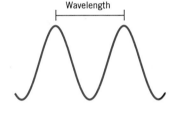

Wavelength

31. Light or electric and magnetic fields (such as X-rays and radio waves) can be thought of as radiation waves moving through space at the speed of light— $3.00 \cdot 10^5$ km (or 186,000 miles) per second. The distance between the crest of one wave to the crest of the next is called the wavelength.

    These radiation waves are referred to as *electromagnetic radiation*. We can draw the entire *electromagnetic spectrum* on an order of magnitude (or logarithmic) scale that shows the wavelengths in centimeters from cosmic rays to long wave radio waves.

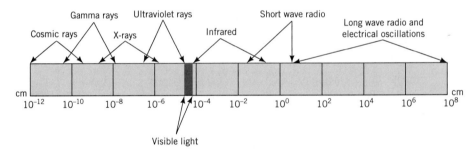

The electromagnetic spectrum.

    a. Within the range of X-ray wavelengths, what is a reasonable size for an "average" wavelength of an X-ray?
    b. What is a reasonable size for an "average" long wave radio wave?
    c. How many orders of magnitude larger is the radio wave than the X-ray?
    d. Very short wavelengths are often measured in Angstroms. One Angstrom, written 1 Å, equals 10^{-8} cm. What is the approximate range of wavelengths of ultraviolet rays in Angstroms? (Sunscreen or sunglasses are commonly used to provide protection against ultraviolet light.)
    e. What is the approximate range of values for wavelengths of infrared radiation in Angstroms? (Although human vision cannot detect infrared radiation, night vision scopes like those used by the military can pick up infrared signals emitted by warm bodies.)

32. a. Read the chapter entitled "The Cosmic Calendar" from Carl Sagan's book *The Dragons of Eden*.

b. Carl Sagan tried to give meaning to the cosmic chronology by imagining the 15-billion-year lifetime of the universe compressed into the span of one calendar year. To get a more personal perspective, consider your date of birth as the time the Big Bang took place. Map the following five cosmic events onto your own life span.

   The Big Bang

   Creation of Earth

   First Life on Earth

   First *Homo sapiens*

   American Revolution

Once you have done the necessary mathematical calculations and placed your results on either a chart or a timeline, then form a topic sentence and write a playful paragraph about what you were supposedly doing when these cosmic events took place. Hand in your calculations along with your writing.

33. Rewrite the following statements using logs.

   a. $10^2 = 100$ **b.** $10^7 = 10,000,000$ **c.** $10^{-2} = 0.01$

Rewrite the following statements using exponents

   d. $\log 10 = 1$ **e.** $\log 10,000 = 4$ **f.** $\log 0.000\,1 = -4$

34. Do the following exercises without a calculator.

   a. Find the following values.

   $\log 100$

   $\log 1000$

   $\log 10,000,000$

   What is happening to the values of $\log x$ as x gets larger?

   b. Find the following values.

   $\log 0.1$

   $\log 0.001$

   $\log 0.000\,01$

   What is happening to the values of $\log x$ as x gets closer to 0?

   c. What is $\log 0$?

   d. What is $\log(-10)$? What do you know about $\log x$ when x is any negative number?

35. Without using a calculator for each number in the form $\log x$, find some numbers a and b such that $a < \log x < b$. Justify your answer. Then verify your answers with a calculator.

   a. $\log 11$ **b.** $\log 12,000$ **c.** $\log 0.125$

36. Use a calculator to determine the following logs. Double check each answer by writing down the equivalent expression using exponents, and then verify using a calculator.

   a. $\log 11$ **b.** $\log 12,000$ **c.** $\log 0.125$

37. On a logarithmic scale, what would correspond to moving over:
    a. 0.001 unit   **b.** 1/2 a unit   **c.** 2 units   **d.** 10 units

38. The difference in the noise level of two sounds is measured in decibels, where decibels $= 10 \log (I_2/I_1)$ and I_1 and I_2 are the intensities of the two sounds. Calculate the difference in noise level when $I_1 = 10^{-15}$ watts/cm^2 and $I_2 = 10^{-8}$ watts/cm^2.

39. The concentration of hydrogen ions in a water solution typically ranges from 10 M to 10^{-15} M. (We have that 1 M equals $6.02 \cdot 10^{23}$ molecules per liter or 1 mole per liter.)[10] Because of this wide range, chemists use a logarithmic scale, called the pH scale, to measure the concentration. The pH is given by the equation pH $= -\log [H^+]$, where $[H^+]$ denotes the concentration of hydrogen ions. Chemists use the symbol H^+ for hydrogen ions and the brackets [ ] mean "the concentration of."

    a. Pure water at 25°C has a hydrogen ion concentration of 10^{-7} M. What is the pH?

    b. In orange juice, $[H^+] \approx 1.4 \cdot 10^{-3}$ M. What is the pH?

    c. Household ammonia has a pH of about 11.5. What is $[H^+]$?

    d. Does a higher pH indicate a lower or a higher concentration of hydrogen ions?

    e. A solution with a pH > 7 is called basic; pH $= 7$ is called neutral; and pH < 7 is called acidic. Identify pure water, orange juice, and household ammonia as either acidic, neutral, or basic. Then plot their positions on the following scale, which shows both the pH and the hydrogen ion concentration.

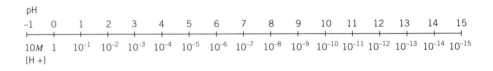

---

[10] You may recall from Algebra Aerobics 7.2 that $6.02 \cdot 10^{23}$ is called Avogadro's number. A mole of a substance is defined as Avogadro's number of molecules of that substance. M is called a molar unit.

EXPLORATION 7.1

The Scale and the Tale of the Universe

Objective

- gain an understanding of the relative size and relative age of objects in the universe using scientific notation and unit conversions

Materials/Equipment

- tape, pins, paper, and string to generate a large wall graph (optional)
- enclosed worksheet and conversion table

Related Readings

Excerpts from *Powers of Ten* and "The Cosmic Calendar" from *The Dragons of Eden*

Related Videos

Powers of Ten and "The Cosmic Calendar" in the PBS series *Cosmos*

Related Software

"E1: Tale and Scale of the Universe" in *Exponential & Log Functions*

Procedure

Work in small groups. Each group should work on a separate subset of objects on the enclosed worksheet.

1. Convert the age and size of objects so they can be compared. You can refer to the conversion table that shows equivalences between English and metric units (Table 7.1 on page 215) and the footnote that lists the meaning of the metric prefixes such as kilo-, giga-, and tera- (note 4 on page 218.)

2. Generate on the blackboard or on the wall a blank graph whose axes are marked off in orders of magnitude (integer powers of 10), with the units on the vertical axis representing age of object, ranging from 10^0 to 10^{10} years, and the units on the horizontal axis representing size of object, ranging from 10^{-12} to 10^{27} years.

3. Each small group should plot the approximate coordinates of their selected objects (age in years, size in meters) on the graph. You might want to draw and label a small picture of your object to plot on your graph.

Discussion/Analysis

- Scan the plotted objects from left to right, looking only at relative size. Does your graph make sense in terms of what you know about the size of these objects?
- Now scan the plotted objects from top to bottom, only considering relative age. Does your graph make sense in terms of what you know about the relative ages of these objects?
- Compare the largest and smallest objects in the table. By how many orders of magnitude do they differ? Compare the sizes of other pairs of objects in the table. Compare the ages of the oldest and youngest objects in the table. By how many orders of magnitude do they differ?

Object	Age (in years)	Size (of radius)	In Scientific Notation Age (in years)	Size (in meters)
Observable universe	15 billion (?)	10^{26} meters		
Surtsey (Earth's newest land mass)	35 years	0.5 miles		
Pleiades (a galactic cluster)	100 million	32.6 light years		
First living organisms on Earth	3.5 billion	0.000 05 meters		
Pangaea (Earth's prehistoric supercontinent)	200 million	4500 miles		
First *Homo sapiens sapiens*	100 thousand	100 centimeters		
First *Tyrannosaurus rex*	200 million	20 feet		
Eukaryotes (first cells with nuclei)	2 billion	0.000 05 meters		
Earth	5 billion	6400 kilometers		
Milky Way galaxy	14 billion	50,000 light years		
First atoms	15 billion	0.000 000 000 1 meters		
Our sun	5 billion	1 gigameter		
Our solar system	5 billion	1 terameter		

EXPLORATION 7.2

Patterns in the Position and Motion of the Planets

Objective

- explore patterns in the positions and motions of the planets and discover Kepler's law

Introduction and Procedure

Four hundred years ago, before Newton's mechanics, Johannes Kepler discovered a law that relates the periods of planets with their average distance from the sun. (A period of a planet is the time it takes a planet to complete one orbit of the sun.) Kepler's strong belief that the solar system was governed by harmonious laws drove him to try to discover hidden patterns and correlations among the positions and motions of the planets. He used the trial-and-error method and continued his search for years.

At the time of his work, Kepler did not know the distance from the sun to each planet in terms of measures of distance such as the kilometer. But he was able to determine the distance from each planet to the sun in terms of the distance from Earth to the sun, now called the astronomical unit or A.U. for short. The A.U. is a unit in which the distance from Earth to the sun is equal to 1. The first column in the table below gives the average distance from the sun to each of the planets in astronomical units or A.U.

Patterns in the Position and Motion of the Planets: Kepler's Discovery

Fill in the following table and look for the relationship that Kepler found.

Kepler's Third Law

		The First Planet Table (Inner Planetary System)		
Planet	Average Distance from Sun (A.U.)*	Cube of the Distance (A.U.^{3})	Orbital Period (years)	Square of the Orbital Period (years^2)
Mercury	0.3870		0.2408	
Venus	0.7232		0.6151	
Earth	1.0000		1.0000	
Mars	1.5233		1.8807	
Jupiter	5.2025		11.8619	
Saturn	9.5387		29.4557	

* 1 A.U. = 149.6 · 10^6 km; 1 year = 365.26 days.

Data from S. Parker and J. Pasachoff, *Encyclopedia of Astronomy,* 2nd ed. Table 1. Elements of Planetary Orbits. Copyright © 1993. McGraw-Hill, Inc. Reprinted with permission.

The planets Uranus, Neptune, and Pluto were discovered after Kepler made his discovery. Check to see whether the relationship you found above holds true for these three planets.

The Second Planet Table (Outer Planetary System)				
Planet	Average Distance from Sun (A.U.)	Cube of the Distance (A.U.^{3})	Orbital Period (years)	Square of the Orbital Period (years^2)
Uranus	19.1911		84.0086	
Neptune	30.0601		164.7839	
Pluto	39.5254		248.5900	

Data from S. Parker and J. Pasachoff, *Encyclopedia of Astronomy,* 2nd ed. Table 1. Elements of Planetary Orbits. Copyright © 1993. McGraw-Hill, Inc. Reprinted with permission.

Summary

- Express your results in words.
- Construct an equation showing the relationship between distance from the sun and orbital period. Solve the equation for distance from the sun. Then solve the equation for orbital period.
- Do your conclusions hold for all of the planets?

Growth and Decay

Overview

For a biologist it is important to understand how populations grow. Determining the best strategy for managing lobster fisheries off the coast of New England requires an understanding of the growth or decline of the lobster population. Controlling insect pests, such as mosquitoes and ticks, or maximizing the yield of penicillin from the fungus *Penicillium notatum* requires a knowledge of population dynamics.

We have examined linear models that are used to describe phenomena that grow or decline at a constant rate of change. This chapter introduces the family of *exponential functions* used to describe phenomena that grow or decay faster and faster over time. Exponential functions can be used to model a bacterial population that doubles every 20 minutes or a radioactive element that decays by 50% every 3000 years. They can also be used to model compound interest rates, musical pitch, inflation, and many other phenomena.

The Explorations help you explore the mathematical properties of exponential functions and give you experience fitting a model to laboratory data on the growth of *E. coli* bacteria under differing conditions.

After reading this chapter you should be able to:

- recognize the properties of exponential functions
- understand the differences between exponential and linear growth
- use exponential functions to model growth and decay
- plot exponential functions on a semi-log graph

8.1 MODELING GROWTH

The Growth of *E. coli* Bacteria

In Exploration 8.1 you can study the growth of E. coli *under varying conditions.*

Measuring and predicting growth is of concern to population biologists, ecologists, demographers, economists, and politicians alike. By studying the growth of a population of the bacterium *E. coli*, we can construct a simple model that can be generalized to describe the growth of other phenomena such as cells, countries, or capital.

Bacteria are very tiny, single-celled organisms that are by far the most numerous organisms on Earth. One of the most frequently studied bacteria is *E. coli*, a rod-shaped bacterium approximately 10^{-6} meter (or 1 micron) long that inhabits the intestinal tracts of humans.[1] The cells of *E. coli* reproduce by a process called fission: the cell splits in half, forming two "daughter cells."

The rate at which fission occurs depends on several conditions, in particular, available nutrients and temperature. Under ideal conditions *E. coli* can divide every 20 minutes. If we start with an initial population of 100 *E. coli* bacteria that doubles during every 20-minute time period, we generate the data in Table 8.1. The initial 100 bacteria double to become 200 bacteria at the end of the first time period, double again to become 400 at the end of the second time period, and keep on doubling to become 800, 1600, 3200, etc., at the ends of successive time periods. At the end of the twenty-fourth time period (at $24 \cdot 20$ minutes $= 480$ minutes or 8 hours), the initial 100 bacteria in our model have grown to over 1.6 billion bacteria!

Because the numbers become astronomically large so quickly, we run into the problems we saw in Chapter 7 when graphing numbers of widely different sizes.

Figure 8.1 shows a graph of the data in Table 8.1 for only the first 10 time periods. We can see from the graph that the relationship between number of bacteria and time is not linear. The number of bacteria seems to be increasing more and more rapidly over time.

Table 8.1

Growth of *E. coli* Bacteria

Time Periods (of 20 minutes each)	No. of *E. coli* Bacteria
0	100
1	200
2	400
3	800
4	1,600
5	3,200
6	6,400
7	12,800
8	25,600
9	51,200
10	102,400
11	204,800
12	409,600
13	819,200
14	1,638,400
15	3,276,800
16	6,553,600
17	13,107,200
18	26,214,400
19	52,428,800
20	104,857,600
21	209,715,200
22	419,430,400
23	838,860,800
24	1,677,721,600

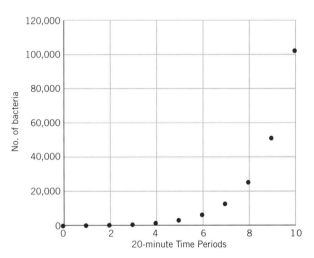

Figure 8.1   Growth of *E. coli* bacteria.

[1] Most types of *E. coli* are beneficial to humans, aiding in human digestion. A few types are lethal. You may have read about deaths resulting from people eating certain deadly strains of *E. coli* bacteria in undercooked hamburgers.

A mathematical model for E. coli growth   If we examine the values in Table 8.1, we see that the initial number of 100 bacteria is repeatedly doubled or multiplied by 2. If we record in a third column (Table 8.2) the number of times we multiply 2 times the original value of 100, we can begin to see a pattern emerge.

Time Period	Number of E. coli Bacteria	Generalized Expression
		Table 8.2
		Pattern in *E. coli* Growth
0	100	$100 = 100 \cdot 2^0$
1	200	$100 \cdot 2 = 100 \cdot 2^1$
2	400	$100 \cdot 2 \cdot 2 = 100 \cdot 2^2$
3	800	$100 \cdot 2 \cdot 2 \cdot 2 = 100 \cdot 2^3$
4	1,600	$100 \cdot 2 \cdot 2 \cdot 2 \cdot 2 = 100 \cdot 2^4$
5	3,200	$100 \cdot 2 \cdot 2 \cdot 2 \cdot 2 \cdot 2 = 100 \cdot 2^5$
6	6,400	$100 \cdot 2 \cdot 2 \cdot 2 \cdot 2 \cdot 2 \cdot 2 = 100 \cdot 2^6$
7	12,800	$100 \cdot 2 \cdot 2 \cdot 2 \cdot 2 \cdot 2 \cdot 2 \cdot 2 = 100 \cdot 2^7$
8	25,600	$100 \cdot 2 \cdot 2 \cdot 2 \cdot 2 \cdot 2 \cdot 2 \cdot 2 \cdot 2 = 100 \cdot 2^8$
9	51,200	$100 \cdot 2 \cdot 2 \cdot 2 \cdot 2 \cdot 2 \cdot 2 \cdot 2 \cdot 2 \cdot 2 = 100 \cdot 2^9$
10	102,400	$100 \cdot 2 \cdot 2 \cdot 2 \cdot 2 \cdot 2 \cdot 2 \cdot 2 \cdot 2 \cdot 2 \cdot 2 = 100 \cdot 2^{10}$

Remembering that 2^0 equals 1 by definition, we can describe the relationship by

$$\text{number of } E.\ coli \text{ bacteria} = 100 \cdot 2^{\text{time period}}$$

If we let N = number of bacteria and t = time period, we can write the equation more compactly as

$$N = 100 \cdot 2^t$$

Since each value of t determines one and only one value for N, the equation represents N as a function of t. If we call the function f, we can write $N = f(t)$ where $N = 100 \cdot 2^t$. The domain of the function, as a model of bacteria growth, includes values of t between 0 and some unspecified positive number since in theory the number of bacteria could increase without bound. This function is called *exponential* since the independent variable, t, occurs in the exponent of the *base* 2. The number 100 is the *initial bacteria population*. We also call 2 the *growth factor*, or the multiple by which the population grows during each time period. So the bacteria population is doubling or, stated another way, is growing by 100% during each time period.

The program "E2: Exponential Growth & Decay" in Exponential & Log Functions *offers a dramatic visualization of growth and decay.*

Linear versus Exponential Growth

The bacteria population grows faster and faster as time goes on, in a behavior typical of exponential functions. Exponential growth is *multiplicative*, which means that for each unit increase in the independent variable, we *multiply* the value of the dependent variable by the growth factor. Contrast this with linear growth, which is *additive*. Linear growth means that for each unit increase in the independent variable, we must *add* a fixed amount to the value of the dependent variable. In the exponential function, $N = 100 \cdot 2^t$, for each increase of one unit in time, t, we *multiply* the previous value of N by 2; but in the linear function, $N = 100 + 2t$, for each increase of one unit in time, t, we *add* 2 to the previous value of N. Table 8.2 shows the multiplicative nature of the exponential function $N = 100 \cdot 2^t$; Table 8.3 shows the additive nature of the linear function $N = 100 + 2t$.

| | | Table 8.3 | |
| | | **Additive Pattern of a Linear Function** | |
t	$N = 100 + 2t$	**Generalized Expression**	
0	100	$100 = 100 + 2 \cdot 0$	
1	102	$100 + 2 = 100 + 2 \cdot 1$	
2	104	$100 + 2 + 2 = 100 + 2 \cdot 2$	
3	106	$100 + 2 + 2 + 2 = 100 + 2 \cdot 3$	
4	108	$100 + 2 + 2 + 2 + 2 = 100 + 2 \cdot 4$	
5	110	$100 + 2 + 2 + 2 + 2 + 2 = 100 + 2 \cdot 5$	
6	112	$100 + 2 + 2 + 2 + 2 + 2 + 2 = 100 + 2 \cdot 6$	
7	114	$100 + 2 + 2 + 2 + 2 + 2 + 2 + 2 = 100 + 2 \cdot 7$	
8	116	$100 + 2 + 2 + 2 + 2 + 2 + 2 + 2 + 2 = 100 + 2 \cdot 8$	
9	118	$100 + 2 + 2 + 2 + 2 + 2 + 2 + 2 + 2 + 2 = 100 + 2 \cdot 9$	
10	120	$100 + 2 + 2 + 2 + 2 + 2 + 2 + 2 + 2 + 2 + 2 = 100 + 2 \cdot 10$	

Exponential growth is more rapid as time goes on, as can be seen in Table 8.4. In both functions when $t = 0$, $N = 100$. After 10 time periods, the values for N are strikingly different: 102,400 for the exponential function versus 120 for the linear function. After 10 time periods, the initial value of 100 has been multiplied by 2 ten times in the exponential function to get $100 \cdot 2^{10}$. In the linear function, 2 has been added to 100 ten times, to get $100 + 2 \cdot 10$.

| | | Table 8.4 | | |
| | | **Comparing Additive Pattern of Linear Growth to Multiplicative Pattern of Exponential Growth** | | |
Time, t	**Linear Function, $N = 100 + 2t$ Pattern**		**Exponential Function, $N = 100 \cdot 2^t$ Pattern**	
0	$100 + 2 \cdot 0$	$= 100$	$100 \cdot 2^0 =$	100
1	$100 + 2 \cdot 1$	$= 102$	$100 \cdot 2^1 =$	200
2	$100 + 2 \cdot 2$	$= 104$	$100 \cdot 2^2 =$	400
3	$100 + 2 \cdot 3$	$= 106$	$100 \cdot 2^3 =$	800
4	$100 + 2 \cdot 4$	$= 108$	$100 \cdot 2^4 =$	1,600
5	$100 + 2 \cdot 5$	$= 110$	$100 \cdot 2^5 =$	3,200
6	$100 + 2 \cdot 6$	$= 112$	$100 \cdot 2^6 =$	6,400
7	$100 + 2 \cdot 7$	$= 114$	$100 \cdot 2^7 =$	12,800
8	$100 + 2 \cdot 8$	$= 116$	$100 \cdot 2^8 =$	25,600
9	$100 + 2 \cdot 9$	$= 118$	$100 \cdot 2^9 =$	51,200
10	$100 + 2 \cdot 10$	$= 120$	$100 \cdot 2^{10} =$	102,400

If we compare any linear growth function (whose graph will be a straight line with a positive slope) to any exponential growth function (whose graph will curve upward), it's not hard to see that sooner or later the exponential curve will permanently lie above the linear graph and continue to grow faster and faster (Figure 8.2). We say that the exponential function eventually dominates the linear function.

> Every exponential growth function will eventually dominate every linear growth function.

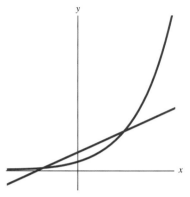

Figure 8.2 Graph comparing linear and exponential growth.

Algebra Aerobics 8.1a

1. Fill in the following table.

t	$N = 10 + 3t$	$N = 10 \cdot 3^t$
0		
1		
2		
3		
4		

2. Sketch a graph of the linear function $N = 10 + 3t$ for $0 \le t \le 10$. Is the slope constant? Justify your answer. What is the vertical intercept?

3. Sketch the graph of the exponential function $N = 10 \cdot 3^t$ for $0 \le t \le 4$. Compare this graph to that of the linear function in Problem 2.

4. On one coordinate plane, sketch the functions in Problem 2 and Problem 3 for $0 \le t \le 2$. Compare the graphs.

The Average Rate of Change Between Points for Linear versus Exponential Functions

Another way to compare linear and exponential functions is to examine average rates of change. We can ask how each function changes over time. Recall that if N is a function of t, then

$$\text{average rate of change} = \frac{\text{change in } N}{\text{change in } t} = \frac{\Delta N}{\Delta t}$$

Examine Table 8.5 which contains the average rate of change for both the linear and exponential function, where $\Delta t = 1$. For all linear functions, we know that the average rate of change is constant. For the linear function $N = 100 + 2t$, we can tell from the equation, Table 8.5, and Figure 8.3 that the (average) rate of change is constant at 2 units. Average rates of change for the exponential function $N = 100 \cdot 2^t$ are calculated in Table 8.5 and graphed in Figure 8.4. These suggest that the average rates of change of an exponential growth function grow *exponentially*.

Table 8.5

Comparing Average Rate of Change Calculations

	Linear Function		Exponential Function	
	$N = 100 + 2t$	Average Rate of Change (Between t and $t - 1$)	$N = 100 \cdot 2^t$	Average Rate of Change (Between t and $t - 1$)
t	N		N	
0	100	n.a.	100	n.a.
1	102	2	200	100
2	104	2	400	200
3	106	2	800	400
4	108	2	1,600	800
5	110	2	3,200	1,600
6	112	2	6,400	3,200
7	114	2	12,800	6,400
8	116	2	25,600	12,800
9	118	2	51,200	25,600
10	120	2	102,400	51,200

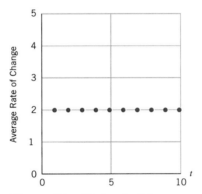

Figure 8.3   The graph of the average rates of change between several pairs of points on the linear function $N = 100 + 2t$.

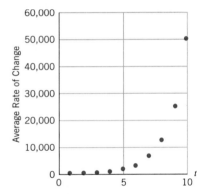

Figure 8.4   The graph of the average rates of change between several pairs of points on the exponential function $N = 100 \cdot 2^t$.

The Exponential Function in the Abstract

We can detach the function $N = f(t)$ where $N = 100 \cdot 2^t$, from any physical context and treat it not as a model relating *E. coli* and time, but as a purely mathematical relationship between two abstract variables, t and N. The domain (the set of possible values for t) is then all the real numbers instead of only values of $t \geq 0$ used in the model. Table 8.6 shows some values for $N = f(t)$ when $-10 < t < +10$.

Remember that

$$2^{-2} = \frac{1}{2^2} = \frac{1}{4} = 0.25 \qquad \text{so} \qquad f(-2) = 100(2^{-2}) = 100\left(\frac{1}{4}\right) = 25$$

$$2^{-4} = \frac{1}{2^4} = \frac{1}{16} = 0.0625 \qquad \text{so} \qquad f(-4) = 100(2^{-4}) = 100\left(\frac{1}{16}\right) = 6.25$$

t	$N = 100 \cdot 2^t$	t	$N = 100 \cdot 2^t$
−10	0.10	2	400.00
−8	0.39	4	1,600.00
−6	1.56	6	6,400.00
−4	6.25	8	25,600.00
−2	25.00	10	102,400.00
0	100.00		

Table 8.6

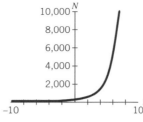

Figure 8.5 Graph of $N = f(t) = 100 \cdot 2^t$.

As values of t move from right to left (decreasing from 0, to −10, to −100, and so on), values of 2^t and therefore of $100 \cdot 2^t$ remain positive but are getting closer and closer to zero. Hence the graph always lies above the t-axis, coming extremely close to it, but never touching it. We say the graph is *asymptotic* to the t-axis. The graph of $f(t)$ in Figure 8.5 may appear to touch the t-axis, but that's only because the values of N become tiny relative to the scale on the N-axis.

The value $f(0)$ (the value of N when $t = 0$) is often referred to as the *initial quantity*. From Table 8.6 we can see that in this case the initial quantity is 100. The base 2 represents the *growth factor*. Each time t increases by 1, the value of N doubles (is multiplied by 2 or increases by 100%).

Algebra Aerobics 8.1b

Make a table and sketch the graph of the function $f(x) = 3^x$ for $-3 \leq x \leq 3$. Is $f(x)$ ever negative? Does the graph ever touch the x-axis, that is, is $f(x)$ ever zero?

Describe the behavior of $f(x)$ as x increases beyond 3. As x decreases beyond −3.

Looking at Real Growth Data for *E. coli* Bacteria

The idealized model for *E. coli* bacteria (in Table 8.1) showed phenomenal growth. It is not surprising that when we actually go to a biology lab and watch real bacteria grow, the growth is substantially slower. Conditions are rarely ideal; bacteria die, nutrients are used up, and temperatures may not be optimal. Table 8.7 and Figure 8.6 on the next page show some real data collected by a graduate student in biology. A solution was inoculated with an indeterminate amount of bacteria at time $t = 0$, and then the number of *E. coli* cells per milliliter was measured at regular time periods of 20 minutes each. For the first two time periods, even though there were probably millions of bacteria per milliliter, the amount was too small to measure.

Using a curve-fitting program to find a best fit exponential function to model the *E. coli* growth, we get

$$N = (1.37 \cdot 10^7) \cdot 1.5^t$$

where t represents the number of 20-minute time periods and $1.37 \cdot 10^7$ (or 13.7 mil-

Table 8.7

Growth of Real *E. coli* for 12 Time Periods

Time Period (20 min each)	Number of Cells/ml
0	n.a.
1	n.a.
2	n.a.
3	$3.00 \cdot 10^7$
4	$7.00 \cdot 10^7$
5	$1.00 \cdot 10^8$
6	$1.80 \cdot 10^8$
7	$2.70 \cdot 10^8$
8	$4.25 \cdot 10^8$
9	$5.80 \cdot 10^8$
10	$8.00 \cdot 10^8$
11	$1.00 \cdot 10^9$
12	$1.16 \cdot 10^9$

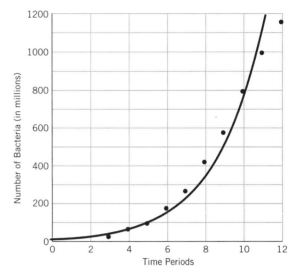

Figure 8.6   Real *E. coli* growth for 12 time periods.

lion) is the estimated number of *E. coli* cells at $t = 0$, when the bacteria were initially placed in the broth. The base of the exponential function, 1.5, is the *growth factor*. For each increase of 1 unit in time t, N is multiplied by 1.5, which increases the value of N by 50%. We call 50%, the percentage increase over each unit of time, the *growth rate*. So in the real experiment, the number of bacteria actually increased by 50% each 20-minute time period, instead of doubling or increasing by 100% as in our first idealized example.

Limitations of the Model

Clearly our model described the potential, not the actual, growth of *E. coli* bacteria. If *E. coli* bacteria really doubled every 20 minutes, then the offspring from a single cell could cover the entire surface of Earth with a layer a foot deep in fewer than 36 hours! The bacterial growth rate in the lab was half of that of the idealized example, but even this growth rate can't be sustained for long. At the slower rate of a 50% increase every 20 minutes, the offspring of a single cell would still cover the surface of Earth with a foot-deep layer in fewer than 3 days. Under most circumstances exponential growth is quickly restricted by environmental limits imposed by shortages of food, space, or oxygen; by predation; or by the accumulation of waste products.

So, while the growth may be exponential at first, it eventually slows down. Figure 8.7 shows the growth of *E. coli* over 24 time periods (or 8 hours), instead of the original 12 time periods (or 4 hours). The first part of the curve (up to $t = 12$) repeats the data points from Table 8.7 and Figure 8.6 with their exponential growth pattern. Then the growth slows down and the curve flattens out as N approaches a maximum population size, called the *carrying capacity*, indicated by the dotted line. Hence the growth rate, the percentage increase for each time period, must be decreasing. This S-shaped curve, typical of real population growth, is called a *sigmoid*, for sigma, the Greek letter S. Our exponential model is applicable to only a part of the growth curve, for values of t roughly between 0 and 12 time periods. The arithmetic of exponentials leads us to the inevitable conclusion that in the long term, the *growth rate* of populations—whether they consist of bacteria or mosquitoes or humans—must approach zero.

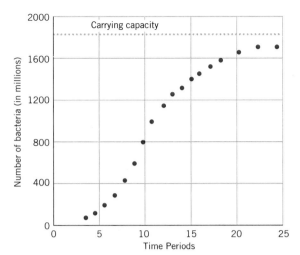

Figure 8.7   Plot of *E. coli* growth for 24 time periods, showing an initial exponential growth that tapers off.

See Excel or graph link file ECOLI.

8.2 VISUALIZING EXPONENTIAL FUNCTIONS

The bacterial growth model belongs to a class of functions called exponential.

An *exponential function* $y = f(x)$ can be represented by an equation of the form

$$y = C \cdot a^x \qquad (a > 0 \text{ and } a \neq 1)$$

where C is the initial value of y (when $x = 0$) and a is the factor by which y is multiplied when x increases by one.

In exponential functions the independent variable is the exponent that is applied to a fixed base. Exponential functions have the form

$$\text{dependent variable} = C \cdot a^{\text{independent variable}}$$
$$y = C \cdot a^x$$

Something to think about

What would the graph of the function $y = C \cdot a^x$ look like if $a = 1$? What kind of function would you have?

Predicting What the Graph of an Exponential Function Will Look Like

Given any exponential function in the form

$$y = C \cdot a^x \qquad \text{where } a > 0$$

we can predict the general shape of its graph. By examining values for a and C it is possible to tell the direction and relative steepness of the curve, and where the curve crosses the y-axis.

You may want to start Exploration 8.2 "Properties of Exponential Functions" now.

The Effect of the Base a

Case 1: $a > 1$ We compare the graphs of three different exponential functions. Each function has $C = 1$, but different values for a, namely $a = 2$, $a = 3$, and $a = 4$.

$$y = 2^x \qquad y = 3^x \qquad y = 4^x$$

To find specific values for the function $y = 2^x$, we need to apply the rules of exponents. For example, when $y = 2^x$ and $x = 3$, then $y = 2^3 = 2 \cdot 2 \cdot 2 = 8$.

When $y = 2^x$ and $x = -2$, then

$$y = 2^{-2} = \frac{1}{2^2} = \frac{1}{4}$$

You can use the program "E3: $y = Ca^x$ Sliders" in Exponential & Log Functions to explore the effects of a and C.

Examine Table 8.8 and Figure 8.8 for the values and graphs of the functions $y = 2^x$, $y = 3^x$, and $y = 4^x$.

Table 8.8			
Values for Three Exponential Functions Where $a > 0$			
x	$y = 2^x$	$y = 3^x$	$y = 4^x$
-2	1/4	1/9	1/16
-1	1/2	1/3	1/4
0	1	1	1
1	2	3	4
2	4	9	16
3	8	27	64
4	16	81	256

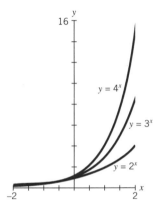

Figure 8.8   When $a > 1$.

What happens to the values for y as x increases? For each of these curves, as the value of x increases, the value of y also increases and the curve slopes upward. Thus, exponential functions where the base $a > 1$, are used to model *growth*.

For each function, y is greater than zero for any value of x, so the graphs lie completely above the x-axis, never touching it. If we move from right to left along the graph, the curves come closer and closer to the x-axis. So each function is asymptotic to the x-axis.

Comparing curves, we see that the larger the value for a, the steeper the graph. Thus when $x > 0$, the graph of $y = 4^x$ lies above the graphs of $y = 2^x$ and $y = 3^x$. When $x < 0$, the graph of $y = 4^x$ lies below the graphs of $y = 2^x$ and $y = 3^x$. This is easier to see in Table 8.8 than in Figure 8.8.

The graph of $y = 3^x$ lies between the graphs of $y = 2^x$ and $y = 4^x$ except at $x = 0$. When $x = 0$, then $y = 1$ for all three functions. Thus the three graphs intersect at $(0, 1)$.

Case 2: $0 < a < 1$ Again, to isolate the effect of a, we examine the following three functions, in which $C = 1$, and a lies between 0 and 1.

$$y = 0.1^x \qquad y = 0.2^x \qquad y = 0.5^x$$

Something to think about

For what values of x is $2^x < 3^x < 4^x$? For what values of x is $2^x > 3^x > 4^x$?

Examine Table 8.9 and Figure 8.9. Again, y is greater than zero for every value of x, so the graphs lie above the x-axis. In Figure 8.9 the graphs may appear to touch the x-axis, but that's only because the values of y become very small as x increases.

Table 8.9			
Values for Three Exponential Functions Where $0 < a < 1$			
x	$y = 0.1^x$	$y = 0.2^x$	$y = 0.5^x$
-2	100	25	4
-1	10	5	2
0	1	1	1
1	0.1	0.2	0.5
2	0.01	0.04	0.25
3	0.001	0.008	0.125
4	0.0001	0.0016	0.0625

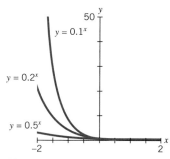

Figure 8.9   When $0 < a < 1$.

Notice that now the values for y *decrease* as x increases in value. Each time x increases by 1, y is multiplied by the base a. Since a is a positive number less than 1, multiplying by a reduces the value of y. Thus exponential functions where $0 < a < 1$ are used to model *decay*.

Comparing these curves, we can see that the smaller the value of a, the more rapidly the function declines. Thus the graph of $y = 0.1^x$ falls faster toward the horizontal axis than the graphs of $y = 0.2^x$ or $y = 0.5^x$.

The graph of $y = 0.2^x$ lies between the graphs of $y = 0.1^x$ and $y = 0.5^x$, except where all three intersect when $x = 0$ and $y = 1$.

The Effect of the Coefficient C

When examining the effect of a, we compared graphs of functions with the same C value, but different values for a. Now to learn the effect of C, we examine functions with the same a value and different values for C. In the following three functions $a = 2$ and C is given different positive values of 1, 3, and 1/4.

$$y = 1 \cdot 2^x \qquad y = 3 \cdot 2^x \qquad y = (1/4) \cdot 2^x$$

From examining Table 8.10 and Figure 8.10 on page 274, we can see that C has two major effects. First, changing C changes the y-intercept. Given any function of the form

$$y = Ca^x \qquad \text{where } a > 0$$

when $x = 0$, then $\qquad y = Ca^0$

since $a^0 = 1$ $\qquad y = C \cdot 1$

so $\qquad y = C$

When $x = 0$, $y = C$, so C is the y-intercept. C is often referred to as the *initial value* or *initial quantity*. Recall that in our *E. coli* model $N = (1.37 \cdot 10^7) \cdot 1.5^t$, the value $1.37 \cdot 10^7$ represented an estimate of the initial bacteria count, the number of bacteria at the start of the experiment when $t = 0$.

Table 8.10

**Values for Three Exponential Functions
with Different Values of C**

x	$y = 2^x$	$y = (3) \cdot 2^x$	$y = (1/4) \cdot 2^x$
-2	1/4	3/4	1/16
-1	1/2	3/2	1/8
0	1	3	1/4
1	2	6	1/2
2	4	12	1
3	8	24	2
4	16	48	4

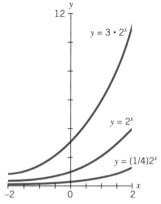

Figure 8.10   Changing the value for C.

Second, an increase in positive values for C corresponds to larger values for y. If $C > 1$, then the graph of $y = C \cdot a^x$ lies above that of $y = a^x$. If $0 < C < 1$, then the graph of $y = C \cdot a^x$ lies between that of $y = a^x$ and the x-axis.

Algebra Aerobics 8.2

1. Identify each of the following exponential functions as representing either growth or decay and specify the initial value.

   a. $y = 4^x$ **b.** $f(x) = 10 \cdot (0.99)^x$

   c. $y = 500\left(\dfrac{2}{3}\right)^x$

2. Compare the graphs of $y = 3^x$ and $5 \cdot 3^x$.

8.3 EXPONENTIAL GROWTH OR DECAY EXPRESSED IN PERCENTAGES

We learned in Section 8.2 how to recognize whether an exponential function represents growth or decay.

> Given an *exponential function* $y = f(x)$ where
> $$y = C \cdot a^x$$
> if $a > 1$, we have *exponential growth*
> if $0 < a < 1$, we have *exponential decay*

Exponential functions can be used to model any phenomenon that has a constant growth (or decay) factor, that is, any phenomenon that increases (or decreases) by a fixed *multiple* at regular intervals. If the growth factor is 2, then the function is in the form $y = C \cdot 2^x$, so the value for y is multiplied by 2 each time x increases by 1. If the growth factor is 1.05, then the function looks like $y = C \cdot (1.05)^x$, and the value for y is multiplied by 1.05 each time x increases by 1.

All exponential functions can be expressed in terms of percentages, called *growth* or *decay rates*. Multiplying by 2 is equivalent to a 100% increase. Multiplying by 1.05 is equivalent to a 5% increase. For example,

$$1.05 \cdot (\$1000) = (1.00 + 0.05)(\$1000)$$
$$= (1.00 \cdot \$1000) + (0.05 \cdot \$1000)$$
$$= \$1000 + (5\% \text{ of } \$1000)$$

The fixed *percentage increase* (or *decrease*) is called the *growth* (or *decay*) *rate*. In common usage, exponential growth is usually described in terms of its growth rate expressed as a percentage. A typical example is a statement from a federal report estimating that the cost of Medicaid (the federal health care program for the poor) will grow by 9.7% annually until 2002.

The terms "growth rate" and "growth factor" can be very confusing. Understanding the mathematical relationship between them can help. A growth *rate* of 10% (or 0.10 in decimal form) corresponds to a growth *factor* of $1 + 0.10$ or 1.10. So if r denotes the growth rate (in decimal form), then $1 + r$ is the growth factor.

A decay *rate* of 7% (or 0.07 in decimal form) corresponds to a decay *factor* of $1 - 0.07$ or 0.93. So if r denotes the decay rate (in decimal form), then $1 - r$ is the decay factor.

If r is the *growth rate* (the percentage increase in decimal form), then the *growth factor*, a, equals $1 + r$ and

$$y = C \cdot a^x = C(1 + r)^x$$

For example, if r is a growth rate equal to 0.25, then $a = 1.25$.

If r is the *decay rate* (the percentage decrease in decimal form), then the *decay factor*, a, equals $1 - r$ and

$$y = C \cdot a^x = C(1 - r)^x$$

For example, if r is a decay rate equal to 0.25, then $a = 0.75$.

John LaRosa, President of Marketdata Enterprises, Inc. in Tampa, Florida, stated that in 1996 sales by commercial weight-loss centers (such as Weight Watchers or Jenny Craig) were \$1.86 billion. He predicted that they would grow 5.5% annually for the next 4 years. Identify the annual growth rate and growth factor, and generate an exponential growth model. What is the domain of the model? What are the projected sales for the year 2000?

Example   1

SOLUTION

The annual growth rate is 5.5% or 0.055 in decimal form. The annual growth factor is 1 + 0.055 or 1.055. If n = number of years since 1996, and S = annual sales in billions of dollars, then $1.86 billion is the initial amount (the sales when $n = 0$). An exponential model would be

$$S = 1.86 \cdot (1.055)^n$$

Since the predictions were only for 4 years beyond 1996, we must restrict the domain to

$$0 \le n \le 4$$

The year 2000 is 4 years beyond 1996, so n is 4, which lies within our domain. The projected sales S for the year 2000 are

$$1.86 \cdot (1.055)^4 \approx 1.86 \cdot 1.239 \approx \$2.30 \text{ billion}$$

Algebra Aerobics 8.3

1. Determine which of the following functions are exponential. Identify each exponential function as representing growth or decay and find the vertical intercept.
   a. $A = 100(1.02^t)$ d. $y = 100x + 3$
   b. $f(x) = 4(3^x)$ e. $M = 2^P$
   c. $g(x) = 0.3(10^x)$ f. $y = x^2$

2. The total diet market consists of sales of all diet-related products—diet foods, weight-loss centers, health clubs, weight-loss drugs, etc. Almost half of the huge market consists of sales of diet soft drinks. The president of Marketdata Enterprises predicted that the 1996 U.S. diet market of 34.8 billion dollars (!) would rise 4.5% annually for the next 4 years. Identify the growth rate and growth factor. Construct an exponential model, specify the variables and the domain, and predict the size of the diet market in each year from 1996 to 2000.

3. Fill in the following table.

Exponential Function	Initial Value	Growth or Decay?	Growth or Decay Factor	Growth or Decay Rate
$A = 4(1.03)^t$				
$A = 10(0.98)^t$				
	1000		1.005	
	30		0.96	
	$50,000	Growth		7.05%
	200 grams	Decay		49%

8.4 EXAMPLES OF EXPONENTIAL GROWTH AND DECAY

Medicare Costs

The costs for almost every aspect of health care in America have risen dramatically over the last 30 years. One of the central issues in the ongoing debate about containing health care costs is the amount of federal dollars spent on Medicare. Medicare is a federal program that since July 1966 has provided two coordinated plans for nearly all people age 65 and over: a hospital insurance plan and a voluntary supplementary medical insurance plan. Table 8.11 and the graph in Figure 8.11 show annual Medicare expenses (in billions of dollars) since 1970.

Table 8.11	
Medicare Costs	
Year	**Medicare Expenses ($ billions)**
1970	7.7
1971	8.5
1972	9.4
1973	10.8
1974	13.5
1975	16.4
1976	19.8
1977	23.0
1978	26.8
1979	31.0
1980	37.5
1981	44.9
1982	52.5
1983	59.8
1984	66.5
1985	72.2
1986	76.9
1987	82.3
1988	89.4
1989	102.5
1990	112.1
1991	123.0
1992	138.7
1993	151.7
1994	169.2

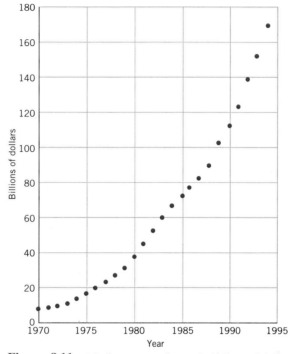

Figure 8.11   Medicare expenditures (in billions of dollars).

Source: U.S. Bureau of the Census, in *The American Almanac 1996–1997: Statistical Abstract of the United States,* 1996.

See Excel and graph link file MEDICARE.

To simplify computations, it is convenient to make the base year, 1970, correspond to $x = 0$. Then 1971 would be represented by $x = 1$; 1972 by $x = 2$; 1980 by $x = 10$, etc.[2] A curve fitting program gives a best-fit exponential function as

$$y = 8.7 \cdot 1.14^x$$

where x = years since 1970 and y = Medicare expenses in billions of dollars.

---

[2] If we did not set $x = 0$ for our baseline year and generated a best fit exponential function for the original data, we'd get an equation something like $y = (3.15 \cdot 10^{-114}) \cdot 1.14^x$. The base, or growth factor, is the same, namely 1.14. The coefficient of $3.15 \cdot 10^{-114}$ looks puzzling, however. $3.15 \cdot 10^{-114}$ is a very small number. It's a decimal place followed by 113 zeroes, and then the digits 3, 1, and 5. How could this minuscule number represent our initial costs? Remember that in the data, our x values started at 1970. The function graphing program is projecting backward to when $x = 0$, almost 2000 years ago to 0 A.D. The value $3.15 \cdot 10^{-114}$ is meaningless. The two graphs would appear identical except for the labels on the x-axis.

The function is graphed in Figure 8.12. The initial quantity of 8.7 billion is the estimated value for Medicare expenses for 1970 in our *model*. It does not represent the exact value of 7.7 billion in Table 8.11. The base of 1.14 is the growth factor. That means that according to our model, multiplying each year's Medicare expenses by 1.14 would approximate the next year's expenses. Multiplying by 1.14 is equivalent to calculating 114%. This represents an annual growth rate of 14%.

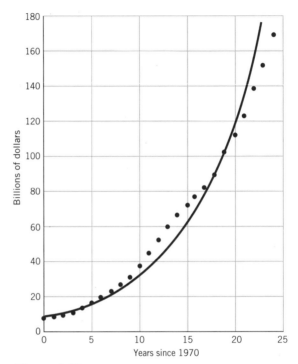

Figure 8.12   Best fit exponential function for Medicare expenses.

Predicting the Future: Two Basic Models, Linear versus Exponential

The San Francisco Chronicle (Thursday, January 5, 1995, p. A6) reported that the number of poor children in California under 2 years of age grew from 184,753 in 1980 to 275,466 in 1990. If this growth were to continue at the same pace, how many poor children under 2 would there be in California in 2000? In 2010?

The phrase "continue at the same pace" is ambiguous. Recall that in Chapter 3 we talked about different ways of describing change over time. One strategy was to subtract one size from another. In this case, we could subtract 184,753 from 275,466 to get 90,713 children, the increase in the number of poor children between 1980 and 1990. We can use this number to construct two quite different models for future growth.

Linear model   We could assume that every decade the number of young poor children will increase by 90,713. Since we are *adding* 90,713 children each decade, that implies that the growth is linear with a constant rate of change of 90,713 children per decade. So if we let d = number of decades since 1980, and n = number of poor children (under 2), the following linear equation will model the growth.

$$\binom{\text{number of}}{\text{poor children}} = (\text{no. in } 1980) + \binom{\text{rate of change}}{\text{per decade}} \cdot (\text{no. of decades})$$

$$n = \quad 184{,}753 \quad + \quad 90{,}713 \cdot d$$

Exponential model   We could assume that every decade, the number of children is increased by 49%, since 90,713 equals approximately 49% of 184,753. Increasing a number by 49% is equivalent to multiplying that number by 1.49. So if we *multiply* the number of children at the start of each decade by 1.49, that implies that the growth is exponential (with a constant growth rate of 0.49 or growth factor of 1.49). If we again let d = number of decades since 1980 and n = number of poor children, the following exponential equation will model the growth.

Exploration 8.3 offers an alternate strategy using ratios to construct exponential functions.

$$\binom{\text{number of}}{\text{poor children}} = (\text{no. in } 1980) \cdot (1 + \text{growth rate per decade})^{(\text{no. of decades})}$$

$$n = \quad 184{,}753 \quad \cdot (1 + 0.49)^d$$

or

$$n = \quad 184{,}753 \quad \cdot (1.49)^d$$

We can compare the predictions of both models in Table 8.12 and Figure 8.13. The numbers for the exponential model are slightly off, since we rounded off the percentage growth to 49%. For example at the end of the first decade, in 1990, our exponential model gives 275,282 instead of 275,446. But we would not expect anything like 6-digit accuracy in our predictions.

Table 8.12

Predictions for Number of Poor Children

No. of Decades Since 1980 d	No. of Poor Children	
	Linear Model $n = 184{,}753 + 90{,}713d$	Exponential Model $n = 184{,}753(1.49)^d$
0	184,753	184,753
1	275,446	275,282
2	366,179	410,170
3	456,892	611,154
4	547,605	910,619
5	638,318	1,356,822

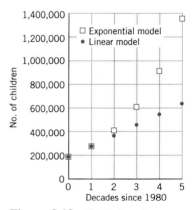

Figure 8.13   Predictions for number of poor children (under age 2) in California using linear and exponential models.

If we had more data points, we would have more information to help decide whether a linear, exponential, or some other model is the most appropriate. The numbers for five decades after 1980 in the year 2030 are highly questionable, but the difference in the projections of the two models is striking. The exponential model predicts more than twice as many poor children (about 1.4 million versus 640,000) as the linear model. Both models should be considered as generating only crude future estimates—and obviously the further out beyond 1990 we try to predict, the more unreliable the estimates become.

Radioactive Decay

One of the toxic radioactive byproducts of nuclear fission is strontium-90. A nuclear accident, like the one in Chernobyl, can release clouds of gas containing strontium-90. The clouds deposit the strontium-90 onto vegetation eaten by cows. Humans ingest strontium-90 from the cows' milk. Strontium-90 replaces calcium in bones, causing cancer. Strontium-90 is particularly insidious because it has a *half-life* of approximately 28 years. That means that every 28 years about half (or 50%) of the existing strontium-90 has decayed into nontoxic, stable zirconium-90, but the other half is still in your bones. The following function, f, gives the amount of remaining strontium-90, after T time periods, where a time period represents 28 years. The number 100 represents the initial amount of strontium-90 in milligrams.

$$f(T) = 100\left(\frac{1}{2}\right)^T$$

We usually measure time just in years, not in periods of 28 years. If we let t = years, then t is directly proportional to T. So $t = k \cdot T$ where k is some constant. When $T = 1$, then $t = 28$. So,

$$t = 28T$$

$$\text{or equivalently } T = \frac{t}{28}$$

We can then rewrite f as a function of t instead of T.

$$f(t) = 100\left(\frac{1}{2}\right)^{(t/28)}$$

Table 8.13 and Figure 8.14 show the decay of 100 mg of strontium-90 over time.

Table 8.13

Decay of Strontium-90

Time in Years, t	Strontium-90 (mg)
0	100.000
28	50.000
56	25.000
84	12.500
112	6.250
140	3.125

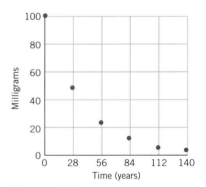

Figure 8.14   Amount of strontium-90.

It takes about 28 years for half of the strontium-90 to decay, but half still remains. It takes an additional 28 years for the remaining amount to halve again, still leaving 25 grams out of the original 100.

Every exponential function has a fixed half-life or doubling time. For example, strontium-90 has a fixed half-life of 28 years. The *E. coli* in our initial model have a fixed doubling time of 1 time period or 20 minutes.

The *doubling time* of an exponentially growing quantity is the time required for the quantity to double in size.

The *half-life* of an exponentially decaying quantity is the time required for one-half of the quantity to decay.

Exponential functions are the only functions with constant doubling or halving times.

The "Rule of 70": A Rule of Thumb for Calculating Doubling or Halving Times

A simple way to understand the significance of percentage growth rates is to compute the doubling time. It has been observed that if a quantity is growing at R percent per year, then its doubling time is approximately $70/R$ years. If the quantity is growing at R percent per *month*, then $70/R$ gives its doubling time in *months*. The same reasoning holds for any unit of time.

   The rule of 70 is easy to apply. We just need to remember to use values for R in percentage form, such as 5% or 10.3%, not their equivalents in decimal form, such as 0.05 or 0.103.

   For now we'll take the rule of 70 on faith. It provides good approximations, especially for smaller values of R (those, say, under 10%). When we return to logarithms in Chapter 12, we'll find out why.

Something to think about

Using a graphing calculator or a spreadsheet, test the rule of 70. Can you find a value for R for which this rule does not work so well?

Example 1

Suppose that at age 25, you invest $1000 in a retirement account, which grows at 5% per year. Since $70/5 = 14$, your investment will double approximately every 14 years. If you retire 42 years later at age 67, then three doubling periods will have elapsed ($3 \cdot 14 = 42$). So your $1000 investment (disregarding inflation) will have increased in value by a factor of 2^3 and be worth $8000.

Example 2

According to an article in the *Los Angeles Times*, Mexico's National Institute of Geography, Information and Statistics concluded from 1995 census data that almost half of Mexico's 91.1 million people now live in overcrowded cities and that the population nationwide will continue to grow an average of 1.8% each year. A private consultant claimed that the growth rate alone "means that the population will double in 40 years." Is the consultant right?

   Using the rule of 70, the approximate doubling time for a 1.8 annual growth rate would be $70/1.8 \approx 38.9$ years. So 40 years seems like a reasonable estimate for the doubling time.

• • •

   The rule of 70 also applies when R represents the percentage at which some quantity is decaying. In these cases, $R/70$ equals the half-life.

Example 3

Something to think about

An atomic weapon is usually designed with a 1% mass margin. That is, it will remain functional until the original fuel has decayed by more than 1%, leaving less than 99% of the original amount. Estimate how many years a plutonium bomb would remain functional.

Plutonium, the fuel for atomic weapons, has an extraordinarily long half-life of about 24,400 years. Once the radioactive element plutonium is created from uranium, 25,000 years later almost half the original amount will remain. You can see why there is concern over stored caches of atomic weapons. Using the rule of 70, we can estimate R, plutonium's annual decay rate.

the rule of 70 gives	$70/R = 24,400$
multiply by R	$70 = 24,400R$
divide and simplify	$R = 70/24,400$
	≈ 0.003

Note that R is already in percentage, not decimal form. So the annual decay rate is a tiny 0.003%.

Algebra Aerobics 8.4a

1. Estimate the doubling time using the rule of 70 when
    a. $g(x) = 100(1.02)^x$ where x is in years
    b. $M = 10,000(1.005)^t$ where t is in months
    c. the annual growth rate is 8.1%
    d. the annual growth factor is 106.5
2. Use the rule of 70 to approximate the growth rate when the doubling time is

    a. 10 years   b. 5 minutes   c. 25 seconds
3. Estimate the time it will take an initial quantity to drop to half its value when
    a. $h(x) = 10(0.95)^x$ where x is in months
    b. $K = 1000(0.75)^t$ where t is in seconds
    c. the annual decay rate is 35%

White Blood Cell Counts

See Excel or graph link file
CELCOUNT

On September 27, 1996, a patient was admitted to Brigham and Women's Hospital in Boston for a bone marrow transplant. The transplant was needed to cure myelodysplastic syndrome, in which the patient's own marrow fails to produce enough white blood cells to fight infection; both Carl Sagan and Paul Tsongas had this disease and underwent this treatment.

The patient's own bone marrow was intentionally destroyed using chemotherapy and radiation, and on October 3 the donated marrow was injected. Each day, the hospital carefully monitored the patient's white blood cell count, to detect when the transplant took hold and the new marrow became active. The patient's counts are listed in Table 8.14 and plotted in Figure 8.15. Normal counts for a healthy individual are between 4000 and 10,000 cells per milliliter.

Table 8.14

White Blood Cell Counts

Date	Cell Count (per ml)	Date	Cell Count (per ml)
27 Sept	1,390	18 Oct	290
28 Sept	2,570	19 Oct	310
29 Sept	2,290	20 Oct	450
30 Sept	2,660	21 Oct	580
1 Oct	1,720	22 Oct	740
2 Oct	1,290	23 Oct	1,070
3 Oct	500	24 Oct	1,210
4 Oct	160	25 Oct	1,870
5 Oct	110	26 Oct	2,030
6 Oct	120	27 Oct	2,540
7 Oct	130	28 Oct	3,350
8 Oct	120	29 Oct	5,460
9 Oct	60	30 Oct	6,940
10 Oct	60	31 Oct	8,640
11 Oct	40	1 Nov	7,650
12 Oct	60	2 Nov	7,430
13 Oct	70	3 Nov	8,790
14 Oct	60	4 Nov	10,100
15 Oct	70	5 Nov	9,620
16 Oct	170	6 Nov	9,420
17 Oct	180		

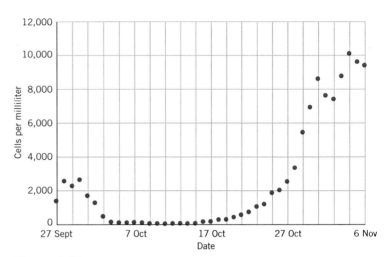

Figure 8.15   White blood cell count.

The white blood cell count dropped steadily until about October 9. Between October 9 and October 15, the count fluctuated below 100. October 15 was the beginning of a dramatic increase. The values rose steadily until October 31. After that, the counts fluctuated as the patient's body began to regulate itself. Several months after the transplant, the count stabilized at about 6200.

Clearly the whole data set does not represent exponential growth, but we can reasonably model the data between say October 15 and October 31 with an exponential function. Table 8.15 (page 284) records the number of days after October 15 and the corresponding counts. So October 15 corresponds to 0, October 16 corresponds to 1, etc. This subset of the original data is plotted in Figure 8.16 (page 284) along with a computer-generated best fit exponential function,

$$y = 105(1.32)^x$$

where x = number of days after October 15 and y = white blood cell count.

The fit looks quite good. The initial quantity of 105 is the white blood cell count for the model, not the actual white blood cell count of 70 measured by the hospital. This discrepancy is not unusual—remember that a best fit function may not necessarily pass through any of the specific data points. The growth factor is 1.32, so the growth rate is 0.32. So between October 15 and October 31, the number of white blood cells were increasing at a rate of 32% a day. According to the rule of 70, the number of white blood cells were doubling roughly every $70/32 \approx 2.2$ days!

Table 8.15

Counts Between October 15th and October 31

Days *after* October 15	Count (per ml)
0	70
1	170
2	180
3	290
4	310
5	450
6	580
7	740
8	1070
9	1210
10	1870
11	2030
12	2540
13	3350
14	5460
15	6940
16	8640

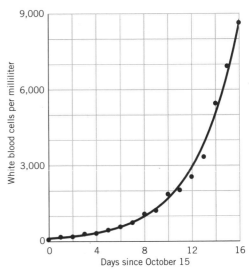

Figure 8.16   White blood cell counts between October 15 and October 31.

Compound Interest

Short-term returns   One of the most common examples of exponential growth is compound interest. Suppose you have $100 that you could put either in a checking account that earns no interest, a NOW account that earns 3% compounded annually, a savings account that earns 5% compounded annually, or a certificate of deposit (CD) that earns 7% compounded annually. How would your results compare after 10 years? After 25 years?

Assuming you make no withdrawals, the money in the checking account will remain constant at $100. The money in the interest-bearing accounts will obviously increase. But by how much?

An interest rate of 3% compounded annually means that at the end of each year, you earn 3% on the current value of your account. The interest is automatically deposited in your account. From then on you earn 3% not only on your *principal* (the initial amount you invested), but also on the interest you have already earned. The growth rate is 3% or 0.03 in decimal form, and the growth factor is 1.03. So each year the current value of the account will be multiplied by 1.03. The functions representing P_r, the current value of your account earning interest r, as a function of n, the number of years, would be

$$P_{0.03} = 100 \cdot (1.03)^n$$
$$P_{0.05} = 100 \cdot (1.05)^n$$
$$P_{0.07} = 100 \cdot (1.07)^n$$

In general,

If

P_0 = original investment

r = interest rate (in decimal form)

n = time periods at which the interest rate is compounded

the resulting value, P_r, of the investment after n time periods is given by the formula

$$P_r = P_0 \cdot (1 + r)^n$$

The interest rate, r, in decimal form is the growth rate and $1 + r$ is the growth factor.

Table 8.16 and Figure 8.17 compare the values of your account over 10 years. At 10 years, the $100 in the 3% account has risen to $134.39, while the $100 in the 7% account has almost doubled to $196.72. The constant value of the $100 in the checking account is not listed in the table but appears on the graph as a flat line.

Table 8.16		
Compound Interest over 10 Years		
Years	Value of $100 at 3%	Value of $100 at 5%	Value of $100 at 7%
0	$100.00	$100.00	$100.00
1	$103.00	$105.00	$107.00
2	$106.09	$110.25	$114.49
3	$109.27	$115.76	$122.50
4	$112.55	$121.55	$131.08
5	$115.93	$127.63	$140.26
6	$119.41	$134.01	$150.07
7	$122.99	$140.71	$160.58
8	$126.68	$147.75	$171.82
9	$130.48	$155.13	$183.85
10	$134.39	$162.89	$196.72

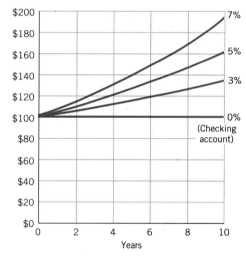

Figure 8.17 $100 invested for 10 years at 3%, 5%, and 7% compounded annually.

Long-term returns   The longer the time period over which you invest, the more dramatic the results. Over time small differences in interest rates can produce enormous differences in returns. Table 8.17 and Figure 8.18 on page 286 show the value of $100 invested at 3%, 5%, and 7% over 40 years. (Note the difference in scales in Figure 8.17 and Figure 8.18.)

If you invested $100 at age 25 for your retirement 40 years later at 3% compounded annually you'd end up with $326; at 5% you'd have $704; and at 7% you'd have almost $1500 or nearly 15 times as much as you started with. So if you were able to invest $10,000 at 7% for 40 years, you'd have about $150,000; if you won the lottery and could invest $100,000, you'd have almost $1.5 million dollars!

	Table 8.17		
	Compound Interest over 40 Years		
Years	**Value of $100 at 3%**	**Value of $100 at 5%**	**Value of $100 at 7%**
0	$100	$100	$100
10	$134	$163	$197
20	$181	$265	$387
30	$243	$432	$761
40	$326	$704	$1497

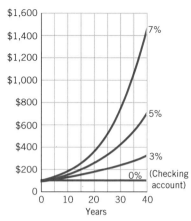

Figure 8.18 $100 invested for 40 years.

Too good to be true?   At decent interest rates, if we could all put aside a few thousand dollars each year, it seems we could retire in comfort. Is the picture as rosy as it seems? Not quite. We haven't taken inflation into account.

Compound interest calculations are the same whether you are dealing with inflation or investments. For example, a 5% annual inflation rate would mean that what cost $1 today would cost $1.05 one year from today.

If we think of the three percentages in Table 8.16 as representing inflation rates, then how much would something that costs $100 today cost in 10 years? It would cost $134.39, $162.89, or $196.72, if the annual inflation rate were 3%, 5%, or 7%, respectively. So if you *invest* $100 at 5% for 10 years, it will be worth $162.89. But inflation will drive up costs. If during those 10 years inflation is also 5% a year, then what originally cost $100 will now cost $162.89. In terms of purchasing power you will come out even, with no net gain or loss. So if the inflation rate equals the interest rate, you are not any better off. If the inflation rate were higher than the interest rate, you would actually lose money on your investment. Because of the erosive nature of inflation, most economists usually use *real* or *constant dollars,* which are dollars adjusted for inflation.

On March 13, 1997, a resident of Melrose, Massachusetts, "scratched" a $5 state lottery ticket and won $1,000,000. The Massachusetts State Lottery Commission notified her that after deducting taxes, they would send her a check for $33,500 every March 13th for 20 years. Is this fair?

Inflation and the Diminishing Dollar

Inflation erodes the purchasing power of the dollar. If you have $1.00 today and annual inflation is 5%, how much will the dollar be worth in a year? In a year, what cost $1.00 today, will cost $1.05. Since $1.00/1.05 = 0.952$, then $1.00 = 0.952 \cdot ($1.05)$. So in a year $1.00 will be worth only 95.2¢. The decay factor is 0.952. The exponential function

$$D = 1 \cdot (0.952)^n$$

gives D, the value (or purchasing power) of today's dollar, at year n in the future if there is a steady inflation rate of 5%. In economists' terms, it gives the real purchasing power in today's dollars. Table 8.18 and Figure 8.19 give an indication of how rapidly the value of the dollar declines. (Dollar values are rounded to the nearest penny.)

Every 14 years, the purchasing power of your money is cut in half. We can think of 14 years as the half-life of the dollar. So, in 14 years, your dollar is worth 50¢. In 28 years, your dollar is worth only 25¢. This is obviously a serious problem faced by those who retire on a fixed pension income.

Table 8.18	
Inflation's Erosion of the Dollar	
Year	Value of Dollar (with 5% inflation)
0	$1.00
1	$0.95
2	$0.91
3	$0.86
4	$0.82
5	$0.78
6	$0.74
7	$0.71
8	$0.67
9	$0.64
10	$0.61
11	$0.58
12	$0.55
13	$0.53
14	$0.50
15	$0.48
16	$0.46
17	$0.43
18	$0.41
19	$0.39
20	$0.37

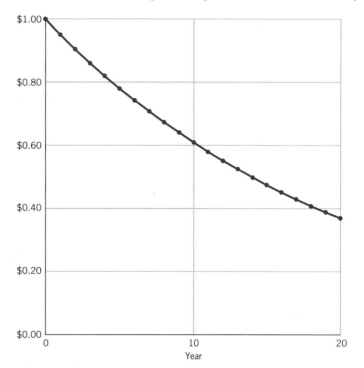

Figure 8.19   Value of a dollar with inflation at 5%.

Algebra Aerobics 8.4b

1. Approximately how long would it take for your money to double if the interest rate, compounded annually, were

   a. 3%   **b.** 5%   **c.** 7%

2. Suppose you are planning to invest a sum of money. Determine the rate that you need so that your investment doubles in

   a. 5 years   **b.** 10 years   **c.** 7 years

3. Construct a function that would represent the resulting value if you invested $1000 for n years at an annually compounded interest rate of

   a. 4%   **b.** 11%   **c.** 110%

4. In the early 1980s Brazil's inflation was running rampant at about 10% per *month*. If this inflation rate continued unchecked, construct a function to describe the purchasing power of 100 cruzeiros after n months. (A cruzeiros is a Brazilian monetary unit.) What would 100 cruzeiros be worth after 3 months? 6 months? A year?

   Note: In such cases sooner or later the government usually intervenes. In 1985 the Brazilian government imposed an anti-inflationary wage and price freeze. When the controls were dropped, inflation soared again reaching a high in March 1990 of 80% per month! At this level, it begins to matter whether you buy groceries in the morning or wait until that night. On August 2, 1993, the government devalued the currency by defining a new monetary unit, the cruzeiros real, equal to 1000 of the old cruzeiros. Inflation still continued, and on July 1, 1994, yet another unit, the real, was defined equal to 2740 cruzeiros real. By 1997, inflation had slowed considerably to about 0.1% per month.

Musical Pitch

Exponential functions can also describe the relationship between musical octaves and vibration frequency. The vibration frequency of middle C is 263 cycles per second (or 263 hertz). The vibration frequency for each subsequent octave above middle C is double that of the previous one. Let N, the independent variable, be the number of octaves above middle C and F, the dependent variable, be the vibration frequency in hertz. Since the frequency doubles at each octave, the growth factor is 2. The initial frequency is 263 hertz. The function is then

$$F = 263 \cdot 2^N$$

Table 8.19 and the graph in Figure 8.20 show a few of the values for the vibration frequency.

If you have access to a computer, you may enjoy playing with a multimedia demonstration of this function in "E6: Musical Keyboard Frequencies" in Exponential & Log Functions.

Table 8.19

Musical Octaves

Number of Octaves above Middle C, N	Vibration Frequency, F (in hertz)
0	263
1	526
2	1052
3	2104
4	4208
5	8416

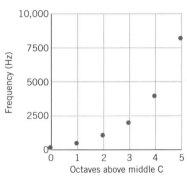

Figure 8.20   Octaves versus frequency.

The Malthusian Dilemma

The most famous attempt to predict growth mathematically was made by a British economist and clergyman, Thomas Robert Malthus, in an essay published in 1798. He argued that the growth of the human population would overtake the growth of food supplies, because the population size was *multiplied* by a fixed amount each year, whereas food production only increased by *adding* a fixed amount each year. In other words, he assumed populations grew exponentially and food supplies grew linearly as functions of time. He concluded that humans were condemned always to breed to the point of misery and starvation, unless the population were reduced by other means, including war or disease.

We can frame his arguments algebraically by letting P_0 represent the original population size, t the time in years, and a the annual growth factor. Then P, the population at time t, is given by

$$P = P_0 \cdot a^t$$

If F_0 represents the amount of food at time $t = 0$, and q the constant quantity added to the food supply each year, then F, the total amount of food, is

$$F = F_0 + q \cdot t$$

Malthus believed the population of Britain, then about 7,000,000, was growing by 2.8% per year ($P_0 = 7,000,000$ and $a = 1.028$). He counted food supply in units that he defined to be enough food for one person for a year. At that time food supply was adequate, so he assumed that Britons were producing 7,000,000 food units. He thought they could increase food production by about 280,000 units a year ($F_0 = 7,000,000$ and $q = 280,000$). So the two functions are

$$P = 7,000,000 \cdot (1.028^t) \qquad F = 7,000,000 + 280,000 \cdot t$$

Table 8.20 and Figure 8.21 reveal that if the formulas were good models, then after about 25 years population would start to exceed food supply, and some people would starve.

Table 8.20 Growth in Population versus Food		
Year	Population (millions)	Food Units (millions)
0	7.00	7.00
5	8.04	8.40
10	9.23	9.80
15	10.59	11.20
20	12.16	12.60
25	13.96	14.00
30	16.03	15.40
35	18.40	16.80
40	21.13	18.20
45	24.25	19.60
50	27.85	21.00
100	110.77	35.00

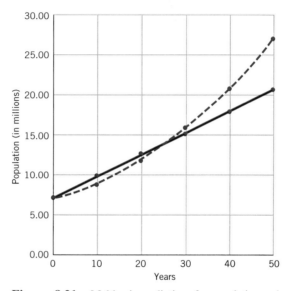

Figure 8.21   Malthus's predictions for population and food.

The two centuries since Malthus published his famous essay have not been kind to his theory. The population of the British Isles in 1995 was about 58 million whereas Malthus' model predicted over 100 million people before the year 1900. Improved food production techniques and the opening of new lands to agriculture have kept food production in general growing faster than the population. The distribution of food is a problem and famines still occur with unfortunate regularity in parts of the world, but the mass starvation Malthus predicted has not come to pass.

8.5 SEMI-LOG PLOTS OF EXPONENTIAL FUNCTIONS

With exponential growth functions, we very often have the same problem that we did in Chapter 7 when we compared the size of atoms to the size of human beings and to the size of the solar system: The numbers go from very small to very large. For example, in our *E. coli* experiment, how can we determine from a graph whether the growth from 100 to 200 cells follows the same rule as the growth from 100 million to 200 million? It is virtually impossible to display the entire data set on a standard graph.

One solution is to use a logarithmic or order of magnitude scale on the vertical axis. Recall that in Chapter 7 we used a logarithmic scale on a single horizontal axis. Moving a fixed distance up or down on a vertical logarithmic scale, corresponds to *multiplying* the variable by a constant factor, rather than to *adding* a constant as on a linear scale. In exponential growth the dependent variable is multiplied by a constant factor each time a fixed constant is added to the independent variable. Exponential growth always appears as a line of constant slope when a logarithmic scale is used on the vertical axis and a standard "linear" scale is used on the horizontal axis. We'll take a closer look at why this is true in Chapter 12. Since most graphing software easily does such *log-linear* or *semi-log* plots, this is one of the easiest and most reliable ways to recognize exponential growth in a data set.

Table 8.21 repeats the *E. coli* data from Table 8.1. Figure 8.22 shows a graph of the data plotted on a semi-log graph. Recall that the data fit the exponential function $N = 100 \cdot 2^t$. On the vertical axis, successive powers of 10 appear at equally spaced intervals: 100 million is just as "far" from 10 million as 100 is from 10. We can display the entire data set on a single graph, and its straight-line shape immediately tells us that it represents exponential growth.

Table 8.21	
Growth of *E. coli*	
t (20-Minute Time Periods)	$N = 100 \cdot 2^t$ (No. of *E. coli* Bacteria)
0	100
1	200
2	400
3	800
4	1,600
5	3,200
6	6,400
7	12,800
8	25,600
9	51,200
10	102,400
11	204,800
12	409,600
13	819,200
14	1,638,400
15	3,276,800
16	6,553,600
17	13,107,200
18	26,214,400
19	52,428,800
20	104,857,600
21	209,715,200
22	419,430,400
23	838,860,800
24	1,677,721,600

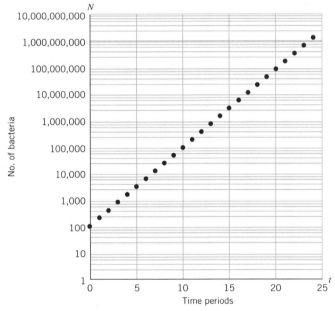

Figure 8.22   Semi-log plot of $N = 100 \cdot 2^t$: Number of *E. coli* bacteria.

When an exponential function is plotted using a standard linear scale on the horizontal axis and a logarithmic scale on the vertical axis, its graph is a straight line. This type of graph is called a *log-linear* or *semi-log* plot.

Algebra Aerobics 8.5

Using the graph in Figure 8.22, estimate the time interval it takes for the population to increase by a factor of 10. From the original expression for the population, $N = 100 \cdot 2^t$, over what time interval does the population increase by a factor of 8? By a factor of 16? Are these three answers consistent with each other?

CHAPTER SUMMARY

An *exponential function* has the form

$$y = C \cdot a^x \qquad (a > 0, \qquad a \neq 1)$$

where C is the *y-intercept* or *initial value* (when $x = 0$) and a is the factor by which y is multiplied when x increases by 1. If $a > 1$, we have *exponential growth*, if $0 < a < 1$, we have *exponential decay*.

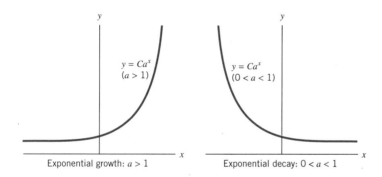

Exponential growth: $a > 1$ Exponential decay: $0 < a < 1$

An exponential function can be expressed in terms of percentages. The fixed percentage increase or decrease is called the growth or decay rate. If r is the *growth rate* (the percentage increase in decimal form), then the *growth factor, a,* equals $1 + r$ and

$$y = C \cdot a^x = C(1 + r)^x$$

If r is the *decay rate* (the percentage decrease in decimal form), then the *decay factor, a,* equals $1 - r$ and

$$y = C \cdot a^x = C(1 - r)^x$$

Exponential functions can be used to model any phenomenon that has a constant growth (or decay) factor, that is, any phenomenon that increases (or decreases) by a fixed multiple at regular intervals.

Exponential functions are commonly used to model the growth of populations. For example, the exponential function $N = 100 \cdot 2^t$ is an idealized model of the growth of *E. coli* bacteria, where 100 is the number of initial bacteria, and 2, the base, is the growth factor. The bacteria are assumed to double every time period, t, of 20 minutes. Populations cannot sustain exponential growth forever. Their long-term growth often displays an S-shaped curve called a *sigmoid,* which is close to exponential in its early phases and then flattens out as it approaches a maximum population size called the

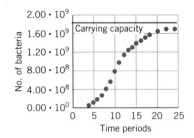

carrying capacity. The adjacent graph shows the results of monitoring the growth of *E. coli* in a lab for 8 hours (24 time periods) that exhibits this **S**-shaped pattern.

The *doubling time* of an exponentially growing quantity is the time required for the quantity to double in size.

The *half-life* of an exponentially decaying quantity is the time required for one-half of the quantity to decay.

Exponential functions are the only functions with constant doubling or halving times.

The "rule of 70" offers a simple way to estimate the doubling time or the half-life. If a quantity is growing at R percent per year, then its doubling time is approximately $70/R$ years. If a quantity is growing at R percent per month, then $70/R$ gives its doubling time in months. Similarly if a quantity is decaying at R percent per year, then its half-life is approximately $70/R$ years.

When an exponential function is plotted using a standard scale on the horizontal axis and a logarithmic scale on the vertical axis, its graph is a straight line. This is called a *log-linear* or *semi-log* plot.

EXERCISES

1. The following exponential functions are of the form $y = C \cdot a^x$. In parts (a), (b), and (c):

   i. Identify C and a for each function.

   ii. Specify whether each function represents growth or decay. In particular, for each unit increase in x what happens to y?

   iii. For each function generate a small table of values.

   iv. Make predictions about the relative shapes of the graphs of the three functions and then check your predictions by graphing the functions on the same grid.

   v. Summarize what you have learned from comparing the three functions.

   a. $y = 2^x$ **b.** $y = (0.5) \cdot 2^x$ **c.** $y = 3^x$
   $\qquad y = 5^x$ $y = 2 \cdot 2^x$ $y = (1/3)^x$
   $\qquad y = 10^x$ $y = 5 \cdot 2^x$ $y = 3 \cdot (1/3)^x$

2. Given an initial value of 50 units for a quantity Q for each part (a)–(d) below, write a function that represents Q as a function of time t. Assume that when t increases by 1,

   a. Q doubles

   b. Q increases by 5%

   c. Q increases by 10 units

   d. Q is multiplied by 2.5

See Excel or graph link file
WORLDPOP.

3. (Requires graphing calculator or computer.)
   Estimates for world population vary, but the data in the table on page 293 are reasonable estimates of the world population from 1800 to 1994.

   a. Either enter the data table into the calculator or computer (you may wish to enter 1800 as 0, 1850 as 50, etc.), or use the data file on the CD-ROM.

b. Generate a best fit exponential function. Record the equation and print out the graph if you can.

c. Interpret each term in the function, and specify the domain and range of the function.

d. What does your model estimate for the growth rate?

e. Using the graph of your function, estimate the following:

    i. The world population in 1750, 1920, 2025, and 2050.

    ii. The approximate years in which world population attained or will attain 1 billion (i.e., 1000 million), 3.2 billion, 4 billion, and 8 billion.

    iii. The length of time your model predicts it takes for the population to double in size.

World Population	
Year	Population (millions)
1800	910
1850	1130
1900	1600
1950	2510
1970	3702
1980	4456
1990	5293
1993	5555
1996	5771

4. Cosmic ray bombardment of the atmosphere produces neutrons, which in turn react with nitrogen to produce radioactive carbon-14. Radioactive carbon-14 enters all living tissues through carbon dioxide (via plants). As long as a plant or animal is alive, carbon-14 is maintained in the organism at a constant level. Once the organism dies however, carbon-14 decays exponentially into carbon-12. By comparing the amount of carbon-14 to the amount of carbon-12, one can determine approximately how long ago the organism died. Willard Libby won a Nobel Prize for developing this technique for use in dating archaeological specimens.

The half-life of carbon-14 is about 5730 years. Assume that the initial quantity of carbon-14 is 500 milligrams.

a. Construct an exponential function that describes the relationship between C, the amount of carbon-14 and t, time. Be sure to specify the units in which you are measuring C and t.

b. Generate a table of values and plot the function. Choose a reasonable set of values for the domain. Remember that the objects we are dating may be up to 50,000 years old.

c. From your graph or equation estimate how many milligrams are left after 4000 years? 15,000 years? 45,000 years?

5. Between 1960 and 1990, the United States grew from about 180 million to 250 million, an increase of approximately 40% in this 30-year period. If the population continues to expand by 40% every 30 years, what would the U.S. population be in the year 2020? In 2050? Use your calculator to estimate when the population of the United States would reach 1 billion.

6. Describe how a 6% inflation rate will erode the value of a dollar over time. Approximately when would a dollar be worth only 50¢? This might be called the half-life of the dollar's buying power under 6% inflation.

7. Belgrade, Yugoslavia:

> *A 10 billion dinar note hit the streets today. . . . With inflation at 20% per day, the note will soon be as worthless as the 1 billion dinar note issued last month. A year ago the biggest note was 5,000 dinars. . . . In addition to soaring inflation, which doubles prices every 5 days, unemployment is at 50%. (from* USA Today, *September 21, 1993)*

In the excerpt, inflation is described in two very different ways. Identify these two descriptions in the text and determine whether they are equivalent. Justify your answer.

8. Tritium, the heaviest form of hydrogen, is a critical element in a hydrogen bomb. It decays exponentially with a half-life of about 12.3 years. Any nation wishing to maintain a viable hydrogen bomb has to replenish its tritium supply roughly every 3 years, so world tritium supplies are closely watched. Construct an exponential function that shows the remaining amount of tritium as a function of time as 100 grams of tritium decays (about the amount needed for an average size bomb).

9. A bank compounds interest annually at 4%.

   a. Write an equation for the value V of 100 dollars in t years.

   b. Write an equation for the value V of 1000 dollars in t years.

   c. After 20 years will the total interest earned on $1000 be 10 times the total interest earned on $100? Why or why not?

10. Below are sketches of the graphs of four exponential functions. Match each sketch with the function that best describes that graph.

$$P = 5 \cdot (0.7^x) \qquad R = 10 \cdot (1.8^x)$$
$$Q = 5 \cdot (0.4^x) \qquad S = 5 \cdot (3^x)$$

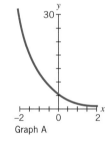
Graph A

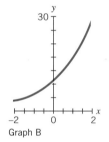
Graph B

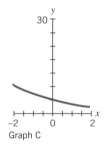
Graph C

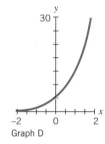
Graph D

11. a. Will the graphs of $y = (1/10)^x$ and $y = 10^x$ intersect? If so, where will they intersect? Explain your answer.

    b. Will the graphs of $y = 5 \cdot (1/10)^x$ and $y = 5 \cdot 10^x$ intersect? If so, where will they intersect? Explain your answer.

12. (Requires graphing calculator or function graphing program.)
    Reliable data about Internet use are hard to come by. Although it is difficult to determine how many individuals use the Internet, it is easy to count the number of Internet hosts, since hosts must be registered. One host, however, could be a single user or an entire service such as America Online, to which several million people subscribe. Some of the best data, published by Network Wizards in Menlo Park, California, show there were 1.2 million hosts in January 1993, 4.9 million in January 1995, and 9.5 million in January 1996.

    a. Give a rough estimate for the doubling time of the number of hosts.

    b. Enter the data on a graphing calculator and generate a best fit exponential function. (Set the year 1993 as year 0.)

    c. From your best fit function, what is the annual growth factor? The annual growth rate?

    d. If this growth rate continues, approximately how long will it take for the number of Internet hosts to reach 20 million? 100 million? 1 billion?

13. The federal government helps students to finance higher education through a grant program called Pell Grants (awards which students do not have to pay back) and through federal loans (which students do have to pay back). The National Education Association reported that the number of Pell Grant recipients fell from 4.2 million in fiscal year (FY) 1993 to 3.7 million in fiscal year (FY) 1994. (*NEA Almanac of Higher Education*, 1996, p. 90). Using FY 1993 as the base year (or year 0) and FY 1994 as year 1, create a model for each of the following assumptions.

   a. Assume that the number of recipients is declining linearly over time. Find an equation relating number of recipients and years.

   b. Assume that the number of recipients is declining exponentially over time. Find an equation relating number of recipients and years.

   c. Graph your functions. Use your graphs to estimate when each of these models will predict that funding will be half of what it was in FY 1993. Check your estimates by substituting these values in each of your functions.

   d. Use each of the functions to predict what would happen in FY 1995.

   e. See if you can find out the number of recipients in FY 1995 and determine which prediction came the closest to the actual number.

14. You have a chance to invest money in a risky investment at 6% interest compounded annually. Or you can invest your money in a safe investment at 3% interest compounded annually.

   a. Write an equation that describes the value of your investment after n years if you invest $100 at 6% compounded annually. Plot the function.

   b. Write an equation that describes the value of your investment after n years if you invest $200 at 3% compounded annually. Plot the function on the same graph as in part (a).

   c. Looking at your graph, will the amount in the first investment in part (a) ever exceed the amount in the second account in part (b)? If so, approximately when?

15. If you have a heart attack and your heart stops beating, the amount of time it takes paramedics to restart your heart with a defibrillator is critical. According to a medical report on the evening news, each minute that passes decreases your chance of survival by 10%. From this wording it is not clear whether the decrease is linear or exponential. Assume that the survival rate is 100% if the defibrillator is used immediately.

   a. Construct and graph a linear function that describes your chances of survival. After how many minutes would your chances of survival be 50% or less?

   b. Construct and graph an exponential function that describes your chances of survival. Now after how many minutes would your chances of survival be 50% or less?

16. According to the *Arkansas Democrat Gazette* (February 27, 1994),

   Jonathan Holdeen thought up a way to end taxes forever. It was disarmingly simple. He would merely set aside some money in trust for the government and leave it there for 500 or 1000 years. Just a penny, Holdeen calculated, could grow to trillions of dollars in that time. But the stash he had in mind would grow much bigger—to quadrillions or quintillions—so big that the government, one day, could pay for all its operations simply from the income. Then taxes could be abolished. And everyone would be better off.

a. Holdeen died in 1967 leaving a trust of $2.8 million that is being managed by his daughter, Janet Adams. In 1994, the trust was worth $21.6 million. The trust is being debated in Philadelphia Orphans' Court. Some lawyers who are trying to break the trust have said that it is dangerous to let it go on, because "it would sponge up all the money in the world." Is this possible?

b. In 500 years, how much would the trust be worth? Would this be enough to pay off the national debt (currently about 6 trillion dollars)? What about after 1000 years? Describe the model you have used to make your prediction.

17. The body eliminates drugs by metabolism and excretion. To predict how frequently a patient should receive a drug dosage, the physician must determine how long the drug will remain in the body. This is usually done by measuring the half-life of the drug, the time required for the total amount of drug to diminish by one-half.

a. Most drugs are considered eliminated after five half-lives, because the amount remaining is probably too low to cause any beneficial or harmful effects. After five half-lives, what percentage of the original dose is left in the body?

b. The accompanying graph shows a drug's concentration in the body over time starting with 100 milligrams.

    i. Estimate the half-life of the drug.

    ii. Construct an equation that approximates the curve.

    iii. How long would it take for five half-lives to occur? Approximately how many milligrams of the original dose would be left then?

    iv. Write a paragraph describing your results to a prospective buyer of the drug.

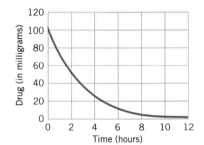

Remaining drug dose in milligrams.

See Excel or graph link file REINDEER.

18. (Requires graphing calculator or function graphing program.)

In 1911 reindeer were introduced to St. Paul Island, one of the Pribilof Islands off the coast of Alaska in the Bering Sea. There was plenty of food and no hunting or reindeer predators. The size of the reindeer herd grew rapidly for a number of years as given in the accompanying table.

Year	Population Size	Year	Population Size	Year	Population Size
1911	17	1921	280	1931	466
1912	20	1922	229	1932	525
1913	42	1923	161	1933	670
1914	76	1924	212	1934	831
1915	93	1925	246	1935	1186
1916	110	1926	254	1936	1415
1917	136	1927	254	1937	1737
1918	153	1928	314	1938	2034
1919	170	1929	339		
1920	203	1930	415		

Source: V.B. Scheffer. The rise and fall of a reindeer herd. *Scientific Monthly,* 73:356–362, 1951.

a. Use the reindeer data file (in Excel or graph link form) to plot the data.

b. Find a best fit exponential function and use it to predict the size of the population in each year. (You may want to set $1911 = 0$.)

c. How does the predicted population from part (b) differ from the observed?

d. Does your answer in part (c) give you any insights into why the model does not fit the observed data perfectly?

e. Estimate the doubling time of this population.

19. (Optional use of graphing calculator or computer.)

According to Carlos Jarque, President of Mexico's National Institute of Geography, Information and Statistics, the population is growing at a rate of 6.46% per year in Mexico's Quintana Roo state, where the tourist industry in Cancun has created a boom economy. The population in Baja California, where many labor-intensive border industries are located, is growing at a rate of 4.29% per year. But the nation's capital, Mexico City (which with more than 21 million inhabitants is already considered the world's largest metropolitan area), is growing at a rate of 0.5% per year.

a. If the growth rates continue, how long will it take for the population of Quintana Roo to double? How long for the population of Baja California to double?

b. Assume the population of Mexico City continues to grow at the same rate.

   i. Write an equation that will predict its future population. (Hint: You may want to set your starting year as $t = 0$.)

   ii. Generate a small table of values.

   iii. Graph the equation.

   iv. Approximately how long will it take for the population to increase by 1 million people? When will the population reach 25 million?

20. (Requires graphing calculator or computer.)

We have seen the accompanying table and graph of the U.S. population a number of times in the text.

See Excel or graph link file USPOP.

Population of the United States 1790 to 1995

Year	Population (in millions)	Year	Population (in millions)
1790	3.9	1910	92.0
1800	5.3	1920	105.7
1810	7.2	1930	122.8
1820	9.6	1940	131.7
1830	12.9	1950	151.3
1840	17.1	1960	179.3
1850	23.2	1970	203.3
1860	31.4	1980	226.5
1870	39.8	1990	248.7
1880	50.2	1992	255.4
1890	63.0	1994	260.7
1900	76.0	1995	263.0

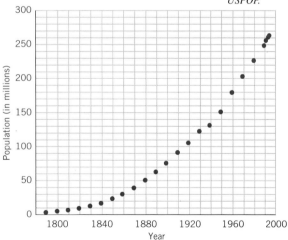

Population of the United States

Source: U.S. Bureau of the Census, Current Population Reports in *The American Almanac 1995–1996 and 1996–1997: Statistical Abstract of the United States,* 1995 and 1996. (Data before 1960 exclude Alaska and Hawaii.)

   a. Find a best fit exponential function.

   b. Graph your function and the actual U.S. population data on the same grid. Describe how the estimated population size differs from the actual population size.

   c. What would your model predict the population to be in the year 2000? In 2020?

   d. In what ways is this exponential function a good model for the data? In what ways is it flawed?

21. Suppose you are offered two similar jobs. The Aerospace Engineering Group offers you a starting salary of $50,000 per year and raises of $1000 every 6 months. The Bennington Corporation offers you an initial salary of $35,000 and a 10% raise every year.

   a. Make a table that shows your salary with each company during the first 8 years.

   b. For each corporation, write a function that gives your salary after t years.

   c. On the same coordinate system, plot the graphs of the functions for years 0 to 20.

   d. In what year does the salary at Bennington exceed the salary at Aerospace?

   e. Using the models from part (b), determine your salary after 10 years and after 20 years at each corporation.

22. The following data show the total government debt for the United States.

Year	1970	1975	1980	1985	1990	1995
National Debt ($ billions)	361	541	909	1827	3266	4961

Data from U.S. Department of the Treasury.

   a. Plot these data by hand and sketch a curve that roughly approximates the data.

   b. Is this a growth or a decay phenomenon? What generic formula might approximate these data?

   c. For 1975 to 1995 calculate the ratio of each debt to the one 5 years earlier, such as (1975 debt)/(1970 debt). This is the growth factor for each particular 5-year period. Is the growth factor staying relatively constant? Average the ratios to get an average growth factor G for 5-year time periods.

   d. Using your value for G in part (c), plot $y = 361G^x$ where $0 \le x \le 5$. Here x represents the number of half-decades since 1970. So $x = 0$ corresponds to 1970; $x = 1$ corresponds to 1975, etc.

   e. Find an estimate for G such that $4961 = 361G^5$. Use this new value of G to plot $y = 361G^x$ where again $0 \le x \le 5$.

   f. Which formula is the better fit to the data? Use this formula to estimate national debts for 2000 and 2005.

   g. Do you think your exponential model is an appropriate model for describing the growth of the national debt? Why?

23. MCI, a phone company that provides long-distance service, introduced a marketing strategy called "Friends and Family." Each person who signed up received a discounted calling rate to 10 specified individuals. The catch was that the 10 people also had to join the "Friends and Family" program.

a. Assume that one individual agrees to join the "Friends and Family" program, and that this individual recruits 10 new members, who in turn each recruit 10 new members, and so on. Write a function to describe the number of people who have signed up for "Friends and Family" after *n* rounds of recruiting.

b. How many "Friends and Family" members, stemming from this one person, will there be after five rounds of recruiting? After 10 rounds?

c. Write a 60 second summary of the pros and cons of this recruiting strategy.

24. (Requires function graphing program with semi-log plot capability.)

See Excel or graph link files CELCOUNT and ECOLI.

a. Load the file CELCOUNT which contains all of the white blood cell counts for the bone marrow transplant patient. Now graph the data on a semi-log plot. Which section(s) of the curve represent exponential growth or decay? Explain how you can tell.

b. Load the file ECOLI which contains the *E. coli* counts for 24 time periods. Graph the *E. coli* data on a semi-log plot. Which section of this curve represents exponential growth or decay?

25. The following graph of the Dow Jones industrial average appeared in the Business section of *The New York Times* on Sunday, February 16, 1997, p. F3. The Dow Jones reflects the performance of 100 mainstream industrial corporations whose stock is traded on the New York Stock Exchange. It had been experiencing phenomenal growth for a number of years and had just reached 7,000 points for the first time ever.

A Different Perspective

This chart, plotted on a logarithmic scale, gives the same visual weight to comparable percentage changes. A 100-point rise when the Dow Jones Industrial Average is at 1,000 looks the same here as a 700-point rise when the Dow is at 7,000. Through this lens, the Dow's recent run-up looks less extraordinary.

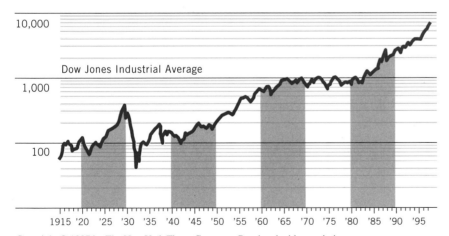

Copyright © 1997 by The New York Times Company. Reprinted with permission.

a. What kind of plot is this?

b. Approximately what was the Dow Jones industrial average in 1929? In 1933? What was happening in American society during this time?

c. In what year did the Dow Jones first reach 1,000? 3,000? 6,000?

d. This plot makes the climb in values appear relatively modest by compressing the vertical scale in order to plot numbers of widely different sizes. On this type of graph, when the plot appears to lie roughly along a straight line, what does this tell you about the original data during that time period?

e. The data between 1980 and 1997 lie roughly along a straight line in this graph. Estimate the annual percentage increase in the Dow Jones during those years.

f. The data between 1933 and 1997 appear to be roughly linear on this graph. Estimate the annual percentage increase in the Dow Jones during this 64-year period.

EXPLORATION 8.1

Devising a Model for the Growth of Bacteria[3]

Objective

- develop a model for bacterial growth by analyzing data on the growth of *E. coli* cells.

Materials/Equipment

- function graphing program with best fit function capabilities or graphing calculators with best fit function capabilities.

This exercise can be carried out in two ways. First, you could analyze data collected by a biology graduate student at UMass/Boston. The data are listed in a table in this Exploration and provided in Excel and graph link files named ECOLIA and ECOLIB. Descriptions of the electronic files are contained in the list of files in the beginning of the Graphing Calculator Workbook. Alternatively, you could measure bacterial growth in a laboratory experiment, then analyze the data generated from your experiment. For this experimental approach, equipment and materials are described on the CD-ROM.

Introduction

In this laboratory, you will analyze the growth of a population of bacteria and try to predict what kind of function defines the dynamics of population growth. Bacteria population growth, which is easy to measure, can be used as a model to characterize population growth in general.

The particular bacterium we use is called *Escherichia coli*, or *E. coli* for short. *E. coli*, one of the most thoroughly studied of all organisms, is a rod-shaped bacterium approximately 1 micrometer (10^{-6} meter) long. Normally, this organism inhabits the intestinal tract of humans. Only certain types of *E. coli* cause problems for the human host. Most types, in fact, aid in our digestion.

Cells of *E. coli* reproduce by simply splitting in half, so that one cell forms two "daughter cells." This form of cell division is called *fission*. It occurs at a rate that depends on the nutrients that are available. When conditions are ideal, *E. coli* can reproduce very quickly. But in a restrictive environment, one in which the food supply is scarce or conditions are poor in other ways, cell division may slow down or stop altogether. To give you an idea of how fast *E. coli* can grow when conditions are unrestricted, consider a population of *E. coli* cells growing under ideal conditions with an unlimited supply of food. If we were to start with one cell, and cell division were to continue uninterrupted under these conditions, then after *only 36 hours*, there would be enough bacterial cells to cover Earth with a layer 1 foot deep. Clearly growth conditions in nature are never so ideal.

---

[3] Developed by Rachel C. Skvirsky, Professor of Biology, University of Massachusetts, Boston.

In this laboratory, we investigate the effects of two environmental variables, nutrients and temperature, on the growth of a bacterial population. We compare the growth of the bacteria using two different nutrient solutions—one is called LB, which is a standard nutrient broth, and the other is Superbroth, which is an enriched nutrient broth. In addition, we compare two temperature conditions. One is body temperature (37°C), and the other (30°C) is between body temperature and room temperature. Remember that *E. coli* normally grows inside the bodies of humans (or other mammals), hence at 37°C.

To monitor the growth of a population of bacteria, we must estimate the number of bacterial cells in the population at different times. Individual bacterial cells are too small to be seen with the naked eye. Therefore we can't simply count them, unless we examine a very small sample with a powerful microscope. Although counting is possible, we use a much faster method. We take advantage of the fact that as bacteria multiply, they increasingly cloud the solution in which they are growing. The more bacteria, the cloudier the solution becomes. This cloudiness occurs because each tiny bacterium absorbs and/or scatters light to a tiny extent. We use this fact in our experiment: We measure the cloudiness of a sample of well-mixed solution by measuring the extent to which the solution absorbs light. Based on this, we can chart the increase in the concentration of bacteria in the solution.

For these measurements, we use a *spectrophotometer*. This is an instrument that measures the amount of light which is absorbed by a solution. Absorbance (or optical density) is related to the number of particles absorbing light. Specifically, optical density is equal to a constant times the concentration of particles. Therefore we use absorbance as a measure of the concentration of particles (bacteria) in the solution.

Biologists have found that an optical density of 0.01 is obtained with a concentration of 1 million bacteria per 5-milliliter (ml) sample. If the optical density were 0.02, then how many bacteria would there be per 5-milliliter?

Procedure (Based on Provided Data)

The following are data from actual measurements of bacterial populations. To carry out this experiment, flasks containing 30 milliliters of either LB or Superbroth medium were inoculated with 0.3 milliliters of an overnight culture of *E. coli* cells (strain HB101). Measurements were taken with a Klett-Summerson colorimeter, which is a simple type of spectrophotometer that measures the optical density or turbidity of the culture. Remember that optical density, represented in this case by a Klett value, is equal to a constant times the concentration of bacteria.

| | LB Cultures | | | | Superbroth Cultures | | | |
| | Temp = 30°C | | Temp = 37°C | | Temp = 30°C | | Temp = 37°C | |
Elapsed Time (min)	Klett Reading*	No. of Cells/ml	Klett Reading*	No. of Cells/ml	Klett Reading*	No. of Cells/ml	Klett Reading*	No. of Cells/ml
20	5	2.50×10^7	2	1.0×10^7	2	1.00×10^7	2	1.00×10^7
50	6	3.00×10^7	6	3.0×10^7	7	3.50×10^7	7	3.50×10^7
80	7	3.50×10^7	11	5.5×10^7	7	3.50×10^7	15	7.50×10^7
110	17	8.50×10^7	27	1.35×10^8	10	5.00×10^7	30	1.50×10^8
130	22	1.10×10^8	37	1.85×10^8	16	8.00×10^7	53	2.65×10^8
165	30	1.50×10^8	62	5.10×10^8	20	1.00×10^8	87	4.35×10^8
180	35	1.75×10^8	75	3.75×10^8	27	1.35×10^8	125	6.25×10^8
200	40	2.00×10^8	87	4.35×10^8	38	1.90×10^8	163	8.15×10^8
230	57	2.85×10^8	106	5.30×10^8	53	2.65×10^8	188	9.40×10^8

* Klett 100 = 5×10^8 cells/ml.

However, a Klett colorimeter uses a different scale from other spectrophotometers. You can assume that a Klett value of 100 corresponds to 5×10^8 cells/ml.

Use these data as the basis for your analysis of bacterial growth. First, make a graph that plots either the Klett reading or the extrapolated cell density (this is the dependent variable and goes on the *y*-axis) versus time (this is the independent variable and goes on the *x*-axis). Note the maximum Klett value or cell density that was measured, as well as the maximum number of minutes elapsed, and use these numbers as guides in setting up the scales of the graph. Use different symbols and/or different colors for the four different cultures, plotting them all on the same graph.

Questions for Thought

Reread the introduction, and be sure you understand the relationship between optical density and concentration of bacteria. You will prepare a report on this lab, which should focus on two issues: (1) The nature of population growth in your cultures of *E. coli*, and (2) a comparison of population growth of *E. coli* in the four different environmental conditions. Be sure to address questions (a)–(g).

a. How does the concentration of bacteria in a 5-ml sample relate to the number of bacteria in the total population? Why is it valid to monitor population growth by measuring 5-ml samples?

b. Does the population size of *E. coli* in the flasks change as time passes? How do you know?

c. Does the growth rate of the *E. coli* population in the flasks change over time? How do you know?

d. The nutritional requirements of various types of bacteria are different; consequently, microbiologists use many different types of media to grow bacteria. LB medium supports the growth of a wide array of microorganisms. In this exercise, we use data from cells grown in LB medium, as well as in a nutrient broth that contains extra nutrients—Superbroth. In which of these growth media do *E. coli* cells grow better? Examine the recipes for these two growth media shown in the Experimental Procedures section on the CD-ROM. What are the differences between these two media?

e. Are the "test tube" environments limited (so that the *E. coli* populations show restricted growth), or unlimited (so that *E. coli* shows unrestricted growth)?

f. How are the shapes of these growth curves likely to change if the experiment is continued for a substantially longer period of time? Why?

g. Does inspection of your graphs reveal differences in growth of *E. coli* populations under different environmental conditions? Do these differences make sense in terms of what you know of the natural habitat of *E. coli*?

Lab Report

Below is the format and some guidelines for preparing your lab report.

Objectives
Describe the purpose of the experiment. What were you trying to accomplish? How were you using data on flasks of growing bacteria to accomplish your objectives?

Materials and Methods
Describe the procedures that you followed, using the past tense to explain what was done. For example, what type of bacteria was used? Under what environmental conditions were the bacteria grown? At what intervals were measurements taken? If you followed the methods outlined

in the Experimental Procedures section on the CD-ROM to take your own measurements of bacterial growth, then additional issues should be addressed. For example, how were the measurements taken? You may reference this Exploration, rather than repeating each and every detail. Perhaps most importantly, this section should describe any ways in which your procedure departed from the printed instructions, or any mistakes that were made. For example, did any flask of cells stay out of the water bath longer than planned? Do you have reason to believe that certain measurements were taken incorrectly? Often a great deal can be learned from the recording and analysis of mistakes or unexpected events.

Results

This section should include all of your data, presented in the form of a table. Enter your data in your function graphing program or graphing calculator and prepare a graph. Plot optical density or cell density, the dependent variable, on the y-axis and time (in minutes), the independent variable, on the x-axis. Remember that optical density is an expression of cell density.

Analysis of Data

To analyze your data, use as a guide the questions asked here and the "Questions for Thought" listed earlier in this Exploration.

- Can you define a function that describes the growth of a population undergoing unrestricted growth? To develop your function, first do the following exercise: Pretend that you start with one bacterial cell. Remember that this cell reproduces by splitting in half. Now pick an arbitrary amount of time (such as 1 hour) that it takes this cell to divide. You can call this amount of time the "doubling time." If you start with one cell, how many cells would you have after one doubling time? How about after 2 or 3 or 4 doubling times? Can you derive an expression that describes how many cells you will have after x hours or doubling times?

- Which environmental conditions best supported the growth of *E. coli*?

- Under the conditions that were used to generate the data, or the conditions that you actually used in the lab, was the growth rate of the population constant? If not, how would you describe the change in growth rate? What do you think causes the change in growth rate? Can you predict what would happen to the population in these flasks if the experiment lasted another whole day? Justify your answer.

- Recall the basic difference between population growth in a restricted environment and growth in an unrestricted environment. In an unrestricted environment, growth rate is constant, whereas in an environment that is restricted by limited nutrients, growth rate declines as the population becomes more dense. How could we determine whether the specific environments in which you grew *E. coli* were restricted or not?

EXPLORATION 8.2

Properties of Exponential Functions

Objective

- explore the effects of a and C on the graph of the exponential function in the form

$$y = Ca^x \qquad \text{where } a > 0 \text{ and } a \neq 1$$

Materials/Equipment

- computer and software on disk, "E3: $y = Ca^x$ Sliders" in *Exponential & Log Functions,* or graphing calculator
- graph paper

Procedure

We start by choosing values for a and C and graphing the resulting equations by hand. From these graphs we make predictions about the effects of a and C on the graphs of other equations. Take notes on your predictions and observations so you can share them with the class. Work in pairs and discuss your predictions with your partner.

Making Predictions

1.   Start with the simplest case, where $C = 1$. The equation will now have the form

$$y = a^x$$

Make a data table and by hand sketch on the same grid the graphs for $y = 2^x$ (here $a = 2$) and $y = 3^x$ (here $a = 3$). Use both positive and negative values for x. Predict where the graphs of $y = 2.7^x$ and $y = 5^x$ would be located on your graph. Check your work and predictions with your partner.

x	$y = 2^x$	$y = 3^x$
-2		
-1		
0		
1		
2		
3		
4		

How would you describe your graphs? Do they have a maximum or a minimum value? What happens to y as x *increases*? What happens to y as x *decreases*? Which graph shows y changing the fastest compared to x?

2.   Now create three functions in the form $y = a^x$ where $0 < a < 1$. Create a data table and graph your functions on the same grid as before. Make predictions for other functions where $C = 1$ and $0 < a < 1$.

3. Now consider the case where C has a value other than 1 for the general exponential function

$$y = Ca^x$$

Create a table of values and sketch the graphs of $y = 0.3(2^x)$ (in this case $C = 0.3$ and $a = 2$) and $y = 5(2^x)$ (in this case $C = 5$ and $a = 2$). What do all these graphs have in common? What do you think will happen when $a = 2$ and $C = 10$? What do you think will happen to the graph if $a = 2$ and $C = -5$? Check your predictions with your partner.

x	$y = 0.3(2^x)$	$y = 5(2^x)$	$y = -5(2^x)$
-2			
-1			
0			
1			
2			
3			
4			

How would you describe your graphs? Do they have a maximum or a minimum value? What happens to y as x *increases*? What happens to y as x *decreases*? Which graph shows y changing the fastest compared to x? What is the y-intercept for each graph?

Testing Your Predictions

Now test your predictions by using a program called "E3: $y = Ca^x$ Sliders" in the *Exponential & Log Functions* software package or by creating graphs using technology.

1. What effect does a have?

Make predictions when $a > 1$ and when $0 < a < 1$, based on the graphs you constructed by hand. Explore what happens when $C = 1$ and you choose different values for a. Check to see whether your observations about the effect of a hold true when $C \neq 1$.

How does changing a change the graph? When does $y = a^x$ describe growth? When does it describe decay? When is it flat? Write a rule that describes what happens when you change the value for a. You only have to deal with cases when $a > 0$.

2. What effect does C have?

Make a prediction based on the graphs you constructed by hand. Now choose a value for a and create a set of functions with different C values. Graph these functions on the same grid.

How does changing C change the graph? What does the value of C tell you about the graph of functions in the form $y = Ca^x$? Describe your graphs when $C > 0$ and when $C < 0$. Use technology to test your generalizations.

Exploration-Linked Homework

Write a 60 second summary of your results, and present it to the class.

EXPLORATION 8.3

Recognizing Exponential Patterns in Data Tables

Objectives

- use ratios to identify exponential growth or decay and to construct exponential functions

Materials/Equipment

- calculator or computer with spreadsheet program (optional)

Procedure

Working in pairs

1. The following tables contain data about the exponential growth of *E. coli* (from Table 8.1) and the exponential decay of the dollar through inflation (from Table 8.18).

Growth of *E. coli* Bacteria

t (Time Periods)	$f(t) = 100 \cdot 2^t$ (No. of Bacteria)	$f(t)/f(t-1)$	$f(t+1)/f(t)$
0	100	n.a.	
1	200		
2	400		
3	800		
4	1,600		
5	3,200		
6	6,400		
7	12,800		
8	25,600		
9	51,200		
10	102,400		n.a.

Erosion of the Dollar through Inflation

n (Year)	$f(n) = 1 \cdot (0.952)^n$ (Value of Dollar)	$f(n)/f(n-1)$	$f(n+1)/f(n)$
0	$1.00	n.a.	
1	$0.95		
2	$0.91		
3	$0.86		
4	$0.82		
5	$0.78		
6	$0.74		
7	$0.71		
8	$0.67		
9	$0.64		
10	$0.61		n.a.

a. In the third column of the *E. coli* data, fill in the values of the ratios $f(t)/f(t-1)$. Similarly in the third column of the inflation data, fill in values for the ratios $f(n)/f(n-1)$.

b. What do you notice about the ratios in each case? What do the ratios tell you about the functional values as the independent variable (*t* or *n*, respectively) increases by 1 unit?

c. Fill in the fourth column with values for the ratios $f(t+1)/f(t)$ and $f(n+1)/f(n)$. Compare the values in columns 3 and 4 in each table. Explain their relationship.

2. Generalizing your results

a. Using the general form for an exponential function, $f(x) = C \cdot a^x$, write an expression for:

$f(k)$, the value of the function when $x = k$.

$f(k+1)$, the value of the function when $x = k + 1$.

$f(k-1)$, the value of the function when $x = k - 1$.

b. Calculate the ratios: $f(k)/f(k-1)$ and $f(k+1)/f(k)$

For exponential functions, these ratios are called *growth or decay factors*. Is this definition consistent with the definitions of growth and decay factors given in Sec. 8.3?

c. How do these ratios relate to the exponential function $f(x) = C \cdot a^x$? State your results as a general rule.

Using Your Results

Use the generalizations for exponential functions that you have found to analyze each of the following data tables.

a. By calculating ratios, determine whether the data in each of the following tables exhibit exponential growth or decay. If the data are exponential, determine the growth or decay factor.

b. For each table explain in your own words how to find $f(x)$ in terms of *x*.

i.

x	$f(x)$
0	1
1	10
2	100
3	1,000
4	10,000

ii.

x	$f(x)$
0	-3
1	-6
2	-12
3	-24
4	-48

iii.

x	$f(x)$
0	25
1	100
2	400
3	1600
4	6400

iv.

x	$f(x)$
0	1.0
1	0.1
2	0.01
3	0.001
4	0.0001

v.

x	$f(x)$
0	15
1	12
2	9
3	6
4	3

vi.

x	$f(x)$
0	0
1	1
2	4
3	9
4	16

c. For each data set that exhibits exponential growth or decay, construct an equation to describe the data. Does your equation predict the values in your data table?

d. Using the equations you generated in part (c), extend each of the exponential data tables to include negative values for *x*.

Functioning with Powers

Overview

Why is the most important rule in scuba diving "never hold your breath"? Why do small animals have faster heartbeats and higher metabolic rates than large ones? Why does the moon exert more force than the sun on Earth's tides?

In this chapter we introduce the family of power functions to help us answer these questions. We examine the properties of different kinds of power functions, analyze their graphs, and compare them to other functions. Sums and differences of certain power functions lead us to polynomial functions.

The Explorations provide an opportunity to construct power functions with specific properties, to visualize the effect of the exponent and the coefficient on graphs of power functions, and to construct scaled figures.

After reading this chapter you should be able to:

- recognize the properties of power functions
- construct and interpret graphs of power functions
- understand direct and inverse proportionality
- compare power, linear, and exponential functions
- identify and evaluate polynomial functions

9.1 THE TENSION BETWEEN SURFACE AREA AND VOLUME

Why do small animals have faster heartbeats and higher metabolic rates than large ones? Why are the shapes of the bodies and organs of large animals often quite different from those of small ones? Why are wood logs hard to set on fire, whereas small sticks and chips burn easily, and sawdust can explode? To find answers to these questions we examine how the relationship between surface area and volume changes as objects increase in size.

Scaling Up a Cube

Let's examine what happens to the surface area and volume of a simple geometric figure, the cube, as we increase its size. In Figure 9.1 we have drawn a series of cubes where the lengths of the edges are 1, 2, 3, and 4 units.[1]

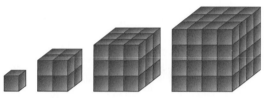

Figure 9.1 Four cubes for which the lengths of the edges are 1, 2, 3, and 4 units, respectively.

Surface area of a cube   If we were painting a cube, the surface area would tell us how much area we would have to cover. Each cube has six identical faces, so

$$\text{surface area} = 6 \cdot \text{area of one face}$$

Each little square on the faces of the cubes in Figure 9.1 has an edge length of 1 unit, and an area of 1 square unit. On a face with the length of edge x, there are x rows of x squares each, so the area of the face equals $x \cdot x$ or x^2. The function $s(x)$ shows how the surface area depends on the length of its edge, x.

$$s(x) = 6x^2$$

The second column of Table 9.1 contains the surface areas of cubes for which the length of the edge is 1, 2, 3, 4, 6, 8, and 10 units, and Figure 9.2 shows a graph of the function $s(x)$.

Since the surface area of a cube with edge length x can be represented as:

$$s(x) = \text{constant} \cdot x^2$$

we say that the surface area is *directly proportional* to the *square* (or second power) of the length of its edge. In Table 9.1 observe what happens to the surface area when we double the length of an edge. If we double the length from 1 to 2 units, the surface

---

[1] It doesn't matter which unit we use, but if you prefer, you may think of "unit" as being "centimeter" or "foot."

Table 9.1		
Surface Area and Volume of Cube with Edge Length x		
Edge length x (unit)	Surface area $s(x) = 6x^2$ (sq. unit)	Volume $v(x) = x^3$ (cubic unit)
1	6	1
2	24	8
3	54	27
4	96	64
6	216	216
8	384	512
10	600	1000

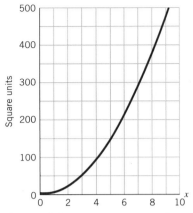

Figure 9.2   Graph of $s(x) = 6x^2$.

area becomes four times larger, increasing from 6 to 24 square units. If we double the length from 2 to 4 units, the surface area is again four times larger, increasing this time from 24 to 96 square units. In general, if we double the length of the edge from x to $2x$, the surface area will increase by a factor of 4.

if	$s(x) = 6x^2$
then	$s(2x) = 6(2x)^2$
apply exponent	$= 6 \cdot 2^2 \cdot x^2$
simplify and rearrange terms	$= 4(6x^2)$
substitute $s(x)$ for $6x^2$	$s(2x) = 4 \cdot s(x)$

Volume of a cube   Now, consider the volume of each cube. Each little cube has an edge length of 1 unit, and a volume of 1 unit cubed. So the volume of a bigger cube equals the number of little cubes that fills it up. Counting cubes is harder than counting squares, because the cubes on the inside are hidden from our view. The cube with 3 units on a side has 3 layers; each layer has 9 cubes laid out in a 3×3 pattern. So the total number of little cubes is 3 layers times 9 cubes per layer or $3 \times 9 = 27$. For a cube with edge length x, there are x layers, each of which is a square arrangement of $x \cdot x$ cubes. So the total volume can be represented by the equation

$$\text{volume} = (\text{number of layers}) \cdot (\text{number of cubes in each layer})$$

or using functional notation as

$$v(x) = x \cdot x^2$$
$$= x^3$$

The third column of Table 9.1 contains the volume for cubes of certain sizes, and Figure 9.3 is the graph of the volume function, $v(x)$.
    Since

$$\text{volume} = \text{constant} \cdot x^3 \qquad \text{(the constant in this case is 1)}$$

we say that the volume is *directly proportional* to the *cube* (or the third power) of the length of its edge.
    What happens to the volume when we double the length? In Table 9.1 if we dou-

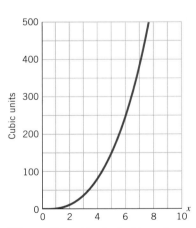

Figure 9.3   Graph of $v(x) = x^3$.

ble the length from 1 to 2 units, the volume increases by a factor of 2^3 or 8, increasing from 1 to 8 cubic units. If we double the length from 2 to 4 units, the volume again becomes 8 times larger, increasing from 8 to 64 cubic units. In general, if we double the edge length from x to $2x$, the volume will increase by a factor of 8.

if	$v(x) = x^3$
then	$v(2x) = (2x)^3$
apply exponent	$= 2^3 \cdot x^3$
simplify	$= 8x^3$
substitute $v(x)$ for x^3	$v(2x) = 8 \cdot v(x)$

Surface area/volume   We have seen that when we double the length of the edge, the surface area increases by a factor of 4 but the volume increases by a factor of 8. If we increase the length of the edge, the volume eventually grows faster than the surface area. So the ratio

$$\frac{\text{surface area}}{\text{volume}}$$

decreases as the side length increases. See the third column of Table 9.2.

Table 9.2
Ratio of Surface Area to Volume of Cube with Length of Edge x

Edge length x	Surface Area $s(x) = 6x^2$	Volume $v(x) = x^3$	Surface Area / Volume $\frac{s(x)}{v(x)} = \frac{6x^2}{x^3} = \frac{6}{x}$
1	6	1	6.00
2	24	8	3.00
3	54	27	2.00
4	96	64	1.50
6	216	216	1.00
8	384	512	0.75
10	600	1000	0.60

Size and Shape

In Exploration 9.2 you can study further the effects of scaling up an object.

What we learned about the cube is true for any object, no matter what the shape. In general:

For any shape, as an object becomes larger while keeping the same shape, the ratio of its surface area to its volume decreases.

Thus, for the larger object, there is relatively less surface area. This fact allows us to understand some basic principles of biology and to answer the questions we asked at the beginning of this section.

Biological functions such as respiration and digestion depend upon surface area but must service the body's entire volume.[2] The biologist J. B. S. Haldane wrote that "comparative anatomy is largely the story of the struggle to increase surface in proportion to volume." This is why the shapes of the bodies and organs of large animals are often quite different from those of small ones. Many large species have adapted by developing complex organs with convoluted exteriors, thus greatly increasing the organs' surface areas. Human lungs, for instance, are heavily convoluted to increase the amount of surface area, thereby increasing the rate of exchange of gases. Stephen Jay Gould writes "the villi of our small intestine increase the surface area available for absorption of food (small animals neither have nor need them)."

Body temperature depends upon the ratio of surface area to volume. Animals generate the heat needed for their volume by metabolic activity, and lose heat through their skin surface. Small animals have more surface area in proportion to their volume than do large animals. Since heat is exchanged through the skin, small animals lose heat proportionately faster than large animals and have to work harder to stay warm. Hence their heartbeats and metabolic rates are faster. As a result, smaller animals burn more energy per unit mass than larger animals.

The surface area/volume tension relates to other physical phenomena as well. To set an object on fire, some part of the volume must be raised to the ignition point. Heat is absorbed through the surface, so large pieces, which have relatively less surface area than smaller ones, are more difficult to ignite. That is why small pieces of wood called tinder are used to light log fires. Tiny dust particles floating in the air can be quite flammable and are responsible for a number of dramatic explosions of grain elevators in the Midwest.

Stephen Jay Gould's essay "Size and Shape" in Ever Since Darwin: Reflections in Natural History *offers an interesting perspective on the relationship between the size and shape of objects.*

Algebra Aerobics 9.1

1. The surface area and volume of a sphere are both functions of the radius. They can be described by the functions:

$$v(r) = (4/3)\pi r^3$$

where $v(r)$ represents the volume of a sphere with radius r; and

$$s(r) = 4\pi r^2$$

where $s(r)$ represents the surface area of a sphere with radius r.

a. What happens to the surface area and volume of a sphere when you double the radius?

b. Which eventually grows faster, the surface area or the volume? As a result, what happens to the surface area/volume ratio as the radius increases?

[2] For those who want to investigate how species have adapted and evolved over time, see D'Arcy Wentworth Thompson, *On Growth and Form* (New York: Dover Publications, Inc., 1992) and Thomas McMahon and John Bonner, *On Size and Life* (New York: Scientific American Books, Inc., 1983).

9.2 POWER FUNCTIONS WITH POSITIVE POWERS

In the last section, we saw that the surface area function, $s(x)$, and the volume function, $v(x)$, of a cube can be represented by the functions:

$$s(x) = 6x^2$$

and

$$v(x) = x^3$$

where x = edge length of the cube. These functions are examples of *power functions.* The general form of an equation for a power function is

$$\text{dependent variable} = (\text{constant}) \cdot (\text{independent variable})^{\text{power}}$$

> A *power function* $y = f(x)$ can be represented by an equation in the form
>
> $$y = kx^p$$
>
> where k and p are any constants.

Something to think about

How would you describe a power function with an exponent of 0?

For example, the functions

$$v = (4/3)\pi r^3 \qquad F = 0.2d^{-2} \qquad A = 25M^{1/2} \qquad \text{and} \qquad y = 5x$$

are all power functions with powers of 3, -2, 1/2, and 1, respectively.

The next few sections focus on power functions with positive integer exponents. In Section 9.6 we study power functions with negative integer exponents. At the end of Chapter 12 we return to power functions, focusing on fractional powers.

Example   1

The radius of the core of Earth is over half the radius of Earth as a whole, yet the core is only about 16% of the total volume of Earth. How is this possible?

SOLUTION
We can think of both Earth and its core as approximately spherical in shape. The function

$$v = \frac{4}{3}\pi r^3$$

describes the volume, v, of a sphere as directly proportional to the cube of its radius, r. So if the core of Earth has radius R, then the core's volume is approximately

$$v_1 = (4/3)\pi(R)^3$$

If the radius of Earth were $2R$ or exactly twice that of its core, Earth's volume would be

$$v_2 = (4/3)\pi(2R)^3$$

rewrite $(2R)^3$
$$= (4/3)\pi(2)^3R^3$$

rearrange terms
$$= (2^3)(4/3)\pi R^3$$

substitute v_1 for $(4/3)\pi R^3$ and 8 for 2^3
$$v_2 = 8v_1$$

Comparing v_2 to v_1 shows v_2 is 2^3 or 8 times larger than v_1. So the volume of a sphere with radius $2R$ is 2^3 or 8 times larger than the volume of a sphere with radius R. If the radius of Earth were exactly twice the radius of its core, then the volume of the core of Earth would be 1/8 or 12.5% of the volume of the whole Earth. Since the radius of the core of Earth is a little larger than half the radius of Earth as a whole, 16% of the whole Earth seems a reasonable estimate for the volume of Earth's core.

Direct Proportionality

In Chapter 4, for linear functions, we said that y is *directly proportional* to x if y equals a constant times x. For example, if $y = 4x$, then y is directly proportional to x. We can extend the same concept to any power function with positive exponents. If $y = kx^p$ and p is positive, we say that y *is directly proportional* to x^p.

In Section 9.1, we saw that the surface area of a cube is directly proportional to the square of its edge length, and that the volume is directly proportional to the cube of its edge length. The symbol $\propto$ is used to indicate direct proportionality.

> If
>
> $$y = kx^p$$
>
> where k is a constant and $p > 0$, we say that y *is directly proportional to* or *varies directly with* x^p. We write this as
>
> $$y \propto x^p$$
>
> k is called the constant of proportionality.

Something to think about

Both the mechanical strength (resistance to breaking) and the muscular strength of a limb are directly proportional to its cross-sectional area. That's why a large animal such as an elephant needs thick heavy legs to support its weight. By contrast, the thin legs of a small animal, such as an antelope, not only support it, but provide enough strength for the animal to run and leap.

Example   2

$A = \dfrac{t^2}{40}$ or, equivalently, $A = \left(\dfrac{1}{40}\right)t^2$ is the equation derived in Section 2.4 that describes the area of a *Sordaria finicola* fungus colony. The area is directly proportional to time squared. 1/40 is the constant of proportionality.

a.   What happens to A, if t is increased by a factor of 5 (multiplied by 5)?

It is probably easiest to see what is happening if we use functional notation, where

$$f(t) = A = \left(\dfrac{1}{40}\right)t^2$$

evaluate f at $5t$	$f(5t) = (1/40)(5t)^2$
apply exponent	$= (1/40)25t^2$
rearrange terms	$= 25(1/40)t^2$
substitute $f(t)$ for $(1/40)t^2$	$f(5t) = 25f(t)$

The value of A becomes 25 times larger (increased by a factor of 25) when t is increased by a factor of 5.

b. What happens to the value of A if t is divided by 5?

evaluate f at $t/5$ $\qquad$ $f(t/5) = (1/40)(t/5)^2$

apply exponent $\qquad$ $= (1/40)\dfrac{t^2}{25}$

rearrange terms $\qquad$ $= (1/25)(1/40)t^2$

substitute $f(t)$ for $(1/40)t^2$ $\qquad$ $f(t/5) = (1/25)f(t)$

The value of A is reduced to $1/25$ of its previous value when t is divided by 5.

c. Rewrite the equation, solving for t. Is t directly proportional to A?

Given $\qquad$ $A = \left(\dfrac{1}{40}\right)t^2$

multiply each side by 40 $\qquad$ $40A = t^2$

take square root of each side
of equation and switch sides $\qquad$ $t = \sqrt{40A}$

$\qquad$ $= \sqrt{40}A^{1/2}$

evaluate $\sqrt{40}$ $\qquad$ $t \approx 6.32A^{1/2}$

So t is directly proportional to $A^{1/2}$. The constant of proportionality is 6.32.

Example 3

In Chapter 7 we encountered the following formula used by police. It gives the speed, S, at which a car must have been traveling given the distance, d, the car skids on a dry tar road after the brakes have been applied.

$$S = \sqrt{30d} \approx 5.48d^{1/2}$$

Speed, S, is in miles per hour and distance, d, is in feet. We can think of S as a function of d.

a. Use the language of proportionality to describe the relationship between S and d.

S is directly proportional to $d^{1/2}$ and the constant of proportionality is 5.48.

b. What are the domain and range of the function?

The implied domain is values of d greater than or equal to 0. The range is the set of all real numbers greater than or equal to 0.

c. What happens to S if d doubles? Quadruples?

We consider S as a function of d and write

$$S = f(d)$$

where $\qquad$ $f(d) = 5.48d^{1/2}$

If we evaluate f at $2d$ $\qquad$ $f(2d) = 5.48(2d)^{1/2}$

apply exponent $\qquad$ $= 5.48 \cdot 2^{1/2}d^{1/2}$

rearrange terms $\qquad$ $= 2^{1/2} \cdot 5.48d^{1/2}$

evaluate $2^{1/2} = \sqrt{2}$ $\qquad$ $\approx 1.414 \cdot (5.48d^{1/2})$

substitute $f(d)$ for $5.48d^{1/2}$ $\qquad$ $f(2d) \approx 1.414f(d)$

Hence, if d doubles, the value of S increases by a factor of $\sqrt{2}$ or approximately 1.414.

If we evaluate f at $4d$	$f(4d) = 5.48(4d)^{1/2}$
apply exponent	$= 5.48 \cdot 4^{1/2} d^{1/2}$
rearrange terms	$= 4^{1/2}(5.48 d^{1/2})$
simplify	$= 2(5.48 d^{1/2})$
substitute $f(d)$ for $5.48 d^{1/2}$	$f(4d) = 2f(d)$

Hence, if d quadruples, S doubles.

d. Solve for d. Is d directly proportional to S?

Given	$S = \sqrt{30d}$
square both sides of the equation	$S^2 = 30d$
divide both sides by 30	$S^2/30 = d$
or	$d = \dfrac{S^2}{30}$

So d is directly proportional to S^2.

Something to think about

Why is $\sqrt{30d} \cdot \sqrt{30d} = 30d$?

Direct Proportionality with More Than One Variable

When a quantity depends on more than one other quantity, we no longer have a simple power function. For example, the volume, V, of a right circular cylinder depends on both the radius, r, of the base and the height, h. The function describing this relationship is

$$V = \text{area of base} \cdot \text{height}$$
$$V = \pi r^2 h.$$

We say V is directly proportional to both r^2 and h.

According to the Stefan–Boltzmann law in physics, dark objects are good emitters as well as good absorbers of radiant energy. The total power, P, emitted by an object at absolute temperature T is given by the following equation:

$$P = kAT^4$$

Example   4

where A is the surface area of the body measured in square units and T is the temperature in degrees Kelvin.[3] The constant k depends on how "black" the object is.

a. Express this relationship in terms of direct proportionality.

P is directly proportional to both A and to T^4.

---

[3] The temperature measured in degrees Kelvin is equal to the temperature in degrees Centigrade plus 273.2°C. The temperature -273.2 degrees Centigrade or 0° Kelvin is called "absolute zero" because at that temperature all motion (of molecules, atoms, etc.) stops.

b. For a particular black body, what happens to the value of P when the temperature increases by a factor of 5?

We can consider the total power, P, as a function of temperature, T, and write

$$P = g(T)$$

where	$g(T) = kAT^4$
evaluate g at $5T$	$g(5T) = kA(5T)^4$
apply exponent	$= kA(625T^4)$
rearrange terms	$= 625kAT^4$
substitute $g(T)$ for kAT^4	$g(5T) = 625g(T)$

So when T increases by a factor of 5, the value of P increases by a factor of 625.

c. How should the temperature T, in degrees Kelvin, of a black body change so the power emission is 1/16 of its original value? Is 1/10,000 of its original value?

To reduce P by a factor of 1/16, T needs to be multiplied by 1/2 since $(1/2)^4 = 1/16$.

To reduce P by a factor of 1/(10,000), T needs to be multiplied by 1/10 since $(1/10)^4 = 1/(10,000)$.

d. Find an equivalent equation by solving for A.

given	$P = kAT^4$
divide both sides of the equation by kT^4	$\dfrac{P}{kT^4} = A$
rewrite exponent and switch sides	$A = \dfrac{PT^{-4}}{k}$

Algebra Aerobics 9.2a

1. Identify which (if any) of the following equations represent direct proportionality.
  a. $y = 5.3x^2$　　　**c.** $y = 5.3x^{-2}$
  b. $y = 5.3x^2 + 10$

2. Given $g(x) = 5x^3$
  a. Calculate $g(2)$ and compare this value to $g(4)$.
  b. Calculate $g(5)$ and compare this value to $g(10)$.
  c. What happens to the value of $g(x)$ if x doubles in value?
  d. What happens to $g(x)$ if x is divided by 2?

3. Express in your own words the relationship between y and x in the following functions:
  a. $y = 3x^5$　　**b.** $y = 2.5x^3$　　**c.** $y = \dfrac{x^5}{4}$

4. Now express each of the relationships in Problem 3 in terms of direct proportionality.

5. **a.** If $P = aR^2$　　　Solve for R
  b. If $V = (1/3)\pi r^2 h$　　Solve for h, then solve for r
  c. If $Y = Z(a^2 + b^2)$　　Solve for Z, then solve for a

Moving to the Abstract

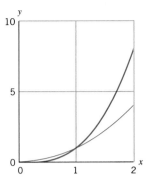

Figure 9.4 Graph of $f(x) = x^2$ and $g(x) = x^3$.

What happens when $x > 0$ and approaches $+\infty$? If we examine what happens to the abstract functions $f(x) = x^2$ and $g(x) = x^3$ when x is positive, it is clear that as x increases, $f(x)$ and $g(x)$ both increase, and as x grows infinitely large, x^2 and x^3 tend toward positive infinity. In symbols we write, as $x \to +\infty$, then $x^2 \to +\infty$ and $x^3 \to +\infty$. If $x = 0$, then x^2 and x^3 are both equal to 0, and if $x = 1$, x^2 and x^3 are both equal to 1. So $f(x)$ and $g(x)$ intersect at $(0, 0)$ and $(1, 1)$. We can see from the graph in Figure 9.4 that when $x > 1$, then $x^3 > x^2$, and when $0 < x < 1$, then $x^3 < x^2$.

What happens when $x < 0$ and approaches $-\infty$? What happens to the functions $f(x) = x^2$ and $g(x) = x^3$ when x is negative? We can see in Table 9.3 and Figure 9.5, the functions exhibit quite different behaviors. When x is negative, x^2 is positive, and as x increases (for example, from -4 to -1), x^2 decreases. As $x \to -\infty$, then $x^2 \to +\infty$. When x is negative, x^3 is negative, and as x increases, x^3 increases. As $x \to -\infty$, then $x^3 \to -\infty$. The range for $f(x)$ is the set of nonnegative real numbers. The range for $g(x)$ is the set of all real numbers. The two graphs in Figure 9.5 show the different behaviors of $f(x) = x^2$ and $g(x) = x^3$.

Table 9.3		
x	$f(x) = x^2$	$g(x) = x^3$
-4	16	-64
-3	9	-27
-2	4	-8
-1	1	-1
0	0	0
1	1	1
2	4	8
3	9	27
4	16	64

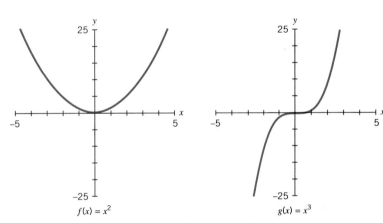

Figure 9.5 The graphs of $f(x) = x^2$ and $g(x) = x^3$.

Consider the functions that represent the volume and the surface area of a cube as two abstract functions. Let $s(x) = 6x^2$ and $v(x) = x^3$.

Example 5

a. Identify the domains and ranges, and sketch the graphs of both functions.

b. Which function eventually dominates when x is positive?

c. For what values of x does $v(x) = s(x)$? For what values of x is $v(x) > s(x)$?

SOLUTION

a. The domain for both abstract functions, $s(x)$ and $v(x)$, is all the real numbers. The range for $s(x)$ is all nonnegative real numbers. The range for $v(x)$ is all real numbers.

b. $v(x)$ eventually dominates $s(x)$.

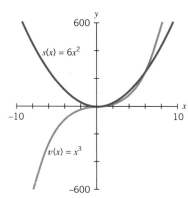

Figure 9.6   Graph of the abstract functions $s(x) = 6x^2$ and $v(x) = x^3$.

c. From the graph in Figure 9.6, the curves appear to intersect when $x = 0$ and when $x = 6$. As with linear systems of equations, we can set the two functions $v(x)$ and $s(x)$ equal to each other to find the value of x where the curves intersect.

$$v(x) = s(x)$$
$$x^3 = 6x^2 \qquad (1)$$

Clearly $x = 0$ is one solution. We have $s(0) = 0$ and $v(0) = 0$, so the graphs intersect at $(0, 0)$.

If $x \neq 0$, we can divide both sides of Equation (1) by x^2 to get $x = 6$. If we evaluate $v(x)$ when $x = 6$

we have $v(6) = 6^3 = 216$

double checking in $s(x)$ $s(6) = 6(6)^2 = 216$

So the graphs of $v(x)$ and $s(x)$ intersect at $(6, 216)$ and $(0, 0)$. When $x > 6$, then $v(x) > s(x)$.

Algebra Aerobics 9.2b

1. a. If $f(x) = 4x^3$, evaluate the following:

$f(2) \qquad f(-2) \qquad f(s) \qquad f(3s)$

b. If $g(t) = -4t^3$, evaluate the following:

$g(2) \qquad g(-2) \qquad g\left(\dfrac{1}{2}t\right) \qquad g(5t)$

2. In each case, indicate whether or not the function is a power function. If it is, identify the independent and dependent variables, the constant of proportionality and the power.

a. $A = \pi r^2$ **c.** $z = w^5 + 10$ **e.** $y = 3x^5$
b. $y = z^5$ **d.** $y = 5^x$

3. a. For both of the following functions, generate small tables, including positive and negative values for x, and graph the functions on the same grid.

$$f(x) = 4x^2 \qquad \text{and} \qquad g(x) = 4x^3.$$

b. What happens to $f(x)$ and to $g(x)$ as $x \to +\infty$?
c. What happens to $f(x)$ and to $g(x)$ as $x \to -\infty$?
d. Specify the domain and range of each function.
e. Where do the graphs of these functions intersect?
f. For what values of x is $g(x) > f(x)$?

9.3 VISUALIZING POSITIVE INTEGER POWERS

Odd Versus Even Powers

Exploration 9.1 can help re-inforce your understanding of power functions with positive integer exponents.

Let n be a positive integer. If x is positive, then x^n is positive for all values of n. But if x is negative, we have to consider whether n is even or odd. If n is even, then x^n is positive, since

$$\text{(negative number)}^{\text{even power}} = \text{positive number.}$$

If n is odd, then x^n is negative, since

$$\text{(negative number)}^{\text{odd power}} = \text{negative number.}$$

Whether the exponent of a power function is odd or even will affect the shape of the graph.

If we graph the simplest power functions: $y = x$, $y = x^2$, $y = x^3$, $y = x^4$, $y = x^5$, $y = x^6$, and so on, we can quickly see that the graphs fall into two groups: the odd powers and the even powers (Figure 9.7).

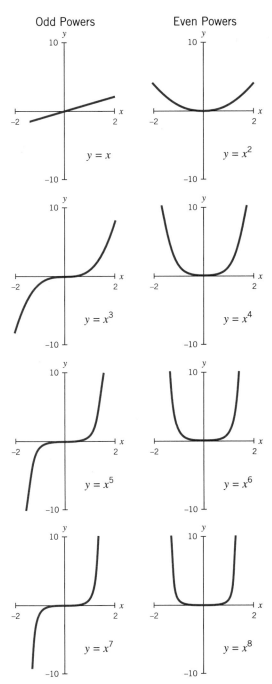

Figure 9.7 Graphs of odd and even power functions.

Something to think about

Take a moment and try to describe in your own words the differences and similarities between the graphs of odd and even power functions. Do the differences between the graphs make sense in terms of what you know about x^n when n is a positive odd integer? What happens when n is a positive even integer?

The graphs of all power functions with either odd or even positive integer exponents go through the origin.

For the odd positive powers, x^1 (usually written as just x), x^3, x^5, x^7, . . . , as x increases, y increases. The graphs are rotationally symmetrical about the origin; that is, if you hold the graph fixed at the origin, and then rotate it 180°, you end up with the same graph. All the graphs of odd powers with exponents greater than 1 appear to "bend" near the origin.

The graphs of even powers, x^2, x^4, x^6, x^8, . . . , are U-shaped. For all positive power functions of even degree, if we start with negative values of x and increase x, y first decreases and then, as x becomes positive, y increases. The graphs are symmetric about the y-axis. If you think of the y-axis as a dividing line, the "left" side of the graph is a mirror image, or reflection, of the "right" side.

The Effect of the Coefficient k

The program "P1: k & p Sliders" in Exponential & Log Functions *can help you visualize the graphs of $y = kx^p$ for different values of k and p.*

The power functions $y = x$, $y = x^2$, $y = x^3$, $y = x^4$, $y = x^5$, and $y = x^6$ all have a coefficient of 1. We now consider the effect of different values for the coefficient k on graphs of power functions in the form

$$y = kx^p \qquad \text{where } p \text{ is a positive integer.}$$

Case 1: $k > 0$ Comparing $y = kx^p$ to $y = x^p$ When k is Positive.

We know from our work with power functions of degree 1, that is, linear functions of the form $y = kx$, that k affects the steepness of the line. For values of $k > 1$, the larger the value for k, the more vertical the graph of $y = kx^p$ becomes compared to $y = x^p$. As k increases, the steepness of the graph increases. The same is true for general power functions of the form $y = kx^p$. We say the graph is *stretched vertically*.

When $0 < k < 1$, the graph of $y = kx^p$ is flatter than the graph of $y = x^p$, and lies closer to the x-axis. We say the graph is *compressed vertically*.

In general, for any positive k, the larger the value for k, the steeper the graph of the power function $y = kx^p$. The graphs in Figure 9.8 illustrate this effect for power functions of degrees 3 and 4.

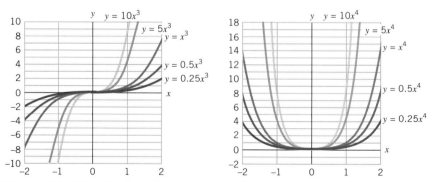

Figure 9.8 When $k > 0$, the larger the value of k, the more closely the power function $y = kx^p$ "hugs" the y-axis for both odd and even powers.

Case 2: $k < 0$ Comparing $y = kx^p$ to $y = x^p$ When k is Negative.

We know from our work with linear power functions of the form $y = kx$ that when k is negative, multiplying x by k not only changes the steepness of the line, but also "flips" or "reflects" the line across the horizontal axis. The graphs of $y = kx$ and $y = -kx$ are mirror images across the x-axis. Similarly, the graphs of $y = kx^p$ and $y = -kx^p$ are mirror images of each other across the x-axis. For example, $y = -3x$ is the mirror image of $y = 3x$, and $y = -7x^3$ is the mirror image of $y = 7x^3$. Figure 9.9 shows various pairs of power functions of the type $y = kx^p$ and $y = -kx^p$.

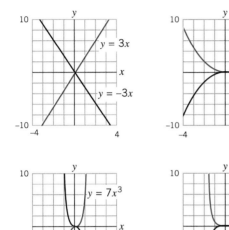

Figure 9.9   In each case, the graphs of $y = kx^p$ and $y = -kx^p$ (shown in blue and black respectively) are mirror images of each other across the x-axis.

Algebra Aerobics 9.3

1. **a.** Do a quick sketch by hand of the power functions $y = x^9$ and $y = x^{10}$.

   b. If possible, check your work using a graphing calculator or computer. Try switching among various window sizes to compare different sections of the graph.

2. **a.** Do a rough hand sketch of:

   i. $y = x$, $y = 4x$,   and $y = -4x$ (all on the same grid)

   ii. $y = x^4$, $y = 0.5x^4$,   and $y = -0.5x^4$ (all on the same grid)

   b. Check the graphs in part (a) using a graphing calculator or function graphing program.

9.4 COMPARING POWER AND EXPONENTIAL FUNCTIONS

Question

Which eventually grows faster, a power function or an exponential function?

Discussion

Although power and exponential functions may appear to be similar in construction, in each function the independent variable assumes a very different role. For power functions the independent variable, x, is the *base* which is raised to a fixed power. Power functions have the form:

$$\text{dependent variable} = k \cdot (\text{independent variable})^{\text{power}}$$

$$y = kx^p$$

where k and p are any constants. For power functions that describe growth, k and p are both positive.

For exponential functions the independent variable, x, is the *exponent* that is applied to a fixed base. Exponential functions have the form:

$$\text{dependent variable} = C \cdot (a)^{\text{independent variable}}$$

$$y = C \cdot a^x$$

where a and C are constants with $a > 0$ and $a \neq 1$. For exponential growth, $C > 0$ and $a > 1$.

Consider the functions: $y = x^3$, a power function; $y = 3^x$, an exponential function; and $y = 3x$, a linear function. Table 9.4 compares the role of the independent variable, x, in the three functions.

	Table 9.4		
x	**Linear Function,** $y = 3x$ (x **is multiplied by 3)**	**Power Function,** $y = x^3$ (x is the *base* **raised to third power)**	**Exponential Function,** $y = 3^x$ (x is the *exponent* **for base 3)**
0	$3 \cdot 0 = 0$	$0 \cdot 0 \cdot 0 = 0$	$3^0 = 1$
1	$3 \cdot 1 = 3$	$1 \cdot 1 \cdot 1 = 1$	$3^1 = 3$
2	$3 \cdot 2 = 6$	$2 \cdot 2 \cdot 2 = 8$	$3^2 = 9$
3	$3 \cdot 3 = 9$	$3 \cdot 3 \cdot 3 = 27$	$3^3 = 27$
4	$3 \cdot 4 = 12$	$4 \cdot 4 \cdot 4 = 64$	$3^4 = 81$
5	$3 \cdot 5 = 15$	$5 \cdot 5 \cdot 5 = 125$	$3^5 = 243$

Visualizing the Difference   Table 9.4 and Figure 9.10 show that the power function $y = x^3$ and the exponential function $y = 3^x$ both grow very quickly relative to the linear function $y = 3x$.

Yet there is a vast difference between the growth of an exponential function and the growth of a power function. In Figure 9.11 we zoom out on the graphs. Notice that

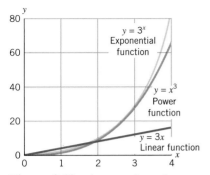

Figure 9.10   A comparison of $y = 3x$, $y = x^3$, and $y = 3^x$.

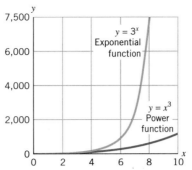

Figure 9.11   "Zooming out" on the graph.

now the scale on the x-axis goes from 0 to 10 (rather than just 0 to 4 as in Figure 9.10) and the y-axis extends to 7500, which is still not large enough to show the value of 3^x once x is slightly greater than 8.

The exponential function $y = 3^x$ clearly dominates the power function $y = x^3$. The exponential function continues to grow so rapidly that its graph appears almost vertical relative to the graph of the power function.

What if we had picked a larger exponent for the power function? Would the exponential function still overtake the power function? The answer is yes. Let's compare, for instance, the graphs of $y = 3^x$ and $y = x^{10}$. If we zoomed in on the graph (see Figure 9.12), we could see that for a while the graph of $y = 3^x$ lies below the graph of $y = x^{10}$. For example, when $x = 2$, then $3^2 < 2^{10}$. But eventually $3^x > x^{10}$. We can see on the graph that somewhere after $x = 30$, the values for 3^x become substantially larger than the values for x^{10}.

In summary,

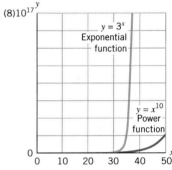

Figure 9.12   Graph of $y = x^{10}$ and $y = 3^x$.

> Any exponential growth function will eventually dominate any power function.

Construct a table using values of $x \geq 0$ for each of the following functions.

$$y = 2^x \qquad \text{and} \qquad y = x^2$$

Then plot the functions on the same grid and answer the following questions.
a.   If $x > 0$, for what values of x is $2^x = x^2$?
b.   Does one function eventually dominate? If so, after what value of x?

Example   1

SOLUTION

a.   For positive values of x, if $x = 2$ or $x = 4$, then $2^x = x^2$.
b.   In Table 9.5 and Figure 9.13, we can see that if $x > 4$, then $2^x > x^2$, so the function $y = 2^x$ will dominate the function $y = x^2$ after $x = 4$.

Table 9.5

x	$y = x^2$	$y = 2^x$
0	0	1
1	1	2
2	4	4
3	9	8
4	16	16
5	25	32
6	36	64

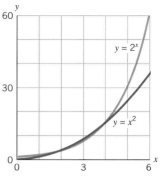

Figure 9.13 Comparison of $y = x^2$ and $y = 2^x$.

Algebra Aerobics 9.4

1. Construct a table using nonnegative values of x for each of the following functions. Then plot the functions on the same grid.

$$y = 4^x \qquad \text{and} \qquad y = x^3$$

Does one function eventually dominate? If so, after approximately what value of x?

2. Which function eventually dominates?
   a. $y = x^{10}$ or $y = 2^x$?
   b. $y = (1.000\ 005)^x$ or $y = x^{1,000,000}$?

9.5 POLYNOMIALS: COMBINING POSITIVE INTEGER POWER FUNCTIONS

Hot Air Balloons

If we fill a balloon with a gas (such as helium or hot air) that is lighter than the surrounding air, it will rise when the buoyant force of the gas is large enough to lift the weight of the balloon. The buoyant force, B, depends on the buoyancy coefficient, B_0, of the gas, and the volume, V, of the balloon's interior. The buoyant force, B, is

$$B = \text{buoyancy coefficient} \cdot \text{volume}$$
$$= B_0(4/3)\pi R^3$$
$$= (4/3)B_0\pi R^3$$

for a spherical balloon of radius R. Replacing $(4/3)B_0\pi$ by a, we can simplify this power function as:

$$B = aR^3$$

The weight, W, of the fabric of which the balloon is made depends on the weight, w, per square unit of the fabric and S, the surface area of the balloon.

weight of fabric = (weight per sq. unit) · (surface area)

$$W = w \cdot S$$
$$= w \cdot 4\pi R^2$$
$$= (4w\pi)R^2$$

By substituting b for $(4w\pi)$, we can simplify this power function to

$$W = bR^2$$

Finally, we may imagine that attached to the balloon is a gondola, a hanging basket in which people can ride, with a constant weight, c. Now we have one force, B, pulling the balloon up, and two forces, W and c, pulling the balloon down. To determine the total force, F, on the balloon, we add together all these components, remembering that buoyancy acts upward, whereas the two weights act downward. The total force, F, is

force = buoyant force − weight of balloon − weight of gondola

$$F = B - W - c$$
$$= aR^3 - bR^2 - c$$

where a, b, and c are real numbers. F is called a *cubic polynomial* in R. The balloon will rise when the radius, R, is large enough to make F positive.

General Polynomials

In general, we can combine one or more positive integer power functions to create a *polynomial function*. For example, if we add together the three power functions $y_1 = 16x^3$, $y_2 = -5x$, and $y_3 = 10$, we create the polynomial function $y = y_1 + y_2 + y_3$, where $y = 16x^3 - 5x + 10$. We say that y is a polynomial of *degree 3*, since the highest power of the independent variable x is 3.

Polynomials are simple functions in the sense that they can be formed from power functions using only the operations of addition, subtraction, and multiplication, with no division or root extractions.

A *polynomial function* $y = f(x)$ can be represented by an equation in the form

$$y = a_n x^n + a_{n-1} x^{n-1} + \cdots + a_1 x^1 + a_0$$

where each coefficient $a_n, a_{n-1} \cdots a_0$ is a constant and n is a positive integer, called the *degree* of the polynomial (provided $a_n \neq 0$).

The constant a_0 is often called "the constant term."

Polynomials of certain degrees have special names:

Polynomials of Degree	Are Called	Example
1	*linear*	$y = -4x - 8$
2	*quadratics*	$y = 3x^2 + 5x - 10$
3	*cubics*	$y = 5x^3 - 4x - 14$
4	*quartics*	$y = -2x^4 - x^3 + 4x^2 + 4x - 14$
5	*quintics*	$y = 8x^5 - 3x^4 + 4x^2 - 4$

Notice that the example for the cubic function, $y = 5x^3 - 4x - 14$, does not include an x^2 term. In this case the coefficient for the x^2 term is zero, since this function could be rewritten as $y = 5x^3 + 0x^2 - 4x - 14$.

The next two chapters focus on the properties and uses of quadratics. Early algebraists believed that higher degree polynomials were not relevant to the physical world and hence were useless: "Going beyond the cube just as if there were more than three dimensions . . . is against nature."[4] It turns out that higher degree polynomials have important applications and we return to them at the end of Chapter 11.

Algebra Aerobics 9.5

1. For the following polynomials, specify the degree of the polynomial, and evaluate each function when $x = -1$.

   a. $f(x) = 11x^5 + 4x^3 - 11$

   b. $y = 4x^3 + 11x^5 - 11$

   c. $g(x) = -2x^4 - 20$

   d. $z = -2x^2 + 3x - 4$

2. Let $f(x) = x^4 - 3x^2$ and $g(x) = 2x^2 + 4$.

   a. Write out the function $h(x)$ if $h(x) = f(x) + g(x)$ and simplify the expression.

   b. What is $h(0)$? $h(2)$? $h(-2)$?

9.6 POWER FUNCTIONS WITH NEGATIVE INTEGER POWERS

Recall that the general form of an equation for a power function is

$$\text{dependent variable} = (\text{constant})(\text{independent variable})^{\text{power}}$$

In Sections 9.1 through 9.5 we focused on functions where the power was a positive integer. We now consider power functions where the power is a negative integer.

Using the rules for negative exponents, we can rewrite power functions in the form

$$y = kx^{\text{negative power}}$$

where k is a constant, as

$$y = \frac{k}{x^{\text{positive power}}}$$

For example, $y = 3x^{-2}$ can be rewritten $y = 3/x^2$. In this form it is easier to make calculations and to see what happens to y as x increases or decreases in value.

[4] Stifel, as cited by Morris Kline in *Mathematical Thought from Ancient to Modern Times* (Oxford: Oxford University Press, 1972).

In Section 9.1 we constructed functions $s(x) = 6x^2$ and $v(x) = x^3$ to describe the surface area and volume of a cube with length of edge x. We can extend our discussion about the ratio of surface area to volume by constructing a new function $r(x)$ where

$$r(x) = \frac{\text{surface area}}{\text{volume}}$$

$$= \frac{s(x)}{v(x)}$$

$$= \frac{6x^2}{x^3}$$

$$= 6/x \text{ or } 6x^{-1}$$

Example 1

Then $r(x) = 6x^{-1}$ is an example of a power function where the power, -1, is a negative integer.

Table 9.6 and Figure 9.14 help us see that as x, the edge length, increases, $r(x)$, the ratio of surface area to volume, decreases.

Table 9.6

Edge Length x	Surface Area / Volume $r(x) = 6/x$
1	6.0
2	3.0
3	2.0
4	1.5
5	1.2
6	1.0
10	0.6

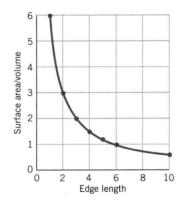

Figure 9.14 The graph of $r(x) = 6/x$ shows that the ratio of surface area/volume decreases as the length of edge, x, increases.

The cardinal rule of scuba diving
The most important rule in scuba diving is, "Never, ever hold your breath." Why?

Example 2

SOLUTION
To answer this question, we need to examine the behavior of gases under pressure. Imagine filling up a balloon with a cubic foot of air and then pulling the balloon deeper and deeper under water. What would you expect to happen? As the balloon moves deeper, the pressure of the surrounding water increases, compressing the air in the balloon. The increase in depth, and the resulting increase in pressure, decreases the volume of air.

Pressure can be measured in units called *atmospheres*. One atmosphere (abbreviated 1 atm) equals 15 lb/in.², which is the atmospheric pressure per square inch at Earth's surface at sea level.

Starting with 1 atm at the surface, each additional 33 feet of water depth increases the pressure by 1 atm. So at 33 feet under water, the pressure is 2 atm, at 66 feet under water, 3 atm, etc. Table 9.7 describes the relationships among depth, pressure, and volume. Figure 9.15 graphs volume versus pressure.

Table 9.7		
Pressure Versus Volume		
Depth (ft)	Pressure (atm)	Volume (ft³)
0	1	1
33	2	1/2
66	3	1/3
99	4	1/4

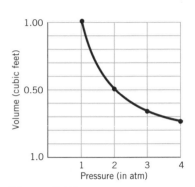

Figure 9.15 Graph of volume versus pressure for a balloon descending under water.

The relationship between volume and pressure can be modeled by the function:

$$V = \frac{1}{P} \qquad \text{or equivalently} \qquad V = P^{-1}$$

where V is the volume in cubic feet and P is the atmospheric pressure measured in atmospheres. V is a power function where the power is -1. It is a special case of Boyle's Law, which states that if the temperature is held fixed, then the volume of a fixed quantity of gas equals a constant divided by the pressure:

$$V = \frac{k}{P} \qquad \text{or equivalently} \qquad V = kP^{-1}$$

The coefficient k depends in part on how much gas you have. In this example, we have 1 cubic foot at 1 atmosphere; thus $k = 1$.

Since Boyle's Law tells us that

$$V = \frac{\text{constant}}{P}$$

we say that volume, V, is *inversely proportional* to the pressure, P. As the pressure increases, the volume decreases. Examine Table 9.7 and Figure 9.15. If we start at 1 atm and double the pressure to 2 atm, the volume of air drops in half to 1/2 ft³. At 4 atm, the volume will drop to one-fourth of the original size (or to 1/4 ft³). Why does this matter to divers?

Suppose you are swimming in a pool, take a lung full of air at the surface, and then dive down to the bottom. As you descend, the buildup of pressure will decrease the volume of air in your lungs. When you ascend back to the surface, the volume of air in your lungs will expand back to its original size and everything is fine.

But when you are scuba diving, you are constantly breathing air that has been pressurized at the surrounding water pressure. If you are scuba diving 33 feet below the surface of the water, the surrounding water pressure is at 2 atm, twice that at the surface. What will happen then if you fill your lungs from your tank, hold your breath, and ascend to the surface? When you reach the surface, the pressure will drop in half, from 2 atm down to 1 atm, so the volume of air in your lungs will double, rupturing your lungs! Hence the first rule of scuba diving: "Never, ever, hold your breath."

Algebra Aerobics 9.6a

1. Assume you are scuba diving at 99 feet below the surface of the water where the pressure is equal to 4 atm. If you use your tank to inflate a balloon with 1 ft^3 of compressed air, by how much will the volume have increased by the time it reaches the surface (assuming it doesn't burst)?

2. Fill in the following data table.

x	$r(x) = 6/x$
0.01	
0.25	
0.50	
1.00	
2.00	
5.00	
10.00	

a. When $x > 0$, what happens to $r(x)$ as x increases in value?

b. When $x > 0$, what happens to $r(x)$ as x gets closer and closer to zero?

c. Is the function $r(x)$ defined for $x = 0$?

In Section 9.2 we saw that for any power function in the form

$$y = \text{constant} \cdot x^p$$

where p is positive, we say that y is *directly proportional* to x^p. For power functions in the form

$$y = \frac{\text{constant}}{x^p}$$

where p is positive, we say that y is *inversely proportional* to x^p. For example, Boyle's Law, $V = \text{constant}/P$, tells us that volume is inversely proportional to pressure when temperature is held fixed. If $y = 8/x^3$, we say that y is *inversely proportional* to x^3.

Let p be a positive number. If

$$y = kx^p$$

we say that y is *directly proportional to* or *varies directly with* x^p. If

$$y = \frac{k}{x^p} = kx^{-p}$$

we say that y is *inversely proportional to* or *varies inversely with* x^p.

When the dependent variable is inversely proportional to the square of the independent variable, the functional relationship is called an *inverse square law.* Inverse square laws are quite common in the sciences. Some examples are shown below.

Example   3

Seeing the light

$I = k/d^2$ is a power function that describes the intensity of light an observer sees from a light source. The intensity of light, I, is inversely proportional to the square of the distance, d, between the light source and the observer. What happens to the intensity of light if the distance between the observer and the source is doubled? Tripled?

SOLUTION

If you are D meters away from a light source and the distance doubles to $2D$ meters, the intensity decreases from $\dfrac{k}{D^2}$ to $\dfrac{k}{(2D)^2} = \dfrac{k}{4D^2} = \dfrac{1}{4}\cdot\left(\dfrac{k}{D^2}\right)$. So the intensity drops to one-fourth of the original intensity as the distance doubles. If the distance triples from D to $3D$, the intensity decreases from $\dfrac{k}{D^2}$ to $\dfrac{k}{(3D)^2} = \dfrac{k}{9D^2} = \dfrac{1}{9}\cdot\left(\dfrac{k}{D^2}\right)$. So the intensity drops to one-ninth of the original intensity as the distance triples.

Example   4

The gravitational force between objects

The gravitational force between you and Earth is inversely proportional to the square of the distance between you and the center of Earth.

a.   Express this relationship as a power function.

$$F(d) = \frac{k}{d^2} = kd^{-2}$$

is a power function that describes the gravitational force $F(d)$ between you and Earth, in terms of a constant k times d^{-2}, where d is the distance between you and the center of Earth.

b.   What happens to $F(d)$ as the distance, d, between you and the center of Earth increases by a factor of 10?

Given $F(d) = \dfrac{k}{d^2}$

if we increase d by a factor of 10 $F(10d) = \dfrac{k}{(10d)^2}$

apply exponent $= \dfrac{k}{100d^2}$

rearrange terms $= \dfrac{1}{100}\cdot\dfrac{k}{d^2}$

substitute $F(d)$ for $\dfrac{k}{d^2}$ $F(10d) = \dfrac{1}{100}\cdot F(d)$

The gravitational force $F(d)$ decreases to 1/100 of its original value if d increases by a factor of 10.

c. What happens to $F(d)$ if the distance d decreases to 1/10 of the original distance?

Evaluate F at $\dfrac{d}{10}$

$$F\left(\frac{d}{10}\right) = \frac{k}{\left(\dfrac{d}{10}\right)^2}$$

apply exponent

$$= \frac{k}{\left(\dfrac{d^2}{100}\right)}$$

simplify

$$= k \cdot \left(\frac{100}{d^2}\right)$$

rearrange

$$= 100 \cdot \frac{k}{d^2}$$

substitute $F(d)$ for $\dfrac{k}{d^2}$

$$F\left(\frac{d}{10}\right) = 100F(d)$$

If d decreases to one-tenth of its original value, the gravitational force increases by a factor of 100.

Why many inverse square laws work

Inverse square laws in physics often depend upon a power source and simple geometry. Imagine a single point as a source of power, emitting perhaps heat, sound, or light. We can think of the power radiating out from the point as passing through an infinite number of concentric spheres. The farther away you are from the point source, the lower the intensity of the power, since it is spread out over the surface area of a sphere that increases in size as you move away from the point source. Therefore the intensity, I, of the power you receive at any point on the sphere will be a function of the distance you are from the point source. The distance can be thought of as the radius, r, of a sphere with the point source at its center.

$$\text{intensity} = \frac{\text{power from a source}}{\text{surface area of sphere}}$$

$$I = \frac{\text{power}}{4\pi r^2}$$

$$= \frac{(\text{power}/4\pi)}{r^2}$$

If the power from the source is constant, we can simplify the expression by substituting a constant $k = \text{power}/(4\pi)$ and rewrite our equation as

$$I = \frac{k}{r^2}$$

The intensity you receive, I, is inversely proportional to your distance from the source if the power from the point source is constant. If you double your distance, the intensity you receive is one-fourth of the original intensity. If you triple your distance, the intensity you receive is one-ninth of the original intensity. In particular, if $r = 1$, then $I = k$; if r doubles to become 2, then $I = k/4 = 0.25k$. The graph of $I = k/r^2$ is sketched in Figure 9.16.

Example 5

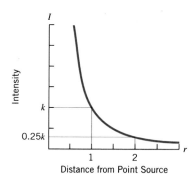

Figure 9.16   Graph of $I = k/r^2$, the relationship between intensity and distance from a point source.

Example   6 **Direct and Inverse Proportionality in the Same Equation: The relative effect of the moon and the sun on the tides**

The force that creates ocean tides on the surface of Earth varies inversely with the cube of the distance from Earth to any other large body in space, and varies directly with the mass of the other body.

a. Construct an equation that describes this relationship.

b. Use this equation to show why the moon has more effect on Earth's tides than the sun.

SOLUTION

a. Using T for the tide-generating force, d for the distance between Earth and another body, and m for the mass of the other body, we have

$$T = \frac{km}{d^3} \qquad \text{or} \qquad T = kmd^{-3}$$

for some constant of proportionality k.

b. Now compare the tide generating force of the moon, T_{moon}, to the tide generating force of the sun, T_{sun}. The sun has a mass approximately 10^7 times larger than the moon's but is about 390 times farther away from Earth than the moon. If we let d = distance from Earth to the moon, then

for the moon
$$T_{moon} = \frac{k \cdot (\text{mass of the moon})}{d^3}$$

for the sun
$$T_{sun} = \frac{k \cdot (\text{mass of the sun})}{(390\,d)^3}$$

substitute for mass of Sun
$$= \frac{k \cdot (\text{mass of the moon}) \cdot 10^7}{(390\,d)^3}$$

apply exponents
$$= \frac{k \cdot (\text{mass of the moon}) \cdot 10^7}{390^3 d^3}$$

rearrange terms
$$= \frac{10^7}{390^3} \cdot \frac{k \cdot (\text{mass of the moon})}{d^3}$$

substitute T_{moon}
$$T_{sun} = \frac{10^7}{390^3} \cdot T_{moon}$$

We can further simplify the coefficient.

use scientific notation
$$T_{sun} = \frac{10^7}{(3.9 \cdot 10^2)^3} \cdot T_{moon}$$

apply exponent and round
$$\approx \frac{10^7}{60 \cdot 10^6} \cdot T_{moon}$$

use law of exponents
$$\approx \frac{10}{60} \cdot T_{moon}$$

so
$$T_{sun} \approx (1/6)T_{moon}$$

Thus, despite the fact that the sun is more massive than the moon, because it is much farther away from Earth than the moon, its effect on Earth's tides is only one sixth that of the moon's.

There are also combined effects of the sun and the moon on the tides. The largest monthly tides are the "spring tides," which occur at the full moon and the new moon, when the sun is aligned in the sky with the moon. The smallest monthly tides are the "neap tides," which occur at the half-moon, when the sun is at right angles to the moon.

Algebra Aerobics 9.6b

1. a. Rewrite each of the following expressions using positive exponents.

    i. $15x^{-3}$ **ii.** $-10x^{-4}$ **iii.** $3.6x^{-1}$

  b. Rewrite each of the following expressions using negative exponents.

    i. $\dfrac{1.5}{x^2}$ **ii.** $-\dfrac{6}{x^3}$ **iii.** $-\dfrac{2}{3x^2}$

2. The time in seconds, t, needed to fill a tank with water is inversely proportional to the square of the diameter, d, of the pipe delivering the water. Write an equation describing this relationship.

3. You are sitting 4 feet from a light reading a book. The light seems too dim, so you move 2 feet closer. What is the change in light intensity?

4. If $g(x) = 3/x^4$, what happens to $g(x)$ when

  i. x doubles?   **ii.** x is divided by 2?

5. a. Fill in the following data table.

x	$f(x) = 1/x^2$	$g(x) = 1/x^3$
0.01		
0.25		
0.50		
1.00		
2.00		
10.00		

  b. Are the abstract functions $f(x) = 1/x^2$ and $g(x) = 1/x^3$ defined when $x = 0$? Are they defined for negative values of x?

  c. Describe what happens to the functions $f(x)$ and $g(x)$ as x approaches positive infinity. What happens to $f(x)$ and $g(x)$ when $0 < x < 1$?

Moving to the Abstract

Let's take a close look at two abstract power functions with negative exponents, $f(x) = x^{-1}$ and $g(x) = x^{-2}$. If we rewrite these functions as $f(x) = 1/x$ and $g(x) = 1/x^2$, it is clear that the domain cannot include zero since the function is undefined when $x = 0$. We say that the functions are *discontinuous* at 0. Now, consider what happens to these functions when x assumes positive or negative values and approaches $+\infty$ or $-\infty$ respectively. Two important questions are: "What happens to these functions as x approaches $+$ or $-\infty$?" and "What happens when x is close to 0?"

What happens when x is positive and approaches $+\infty$? From Table 9.8 and Figure 9.17, when x is positive and increasing, both $f(x)$ and $g(x)$ *decrease*. For both functions as $x \to +\infty$, then $1/x$ and $1/x^2$ grow smaller and smaller, approaching, but never reaching, zero. Both graphs are asymptotic to the x-axis.

Something to think about

Explain why $f(x) = 1/x$ is undefined at $x = 0$.

Table 9.8

	$x \geq 0$	
x	$f(x) = 1/x$	$g(x) = 1/x^2$
0	Undefined	Undefined
1/100	100	10,000
1/4	4	16
1/3	3	9
1/2	2	4
1	1	1
2	1/2	1/4
3	1/3	1/9
4	1/4	1/16
100	1/100	1/10,000

What happens when $0 < x < 1$?   We can see in Table 9.8 and Figure 9.17 that when x is positive, as x grows smaller and smaller and approaches zero, the values for $f(x) = 1/x$ and $g(x) = 1/x^2$ get larger and larger. Let's examine what happens when we use fractions between 0 and 1 to calculate values for $f(x) = 1/x$ and $g(x) = 1/x^2$

$$f\left(\frac{1}{4}\right) = \frac{1}{\left(\frac{1}{4}\right)} = 1 \div \frac{1}{4} = 1 \cdot \frac{4}{1} = 4$$

$$f\left(\frac{1}{100}\right) = \frac{1}{\left(\frac{1}{100}\right)} = 1 \div \frac{1}{100} = 1 \cdot \frac{100}{1} = 100$$

$$g\left(\frac{1}{4}\right) = \frac{1}{\left(\frac{1}{4}\right)^2} = 1 \div \frac{1}{16} = 1 \cdot \frac{16}{1} = 16$$

$$g\left(\frac{1}{100}\right) = \frac{1}{\left(\frac{1}{100}\right)^2} = 1 \div \frac{1}{10,000} = 1 \cdot \frac{10,000}{1} = 10,000$$

If x is positive and approaches zero, then $f(x) = 1/x$ and $g(x) = 1/x^2$ approach positive infinity. So both graphs are also asymptotic to the y-axis.

What happens when x is negative?   When x is negative, the value of $f(x) = 1/x$ is always negative and as $x \rightarrow 0$, then $1/x \rightarrow -\infty$. The function $g(x) = 1/x^2$ is positive for all values of x and whether x is positive or negative as $x \rightarrow 0$, then $1/x^2 \rightarrow +\infty$ (Table 9.9).

The two graphs in Figure 9.18 exhibit different shapes for negative values of x. The graphs of these functions approach the x-axis and y-axis but never reach them for positive and negative values of x. Thus, both curves are *asymptotic* to both axes.

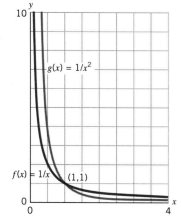

Figure 9.17   Graphs of $f(x) = 1/x$ and $g(x) = 1/x^2$, when $x > 0$.

Table 9.9

	$x < 0$	
x	$f(x) = 1/x$	$g(x) = 1/x^2$
-100	$-1/100$	1/10,000
-4	$-1/4$	1/16
-3	$-1/3$	1/9
-2	$-1/2$	1/4
-1	-1	1
$-1/2$	-2	4
$-1/3$	-3	9
$-1/4$	-4	16
$-1/100$	-100	10,000

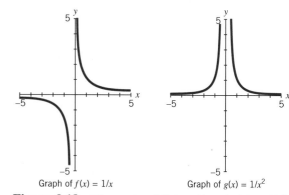

Graph of $f(x) = 1/x$ Graph of $g(x) = 1/x^2$

Figure 9.18   The graphs of $f(x) = 1/x$ and $g(x) = 1/x^2$.

9.7 VISUALIZING NEGATIVE INTEGER POWER FUNCTIONS

If we examine the graphs (Figure 9.19) of the simplest negative power functions $y = x^{-1}$, $y = x^{-2}$, $y = x^{-3}$, $y = x^{-4}$, $y = x^{-5}$, $y = x^{-6}$, we see that the negative power functions, like the positive power functions, fall into two groups: the odd powers and the even powers.

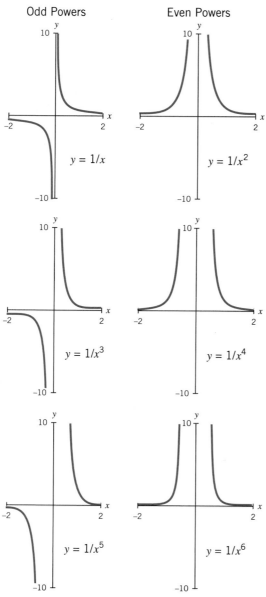

Figure 9.19   Graphs of power functions with even and odd powers.

In Exploration 9.3 you examine the effect of k on the graphs of negative power functions.

For both the even and odd negative powers, the functions are not defined when $x = 0$; hence, the domains never include 0. As $x \to 0$, then either $y \to +\infty$ or $y \to -\infty$. The graphs of these functions approach the y-axis but never reach it. Thus, each curve has the y-axis as a *vertical asymptote*.

For both even and odd negative powers, y is never equal to zero. As $x \to +\infty$ or $x \to -\infty$, the values of y come closer and closer to zero but never reach zero. The graphs of these functions approach the x-axis but never touch it. Thus each curve has the x-axis as a *horizontal asymptote*.

All of the odd negative powers have rotational symmetry about the origin. If you hold the graph fixed at the origin and rotate it 180°, you end up with the same image.

All of the even negative power functions are symmetrical about the y-axis. If you think of the y-axis as the dividing line, the "left" side of the graph is a mirror image, or reflection, of the "right" side.

Algebra Aerobics 9.7

1. Rewrite each of the following expressions with positive exponents and then evaluate each expression for $x = -2$.
    a. x^{-2} c. $4x^{-3}$
    b. x^{-3} d. $-4x^{-3}$

2. Sketch the graph for each equation
    a. $y = x^{-10}$ b. $y = x^{-11}$

3. On the same grid, graph the following equations.
    a. $y = x^{-2}$ and $y = x^{-3}$
        Where do these graphs intersect? How are the graphs similar and how are they different?
    b. $y = 4x^{-2}$ and $y = 4x^{-3}$
        Where do these graphs intersect? How are the graphs similar and how are they different?

CHAPTER SUMMARY

Power functions

A *power function* has the form $y = kx^p$ where k and p are any constants.

Let p be a positive integer.
        If $y = kx^p$, we say that y is *directly proportional to x^p*.

        If $y = \dfrac{k}{x^p} = kx^{-p}$, we say that y is *inversely proportional to x^p*.

k is called the *constant of proportionality*.
Any exponential growth function will eventually dominate any power function.

Positive integer powers:
$$y = kx, \ y = kx^2, \ y = kx^3, \ \ldots \ .$$

If $y = x^p$ where p is a positive integer (and $k = 1$), then for odd powers:

- as x increases, y increases.
- graph has rotational symmetry about the origin and "bends" near the origin.

for even powers:

- as x increases from negative values to positive values, y decreases and then increases as x becomes positive.
- graph is symmetric about the y-axis and is U-shaped.

If $y = kx^p$ where k is a constant and p is a positive integer, then:

If $k > 0$, the larger the value for k, the steeper the graph of $y = kx^p$.

Changing the sign of k "flips" the graph over the x-axis; that is, the graphs of $y = kx^p$ and $y = -kx^p$ are mirror images or reflections of each other across the x-axis.

Negative integer powers: $y = kx^{-1}, \; y = kx^{-2}, \; y = kx^{-3}, \; \ldots$

Power functions in the form

$$y = kx^{\text{negative power}}$$

where k is a constant, can be rewritten as

$$y = \frac{k}{x^{\text{positive power}}}$$

Power functions in this form are not defined when $x = 0$; hence, their domains do not include 0.

Graphs of power functions of the form

$$y = \frac{1}{x^p}$$

where p is a positive integer, are *asymptotic* to both the x- and y-axes.

For odd powers, the graphs have rotational symmetry about the origin.

For even powers, the graphs are symmetrical about the y-axis.

Polynomial functions

A *polynomial function* has the general form

$$f(x) = a_n x^n + a_{n-1} x^{n-1} + \; \cdots \; + a_1 x^1 + a_0$$

where each coefficient $a_n, a_{n-1}, \cdots a_0$ is a constant and n is a positive integer, called the *degree* of the polynomial (provided $a_n \neq 0$). The term a_0 is often referred to as "the constant term."

Polynomials of Degree	Are Called
1	*linear*
2	*quadratics*
3	*cubics*
4	*quartics*
5	*quintics*

EXERCISES

1. Evaluate the following expressions for $x = 2$ and $x = -2$.

   a. $5x^2$ b. $5x^3$ c. $-5x^2$ d. $-5x^3$

2. Identify which of the following are power functions. For each power function, identify the value for k and the value for p, the power.

   a. $y = -3x^2$ b. $y = 3x^{10}$ c. $y = x^2 + 3$ d. $y = 3^x$

3. The radius of a certain protein molecule is $0.000\,000\,000\,1 = 10^{-10}$ meters. Assuming the molecule is roughly spherical, answer the following questions. Express your answers in scientific notation.

   a. Find its surface area, S, in square meters. Remember that $S = 4\pi r^2$.

   b. Find its volume, V, in cubic meters. Remember that $V = \dfrac{4}{3}\pi r^3$.

   c. Find the ratio of the surface area to the volume.

4. The radius of Earth is about 6400 kilometers. Assume that Earth is spherical. Express your answers to the questions below in scientific notation.

   a. Find the surface area of Earth in square meters.

   b. Find the volume of Earth in cubic meters.

   c. Find the ratio of the surface area to the volume.

5. Y is directly proportional to X^3.

   a. Express this relationship as a function where Y is the dependent variable.

   b. If $Y = 10$ when $x = 2$, then find the value of k, the constant of proportionality, in part (a).

   c. If X is increased by a factor of 5, what happens to the value of Y?

   d. If X is divided by 2, what happens to the value of Y?

   e. Rewrite your equation from part (a) in terms of X.

6. If the radius of a sphere is x meters, what happens to the surface area and volume of a sphere when you

   a. quadruple the radius?

   b. multiply the radius by n?

   c. divide the radius by 3?

   d. divide the radius by n?

7. Assume a box has a square base and the length of a side of the base is equal to twice the height of the box.

   a. Write functions for the surface area and the volume which are dependent on the height, h.

   b. Construct a small table and graph for each of the two functions.

   c. When the height triples, what happens to the surface area? What happens to the volume?

   d. As the height increases, what will happen to the ratio surface area/volume?

8. The volume, V, of a cylindrical can, where r is the radius of the base and h is the height, is given by the following equation.

$$V = \pi r^2 h$$

The total surface area, S, of the can is

$$S = \text{area of the curved surface} + 2 \cdot (\text{area of the base})$$
$$= 2\pi rh + 2\pi r^2$$

a. Assume the height is three times the radius. Write the volume and the surface area as functions of the radius.

b. Does the volume or the surface area eventually grow faster as r increases? Show why.

9. From Galileo's experiments we know that the distance, d, a ball travels down an inclined plane is proportional to the square of the total time, t, of the motion.

   a. Express this relationship as a function where d is the dependent variable.

   b. If the ball travels a total of 4 feet in 0.5 seconds, find the value of the constant of proportionality in part (a).

   c. Using your equation in part (a), solve for t.

10. **a.** L is directly proportional to x^5. What is the effect of doubling x?

    b. M is directly proportional to x^p where p is a positive integer. What is the effect of doubling x?

11. In each part sketch the three graphs on the same grid and clearly label each function. Describe how the three graphs are alike and not alike.

    a. $y = x^1$ $y = x^3$ $y = x^5$

    b. $y = x^2$ $y = x^4$ $y = x^6$

    c. $y = x^3$ $y = 2x^3$ $y = -2x^3$

    d. $y = x^2$ $y = 4x^2$ $y = -4x^2$

    If possible, check your results using a graphing calculator or function graphing program.

12. In "Love That Dirty Water," the *Chicago Reader,* April 5, 1996, Scott Berinato interviewed Ernie Vanier, Captain of the towboat *Debris Control.* The Captain said, "We've found a lot of bowling balls. You wouldn't think they'd float, but they do."

    When will a bowling ball float in water? The bowling rule book specifies that a regulation ball must have a circumference of exactly 27 inches. Recall that the circumference, C, of a circle is directly proportional to its radius, r, by the formula $C = 2\pi r$.

    a. What is a regulation bowling ball's radius in inches? In centimeters? (Note: 1 in. = 2.54 cm.)

    b. What is the *volume* of a regulation bowling ball in cubic inches? In cubic centimeters? (Neglect the finger holes.)

    c. What is the weight in grams of a volume of water equivalent in size to a regulation bowling ball? (Water weighs 1 g/cm^3.)

    d. Regulation bowling balls are made in weights between 6 and 16 pounds. What is the range of weights in kilograms? (1 kg = 2.205 lb.)

    e. What is the heaviest weight of a bowling ball that will float in water?

f. Typical men's bowling balls are 15 or 16 pounds. Women commonly use 12-pound bowling balls. What will happen to the men's and to the women's bowling balls? Will they sink or float?

13. Graph the following functions on the same grid. Use a graphing calculator or function graphing program if available.

$$y = x^4 \qquad y = 4^x$$

a. Where do your graphs intersect? Do they intersect more than once? For what values of x are the two functions equal?

b. For positive values of x, describe what happens to the right and left of any intersection points. You may need to change the scales on the axes or change the windows on the graphing calculator in order to see what is happening.

c. Which eventually dominates, $y = x^4$ or $y = 4^x$?

14. a. Which eventually dominates, $y = (0.001)^x$ or $y = x^{1000}$?

b. As the independent variable approaches $+\infty$, which function eventually approaches zero faster, an exponential decay function or a power function with negative integer exponents?

15. Match the following data tables with the appropriate function.

Table 1		Table 2		Table 3		Table 4		Table 5	
x	y	x	y	x	y	x	y	x	y
1	2	1	2	1	1/2	1	2	1	1
2	1	2	4	2	1	2	4	2	4
3	2/3	3	8	3	3/2	3	6	3	9
4	1/2	4	16	4	2	4	8	4	16

a. $f(x) = 2x$ **d.** $f(x) = x^2$

b. $f(x) = \dfrac{x}{2}$ **e.** $f(x) = 2^x$

c. $f(x) = \dfrac{2}{x}$

16. For function (a) in Exercise 15, can you find a specific value x_0 for x such that $f(x_0) = x_0$? Repeat the process for functions (b) through (e).

17. If x is positive, for what values of x will $3 \cdot 2^x < 3 \cdot x^2$? For what values of x will $3 \cdot 2^x > 3 \cdot x^2$?

18. Evaluate the following polynomials for $x = 2$ and $x = -2$ and specify the degree of the polynomial.

a. $y = 3x^2 - 4x + 10$

b. $y = x^3 - 5x^2 + x - 6$

c. $y = -2x^4 - x^2 + 3$

19. Evaluate the following expressions for $x = 2$ and $x = -2$.

a. x^{-3} **b.** $4x^{-3}$ **c.** $-4x^{-3}$ **d.** $-4x^3$

20. The intensity of light from a point source is inversely proportional to the square of the distance from the light source. If the intensity is 4 watts per square meter at a distance of 6 m from the source, find the intensity at a distance of 8 m from the source. Find the intensity at a distance of 100 m from the source.

21. a. Make a table and sketch a graph for each of the following functions. Be sure to include negative and positive values for x, as well as values for x that lie close to zero.

$$y_1 = \frac{1}{x} \qquad y_2 = \frac{1}{x^2} \qquad y_3 = \frac{1}{x^3} \qquad y_4 = \frac{1}{x^4}$$

b. Describe the domain and range of each function.

c. Describe the behavior of each function as x approaches positive infinity and as x approaches negative infinity.

d. Describe the behavior of each function when x is near 0.

22. In each part, sketch the three graphs on the same grid and label each function. Describe how the three graphs are similar and how they are different.

a. $y = x^{-1}$ $\qquad$ $y = x^{-3}$ $\qquad$ $y = x^{-5}$

b. $y = x^0$ $\qquad$ $y = x^{-2}$ $\qquad$ $y = x^{-4}$

c. $y = 2x^{-1}$ $\qquad$ $y = 4x^{-1}$ $\qquad$ $y = -2x^{-1}$

23. Boyle's Law says that if the temperature is held constant, then the volume, V, of a fixed quantity of gas is inversely proportional to the pressure, P. That is, $V = k/P$ for some constant k. What happens to the volume if

a. the pressure triples?

b. the pressure is multiplied by n?

c. the pressure is halved?

d. the pressure is divided by n?

24. a. B is inversely proportional to x^4. What is the effect on B of doubling x?

b. Z is inversely proportional to x^p where p is a positive integer. What is the effect on Z of doubling x?

25. a. Generate tables and graphs for each of the following functions.

$$g(x) = 5x \qquad h(x) = x/5 \qquad t(x) = 1/x \qquad f(x) = 5/x$$

b. Describe the ways in which the graphs in part (a) are alike and the ways in which they are not alike.

26. The frequency, F, (the number of oscillations per unit of time) of an object of mass, m, attached to a spring is inversely proportional to the square root of m.

a. Write an equation describing the relationship.

b. If a mass of 0.25 kg attached to a spring makes 3 oscillations per second, find the constant of proportionality.

c. Find the number of oscillations per second made by a mass of 0.01 kg that is attached to the spring in part (b).

27. A light fixture is mounted on a 10-foot-high ceiling over a 3-foot-high counter. How much will the illumination improve if the light fixture is lowered to 4 feet above the counter?

28. Match each of the following functions with its graph.

a. $y = 3(2^x)$ c. $y = 2x - 3$ e. $y = x^{-3}$

b. $y = 2 - x$ d. $y = x^3$ f. $y = x^{-2}$

Graph 1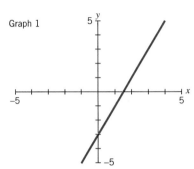

Graph 2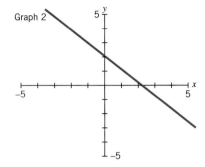

Graph 3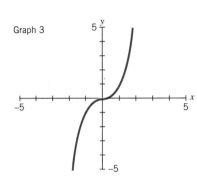

Graph 4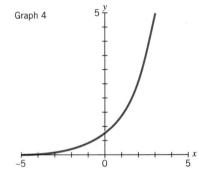

Graph 5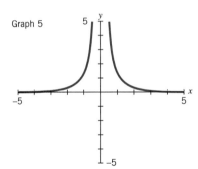

Graph 6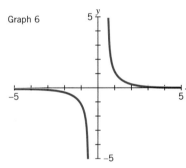

29. A cube of edge length x has a surface area $s(x) = 6x^2$ and a volume $v(x) = x^3$. We constructed the function $r(x) = s(x)/v(x) = 6x^2/x^3 = 6/x$. Consider $r(x)$ as an abstract function. What is the domain? Construct a small table of values including negative values of x, and plot the graph. Describe what happens to $r(x)$ when x is positive and $x \to 0$. What happens to $r(x)$ when x is negative and $x \to 0$?

Water Depth (feet)	Pressure (atm)
0	1
33	2
66	3
99	4

30. The pressure of the atmosphere around us is relatively constant at 15 pounds per square inch at sea level, or 1 atmosphere of pressure (1 atm). In other words, the column of air above one square inch of Earth's surface is exerting 15 pounds of force on that square inch of Earth. Water is considerably more dense. As we saw in Section 9.6, pressure increases at a rate of 1 atm for each additional 33 feet of water. The adjacent table shows a few corresponding values for water depth and pressure.

a. What type of relationship does the table describe?

b. Construct an equation that describes pressure, P, as a function of depth, D.

c. In Section 9.6 we looked at a special case of Boyle's Law for the behavior of gases, $P = 1/V$ (where if $V = 1$ cubic foot, then $P = 1$ atm).

   i. Use Boyle's Law and the equation you found in part (b) to construct an equation for volume, V, as a function of depth, D.

   ii. When $D = 0$ feet, what is V?

   iii. When $D = 66$ feet, what is V?

   iv. If a snorkeler takes a lung full of air at the surface, dives down to 10 feet, and returns to the surface, describe what happens to the volume of air in her lungs.

   v. A large balloon is filled with a cubic foot of compressed air from a scuba tank at 132 feet below water level, sealed tight, and allowed to ascend to the surface. Use your equation to predict the change in volume of air.

31. Waves on the open ocean travel with a velocity that is directly proportional to the square root of their wavelength, the distance from one wave crest to the next. (See W. T. D'Arcy Thompson, *On Growth and Form,* New York: Dover Publications, Inc., 1992.)

a. Since the time between successive waves equals $\dfrac{\text{wavelength}}{\text{velocity}}$, show that time is also directly proportional to the square root of the wavelength.

b. On one day waves crash on the beach every 3 seconds. On the next day, the waves crash every 6 seconds. On the open ocean, how much farther apart do you expect the wave crests to be on the second day than on the first?

32. Suppose you are traveling in your car at speed, S, and you suddenly brake hard, leaving skid marks on the road. A "rule of thumb" for the distance, D, that the car would skid is given by

$$D = \frac{S^2}{30f}$$

where $D = $ the distance the car skidded in feet, $S = $ speed of the car in miles per hour, and f is a number called the coefficient of friction that depends upon the road surface and condition. For a dry tar road, $f \approx 1.0$. For a wet tar road, $f \approx 0.5$. (We saw a variation of this problem in Example 1, Section 7.5 and Example 3, Section 9.2.)

a. What is the function relating distance skidded and speed for a dry tar road? For a wet tar road?

b. Generate a small table of values for both functions, including speeds between 0 and 100 miles per hour.

c. Plot both functions on the same grid.

d. Why do you think the coefficient of friction is less for a wet road than for a dry road? What effect does this have on the graph in part (c)?

e. In the accompanying table, estimate the speed given the following distances skidded on dry and on wet tar roads. Describe the method you used to find these numbers.

Distance Skidded (feet)	Estimated Speed (mph)	
	Dry Tar	Wet Tar
25		
50		
100		
200		
300		

f. If one car is going twice as fast as another when they both jam on the brakes, how much farther will the faster car skid? Explain. Does your answer depend on whether the road is dry or wet?

33. Which of the graphs of the following pairs of equations intersect? If the graphs intersect, find the point or points of intersection.

   a. $y = 2x$ $\qquad\qquad$ $y = 4x^2$

   b. $y = 4x^2$ $\qquad\quad\;$ $y = 4x^3$

   c. $y = x^{-2}$ $\qquad\quad$ $y = 4x^2$

   d. $y = x^{-1}$ $\qquad\quad$ $y = x^{-2}$

   e. $y = 4x^{-2}$ $\qquad\;$ $y = 4x^{-3}$

34. A function is said to be *even* if $f(-x) = f(x)$, and *odd* if $f(-x) = -f(x)$. Use these definitions to answer the questions below.

   a. Show that the even power functions are even.

   b. Show that the odd power functions are odd.

   c. What do you predict about the symmetries in the graphs of even power functions? What do you predict about the symmetries in the graphs of odd power functions? Check your predictions with a function graphing program or graphing calculator.

   d. Show whether each of the following functions is even, odd, or neither.

   $\quad$ **i.** $f(x) = x^4 + x^2$

   $\quad$ **ii.** $u(x) = x^5 + x^3$

   $\quad$ **iii.** $h(x) = x^4 + x^3$

   $\quad$ **iv.** $g(x) = 10 \cdot 3^x$

   e. For each function that you have identified as even or odd, what do you predict about the symmetries of their graphs? Check your predictions with a function graphing program or graphing calculator.

EXPLORATION 9.1

Predicting Properties of Power Functions

Objectives

- construct power functions with positive integer exponents
- find patterns in the graphs of power functions

Materials

- graphing calculator or function graphing program or "P1: k & p Sliders" in *Power Functions*
- graph paper

Procedure

Working in Pairs

We will be constructing power functions in the form $y = kx^p$ where k and p are constants and p is a positive integer. In each case, if possible, check your work using technology.

1. **a.** For each of the following, write the equation of a power function such that:
    - **i.** the graph of your function is symmetric across the y-axis
    - **ii.** the graph of your function is symmetric around the origin
    b. For each of the following, construct the equations for two power functions such that:
    - **i.** the graphs of your two functions are mirror images of each other across the x-axis
    - **ii.** the graphs of your two functions are mirror images of each other across the y-axis

2. Using the same power for each function, construct two different power functions such that:
    a. both functions have even powers and the graph of one function "hugs" the y-axis more closely than the other function.
    b. both functions have odd powers and the graph of one function "hugs" the y-axis more closely than the other function.

3. Using the same value for the coefficient, k, construct two different power functions such that:
    a. both functions have even powers and the graph of one function "hugs" the y-axis more closely than the other function.
    b. both functions have odd powers and the graph of one function "hugs" the y-axis more closely than the other function.

4. a. Choose a value for k that is greater than 1 and construct the following functions.

   i. $y = kx^0 = k$ $y = kx^2$ $y = kx^4$

   ii. $y = kx^1$ $y = kx^3$ $y = kx^5$

   b. Graph the functions in (i) on the same grid. Choose a scale for your axes so you can examine what happens when $0 < x < k$. Now regraph, choosing a scale for your axes so you can examine what happens when $x > k$.

   c. Graph the functions in (ii) on the same grid. Choose a scale for your axes so you can examine what happens when $0 < x < k$. Now regraph, choosing a scale for your axes so you can examine what happens when $x > k$.

   d. Describe your findings.

Class Discussion

Compare your findings. As a class develop a 60 second summary describing the results.

Exploration-Linked Homework

A Challenge Problem

In Chapter 9 we primarily studied power functions with integer exponents. Try extending your analyses to power functions with positive fractional exponents. Start with fractions that in reduced form have 1 in the numerator. What would the graphs of $y = x^{1/2}$ and $y = x^{1/4}$ look like? What are the domains? The ranges? What would the graphs of $y = x^{1/3}$ or $y = x^{1/5}$ look like? What about their domains and ranges? Generalize your results for fractions of the form $1/n$.

EXPLORATION 9.2

Scaling Objects

Objectives

- find and use general formulas for scaling different types of objects

Procedure

1. **Scaling factors**

   When a picture or three-dimensional object is enlarged or shrunk, each linear dimension is multiplied by a constant called the *scaling factor*. For the two squares below, the scaling, F, is 3. That means that any linear measurement (for example, the length of the side or of the diagonal) is 3 times larger in the bigger square.

   Scaling up a square by a factor of 3

   a. What is the relationship between the area of the two squares above? Show that for all squares:

   $$\text{Area scaled} = (\text{original area}) \cdot F^2 \qquad \text{where } F \text{ is the scaling factor}$$

   b. Given a scaling factor, F, find an equation to represent the surface area of a scaled-up cube, S_1, in terms of the surface area of the original object, S_0.

   c. Given a scaling factor, F, find an equation to represent the volume of a scaled-up cube, V_1, in terms of the volume of the original object, V_0.

2. **Representing volumes**

   We can describe the volume, V, of any three-dimensional object as $V = kL^3$, where V depends on any length, L, which describes the size of the object. The coefficient, k, depends upon the shape of the object and the particular length and the measurement units we choose. It does not matter which length we use. For a statue of a deer, for example, L could represent the deer's width or the length of an antler or a tail. If we let L_0 represent the overall height of the deer in Figure 1, then L_0^3 is the volume of a cube with edge length L_0 that contains the entire deer. The actual volume, V_0, of the deer is some fraction k of L_0^3; that is, $V = kL_0^3$ for some constant k.

   Figure 1   Figure 2

3. **Finding the scaling-factor**

If we scale up the deer to a height of L_1 (see Figure 2), then the volume of the scaled-up deer equals kL_1^3. The scaled-up object inside the cube will still occupy the same fraction, k, of the volume of the scaled-up cube.

a. If an original object has volume V_0, and some measure of its length is L_0, then we know $V_0 = kL_0^3$. Rewrite the equation, solving for the coefficient, k, in terms of L_0 and V_0.

b. Write an equation for volume, V_1, of a replica scaled to have length L_1. Substitute for k the expression you found in part (a). Simplify the expression such that V_1 is expressed as some term times V_0.

c. The ratio L_1/L_0 is the *scaling factor.* A scaling factor is the number by which each linear dimension of an original object is *multiplied* when the size is changed. Write an equation for the volume of a scaled object, V_1, in terms of V_0, the volume of the original object and F, the scaling factor.

4. **Using the scaling factor**

a. Draw a circle with a one-inch radius. Then draw a circle with an area four times as large. What is the change in the radius?

b. A photographer wants to blow up a 3-cm by 5-cm photograph. If he wants the final print to be double the area of the original photograph, what scaling factor should he use?

c. A sculptor is commissioned to make a bronze statue of George Washington sitting on a horse. To fit into its intended location, the final statue must be 15 feet long from the tip of the horse's nose to the end of its tail. In her studio, the sculptor experiments with smaller statues that are only 1 foot long. Suppose that the final version of the small statue requires 0.15 cubic feet of molten bronze. When the sculptor is ready to plan construction of the larger statue, how can she figure out how much metal she needs for the full-size one?

d. (Adapted from COMAP, *For All Practical Purposes* W. H. Freeman and Co., New York 1988 p. 370.) One of the famous problems of Greek antiquity was the *duplication of the cube.* Our knowledge of the history of the problem comes down to us from Eratosthenes (circa 284 to 192 B.C.), who is famous for his estimate of the circumference of Earth. According to him, the citizens of Delos were suffering from a plague. They consulted the oracle, who told them that to rid themselves of the plague, they must construct an altar to a particular god. That altar must be the same shape as the existing altar but double the volume. What should the scaling factor be for the new altar?

e. Pyramids have been built by cultures all over the world, but the largest and most famous were built in Egypt and Mexico.

   i. The Great Pyramid of Khufu at Giza in Egypt has a square base 755 feet on a side. It originally rose about 481 feet high (the top 31 feet have been destroyed over time). Find the original volume of the Great Pyramid. The volume V of a pyramid is given by $V = (B \cdot H)/3$, where B is the area of the base and H is its height.

   ii. The third largest pyramid of Giza, the Pyramid of Menkaure, occupies approximately one-quarter of the land area covered by the Great Pyramid. Menkaure's pyramid is the same shape as the original shape of the Great Pyramid but it is scaled down. Find the volume of the Pyramid of Menkaure.

EXPLORATION 9.3

Visualizing Power Functions with Negative Integer Powers

Objectives

- examine the effect of k on negative integer power functions

Materials

- graphing calculator or function graphing program
- graph paper

Related Software

- "P1: k & p Sliders" in *Power Functions*

Procedure

Class Demonstration: Constructing Negative Integer Power Functions

1. A power function has the form $y = kx^p$ where k and p are constants. Consider the following power functions with negative integer exponents where k is 4 and 6 respectively.

$$y = 4x^{-2} \qquad \text{and} \qquad y = 6x^{-2}$$

We can also write these as

$$y = \frac{4}{x^2} \qquad \text{and} \qquad y = \frac{6}{x^2}$$

What are the constraints on the domain and the range for each of these functions?

2. Construct a table for these functions using positive and negative values for x. How do different values for k lead to different values for y for these two functions? Sketch a graph for each function. Check your graphs with a function graphing program or graphing calculator.

   a. Describe the overall behavior of these graphs. In each case, when is y increasing? When is y decreasing? How do the graphs behave for values of x near 0?

   b. Describe how these graphs are similar and how they are different.

Working in Small Groups

In the following exploration you will predict the effect of the coefficient k on the graphs of power functions with negative integer exponents. In each part, compare your findings, and then write down your observations.

1. **a.** Choose a value for p in the function $y = \dfrac{k}{x^p}$ where p is a positive integer.

   Construct several functions where p has the same value but k assumes different *positive* values (as in the example above where $k = 4$ and $k = 6$). What effect do you think k has on the graphs of these equations? Graph the functions. As you choose larger and larger values for k, what happens to the graphs? Try choosing values for k between 0 and 1. What happens to the graphs?

b. Using the same value of p as in part (a), construct several functions with *negative* values for the constant k. What effect do you think k has on the graphs of these equations? Graph the functions. Describe the effect of k on the graphs of your equations. Do you think your observations about k will hold for any value of p?

c. Choose a new value for p and repeat your experiment. Are your observations still valid? Compare your observations with your partner. Have you examined both odd and even negative integer powers?

    In your own words, describe the effect of k on the graphs of functions of the form $y = k/x^p$, where p is a positive integer. What is the effect of the sign of the coefficient k?

2. **a.** Choose a value for k where $k > 1$ and rewrite each function in the form $y = k/x^p$.

i. $y = kx^0 = k$	$y = kx^{-2}$	$y = kx^{-4}$
ii. $y = kx^{-1}$	$y = kx^{-3}$	$y = kx^{-5}$

b. Graph the functions in part (a.i) on the same grid. Choose a scale for your graphs such that you can examine what happens when $0 < x < k$. Now choose scales for your axes so you can examine what happens when $x > k$.

c. Graph the functions in part (a.ii) on the same grid. Choose scales for your graphs such that you can examine what happens when $0 < x < k$. Now choose scales for your axes so you can examine what happens when $x > k$.

d. Describe your findings.

Class Discussion

Compare the findings of each of the small groups. As a class, develop a 60 second summary on the effect of k on power functions in the form $y = \dfrac{k}{x^p}$, where p is a positive integer.

The Mathematics of Motion: A Case Study

Overview

Scientists probe the functioning of our universe with carefully designed and implemented experiments. The process of generating hypotheses, conducting experiments, analyzing results, and forming conclusions, all subject to reverification by others, is called the scientific method. Mathematics provides the language needed to describe scientific findings.

In this chapter we use the laboratory methods of the modern physicist to collect and analyze data for freely falling bodies and then we examine the questions that Galileo asked about bodies in motion. In generating questions that could be answered through the process of experimentation, Galileo discovered the laws of motion for freely falling bodies. The mathematical focus of the chapter is *quadratic functions* (polynomials of degree 2), which provide useful models for describing the motion of moving bodies.

In the first Exploration you analyze free-fall data. In the second Exploration you derive a free-fall height equation using the concept of average velocity.

After reading this chapter you should be able to:

- understand the importance of the scientific method
- describe the relationship between distance and time for freely falling bodies
- derive equations describing the velocity and acceleration of a freely falling body

10.1 THE SCIENTIFIC METHOD

Today we take for granted that scientists study physical phenomena in laboratories, using sophisticated equipment. But in the early 1600s, when Galileo did his experiments on motion, the concept of laboratory experiments was unknown. In his attempts to understand nature, Galileo asked questions that could be tested directly in experiments. His use of observation and direct experimentation and his discovery that aspects of nature were subject to quantitative laws was of decisive importance not only in science but in the broad history of human ideas.

The Greeks and medieval thinkers believed that basic truths existed within the human mind and that these truths could be uncovered through reasoning, not empirical experimentation. Their scientific method has been described as a "qualitative study of nature." Greek and medieval scientists were interested in *why* objects fall. They believed that a heavier object fell faster than a lighter one because "it has weight and it falls to the Earth because it, like every object, seeks its natural place, and the natural place of heavy bodies is the center of the Earth. The natural place of a light body, such as fire, is in the heavens, hence fire rises."[1]

Galileo changed the question from *why* things fall to *how* things fall. This question suggested other questions that could be tested directly by experiment. "By alternating questions and experiments, Galileo was able to identify details in motion no one had previously noticed or tried to observe."[2] His quantitative descriptions of objects in motion led not only to new ways of thinking about motion, but to new ways of thinking about science. His process of careful observation and testing began the critical transformation of science from a qualitative to a quantitative study of nature.[3] Galileo's decision to search for quantitative descriptions "was the most profound and the most fruitful thought that anyone has had about scientific methodology."[4] This approach became known as "The Scientific Method."

10.2 INTERPRETING DATA FROM A FREE-FALL EXPERIMENT

Exploration 10.1 parallels the discussions in Sections 10.2 – 10.4. The software "Q11: Freely Falling Objects" in Quadratic Functions *simulates the free-fall experiment.*

To gain some understanding of how modern science approaches the search for quantitative natural laws, we examine a modern version of Galileo's free-fall experiment. This classic experiment can be performed either in class with a graphing calculator connected to a motion sensor, or in a physics lab with an apparatus that drops a weight and records its position on a tape at small, regularly timed intervals. The sketch of a tape (Figure 10.1) gives data collected by a group of students in a free-fall experiment. Each dot represents how far the object fell in each succeeding 1/60th of a second.

Since the first few dots are too close together to get accurate measurements, we start measurements instead at the sixth dot, which we call dot_0. At this point, the object is already in motion. This dot is considered to be the starting point, and the time, t,

---

[1] Morris Kline, *Mathematics for the Nonmathematician* (Dover Publications, New York, 1967) p. 287.

[2] Elizabeth Cavicchi, "Watching Galileo's Learning" in the Anthology of Readings.

[3] Galileo's scientific work was revolutionary not only in science, but in the politics of the time; his work was condemned by the ruling authorities and he was arrested.

[4] Morris Kline, p. 288.

at dot_0, is set at 0 seconds. The next dot represents the position of the object 1/60th of a second later. Time increases by 1/60th of a second for each successive dot. In addition to assigning a time to each point, we also measure the total distance fallen, d, from the point designated dot_0. For every dot we have two values: the time, t, and the distance fallen, d. At dot_0, we have $t = 0$ and $d = 0$.

The time and distance measurements from the tape are recorded in Table 10.1 and plotted on the graph in Figure 10.2. Time, t, is the independent variable, and distance, d, is the dependent variable. The graph gives a representation of the data collected on distance fallen over time, not a picture of the physical motion of the object.

The graph of the data looks more like a curve than a straight line, so we expect the

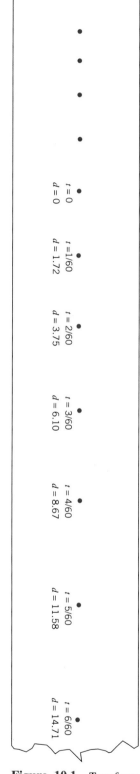

Table 10.1	
Time (sec)	**Total Distance Fallen (cm)**
0.0000	0.00
0.0167	1.72
0.0333	3.75
0.0500	6.10
0.0667	8.67
0.0833	11.58
0.1000	14.71
0.1167	18.10
0.1333	21.77
0.1500	25.71
0.1667	29.90
0.1833	34.45
0.2000	39.22
0.2167	44.22
0.2333	49.58
0.2500	55.15
0.2667	60.99
0.2833	67.11
0.3000	73.48
0.3167	80.10
0.3333	87.05

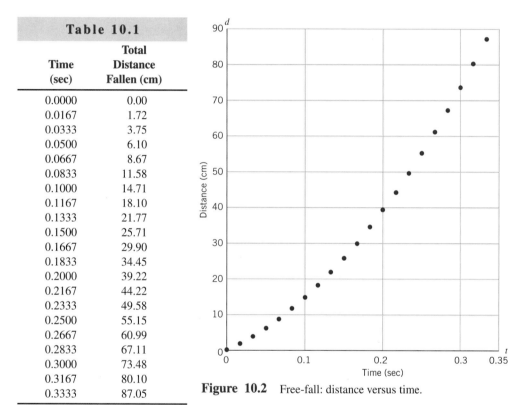

Figure 10.2   Free-fall: distance versus time.

average rates of change between different pairs of points to be different. We know how to calculate the average rate of change between two points, and that it represents the slope of a line segment connecting the two points.

$$\text{average rate of change} = \frac{\text{change in distance}}{\text{change in time}} = \text{slope of line segment}$$

Table 10.2 and Figure 10.3 on page 356 show the increase in the average rate of change over time for three different pairs of points. The time interval nearest the start of the fall shows a relatively small change in the distance per time step and therefore a relatively gentle slope of 188 cm/sec. The time interval farthest from the start of the fall shows a greater change of distance per time step and a much steeper slope of 367 cm/sec.

Figure 10.1   Tape from a free-fall experiment.

See Excel or graph link file, FREEFALL, which contains the data in Table 10.1.

	Table 10.2	
t	d	**Average Rate of Change**
0.0500	6.10	$\dfrac{21.77 - 6.10}{0.1333 - 0.0500} \approx 188$ cm/sec
0.1333	21.77	
0.0833	11.58	$\dfrac{49.58 - 11.58}{0.2333 - 0.0833} \approx 253$ cm/sec
0.2333	49.58	
0.2167	44.22	$\dfrac{87.05 - 44.22}{0.3333 - 0.2167} \approx 367$ cm/sec
0.3333	87.05	

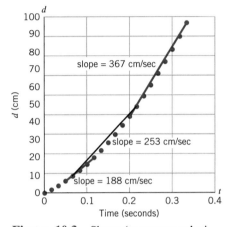

Figure 10.3   Slopes (or average velocities) between three pairs of end points.

In this experiment the average rate of change has an additional important meaning. For objects in motion, the change in distance divided by the change in time is also called the *average velocity* over that time period. For example, in the first set of calculations in Table 10.2, the average rate of change of 188 cm/sec represents the average velocity of the falling object between 0.0500 and 0.1333 seconds.

$$average\ velocity = \frac{change\ in\ distance}{change\ in\ time}$$

Important Questions

Do objects fall at a constant speed?[5] The rate of change calculations and the graph in Figure 10.3 indicate that the average rate of change of position with respect to time, the velocity, of the falling object is not constant. Moreover, the average velocity appears to be increasing over time. In other words, as the object falls, it is moving faster

---

[5] In everyday usage, "speed" and "velocity" are used interchangeably. In physics, "velocity" indicates the direction of motion by the sign of the number, positive for forward, negative for backward. "Speed" means the absolute value or magnitude of the velocity. So, speed is never negative, whereas velocity can be positive or negative.

and faster. Our calculations agree with Galileo's observations. He was the first person to show that the velocity of freely falling objects is not constant.

This finding prompted Galileo to ask more questions. One of these questions was: If the velocity of freely falling bodies is *not constant,* is it increasing at a *constant rate?* Galileo discovered that the velocity of freely falling objects does increase at a constant rate. If the rate of change of velocity with respect to time is constant, then the graph of velocity versus time is a straight line. The slope of that line is constant and equals the rate of change of velocity with respect to time. A theory of gravity has been built around Galileo's discovery of a constant rate of change for the velocity of freely falling bodies. This constant of nature, the gravitational constant of Earth, is denoted by g.

In Section 10.4 we show that for the free-fall data in Table 10.1 the graph of velocity versus time is indeed a straight line and the slope of that line is approximately 980 cm/sec^2, which is the conventional value for g. This tells us that the velocity of the freely falling object increases by 980 cm/sec during each second of fall.

Algebra Aerobics 10.2

1. Complete the table below. What happens to the average velocity of the object as it falls?

Time (sec)	Distance Fallen (cm)	Average Velocity (average rate of change over previous 1/60th of a second)
0.0000	0.00	N/A
0.0333	3.75	$(3.75 - 0.00) / (0.0333 - 0.0000) \approx 113$ cm/sec
0.0667	8.67	$(8.67 - 3.75) / (0.0667 - 0.0333) \approx 147$ cm/sec
0.1000	14.71	
0.1333	21.77	
0.1667	29.90	

10.3 DERIVING AN EQUATION RELATING DISTANCE AND TIME

Galileo wanted to describe mathematically the distance an object falls over time. Using mathematical and technological tools not available in Galileo's time, we can describe the distance fallen over time in the free-fall experiment using a "best fit" function for our data. Galileo had to describe his finding in words. Galileo described the free-fall motion first by direct measurement and then abstractly with a time-squared rule. "This discovery was revolutionary, the first evidence that motion on Earth was subject to mathematical laws."[6]

Using Galileo's finding that distance is related to time by a time-squared rule, we find a best fit polynomial function of degree 2 (a quadratic) for the free-fall data. A

If you are interested in learning more about how Galileo made his discoveries, read Elizabeth Cavicchi's "Watching Galileo's Learning."

[6] Elizabeth Cavicchi, "Watching Galileo's Learning" in Anthology of Readings.

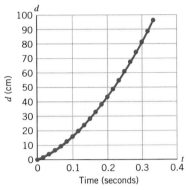

Figure 10.4   Best fit function for distance versus time.

function graphing program gives the following best fit quadratic function for the free-fall data in Table 10.1.

$$d = 487.92t^2 + 98.6t + 0.0027$$

Figure 10.4 shows a plot of the data and the function. If your curve fitting software program or graphing calculator does not provide a measure of closeness of fit, such as the correlation coefficient for regression lines, you may have to rely on a visual judgment. Rounding the coefficients to the nearest unit, we obtain the following equation.

$$d = 488t^2 + 99t + 0$$
$$= 488t^2 + 99t \tag{1}$$

We now have a mathematical model for our free-fall data.

What are the units for each term of the equation? Since d is in centimeters, then each term on the right-hand side of Equation (1) must also be in centimeters. Since t is in seconds, then 488 must be in cm/sec^2.

$$\frac{cm}{\sec^2} \cdot \frac{\sec^2}{1} = cm$$

99 must be in cm/sec and the constant term, 0, in cm.

What if we ran the experiment again? How would the results compare? In one class, four small groups did the free-fall experiment, plotted the data, and found a corresponding best fit second-degree polynomial. The functions are listed below, along with Equation (1). In each case we have rounded the coefficients to the nearest unit. All of the constant terms rounded to 0.

$$d = 488t^2 + 99t \tag{1}$$
$$d = 486t^2 + 72t \tag{2}$$
$$d = 484t^2 + 173t \tag{3}$$
$$d = 486t^2 + 73t \tag{4}$$
$$d = 495t^2 + 97t \tag{5}$$

Examine the coefficients of each of the terms in these equations. All the functions have similar coefficients for the t^2 term, very different coefficients for the t term, and 0 for the constant term. Why is this the case? Using concepts from physics, we can describe what each of the coefficients represents.

The coefficients of the t^2 term found in Equations (1)–(5) are all close to one-half of 980 cm/sec^2 or half of g, Earth's gravitational constant. The data from this simple experiment give very good estimates for 1/2 of g.

The coefficient of the t term represents the initial velocity, v_0, of the object when $t = 0$. In Equation (1), $v_0 = 99$ cm/sec. Recall that we didn't start to take measurements until the sixth dot, the dot we called dot_0. So at dot_0, where we set $t = 0$, the object was already in motion with a velocity of approximately 99 cm/sec. The initial velocities or v_0 values in Equations (2)–(5) range from 72 to 173 cm/sec. Each v_0 represents approximately how fast the object was moving when $t = 0$, the point chosen to begin recording data in each of the various experiments.

The constant term rounded to zero in each of Equations (1)–(5). On the tape when we set $t = 0$, we set $d = 0$. So we expect that in all our equations the constant terms, that represent the distance at time zero, to be approximately 0. If we substitute 0 for t in Equations (1)–(5) above, the value for d is indeed 0. If we looked at additional experimental results, we might encounter some variation in the constant term but all should have values ≈ 0.

Galileo's discoveries are the basis for the following equations relating distance and time.

The general equation of motion of freely falling bodies that relates distance fallen, d, to time, t, is

$$d = (1/2)gt^2 + v_0 t$$

where $v_0 = $ initial velocity and $g = $ acceleration due to gravity on Earth.

For example, in our particular model the equation is $d = 488t^2 + 99t$, where 488 approximates half of g in cm/sec^2, and 99 approximates the initial velocity in cm/sec.

Algebra Aerobics 10.3

1. A freely falling body has an initial velocity of 125 cm/sec. Assume that $g = 980$ cm/sec^2.

   a. Write an equation that relates d, distance fallen in cm, to t, time in seconds.

   b. How far has the body fallen after 1 second? After 3 seconds?

   c. If the initial velocity were 75 cm/sec, how would your equation in part (a) change?

2. In the equation of motion, $d = (1/2)gt^2 + v_0 t$, we specified that distance was measured in centimeters, velocity in centimeters per second, and time in seconds. Rewrite this as an equation that shows only units of measure. Verify that you get cm = cm.

3. The equation $d = (1/2)gt^2 + v_0 t$ could also be written using distances measured in meters. Rewrite the equation showing only units of measure and verify that you get meters = meters.

10.4 DERIVING EQUATIONS FOR VELOCITY AND ACCELERATION

Returning to Galileo's Question

If the velocity for freely falling bodies is not constant, is it increasing at a constant rate? Galileo discovered that the rate of change of the velocity of a freely falling object is constant. In this section we confirm his finding with data from the free-fall experiment.

Velocity: Change in Distance over Time

If the rate of change of velocity is constant, then the graph of velocity versus time should be a straight line. Previously we calculated the average rates of change of distance with respect to time (or average velocities) for three arbitrarily chosen pairs of points. Now, in Table 10.3 we calculate the average rates of change for all the pairs of adjacent points in our free-fall data. The results are in column four. Since each computed velocity is the average over an interval for increased precision, we associate each velocity with the midpoint time of the interval instead of one of the endpoints. In Figure 10.5, we plot velocity from the fourth column against the midpoint times from the third column. The graph is strikingly linear.

	Table 10.3		
t, Time (sec)	d, Distance Fallen (cm)	t, Midpoint Time (sec)	v, Velocity (cm/sec)
0.0000	0.00		
0.0167	1.72	0.0084	103.2
0.0333	3.75	0.0250	121.8
0.0500	6.10	0.0417	141.0
0.0667	8.67	0.0584	154.2
0.0833	11.58	0.0750	174.6
0.1000	14.71	0.0917	187.8
0.1167	18.10	0.1084	203.4
0.1333	21.77	0.1250	220.2
0.1500	25.71	0.1417	236.4
0.1667	29.90	0.1584	251.4
0.1833	34.45	0.1750	273.0
0.2000	39.22	0.1917	286.2
0.2167	44.22	0.2084	300.0
0.2333	49.58	0.2250	321.6
0.2500	55.15	0.2417	334.2
0.2667	60.99	0.2584	350.4
0.2833	67.11	0.2750	367.2
0.3000	73.48	0.2917	382.2
0.3167	80.10	0.3084	397.2
0.3333	87.05	0.3250	417.0
0.3500	94.23	0.3417	430.8

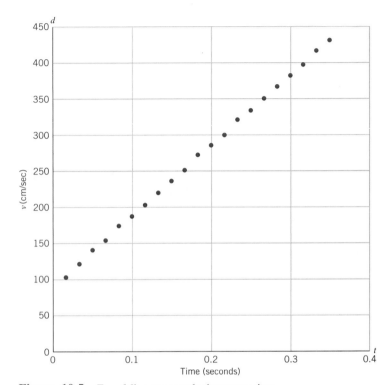

Figure 10.5   Free-fall: average velocity versus time.

Generating a best fit linear function and rounding to the nearest unit, we obtain the following equation:

$$\text{average velocity} = 977t + 98$$

for the average velocity (in cm/sec) at time t where t is in seconds. The graph of this function appears in Figure 10.6. The slope of the line is constant and equals the rate of change of velocity with respect to time. So while the velocity is not constant, its rate of change with respect to time *is* constant.

The coefficient of t, 977, is the slope of the line and in physical terms represents g, the acceleration due to gravity. The conventional value for g is 980 cm/sec^2. So the velocity of the freely falling object increases by about 980 cm/sec during each second of free-fall.

With this equation we can estimate the velocity at any given time t. When $t = 0$, then $v = 98$ cm/sec. This means that the object was already moving at about 98 cm/sec when we set $t = 0$. In our experiment, the velocity when $t = 0$ depends on where we choose to start measuring our dots. If we had chosen a dot closer to the beginning of the free-fall, we would have had an initial velocity smaller than 98 cm/sec. If we had chosen a dot farther away from the start, we would have had an initial velocity larger than 98 cm/sec. Note that 98 cm/sec closely matches the value of 99 cm/sec that we found for the initial velocity for these data in Section 10.3.

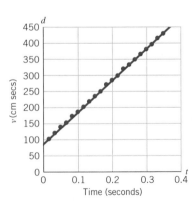

Figure 10.6   Best fit linear function for average velocity versus time.

The general equation that relates v, the velocity of freely falling bodies, to t, time, is

$$v = gt + v_0$$

where v_0 = initial velocity (velocity at time $t = 0$) and g is the acceleration due to gravity.

Something to think about

The graph of distance fallen versus time in Figure 10.1 looks as if it could possibly be an exponential function. Why is the data set not exponential? (Hint: What does the graph of the average rate of change of an exponential function look like?)

Acceleration: Change in Velocity over Time

Acceleration means a change in velocity or speed. If you push the accelerator pedal in a car down just a bit, the speed of the car increases slowly. If you floor the pedal, the speed increases rapidly. The rate of change of velocity with respect to time is called *acceleration*. Calculating the average rate of change of velocity with respect to time gives an estimate of acceleration. For example, if a car is traveling at 20 miles/hr and 1 hour later the car has accelerated to 60 miles/hr, then

$$\frac{\text{change in velocity}}{\text{change in time}} = \frac{60 - 20 \text{ mi/hr}}{1 \text{ hr}} = (40 \text{ mi/hr})/\text{hr} = 40 \text{ mi/hr}^2.$$

In 1 hour, the velocity of the car changed from 20 to 60 miles per hour, so its average acceleration was 40 miles/hr/hr, or 40 mi/hr^2.

$$\textit{average acceleration} = \frac{\text{change in velocity}}{\text{change in time}}$$

Table 10.4 on page 362 uses the average velocity data and midpoint time from Table 10.3 to calculate average accelerations. Figure 10.7 shows the plot of average acceleration in cm/sec^2 (the third column) versus time in seconds (the first column).

The data lie along a roughly horizontal line. The average acceleration values vary between a low of 790.4 and a high of 1301.2 cm/sec^2 with a mean of 983.3. Rounding off we have:

$$\text{acceleration} \approx 980 \text{ cm/sec}^2$$

This expression confirms that for each additional 1 second of fall, the velocity of the falling object increases by approximately 980 cm/sec. The longer it falls, the faster it goes. We have verified a characteristic feature of gravity near the surface of Earth: it causes objects to fall at a velocity that increases every second by about 980 cm/sec. We say that the acceleration due to gravity near Earth's surface is 980 cm/sec^2.

In order to express g in feet/sec^2, we need to convert 980 centimeters into feet. We start with the fact that 1 ft = 30.48 cm. So the conversion factor for centimeters to feet is $\left(\dfrac{1 \text{ ft}}{30.48 \text{ cm}}\right) = 1$. If we multiply 980 cm by $\left(\dfrac{1 \text{ ft}}{30.48 \text{ cm}}\right)$ to convert centimeters to feet, we get

$$980 \text{ cm} = (980 \text{ cm})\left(\frac{1 \text{ ft}}{30.48 \text{ cm}}\right) \approx 32.15 \text{ ft} \approx 32 \text{ ft}$$

So a value of 980 cm/sec^2 for g is equivalent to approximately 32 feet/sec^2.

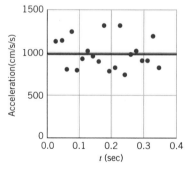

Figure 10.7 Average acceleration for free-fall data.

	Table 10.4	
t, Midpoint Time (sec)	v, Average Velocity (cm/sec)	Average Acceleration (cm/sec^2)
0.0084	103.2	n.a.
0.0250	121.8	1120.5
0.0417	141.0	1149.7
0.0584	154.2	790.4
0.0750	174.6	1228.9
0.0917	187.8	790.4
0.1084	203.4	934.1
0.1250	220.2	1012.0
0.1417	236.4	970.1
0.1584	251.4	898.2
0.1750	273.0	1301.2
0.1917	286.2	790.4
0.2084	300.0	826.3
0.2250	321.6	1301.2
0.2417	334.2	754.5
0.2584	350.4	970.1
0.2750	367.2	1012.0
0.2917	382.2	898.2
0.3084	397.2	898.2
0.3250	417.0	1192.8
0.3417	430.8	826.3

The conventional values for g, the acceleration due to gravity near the surface of Earth, are

$$g = 32 \text{ ft/sec}^2$$

or equivalently

$$g = 980 \text{ cm/sec}^2 = 9.8 \text{ m/sec}^2.$$

The numerical value used for the constant g depends on the units being used for the distance, d, and the time, t. The exact value of g also depends on where it is measured.[7]

---

[7] Because the Earth is rotating, is not a perfect sphere, and is not uniformly dense, there are variations in g according to latitude and elevation. The following are a few examples of local values for g.

Location	North Latitude (°)	Elevation (m)	g (cm/sec^2)
Panama Canal	9	0	978.243
Jamaica	18	0	978.591
Denver, Colo.	40	1638	979.609
Pittsburgh, Pa.	40.5	235	980.118
Cambridge, Mass.	42	0	980.398
Greenland	70	0	982.534

Source: Hugh D. Young, *University Physics,* Vol. I, 8th ed., (Reading, Mass.: Addison Wesley, 1992) p. 336.

The general equations for the motion of freely falling bodies that relate distance, d, and velocity, v, to time, t, are

$$d = (1/2)gt^2 + v_0 t$$
$$v = gt + v_0$$

where v_0 = initial velocity and g = acceleration due to gravity.

Algebra Aerobics 10.4

1. A freely falling object has an initial velocity of 50 cm/sec.

   a. Write two motion equations, one relating distance and time and the other relating velocity and time.

   b. How far has the object fallen and what is its velocity after 1 second? After 2.5 seconds? Be sure to identify units in your answers.

2. A freely falling object has an initial velocity of 20 ft/sec.

   a. Write one equation relating distance fallen (in feet) and time (in sec), and a second equation relating velocity (in ft/sec) and time.

   b. How many feet has the object fallen and what is its velocity after 0.5 seconds? After 2 seconds?

3. If the equation $d = 4.9t^2 + 11t$ represents the relationship between distance and time for a freely falling body, in what units is distance now being measured? How do you know?

10.5 USING THE FREE-FALL EQUATIONS

Applying the Model

Interpreting a free-fall equation

The data from a free-fall tape generate the following equation relating distance fallen in centimeters and time in seconds.

Example 1

$$d = 485.7t^2 + 72.6t$$

a. Give a physical interpretation of each of the coefficients along with their appropriate units of measurement.

The coefficient 485.7 is in cm/sec^2 and approximates half the acceleration due to gravity; 72.6 is in cm/sec and represents the initial velocity of the object.

b. How far has the object fallen after 0.05 seconds? 0.10 seconds? 0.30 seconds?

Table 10.5 shows the value of d for t at 0.05, 0.10, and 0.30 seconds.

	Table 10.5	
t (sec)	$d = 485.7\,t^2 + 72.6\,t$	d (cm)
0.05	$486(0.05)^2 + 73(0.05) \approx$	4.9
0.10	$486(0.10)^2 + 73(0.10) \approx$	12.2
0.30	$486(0.30)^2 + 73(0.30) \approx$	65.6

After 0.1 second, for instance, the object would have fallen 12.2 cm.

Translating units

Example 2   What would the free-fall equation $d = 490t^2 + 90t$ become if d were measured in feet instead of centimeters?

SOLUTION
In the equation $d = 490t^2 + 90t$, the number 490 represents $(1/2)g$ and is in cm/sec^2. Since g is approximately 32 feet/sec^2, then $(1/2)g$ is approximately 16 feet/sec^2. We need to convert the initial velocity of 90 cm/sec to feet/sec. Since 1 cm = 0.0325 feet, then

$$90 \text{ cm} = (90 \text{ cm})(0.0325 \text{ ft/cm}) \approx 3 \text{ feet}.$$

So 90 cm/ft $\approx$ 3 ft/sec.
    We can then rewrite the original equation as $d = 16t^2 + 3t$, where t is still in seconds but d is now in feet.

Translating from distance to height measurements

Example 3   Assume we have the following motion equation relating distance fallen in centimeters and time in seconds

$$d = 490t^2 + 45t$$

and that when $t = 0$ the height, h, of the object was 100 cm above the ground. Until now, we have considered the distance from the point the object was dropped, a value that *increases* as the object falls down. How can we describe a different distance, the *height above ground* of an object, as a function of time, a value that *decreases* as the object falls?

SOLUTION
At time 0, the distance fallen is 0 and the height above the ground is 110 centimeters. After 0.05 seconds, the object has fallen about 3.5 cm, so its height would be $110 - 3.5 = 106.5$ cm. For an arbitrary distance d, we have $h = 110 - d$. Table 10.6 gives associated values for time, t, distance, d, and height, h. The graphs in Figure 10.8 show distance versus time and height versus time.

Table 10.6

t, Time (sec)	d, Distance Fallen (cm) ($d = 490t^2 + 45t$)	h, Height Above Ground (cm) ($h = 110 - 45t - 490t^2$)
0.00	0.0	110.0
0.05	3.5	106.5
0.10	9.4	100.6
0.15	17.8	92.2
0.20	28.6	81.4
0.25	41.9	68.1
0.30	57.6	52.4
0.35	75.8	34.2
0.40	96.4	13.6

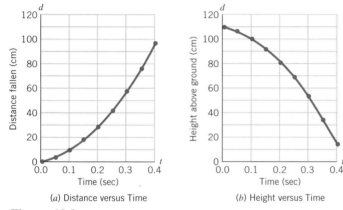

(a) Distance versus Time   (b) Height versus Time

Figure 10.8   Representations of free-fall data

The equation $d = 490t^2 + 45t$ relates distance fallen and time. How can we find an expression relating height and time?

We know that the relationship between height and distance is $h = 110 - d$. We can substitute the expression for d into the height equation.

$$h = 110 - d$$
$$= 110 - (490t^2 + 45t)$$
$$h = 110 - 490t^2 - 45t \tag{1}$$

Since we can switch the order in which we add terms, we could of course write this equation as $h = -490t^2 - 45t + 110$. The constant term, here 110 cm, represents the initial height when $t = 0$. By placing the constant term first as in Equation (1), we emphasize 110 cm as the initial or starting value. Height equations often appear in the form $h = c + bt + at^2$ to emphasize the constant term c as the starting height. This is similar to writing linear equations in the form $y = b + mx$ to emphasize the constant term b as the base or starting value. Rewriting Equation (1), we have

$$h = 110 - 45t - 490t^2$$

Note that in our equation for height, the coefficients of both t and t^2 are negative. If we consider what happens to the height of an object in free-fall, this makes sense.

In Exploration 10.2 you examine alternate strategies for determining the height equation.

As time increases, the height decreases. (See Table 10.6 and Figure 10.8.) When we were measuring the increasing distance an object fell, we did not take into account the direction in which it was going (up or down). We cared only about the magnitudes (the absolute values) of distance and velocity, which were positive. But when we are measuring a decreasing height or distance, we have to worry about directions. In these cases we define downward motion to be negative and upward motion to be positive. In the height equation, $h = 110 - 45t - 490t^2$, the constant term, the initial height, is 110 cm; and the change in height resulting from initial velocity, $-45t$, is negative because the object was moving down when we started to measure it. The change in height caused by acceleration, $-490t^2$, is negative because the effect of gravity is downward motion—it reduces the height of a falling object as time increases.

Once we have introduced the notion that downward motion is negative and upward motion is positive, we can also deal with situations in which the initial velocity is upward and the acceleration is downward.

Example   4

Working with an upward initial velocity

Write an equation relating height and time if an object at an initial height of 87 cm is thrown upward with an initial velocity of 97 cm/sec.

SOLUTION

The initial height is 87 cm when $t = 0$, so the constant term is 87 cm. The coefficient of t, or the initial velocity term, is positive 97 cm/sec since the initial motion is upward. The coefficient of t^2, the gravity term, is negative 490 cm/sec^2, since the effect of gravity is downward motion.

Substituting these values into the equation for height, we get

$$h = 87 + 97t - 490t^2$$

Table 10.7 gives a series of values for heights corresponding to various times.

Table 10.7	
t (sec)	h (cm)
0.00	87.00
0.05	90.63
0.10	91.80
0.15	90.53
0.20	86.80
0.25	80.63
0.30	72.00
0.35	60.93
0.40	47.40
0.45	31.43
0.50	13.00

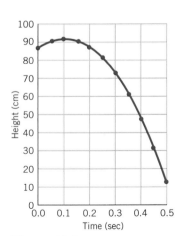

Figure 10.9   Height of a thrown ball.

The graph of the heights at each time in Figure 10.9 should not be confused with the trajectory of a thrown object. The actual motion we are talking about is a purely vertical motion—straight up and straight down. The graph shows that the object travels up for a while before it starts to fall. This corresponds with what we all know from practical experience throwing balls. The upward (positive) velocity is decreased by the pull of gravity until the object stops moving upward and begins to fall. The downward (negative) velocity is increased in magnitude by the pull of gravity until the object strikes the ground.

The velocity equation is

$$v = -gt + v_0$$

where v_0 could be either positive or negative, depending on whether the object is thrown upward or downward, and the sign for the g term is negative because gravity accelerates downward in the negative direction.

In summary:

The general equations of motion of freely falling bodies that relate height, h, and velocity, v, to time, t, are

$$h = h_0 + v_0 t - (1/2)gt^2$$
$$v = -gt + v_0$$

where h = initial height, g = acceleration due to gravity and v_0 = initial velocity (which can be positive or negative).

Algebra Aerobics 10.5

1. In the equation $d = 4.9t^2 + 500t$, time is measured in seconds and distance in meters. What does the number 500 represent?

2. In the height equation $h = 300 + 50t - 4.9t^2$, time is measured in seconds and height in meters.

    a. What does the number 300 represent?

    b. What does the number 50 represent? What does the fact that 50 is positive tell you?

3. The height of an object that was projected vertically from the ground with initial velocity of 200 meters per second is given by the equation $h = 200t - 4.9t^2$, where t is in seconds.

    a. Find the height of the object after 0.1, 2, and 10 seconds.

    b. Sketch a graph of height versus time.

    c. Use the graph to determine the maximum height of the projectile and the approximate number of seconds that the object traveled before hitting the ground.

4. The height of an object that was shot downward from a 200-meter platform with an initial velocity of 50 m/sec is given by the equation $h = -4.9t^2 - 50t + 200$, where t is in seconds. Sketch the graph of height versus time. Use the graph to determine the approximate number of seconds that the object traveled before hitting the ground.

CHAPTER SUMMARY

Galileo discovered that the velocity of freely falling bodies is *not constant,* but it is increasing at a *constant rate.* His process of careful observation and experimentation began the critical transformation of science from a qualitative to a quantitative study of nature. This approach became known as "The Scientific Method."

We can replicate his experiments by taking distance versus time measurements for a freely falling body and examining:

$$average\ velocity = \frac{change\ in\ distance}{change\ in\ time}$$

and

$$average\ acceleration = \frac{change\ in\ velocity}{change\ in\ time}$$

The general equations of motion for freely falling bodies that relate distance fallen, velocity, and time are:

$$d = (1/2)gt^2 + v_0 t$$

and

$$v = gt + v_0$$

where d = distance, t = time, v_0 = initial velocity, and g = acceleration near the surface of Earth due to gravity.

The conventional values for g are 32 ft/sec^2 or equivalently 980 cm/sec^2, which is equal to 9.8 m/sec^2.

The general equations of motion for freely falling bodies that relate height, velocity, and time are:

$$h = h_0 + v_0 t - (1/2)gt^2$$

and

$$v = -gt + v_0$$

where h_0 = initial height, v_0 = initial velocity (which may be positive or negative), and g = acceleration due to gravity. In these equations downward motion is considered negative, upward motion is positive.

Exercises

1. The essay "Watching Galileo's Learning" examines the learning process that Galileo went through to come to some of the most remarkable conclusions in the history of science. Write a summary of one of Galileo's conclusions about motion. Include in your summary the process by which Galileo made this discovery and some aspect of your own learning or understanding of Galileo's discovery.

2. The equation $d = 490t^2 + 50t$ describes the relationship between distance fallen, d, in centimeters, and time, t, in seconds, for a particular freely falling object.
   a. Interpret each of the coefficients and specify its units of measurement.
   b. Generate a table for a few values of t between 0 and 0.3 seconds.
   c. Graph distance versus time by hand. Check your graph using a computer or calculator if available.

3. The equation $d = 4.9t^2 + 1.7t$ describes the relationship between distance fallen, d, in meters, and time, t, in seconds, for a particular freely falling object.
   a. Interpret each of the coefficients and specify its units of measurement.
   b. Generate a table for a few values of t between 0 and 0.3 seconds.
   c. Graph distance versus time by hand. Check your graph using a computer or calculator if available.
   d. Relate your answers to earlier results in this chapter.

4. The equation $d = (1/2)gt^2 + v_0t$ could be written using distance measured in feet. Rewrite the equation showing only units of measure and verify that you get feet = feet.

5. A freely falling object has an initial velocity of 12 ft/sec.
   a. Construct an equation relating distance fallen and time.
   b. Generate a table by hand for a few values of the distance fallen between 0 and 5 seconds.
   c. Graph distance versus time by hand. Check your graph using a computer or graphing calculator if available.

6. Use the information in Exercise 5 to do the following:
   a. Construct an equation relating velocity and time.
   b. Generate a table by hand for a few values of velocity between 0 and 5 seconds.
   c. Graph velocity versus time by hand. If possible, check your graph using a computer or graphing calculator.

7. Let $h = 85 - 490t^2$ be a motion equation describing height, h, in centimeters and time, t, in seconds.
   a. Interpret each of the coefficients and specify its units of measurement.
   b. What is the initial velocity?
   c. Generate a table for a few values of t between 0 and 0.3 seconds.

d. Graph height versus time by hand. Check your graph using a computer or calculator if possible.

8. Let $h = 85 + 20t - 490t^2$ be a motion equation describing height, h, in centimeters and time, t, in seconds.

   a. Interpret each of the coefficients and specify its units of measurement.

   b. Generate a table for a few values of t between 0 and 0.3 seconds.

   c. Graph height versus time by hand. Check your graph using a computer or calculator if available.

9. At $t = 0$, a ball is thrown upward at a velocity of 10 ft/sec, from the top of a building 50 feet high. The ball's height is measured in feet above the ground.

   a. Is the initial velocity positive or negative? Why?

   b. Write the motion equation that describes height, h, at time, t.

10. The relationship between the velocity of a freely falling object and time is given by

$$v = -gt - 66$$

where g is the acceleration due to gravity and the units for velocity are cm/sec.

   a. What value for g should be used in the equation?

   b. Generate a table of values for t and v, letting t range from 0 to 4 seconds.

   c. Graph velocity versus time and interpret your graph.

   d. What was the initial condition? Was the object dropped or thrown upward? Explain your reasoning.

11. The concepts of velocity and acceleration are useful in the study of human childhood development. The accompanying figure shows (*a*) a standard growth curve of weight over time; (*b*) the rate of change of weight over time (the *growth rate* or *velocity*); and (*c*) the rate of change of the growth rate over time, (or *acceleration*). Describe in your own words what each of the graphs shows about a child's growth.

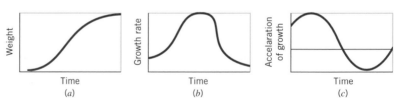

(a)   (b)   (c)

Graphs adapted from Barry Bogin, "The Evolution of Human Childhood," *BioScience,* Vol. 40, p. 16.

One screen in "Q11: Freely Falling Objects" in Quadratic Functions *simulates this activity.*

12. (This exercise requires a free-fall data tape created using a spark timer.)

   a. Make a graph from your tape: cut the tape with scissors crosswise at each spark dot, so you have a set of strips of paper that are the actual lengths of the distances fallen by the object during each time interval. Arrange them evenly spaced in increasing order, with the bottom of each strip on a horizontal line. The end result should look like a series of steps. You could paste or tape them down on a big piece of paper or newspaper.

   b. Use a straight edge to draw a line that passes through the center of the top of each strip. Is the line a good fit? Each separate strip represents the distance the object fell during a fixed time interval, so we can think of the strips as representing change in distance over time, or an average velocity. Interpret the graph of the line you have constructed in terms of the free-fall experiment.

13. A certain baseball is at height, $h = 4 + 64t - 16t^2$ feet, at time, t, in seconds. Compute the average velocity for each of the following time intervals and indicate for which intervals the baseball is rising and for which it is falling. In which interval was the average velocity the greatest?

   a. $t = 0$ to $t = 0.5$

   b. $t = 0$ to $t = 0.1$

   c. $t = 0$ to $t = 1$

   d. $t = 1$ to $t = 2$

   e. $t = 2$ to $t = 3$

   f. $t = 1$ to $t = 3$

   g. $t = 4$ to $t = 4.01$

14. At $t = 0$, an object is in free fall 150 cm above the ground, falling at a rate of 25 cm/sec. Its height, h, is measured in centimeters above the ground.

   a. Is its velocity positive or negative? Why?

   b. Write an equation that describes its height, h, at time, t.

   c. What is the average velocity from $t = 0$ to $t = 1/2$? How does it compare to the initial velocity?

15. A freely falling object falls a distance of d feet in t seconds.

   a. Solve the distance equation, $d = 16t^2$, for t.

   b. What values of d make physical sense?

   c. Write t as a function of d. Specify the domain. Remember that for t to be a function of d, there must be one and only one value of t for each value of d.

   d. Fill in the following table of values for this function.

d	t
0	
16	
32	
64	
96	

   e. Sketch by hand the graph of the function you found in part (c). Note that d is now the independent variable and hence graphed on the horizontal axis. The dependent variable is now t and hence graphed on the vertical axis.

16. *The force of acceleration on other planets*

   We have seen that the function $d = (1/2)g \cdot t^2 + v_0 \cdot t$ (where g is the force of acceleration due to Earth's gravity and v_0 is the object's initial velocity) is a mathematical model for the relationship between time and distance fallen by freely falling bodies near Earth's surface. This relationship also holds for freely falling bodies near the surface of other planets. We would need to replace g, the force of acceleration of Earth's gravitational field, with the force of acceleration for the planet under consideration. The following table gives the force of acceleration due to gravity for planets in our solar system.

	Acceleration Due to Gravity	
	meters/sec^2	feet/sec^2
Mercury	3.7	12.1
Venus	8.9	29.1
Earth	9.8	32.1
Mars	3.7	12.2
Jupiter	24.8	81.3
Saturn	10.4	34.1
Uranus	8.5	27.9
Neptune	11.6	38.1
Pluto	0.6	2.0

Source: The Astronomical Almanac, U.S. Naval Observatory, 1981.

a. Choose units of measurement (meters or feet) and three of the planets (other than Earth). For each of these planets, find an equation for the relationship between distance an object falls and time. Construct a table as shown below. Assume for the moment that the initial velocity of the freely falling object is 0.

Name of Planet	Function Relating Distance and Time	Units for Distance

b. Using a function graphing program or a graphing calculator, plot the three functions, with time on the horizontal axis and distance on the vertical axis. Choose a domain that includes both negative and positive values of *t*.

c. On which of your planets will an object fall the farthest in a given time? On which will it fall the least distance in a given time? What part of the domain makes physical sense for your models? Why?

d. Examine the graphs and think about the *similarities* that they share. Describe their general shape. What happens to *d* as the value for *t* increases? As the value for *t* decreases?

e. Think about the *differences* among the three curves. What effect does the coefficient of the t^2 term have on the shape of the graph? When the coefficient gets larger (or smaller) how is the shape of the curve affected? Which graph shows *d* increasing the fastest compared to *t*?

17. In 1974 in Anaheim, California, Nolan Ryan threw a baseball at just over 100 mph. If he had thrown the ball straight upward at this speed, it would have risen to a height of over 335 feet and taken just over 9 seconds to fall back to Earth. Choose another planet and see what would have happened if he had been able to

throw a baseball straight up at 100 mph on that planet. Use the table for the acceleration due to gravity on other planets in Exercise 16.

18. Suppose an object is moving with constant acceleration, a, and its motion is initially observed at a moment when its velocity is v_0. We set time, t, equal to 0, at this point when velocity equals v_0. Then its velocity t seconds after the initial observation is $V(t) = at + v_0$. (Note that the product of acceleration and time is velocity.) Now suppose we want to find its average velocity between time 0 and time t. The average velocity can be measured in two ways. First, we can find the average of the initial and final velocities by calculating a numerical average or mean: that is, we add the two velocities and divide by 2. So, between time 0 and time t,

$$\text{average velocity} = \frac{v_0 + V(t)}{2} \tag{1}$$

As in Section 10.3, we can also find the average velocity by dividing the change in distance by the change in time. Thus, between time 0 and time t,

$$\text{average velocity} = \frac{\Delta \text{ distance}}{\Delta \text{ time}} = \frac{d - 0}{t - 0} = \frac{d}{t} \tag{2}$$

Since both Equations (1) and (2) are equal to the average velocity, they must be equal to each other.

$$\frac{d}{t} = \frac{v_0 + V(t)}{2} \tag{3}$$

But we also know that $V(t) = at + v_0$. Substitute this expression for $V(t)$ in Equation (3) and solve for d. Interpret your results.

19. An object that is moving on the ground is observed to have (initial) velocity of 60 cm/sec and to be accelerating at a constant rate of 10 cm/sec^2.

   a. Determine its velocity after 5 seconds; after 60 seconds; after t seconds.

   b. Find the average velocity for the object between 0 and 5 seconds.

20. Find the distance traveled by the object described in Exercise 19 above after 5 seconds by using two different methods.

   a. Use the formula distance equals rate times time. For rate, use the value found in Exercise 19(b). For time, use 5 seconds.

   b. Write an equation of motion $d = (1/2)at^2 + v_0 t$ using $a = 10$ cm/sec^2 and $v_0 = 60$ cm/sec and evaluate when $t = 5$. Does your answer agree with part (a)?

21. An object is observed to have (initial) velocity of 200 m/sec and to be accelerating at 60 m/sec^2.

   a. Write an equation for its velocity after t seconds.

   b. Write an equation for the distance traveled after t seconds.

22. You may have noticed that when a basketball player or dancer jumps straight up in the air, in the middle of a blurred impression of vertical movement, the jumper appears to "hang" for an instant at the top of the jump.

   a. If a player jumps 3 feet straight up, how long does it take him to fall back down to the ground from the top of the jump? At what downward velocity does he hit the ground?

   b.   At what initial upward velocity does the player have to leap to achieve a 3-foot high jump? How long does the total jump take from takeoff to landing?

   c.   How much vertical distance is traveled in the first third of the total time the jump takes? In the middle third? In the last third?

   d.   Now explain in words why it is that the jumper appears suspended in space at the top of the jump.

   23.   In the Anthology Reading, "Watching Galileo's Learning," Cavicchi notes that Galileo generated a sequence of odd integers from his study of falling bodies. Show that in general the odd integers can be constructed from the difference of the squares of successive integers, that is, that the terms $(n + 1)^2 - n^2$ (where $n = 1, 2, 3, \ldots$) generate a sequence of odd integers.

EXPLORATION 10.1

Free-Fall Experiment

Objective

- determine the relationship for a freely falling object between time and distance fallen

Equipment/Materials

- graphing calculator with best fit function capabilities or computer with spreadsheet and function graphing program
- notebook for recording measurements and results

- if using precollected data, see Excel or graph link file FREEFALL
- Equipment needed for collecting data in physics laboratory
  - free-fall apparatus
  - meter sticks 2 meters long
  - masking tape
- Equipment needed for collecting data with CBL[T] (Calculator-Based Laboratory System[T])
  - CBL[T] unit with AC-9201 power adapter
  - Vernier CBL[T] ultrasonic motion detector
  - graphing calculator
  - extension cord and some object to drop, such as a pillow or rubber ball

Related Readings

"Watching Galileo's Learning"

Related Software

"Q11: Freely Falling Objects" and "Q12: Average Rates of Change" in *Quadratic Functions*
"C3: Average Velocity & Distance" in *Rates of Change*

Preparation

If collecting data in a physics lab, schedule a time for doing the experiment and have the lab assistant available to set up the equipment and assist with the experiment. If collecting data with a CBL[T] unit with graphing calculators, instructions for using a CBL[T] unit are in the Instructor's Manual and on the CD-ROM.

Procedure

The following procedures can be used for collecting data in a physics lab.[8] If you are collecting data with a CBL[T], collect the data and go to the Results section. If you are using the precollected data in the file FREEFALL, go directly to the Results section.

Collecting the Data

Since the falling times are too short to record with a stopwatch, we use a free-fall apparatus. Every 1/60th of a second a spark jumps between the falling object or "bob" and the vertical metal pole supporting the tape. Each spark burns a small dot on the tape, recording the bob's position. The procedure is to:

1. Position the bob at the top of the column in its holder.
2. Pull the tape down the column so that a fresh tape is ready to receive spark dots.
3. Be sure that the bob is motionless before you turn on the apparatus.
4. Turn on the spark switch and bob release switch as demonstrated by the lab assistant.
5. Tear off the length of tape recording the fall of the bob.

Obtaining and Recording Measurements from the Tapes

The tape is a record of the distance fallen by the bob between each 1/60th of a second spark dot. Each pair of students should measure and record the distance between the dots on the tape. Let d = the distance fallen in centimeters and t = time in seconds.

1. Fasten the tape to the table using masking tape.
2. Inspect the tape for missing dots. Caution: the sparking apparatus sometimes misses a spark. If this happens, take proper account of it in numbering the dots.
3. Position the 2-meter stick on its edge along the dots on the tape. Use masking tape to fasten the meter stick to the table, making sure that the spots line up in front of the bottom edge of the meter stick so you can read their positions off of the stick.
4. Beginning with the sixth visible dot, mark the time for each spot on the tape; i.e., write $t = 0/60$ sec by the sixth dot, $t = 1/60$ by the next dot, $t = 2/60$ by the next dot, and so on until you reach the end of the tape.

    Note: The first five dots are ignored in order to increase accuracy of measurements. One cannot be sure that the object is released exactly at the time of the spark, instead of between sparks, and the first few dots are too close together to get accurate measurements. When the body passes the sixth dot, it already has some velocity, which we call v_0, and this point is arbitrarily taken as the initial time, $t = 0$.

5. Measure the distances (accurate to a fraction of a millimeter) from the sixth dot to each of the other dots. Record each distance by the appropriate dot on the tape.
6. Recheck your measurements.
7. Clean your work area.

---

[8] These procedures are adapted from "Laboratory Notes for Experiment 2: The Kinematics of Free Fall," UMass/Boston, Elementary Physics 181.

Results

Use your notebook to keep a record of your data, observations, graphs, and analysis of the data.

a. Record the data obtained from your measurements on the tape or from using a CBL^T unit. If you are entering your data into a function graphing program or a spreadsheet, you can use a printout of the data and staple it into your lab notework.

Your data should include time, t, and distance fallen, d, as in the following table.

t (sec)	d (cm)
0/60	0
1/60	__
2/60	__
....	
To last record	__

This table assumes regular time intervals of 1/60th of a second. Check your equipment to see whether it uses a different interval size.

b. Note at which dot on the tape you started to make your measurements.

Analysis of Data

1. By hand:
   a. Graph your data, using the vertical axis for distance fallen, d, in centimeters and the horizontal axis for time, t, in seconds. What does your graph suggest about the average rate of change of distance with respect to time?

   b. Calculate the average rate of change for distance, d, with respect to time, t, for three pairs of points from your data table. Show your work.

   $$\text{average rate of change} = \frac{\text{change in distance}}{\text{change in time}} = \frac{\Delta d}{\Delta t}$$

   This average rate of change is called the *average velocity* of the falling object between these two points. Do your calculations support your answer in part (a)?

   c. Jot down your observations from your graph and calculations in your notebook. Staple your graph into your notebook.

2. With graphing calculators or computers:
   a. Now use technology to graph your data for the free-fall experiment. Plot time, t, on the horizontal axis and distance fallen, d, on the vertical axis.

   b. Find a best fit function for distance fallen over time.

   c. Use your spreadsheet or graphing calculator to calculate the average rate of change in distance over each of the small time intervals. This average rate of change is the average velocity over these time intervals.

   d. Plot average velocity versus time, with time on the horizontal axis and average velocity on the vertical axis.

   e. Jot down your observations from your graphs and calculations in your notebook. Be sure to specify the units for any numbers you recorded.

Conclusions

Summarize your conclusions from the experiment.

- Describe what you found out from your graph of distance versus time and your calculations for the average rate of change of distance with respect to time. Is the average rate of change of distance over time the same for each small time interval?

- What does your graph of the average velocity versus time tell you about the average velocity of the freely falling body? Is the average rate of change in velocity from one interval to the next roughly constant?

- In light of the readings and class discussion, interpret your graphs for distance and average velocity and interpret the coefficients in the equation you found for distance.

EXPLORATION 10.2

Constructing Equations of Motion

Objectives

- derive an equation for the height of an object in free fall
- interpret the height and velocity equations for the free-fall experiment

Materials/Equipment

- graph paper
- graphing calculator or computer with spreadsheet and function graphing program
- free-fall data from Exploration 10.1 or from Table 10.1 (see the Excel or graph link file FREEFALL)

Procedure

In this Exploration, we use alternate strategies for deriving equations for the height, h, of a freely falling object. We can use data from Table 10.1 or data that you have collected from a free-fall experiment in Exploration 10.1. To find the height of the object, subtract the distance fallen from the initial height at which the object was dropped. For the data set in Table 10.1 the initial height was approximately 90.5 cm.

1. Enter your data for time and distance either into a graphing calculator or a spreadsheet in two columns, then calculate the height in column three, and the average velocity in column four. Some sample calculations for the data in Table 10.1 are shown below. Note that the average velocities you compute will be negative. This is because we are now considering the direction in which the object is moving. The object is falling down, so its height is decreasing (see Example 3 in Section 10.5).

Time, t (sec)	Distance, d (cm)	Height, h (cm)	Average Velocity (cm/sec) over previous 1/60th of a sec
0.0000	0.00	$(90.5 - 0.00) = 90.50$	n.a.
0.0167	1.72	$(90.5 - 1.72) = 88.78$	$\dfrac{(88.78 - 90.50)}{(0.0167 - 0.000)} \approx -103.0$

2. Construct and interpret graphs for the following:
    a. height versus time
    b. average velocity versus time
3. Find an equation of a line approximating the average velocity over time from your plot in 2 (b). Generate the line and equation either by hand or by using technology to determine a regression line. Call this Equation (1).
4. An equation giving height, h, at time, t, can be derived from the equation for the average velocity from the starting time at 0 to time t. If we let h_0 = height at $t = 0$, then we

have

$$\text{average velocity between time 0 and time } t = \frac{\text{change in height}}{\text{change in time}}$$

$$= \frac{h - h_0}{t - 0}$$

$$= \frac{h - h_0}{t}$$

Solve this equation for h, the height, and call the result Equation (2). We get

$$h = h_0 + t \cdot (\text{average velocity between 0 and } t) \tag{2}$$

5. We now need an expression for average velocity between time 0 and time t. The graph of velocity is a straight line. So it's plausible and justifiable by more advanced means (calculus) to take the average velocity as just the velocity at time $t/2$, halfway between the extremes of 0 and t (see the accompanying figure). Note that $v_0 = $ velocity at $t = 0$, and $v_t = $ velocity at time t. Find the average velocity (v_{average}) from time 0 to time t by using $t/2$ for time in Equation (1). Call this result Equation (3).

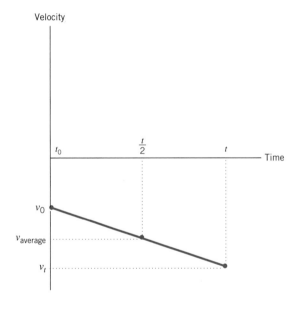

Substitute the expression for average velocity (from time 0 to time t) given by Equation (3) into Equation (2). Express in your own words what this equation tells you about the height, h.

6. Use a graphing calculator or computer to generate a best fit quadratic for height versus time data in Table from part (1). How do your two height equations compare?

Analysis of Results

In your own words explain what you did to find the average velocity from time 0 to time t.

How close did you come to the conventional value for g in both of your height equations?

Interpret the coefficients for each of the terms of the equations you found for height.

Parabolic Reflections and Polynomials

Overview

The free-fall experiment in Chapter 10 introduced quadratic relationships between distance and time. The graph of a quadratic function has a distinctive bowl-like shape and is called a *parabola*. A parabolic curve can be seen by watching a basketball foul shot. *Ball*istics, literally the study of dropped, thrown, or projected balls, gives many variations of quadratic data, all of which appear as parabolic curves when vertical position is plotted versus time. Mirrors in the shape of parabolas have the property of focusing or reflecting parallel rays.

In this chapter we study the general properties of *quadratic functions* (polynomials of degree 2). We develop ways to identify critical properties of quadratics and sketch their graphs. We then extend our study to include polynomials of higher degrees.

The Explorations build an intuitive understanding of quadratic functions and show some of their practical applications.

After reading this chapter you should be able to:

- identify, evaluate, and graph quadratic functions
- determine the vertex, the axis of symmetry, and the intercepts of a quadratic function
- convert quadratic functions from one form to another
- describe the basic properties of higher order polynomial functions

11.1 VISUALIZING QUADRATIC FUNCTIONS

The Definition of a Quadratic Function

In Chapter 10, quadratic functions of the form $d = at^2 + bt + c$ were used to model distance fallen, d, versus time, t, in the free-fall experiment. Quadratics are a subset of the family of polynomial functions. In Chapter 9, we introduced polynomial functions as the sum of one or more power functions that have positive integer powers.

A *quadratic function* or polynomial function of degree 2 has the form $y = f(x)$, where

$$y = ax^2 + bx + c \qquad a, b, \text{ and } c \text{ are constants and } a \neq 0.$$

Important Characteristics of the Graphs of Quadratic Functions

Every quadratic function has certain key features, which we introduce here and study in more depth in the next few sections of this chapter.

The shape of the graph   The graph of a quadratic function is called a *parabola*. All parabolas share a distinctive bowl-like shape, some are wide, some are narrow, some open upward *(concave up)* and some open downward *(concave down)* (Figure 11.1). Although our graphs can only show part of a parabola, the "arms" of a parabola extend indefinitely upward or downward.

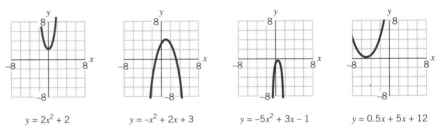

| $y = 2x^2 + 2$ | $y = -x^2 + 2x + 3$ | $y = -5x^2 + 3x - 1$ | $y = 0.5x + 5x + 12$ |

Figure 11.1   Graphs of four different quadratic functions.

When will a quadratic function have a maximum value?

The vertex and axis of symmetry   Each parabola has a *vertex* that is the lowest or the highest point on the curve, depending on whether the parabola opens upward or downward. For example, when a parabola opens upward, the vertex is the lowest point on the parabola. This point has the minimum vertical value and thus represents the minimum value of the function. We say the function has a *minimum* at the vertex.

Now imagine drawing a vertical line through the vertex and folding the parabola along that line. The right half of the curve will fall exactly on the left half. The right half of the curve is a mirror image of the left half. The vertical line of the fold is called the *axis of symmetry* (Figure 11.2). The graph is said to be *symmetrical* about its axis of symmetry. If the vertex has coordinates (d, e), then the equation for the axis of symmetry is $x = d$.

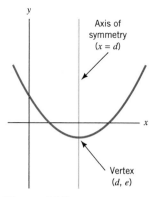

Figure 11.2   Each parabola has a vertex that lies on an axis of symmetry.

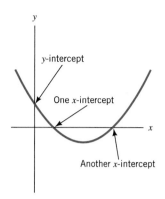

Figure 11.3   Each parabola has a vertical intercept and possibly horizontal intercepts.

Horizontal and vertical intercepts   While each parabola always has a vertical, or *y*-intercept, it may or may not have any horizontal, or *x*-intercepts. (See Figures 11.1 and 11.3.)

Something to think about

How many horizontal intercepts could a parabola have?

Algebra Aerobics 11.1a

Using a graphing tool, plot each of the following quadratic functions. Estimate the coordinates of the vertex for each parabola and give the equation of the axis of symmetry. Is the parabola concave up or concave down? Specify the number of *x*-intercepts and identify the *y*-intercept.

1. $y = -x^2 + 2$
2. $y = x^2 + 2x + 1$
3. $y = 0.5x^2 - 2x + 3$

The Effect of the Coefficient *a*
Case 1: *a* > 0

You may wish to do Exploration 11.1, Part I, in parallel with this section.

Given a quadratic function, $y = ax^2 + bx + c$, when *a* is positive, the parabola opens upward. The vertex is the lowest point on the curve and represents the minimum value of the function. What happens as we increase the size of *a*? We start by examining the simplest quadratic, $y = x^2$, where $a = 1$, and *b* and *c* are 0. Figure 11.4 shows the graph of $y = x^2$ as a parabola that opens upward. Now consider $y = ax^2$ when $a = 2$ and when $a = 4$. Examine the values in Table 11.1. Except when $x = 0$, larger values for the coefficient *a* will produce larger values for *y* for any given value of *x*.

Now compare the graphs of $y = x^2$, $y = 2x^2$, and $y = 4x^2$ in Figure 11.4. Each parabola opens upward, and the larger the value of *a*, the steeper the graph. Thus the graph of $y = 2x^2$ hugs the *y*-axis more closely than the graph of $y = x^2$. The graph of $y = 4x^2$ is even closer to the *y*-axis. Imagine the graph of $y = 3x^2$. It lies between the graphs of $y = 2x^2$ and $y = 4x^2$.

Table 11.1		
Quadratics with $a \geq 1$		
x	$y = x^2$	$y = 2x^2$	$y = 4x^2$
-2	4	8	16
-1	1	2	4
0	0	0	0
1	1	2	4
2	4	8	16

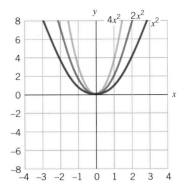

Figure 11.4 Parabolas with $a \geq 1$.

You can use "Q1: a, b, c Sliders" in Quadratic Functions *to explore the effects of a, b, and c.*

Try setting b and c to values other than 0, while keeping values of a greater than or equal to 1. Varying a still has the same effect. The parabola still opens upward and gets narrower as the value of a increases.

For example, suppose we set $b = 3$ and $c = 7$; then when we look at the quadratic function $y = 4x^2 + 3x + 7$, we know, before calculating any values or plotting any points, that since $a = 4$, the graph must open upward and would be narrower, for example, than the graph of $y = 2x^2 + 3x + 7$, since 4 is greater than 2.

What happens to the parabola if a is between 0 and 1? Again we can use the function $y = x^2$ as a basis of comparison. Using Table 11.2 and Figure 11.5, we see that all the parabolas still open upward and the vertex is the lowest point on the curve, but the smaller the value of the coefficient a, the more the parabola flattens out. The parabola for $y = 0.5x^2$ is broader than that for $y = x^2$ and the graph of $y = 0.25x^2$ lies even closer to the x-axis.

Table 11.2		
Quadratics with $0 < a \leq 1$		
x	$y = x^2$	$y = 0.5x^2$	$y = 0.25x^2$
-2	4	2.0	1.00
-1	1	0.5	0.25
0	0	0.0	0.00
1	1	0.5	0.25
2	4	2.0	1.00

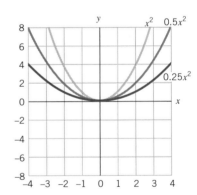

Figure 11.5 Parabolas with $0 < a \leq 1$.

Try setting b and c to values other than 0. Does varying the values of a between 0 and 1 still have the same effect? The answer is yes.

In general, when a is positive the graph of $y = ax^2 + bx + c$ opens upward, the vertex is the lowest point on the curve, and the larger the value of the coefficient a, the narrower the parabola.

Case 2: $a < 0$

Examine the parabolas in Figure 11.6 where $a < 0$.

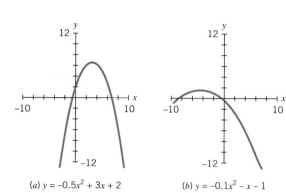

(a) $y = -0.5x^2 + 3x + 2$ (b) $y = -0.1x^2 - x - 1$ (c) $y = -0.25x^2 - 2$

Figure 11.6   Parabolas with $a < 0$.

Something to think about

Using "Q1: a, b, c Sliders" in *Quadratic Functions* in the course software, can you describe the movement of the vertex of the original parabola as you change the value for b, but hold a and c fixed?

When a is negative, the parabola opens downward and the vertex is the highest point on the curve, representing the *maximum* value of the function. When $|a|$, the absolute value of a, increases, the curve narrows.

The Effect of the Constant Term c

Changing c changes only the vertical position of the graph, not its shape (Figure 11.7). The parabola $y = x^2 + 4$ is raised four units above the graph of $y = x^2$. Similarly, the graph of $y = x^2 - 4$ is four units below the graph of $y = x^2$. The constant term c has the same effect for any value of a and b.

Think of c as an "elevator" term. Without changing the shape of the curve, increasing the value of c shifts the parabola up and decreasing the value of c shifts the parabola down.

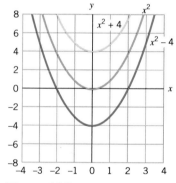

Figure 11.7   Parabolas with different values for c.

The graph of a quadratic function $y = ax^2 + bx + c$

- is called a *parabola*
- has a bowl-like shape
- has a lowest or highest point called the *vertex*
- is symmetric about a vertical line, called the *axis of symmetry,* that runs through the vertex
- opens up if $a > 0$ and down if $a < 0$
- becomes narrower as $|a|$, the absolute value of a, is increased
- maintains its shape, but is shifted up if c is increased and down if c is decreased

Algebra Aerobics 11.1b

1. Without drawing the graph, describe whether the graph of each of the functions in parts (a)–(d)

    i. has a maximum or minimum at the vertex.

    ii. is narrower or broader than $y = x^2$.

    a. $y = 2x^2 - 5$

    b. $y = 0.5x^2 + 2x - 10$

    c. $y = 3 + x - 4x^2$

    d. $y = -0.2x^2 + 11x + 8$

2. Without drawing the graph, list these parabolas in order, from the narrowest to the broadest.

    a. $y = x^2 + 20$

    b. $y = 0.5x^2 - 1$

    c. $y = \dfrac{1}{3}x^2 + x + 1$

    d. $y = 4x^2$

    e. $y = 0.1x^2 + 2$

    f. $y = -2x^2 - 5x + 4$

Without drawing the graphs, in Problems 3–7, compare the graph of (b) to the graph of (a).

3. **a.** $y = x^2 + 2$ **b.** $y = 2x^2 + 2$

4. **a.** $f(x) = x^2 + 3x + 2$ **b.** $g(x) = x^2 + 3x + 8$

5. **a.** $d = t^2 + 5$ **b.** $d = -t^2 + 5$

6. **a.** $f(z) = -5z^2$ **b.** $g(z) = -0.5z^2$

7. **a.** $h = -3t^2 + t - 5$ **b.** $h = -3t^2 + t - 2$

11.2 FINDING THE VERTEX

Why the Vertex Is Important

Maximum and minimum values   Since the vertex is the point at which a parabola reaches a maximum or minimum value, the vertex often assumes particular significance in a model. For example, biologists have discovered a quadratic relationship between species diversity and ocean depth. The graph in Figure 11.8 shows the number of different species as a function of the depth of the water in cold northern oceans. As the depth increases, the number of species initially increases, reaches a maximum at the vertex (at a depth of approximately 1250 meters) and then decreases.[1]

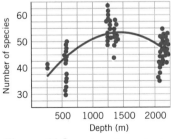

Source: Ron J. Etter and Frederick Grassle, "Patterns of Species Diversity in the Deep Sea as a Function of Sediment Particle Size Diversity," *Nature*, Vol. 360: pp. 576–578, Dec. 10, 1992. Reprinted with permission from Macmillan Magazines Limited.

Figure 11.8   A best fit quadratic function for species diversity as a function of depth in northern regions.

[1] According to the authors of the article, each point represents data from a single core sample with a cross section of 0.25 m². Each sample costs more than $100,000 to collect and analyze!

Urban planners and highway designers are interested in maximizing the number of cars that pass along a section of roadway in a certain amount of time. Observations indicate that the primary variable controlling traffic flow (hence a good choice for the independent variable) is the density of cars on the roadway; the closer each driver is to the car ahead, the more slowly he or she drives. The graph in Figure 11.9 shows a quadratic model of traffic flow rate (the number of cars per hour) as a function of the density (the number of cars per mile). It was derived from observing traffic patterns in the Lincoln Tunnel, which connects New York and New Jersey.

Example 1

Figure 11.9 The quadratic relationship between traffic flow rate and density.

Source: Adapted from G.B. Whitman, *Linear and Nonlinear Waves.* New York: John Wiley, 1974, p. 68.

We can estimate the domain (possible values for the independent variable density) to lie between 0 and 165 cars per mile. The range (possible values for the dependent variable traffic flow) appears to lie between 0 and 1430 cars per hour. The function increases until the density equals about 83 cars per mile, at which point the traffic flow appears to reach a maximum rate of 1430 cars per hour. This point, (83, 1430), corresponds to the vertex of the parabola. Note that at 83 cars per mile or equivalently 83 cars per 5280 feet, cars are spaced about 64 feet apart (since $64 \approx 5280/83$). This spacing apparently maximizes the traffic flow, the number of cars per hour. When the density is either below or above 83 cars per mile, the traffic flow is less than 1430 cars per hour.

Using this model, what predictions would you make about the traffic flow when the density is approximately 165 cars per mile?

Finding the Vertex of a Parabola

Finding the vertex when $b = 0$ If $b = 0$, then a quadratic function is of the form $g(x) = ax^2 + c$. Compare the vertex of $g(x)$ to the vertex of the basic quadratic function $f(x) = x^2$. The vertex of $f(x)$ represents a minimum since a is positive. We also know that $f(x) = x^2$ is greater than or equal to 0 and that $f(0) = 0$. So the minimum value for $f(x)$ is 0, and thus the vertex is at (0, 0), the origin. The graph of $g(x)$ is the graph of $f(x)$ that has been narrowed or flattened by a factor of a, which does not change the vertex, and then elevated c units, which does change the vertex. So the vertex of $g(x)$ is the vertex of $f(x)$ vertically shifted c units: that is, the vertex of $g(x)$ is (0, c). For example, the vertex of $j(x) = -3x^2$ is at (0, 0), the vertex of $g(x) = -3x^2 + 5$ is at (0, 5), and the vertex of $h(x) = 2x^2 - 7$ is at (0, -7).

For cases in which b may not equal 0, we can use a general formula to determine the vertex.

A general formula for the vertex   The following formula can be used to find the vertex of any parabola. Section 11.7 explains why this formula works.

> The *vertex* of the quadratic function
> $$y = ax^2 + bx + c$$
> has coordinates $\left(-\dfrac{b}{2a},\ -\dfrac{b^2}{4a} + c \right)$

Example   2

Find the vertex for the quadratic height function $f(t) = 34 + 32t - 16t^2$, where t is in seconds and height is in feet. What significance does the vertex have? Sketch the graph.

SOLUTION
Remember a is the coefficient of the squared term, which in this case happens to be the third term. So $f(t) = c + bt + at^2$ and $a = -16$, $b = 32$, and $c = 34$. Since a is negative, we know that the graph opens downward, so the vertex represents a maximum value.

Using the formula for the horizontal coordinate of the vertex, we have

$$-\frac{b}{2a} = -\frac{32}{2(-16)} = \frac{-32}{-32} = 1$$

Using the formula for the vertical coordinate of the vertex, we get

$$-\frac{b^2}{4a} + c = -\frac{32^2}{4(-16)} + 34 = \frac{-1024}{-64} + 34 = 16 + 34 = 50$$

So the vertex is at $(1, 50)$.

Since the formula for the vertical coordinate is pretty complicated to remember, you may prefer to find the vertical coordinate of the vertex by evaluating $f(t) = 34 + 32t - 16t^2$ at the appropriate value of the horizontal coordinate, which in this case is $t = 1$. Setting $t = 1$,

$$f(1) = 34 + 32(1) - 16(1)^2$$
$$= 34 + 32 - 16$$
$$= 50$$

As we would expect, we obtain the same result. Figure 11.10 shows the function's graph.

The vertex represents the point at which the object reaches a maximum height. So at 1 second, the object reaches a maximum height of 50 feet.

Note that our height model consists only of the solid part of the curve. Values of negative time or negative height do not have meaning in this problem.

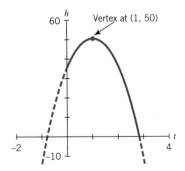

Figure 11.10   Graph of the function $f(t) = 34 + 32t - 16t^2$.

Find the vertex and sketch the graph of $f(x) = x^2 - 10x + 100$.

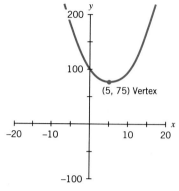

Example 3

SOLUTION

Here $a = 1$, $b = -10$, and $c = 100$. Since the coefficient a is positive, we know that the graph opens upward and the vertex represents a minimum value.

Using the formula for the horizontal coordinate of the vertex, we have

$$x = -\frac{b}{2a} = -\frac{(-10)}{2(1)} = \frac{10}{2} = 5$$

To find the vertical coordinate we can find the value of $f(x)$ when $x = 5$.

given $f(x) = x^2 - 10x + 100$

let $x = 5$ $f(5) = 5^2 - 10(5) + 100$

simplify $f(5) = 75$

The coordinates of the vertex are (5, 75). Figure 11.11 shows a sketch of the graph.

Figure 11.11   Graph of $y = x^2 - 10x + 100$.

Algebra Aerobics 11.2

1. Find the vertex of the graph of each of the following quadratic functions.
   a. $f(x) = 2x^2 - 4$ c. $w = 4t^2 + 1$
   b. $g(z) = -z^2 + 6$

2. Find the vertex of the graph of each of the following functions and then sketch the graphs on the same grid.
   a. $y = x^2 + 3$ b. $y = -x^2 + 3$

3. Find the vertex and do a rough sketch of the graph of each of the following functions.
   a. $y = x^2 + 3x + 2$ c. $g(t) = -t^2 - 4t - 7$
   b. $f(x) = 2x^2 - 4x + 5$

11.3 FINDING THE INTERCEPTS

In a quadratic model, the horizontal and vertical intercepts often convey important information. Consider the parabola shown in Figure 11.12 where the solid part of the graph models the height of a thrown object as a function of time. The graph is similar to Figure 11.10. The vertical intercept tells us the starting height, and the horizontal intercept within the domain of the model tells us when the object has hit the ground.

Finding the y- (or Vertical) Intercept

The y-intercept is the point at which the graph of a function $y = f(x)$ crosses the y-axis. It corresponds to the value for y when $x = 0$. To find the y-intercept of the quadratic function $y = f(x)$ where $f(x) = -0.5x^2 + 3x + 2$, we would evaluate $f(x)$ when $x = 0$.

$$f(0) = -0.5(0)^2 + 3(0) + 2$$
$$f(0) = 2$$

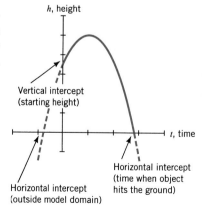

Figure 11.12   Height of a thrown object as a function of time.

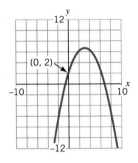

$$y = -0.5x^2 + 3x + 2$$

Figure 11.13   The y-intercept indicates where the function crosses the y-axis.

So the y-intercept is $(0, 2)$. Since the x-coordinate is zero, we often say simply that the y-intercept is 2. The graph is shown in Figure 11.13.

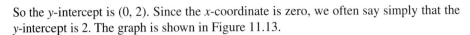

To find a general expression for the vertical intercept for the function $f(x) = ax^2 + bx + c$, we evaluate $f(x)$ when $x = 0$.

$$f(x) = a(0)^2 + b(0) + c$$
$$= c$$

The y-intercept is the point $(0, c)$ but is often simply referred to as c.

Finding the x- (or Horizontal) Intercepts

The x-intercepts are the points where the graph of a function $y = f(x)$ crosses the horizontal axis. Since the y-coordinate of an x-intercept is always 0, we often refer to an x-intercept by its x-coordinate only. For example, if a function has an x-intercept at $(5, 0)$, we often abbreviate this to say the x-intercept is 5. A quadratic function may have no, one, or two x-intercepts, as Figure 11.14 illustrates.

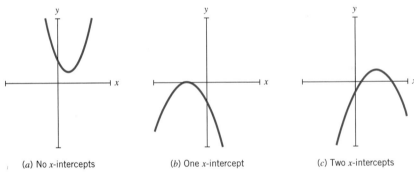

(a) No x-intercepts   (b) One x-intercept   (c) Two x-intercepts

Figure 11.14   Graphs of quadratic functions showing three possible cases for the number of x-intercepts.

Predicting the number of x-intercepts   If we know the vertex and whether a parabola opens upward or downward, we can predict the number of horizontal intercepts. For example, the graph of the function

$$f(x) = x^2 + 4x + 9$$

opens upward since the coefficient of x^2 is positive. From the formulas in Section 11.2 we can compute a vertex of $(-2, 5)$. Since the point $(-2, 5)$ lies above the x-axis and the parabola opens upward, we know that there are no x-intercepts, as the sketch in Figure 11.15 shows.

Estimating x-intercepts with a function graphing program   If you have access to a function graphing program on a computer or calculator, you can use it to approximate the x-intercepts of a quadratic.

For example, if we plot the function $g(x) = -1.7x^2 + 2.3x + 0.9$ (see Figure 11.16), we can see that the resulting graph has two x-intercepts, one between -1 and 0 and another between 1 and 2. By zooming in on the graph or using special features (such as *trace* or *root finder*), a calculator can estimate the two x-coordinates where

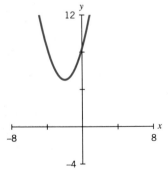

Figure 11.15   Graph of $f(x) = x^2 + 4x + 9$.

Figure 11.16 Graph of $g(x) = -1.7x^2 + 2.3x + 0.9$.

the curve crosses the x-axis. The two values for this graph are approximately $x = -0.32$ and $x = 1.67$.

The technological approach gives good estimates, but they may not be exact. Using equations it is possible to find the precise values of any intercepts.

Algebra Aerobics 11.3a

1. Find the y-intercept of each parabola.
   a. $y = 2x - 3x^2$
   b. $y = 5 - x - 4x^2$
   c. $y = \dfrac{2}{3}x^2 + 6x - \dfrac{11}{3}$

2. Find and interpret the vertical h-intercept for the graph of the given height equations.
   a. $h = -4.9t^2 + 50t + 80$ (h is in meters, and t is in seconds)

2. b. $h = 150 - 80t - 490t^2$ (h is in centimeters, and t is in seconds)

3. In parts (a)–(d) determine the vertex and whether the graph opens upward or downward. Then predict the number of x-intercepts. Graph the function to confirm your answer. Estimate the values of the x-intercepts.
   a. $f(x) = x^2 + 4x - 7$ c. $y = -3x^2$
   b. $y = 4 - x - 2x^2$ d. $y = -2x^2 - 5$

Using Equations to Find the x-Intercepts: The Quadratic Formula

When we are looking for x-intercepts, we are asking "What value(s) of x makes the function $f(x)$ zero?" To find the x-intercepts of $f(x) = ax^2 + bx + c$, we set $f(x) = 0$ and solve for x. The solutions to the equation $0 = ax^2 + bx + c$ are called the *roots* of the equation. We can use the quadratic formula to find the roots. We state the formula here and prove it in Section 11.7.

The Quadratic Formula

For any quadratic equation of the form $0 = ax^2 + bx + c$, the solutions or *roots* of the equation are given by

$$x = \frac{-b \pm \sqrt{b^2 - 4ac}}{2a}$$

The term under the radical sign, $b^2 - 4ac$, is called the *discriminant*.

The symbol $\pm$ lets us write the two roots at

$$x = \frac{-b + \sqrt{b^2 - 4ac}}{2a} \qquad \text{and} \qquad x = \frac{-b - \sqrt{b^2 - 4ac}}{2a}$$

with one formula.

Getting the terminology straight   A value of x that makes the value of the function $f(x)$ equal to 0 is called a *zero of the function*. The zeros of a function $f(x)$ are the *roots of the equation* $f(x) = 0$. Every x-intercept on the graph of $f(x)$ represents a zero of the function. For example, to find the zeros of the function $f(x) = x^2 - 9$, we set $f(x) = 0$ and find the roots of the equation $0 = x^2 - 9$. The roots occur when $x = +3$ and $x = -3$. These roots are the zeros and also the x-intercepts of $f(x)$.

Example 1

a. What are the horizontal intercepts (if any) for the quadratic height function $h = 34 + 32t - 16t^2$ where t is in seconds and h is in feet? We saw this function before in Example 2, Section 11.2.

b. What significance do the intercepts have?

SOLUTION

a. Here $a = -16$, $b = 32$, and $c = 34$. Since a is negative, we know that the graph opens downward. In Section 11.2 we found the vertex to be at $(1, 50)$, which is above the t-axis. So the function must have two t-intercepts. To find these intercepts, we set $h = 0$ and solve the resulting equation $0 = 34 + 32t - 16t^2$. We substitute the values of a, b, and c into the quadratic formula.

Use the quadratic formula
$$t = \frac{-b \pm \sqrt{b^2 - 4ac}}{2a}$$

Substitute for a, b, and c
$$= \frac{-32 \pm \sqrt{32^2 - 4(-16)(34)}}{(2)(-16)}$$

$$= \frac{-32 \pm \sqrt{1024 + 2176}}{-32}$$

(note that the discriminant = 3200, a positive real number)
$$= \frac{-32 \pm \sqrt{3200}}{-32}$$

Evaluate with a calculator
$$t \approx \frac{-32 \pm 56.6}{-32}$$

So there are two roots,

one at
$$\frac{-32 + 56.6}{-32} = \frac{24.6}{-32} \approx -0.77$$

and the other at
$$\frac{-32 - 56.6}{-32} = \frac{-88.6}{-32} \approx 2.77$$

Therefore the parabola crosses the horizontal axis at approximately $(-0.77, 0)$ and $(2.77, 0)$, as shown in Figure 11.17.

b. The t-intercept at $(-0.77, 0)$ lies outside the model, since it represents a negative value for time, t. The positive t-intercept says that when $t = 2.77$ seconds, $h = 0$ feet. In other words, at 2.77 seconds the object hit the ground.

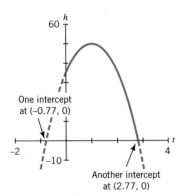

One intercept at (–0.77, 0)

Another intercept at (2.77, 0)

Figure 11.17 Graph of the equation $h = 34 + 32t - 16t^2$, with two horizontal intercepts.

Find the x-intercepts of $f(x) = x^2 + 3x + 2.25$.

Example 2

SOLUTION

Here $a = 1$, $b = 3$, and $c = 2.25$. We can find the x-intercepts by solving the equation $0 = x^2 + 3x + 2.25$. The quadratic formula says that the solutions occur at

$$x = \frac{-b \pm \sqrt{b^2 - 4ac}}{2a}$$

$$= \frac{-3 + \sqrt{(3)^2 - 4(1)(2.25)}}{(2)(1)}$$

$$= \frac{-3 + \sqrt{9 - 9}}{2}$$

$$= \frac{-3 + \sqrt{0}}{2} \qquad \text{(note that the discriminant} = 0)$$

$$= \frac{-3}{2}$$

$$x = -1.5$$

In this case, the quadratic formula produces only one value, 1.5. So there is only one x-intercept at $(-1.5, 0)$, which is also the vertex of the parabola (see Figure 11.18).

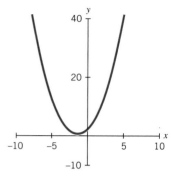

Figure 11.18 Graph of $f(x) = x^2 + 3x + 2.25$, with one x-intercept at the vertex.

Find the x-intercepts of the function $f(x) = -x^2 - 6x - 10$.

Example 3

SOLUTION

We can find the intercepts of the function by using the quadratic formula to solve the equation $0 = -x^2 - 6x - 10$. In this case $a = -1$, $b = -6$, and $c = -10$. Thus, we have

$$x = \frac{-b \pm \sqrt{b^2 - 4ac}}{2a}$$

$$= \frac{-(-6) + \sqrt{(-6)^2 - 4(-1)(-10)}}{(2)(-1)}$$

$$= \frac{6 \pm \sqrt{36 - 40}}{-2}$$

$$x = \frac{6 \pm \sqrt{-4}}{-2}$$

Here the discriminant, -4, is negative, so taking its square root presents a problem. $\sqrt{-4}$ is not a real number since there is no real number, r, such that $r^2 = -4$. Therefore the solutions or roots $\dfrac{6 + \sqrt{-4}}{-2}$ are not real. Since there are no real values for x such that $f(x) = 0$, there are no x-intercepts, as we can see in Figure 11.19.

Imaginary and Complex Numbers

Mathematicians were uncomfortable with the notion that certain quadratic equations

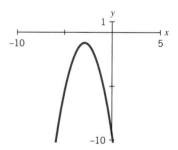

Figure 11.19 Graph of $y = -x^2 - 6x - 10$ with no x-intercepts.

did not have solutions. So they literally invented a number system in which such equations would be solvable. They created new numbers, called *imaginary numbers,* that were used to extend the real number system to a larger system, called the *complex numbers.*

A number such as $\sqrt{-4}$ is called *imaginary.* We can also write $\sqrt{-4}$ as $\sqrt{(4)(-1)} = \sqrt{4} \cdot \sqrt{-1} = 2\sqrt{-1}$. The number $\sqrt{-1}$ is a special imaginary number called *i.* We can write $\sqrt{-4}$ as $2i$. Any *imaginary number* has the form $a + bi$, where a and b are real numbers and $b \neq 0$. For example:

$$-2 + 7i$$
$$4 + \sqrt{-9} = 4 + 3\sqrt{-1} = 4 + 3i$$
$$\sqrt{-25} = 5\sqrt{-1} = 5i = 0 + 5i$$

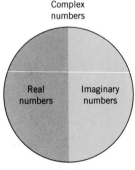

Complex numbers

Real numbers   Imaginary numbers

Figure 11.20   The real numbers together with the imaginary numbers form the set of complex numbers.

are all imaginary numbers.

If we remove the constraint that b be nonzero, then we have the larger set of *complex numbers;* that is, the complex numbers consist of all numbers of the form $a + bi$ where a and b are real numbers. Any real number a can be written as the complex number $a + 0i$. For example, $11 = 11 + 0i$. So the real numbers are a subset of the complex numbers.

Every complex number, $a + bi$, is either real (if $b = 0$) or imaginary (if $b \neq 0$). The complex numbers consist of all the real numbers and all the imaginary numbers. (See Figure 11.20.)

Solutions to quadratic equations may be real or imaginary   In Example 3, the zeros of the function $f(x) = -x^2 - 6x - 10$, that is the values of x for which $f(x) = 0$, are both imaginary numbers. Their values are

$$x = \frac{6 + \sqrt{-4}}{-2} = \frac{6}{(-2)} + \frac{\sqrt{-4}}{(-2)} = -3 - \frac{2i}{2} = -3 - i$$

and

$$x = \frac{6 - \sqrt{-4}}{-2} = \frac{6}{(-2)} - \frac{\sqrt{-4}}{(-2)} = -3 + \frac{2i}{2} = -3 + i$$

So for $f(x) = -x^2 - 6x - 10$, there is no real number x such that $f(x) = 0$, and hence its graph has no x-intercepts. But there are two imaginary numbers, $-3 - i$ and $-3 + i$, such that $f(-3 - i) = 0$ and $f(-3 + i) = 0$. Hence $f(x)$ has no real zeros, but does have 2 imaginary zeros. The number of distinct real zeros determines the number of x-intercepts.

To find the zeros of the function $f(x) = ax^2 + bx + c$, we use the quadratic formula to solve the equation $0 = ax^2 + bx + c$. The discriminant $b^2 - 4ac$, the term under the radical, can be used to determine the number of real roots and hence the number of x-intercepts.

If the discriminant $b^2 - 4ac = 0$, there is only one real root at

$$x = \frac{-b}{2a}$$

Explorations 11.2 and 11.3 will give you practice in working with quadratic functions that model physical phenomena.

and hence the graph has one x-intercept. Since this only happens when the vertex is on the x-axis, $\dfrac{-b}{2a}$ is also the x-coordinate of the vertex. (See Example 2, page 393.)

If the discriminant $b^2 - 4ac > 0$, then $\sqrt{b^2 - 4ac}$ is a real number, which means that there are two real roots, and hence two x-intercepts. (See Example 1.)

If the discriminant $b^2 - 4ac < 0$, then $\sqrt{b^2 - 4ac}$ is not a real number, which means that there are no real roots, and the graph has no x-intercepts. (See Example 3.)

In summary,

To find the x-intercepts of a quadratic function

$$f(x) = ax^2 + bx + c$$

we use the quadratic formula to find the solutions or *roots* of the associated equation

$$0 = ax^2 + bx + c$$

If $b^2 - 4ac > 0$, there are two real roots and the graph has two x-intercepts.

If $b^2 - 4ac = 0$, there is one real root and the graph has one x-intercept.

If $b^2 - 4ac < 0$, there are no real roots and the graph has no x-intercepts.

Algebra Aerobics 11.3b

1. Evaluate the discriminant and then predict the number of x-intercepts for each function. Use the quadratic formula to find all the zeros (real and imaginary) of each function and identify the coordinates of any x-intercept(s).

   a. $y = 4 - x - 5x^2$ **c.** $y = 2x^2 + 5x + 4$

   b. $y = 4x^2 - 28x + 49$

2. Find the vertex and y-intercept of the functions in Problem (1) and sketch the graphs.

3. Return to Figure 11.1 (on p. 382) and for each parabola identify the number of real and the number of imaginary zeros.

The relationship between the quadratic formula and the formula for the vertex The vertex for the quadratic function $f(x) = ax^2 + bx + c$ lies on its axis of symmetry. The vertex formula tells us that the x-coordinate of the vertex is $-b/(2a)$. So the equation for the axis of symmetry is $x = -b/(2a)$. If the discriminant, $b^2 - 4ac$, is positive, we have seen that there are two x-intercepts. The quadratic formula tells us that these intercepts are located at

$$x = \frac{-b \pm \sqrt{b^2 - 4ac}}{2a}$$

We can regroup terms and write

$$x = \frac{-b}{2a} \pm \frac{\sqrt{b^2 - 4ac}}{2a}$$

See "Q6: Roots a-b-c Form"
in Quadratic Functions.

By letting $d = \dfrac{\sqrt{b^2 - 4ac}}{2a}$ we can simplify the x-intercepts of the function as

$$x = \frac{-b}{2a} \pm d$$

The two x-intercepts are located at

$$\frac{-b}{2a} + d \qquad \text{and} \qquad \frac{-b}{2a} - d$$

and hence are equally spaced d units to the left and right of the axis of symmetry, the line at $x = -b/(2a)$.

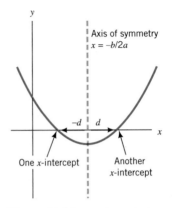

Figure 11.21   Each parabola has a vertex that lies on an axis of symmetry.

If $b^2 - 4ac = 0$, then $d = 0$, so there is only one x-intercept at $(-b/(2a), 0)$ which is also the vertex of the function.

11.4 DIFFERENT FORMS OF THE QUADRATIC FUNCTION

There are three major forms in which quadratic functions can be written. In this and the next two sections we will see that all three of the forms are equivalent, but each emphasizes different characteristics of the function.

The a-b-c Form

A quadratic function written $f(x) = ax^2 + bx + c$ is said to be in a-b-c form. What can we determine easily from a quadratic function written in this way? The value of a (the coefficient of the x^2 term) tells us whether the graph opens upward (if $a > 0$) and therefore has a minimum at the vertex, or opens downward (if $a < 0$) and has a maximum at the vertex. The value of c tells us the y-intercept. We can determine the coordinates of the vertex by calculating the values $\left(-\dfrac{b}{2a}, -\dfrac{b^2}{4a} + c \right)$ and determine the values of any x-intercepts by using the quadratic formula.

The Factored Form

A quadratic function in factored form, such as $f(x) = (2x + 5)(x - 3)$ allows us to identify the x-intercepts (or zeros) quickly. In Section 11.5 we'll see that the x-intercepts occur when $2x + 5 = 0$ and $x - 3 = 0$ or (by solving both equations) when $x = -2.5$ and $x = 3$.

The a-h-k Form

There is another convenient form for writing quadratic functions that allows us easily to identify not only whether a quadratic function has a maximum or minimum at its vertex, but also the specific coordinates of the vertex. A quadratic function written

$$f(x) = a(x - h)^2 + k$$

is said to be in a-h-k form. In Section 11.6 we'll see that the values of h and k tell us the location of the vertex; that is, the vertex has coordinates (h, k).

11.5 THE FACTORED FORM OF THE QUADRATIC FUNCTION

A quadratic function written as the product of two terms both involving the independent variable is said to be in *factored form*. Most quadratic functions cannot be written as the product of factors with integer coefficients. But when a function is factored, the horizontal intercepts are easy to find. For example, the following function is in factored form.

$$f(x) = (x - 4)(x + 5)$$

To find the x-intercepts, set $f(x) = 0$ to get

$$0 = (x - 4)(x + 5) \tag{1}$$

and solve for x.

In Equation (1), we have the product of two terms, $(x - 4)$ and $(x + 5)$, equal to 0. The *zero product rule* tells us that whenever the product of terms equals 0, one or more of the terms must equal 0.

Zero Product Rule

For any two real numbers r and s, if the product $rs = 0$, then r or s or both must equal 0.

We can apply the zero product rule to Equation (1) to get

$$x - 4 = 0 \qquad \text{or} \qquad x + 5 = 0$$
$$x = 4 \qquad\qquad\qquad x = -5$$

The x-intercepts are at 4 and -5, or equivalently the function crosses the x-axis at $(4, 0)$ and $(-5, 0)$.

To change between the *a-b-c* form of the quadratic function and the factored form, some knowledge of factoring and multiplying binomials is necessary. We offer a short review of these skills.

Factoring Review

Multiplying binomials   When we multiply binomials, we use the distributive law. For example, to multiply $(x + 2)(x + 5)$, we need to multiply each term in the first expression by each term in the second expression.

apply distributive law

apply distributive law again

simplify

$$(x + 2)(x + 5) = x(x + 5) + 2(x + 5)$$
$$= x^2 + 5x + 2x + 10$$
$$= x^2 + 7x + 10$$

We say that $x^2 + 7x + 10$ is the *product* of $(x + 2)$ and $(x + 5)$ or that $(x + 2)$ and $(x + 5)$ are *factors* of $x^2 + 7x + 10$.

We can generalize to any two binomials.

$$(a + b)(c + d) = a(c + d) + b(c + d)$$
$$= ac + ad + bc + bd$$

One strategy for remembering all the terms in the product (though in a slightly different order) is to consider those 4 terms as products of the first (F), outside (O), inside (I), and last (L) two terms of the factors.

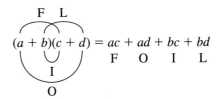

Factoring quadratics   If we know only the product, to find the factors requires thinking, practice, and a few hints. It is often a trial-and-error process. We usually restrict ourselves to finding factors with integer coefficients.

First, look for common factors in all of the terms.

For example, $10x^2 + 2x$ can be factored as $2x(5x + 1)$

Second, if factoring a trinomial, look for two binomial factors.

This is easiest to do when the coefficient of x^2 is 1. For example, to factor $x^2 + 7x + 12$ we want to rewrite it as

$$(x + m)(x + n)$$

for some m and n. Note that the coefficients of both x's in the factors equal 1, since x times x is equal to the x^2 in the original expression.

Now we need to determine values for the constants m and n. We know that when we multiply $m \cdot n$ we need to get 12. So we consider pairs of integers whose product is 12, namely, 1 and 12, or 2 and 6, or 3 and 4. All of the signs are positive, so we know that all the signs in each linear factor are positive. We can then narrow our list of factors of 12 to those whose sum equals 7, the coefficient of the x term. Only the factors 3 and 4 fit this criterion. So substituting 3 for m and 4 for n, we can factor our polynomial.

$$(x + 3)(x + 4) = x^2 + 7x + 12$$

You can check that these factors work by multiplying them out.

Third, when factoring a binomial, look for the special case of the difference of two squares. In this case the middle terms cancel out when multiplying.

$$\begin{aligned} x^2 - 25 &= (x - 5)(x + 5) \\ &= x^2 - 5x + 5x - 25 \\ &= x^2 - 25 \end{aligned}$$

In general,

$$x^2 - n^2 = (x - n)(x + n)$$

We can use these strategies to see if a quadratic function can be easily factored. If a quadratic function is in factored form, we can then easily identify the horizontal intercepts.

Write the function $f(x) = 5700x^2 + 3705x$ in factored form. Identify the x- and y-intercepts. Find the vertex, and graph the function.

Example 1

SOLUTION

given $\qquad f(x) = 5700x^2 + 3705x$
factor out $15x \qquad f(x) = 15x(380x + 247)$

and we have $f(x)$ in factored form.

To find the x-intercepts of the function, set $f(x)$ equal to 0 and solve the resulting equation.

given $\qquad\qquad\qquad f(x) = 15x(380x + 247)$
set $f(x) = 0 \qquad\qquad\qquad 0 = 15x(380x + 247)$

apply zero product rule to get $\qquad 15x = 0 \quad$ or $\quad 380x + 247 = 0$
solve for $x \qquad\qquad\qquad\qquad x = 0 \qquad\qquad 380x = -247$
$$x = -247/380$$
$$x = -0.65$$

The function crosses the x-axis when $x = -0.65$ and when $x = 0$, or equivalently at the points $(-0.65, 0)$ and the origin $(0, 0)$. The origin is also the y-intercept. The parabola opens up (since the coefficient of x^2 is positive) and crosses the x-axis twice, so we can assume that the vertex is below the x-axis. (See Figure 11.22.) The x-coordinate of the vertex equals $\dfrac{-b}{2a} = \dfrac{-3705}{2(5700)} = \dfrac{-3705}{11400} = -0.325$. Note that this value lies on the axis of symmetry exactly halfway between the two x-intercepts at -0.65 and 0. The y-coordinate of the vertex equals

$$f(-0.325) = 5700(-0.325)^2 + (3705)(-0.325) \approx -602.$$

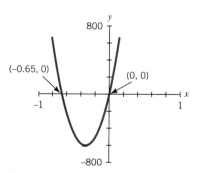

Figure 11.22 Graph of $f(x) = 5700x^2 + 3705x$.

Example 2

Write the function $y = -3x^2 + 12x - 12$ in factored form, identify the x- and y-intercepts and the vertex. Sketch the graph.

SOLUTION

given	$y = -3x^2 + 12x - 12$
pull out common factor of -3	$y = -3(x^2 - 4x + 4)$
factor the remaining trinomial	$y = -3(x - 2)(x - 2)$

and we have y in factored form.

To find the x-intercepts, set $y = 0$ and solve the equation for x.

given	$y = -3(x - 2)(x - 2)$
set $y = 0$	$0 = -3(x - 2)(x - 2)$
apply the zero product rule	$x - 2 = 0$
solve for x	$x = 2$

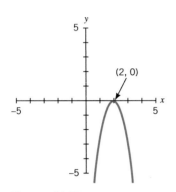

Figure 11.23 Graph of $y = -3x^2 + 12x - 12$.

The function $y = -3x^2 + 12x - 12$ intersects the x-axis only when $x = 2$. In this case there is only one x-intercept at $(2, 0)$ that is also the vertex of the parabola. (See Figure 11.23.) The parabola opens down, since $a = -3$ and it intersects the y-axis once at $(0, -12)$.

• • •

Note that, we can factor any quadratic function using the quadratic formula, though the factors may not have real, much less integer, coefficients. Since the quadratic formula gives the zeros, say r and s, of the function $f(x) = ax^2 + bx + c$ as

$$r = \frac{-b + \sqrt{b^2 - 4ac}}{2a} \qquad \text{and} \qquad s = \frac{-b - \sqrt{b^2 - 4ac}}{2a}$$

we can always factor $f(x)$ as

$$f(x) = a(x - r)(x - s)$$

Algebra Aerobics 11.5

When possible, put the function in factored form with integer coefficients, and find the x-intercepts.

a. $y = -16t^2 + 50t$

b. $y = x^2 + x - 6$

c. $y = 2x^2 + x - 5$

d. $h(t) = 69 - 9t^2$

e. $f(x) = -2x^2 + 12x + 54$

f. $g(x) = 64x^2 + 16x + 4$

11.6 LIFTS AND SHIFTS: UNDERSTANDING THE *a-h-k* FORM

Getting to the *a-h-k* Form: Parabolic Shifts

A quadratic function written as $y = a(x - h)^2 + k$ is said to be in the *a-h-k* form. How can we transform a function in *a-b-c* form into the *a-h-k* form? Why does a function written in *a-h-k* form have a vertex at (h, k)? Do the vertex coordinates (h, k) coincide with the vertex coordinates given earlier in terms of *a*, *b*, and *c*? The answers to these questions lie in vertical and horizontal shifts of a simple quadratic function of the form $f(x) = ax^2$.

You may wish to do Exploration 11.1, Part II, in parallel with this section.

Shifting a parabola horizontally h units   The simplest quadratic functions are power functions, such as

$$f(x) = 2x^2$$

where here $a = 2$, and *b* and *c* are both 0. The graph of this function is a parabola that opens upward (since the coefficient of x^2 is positive) and therefore has a minimum at its vertex. Since $f(x) \geq 0$ and when $x = 0$, we have $f(x) = 0$, then the vertex is at $(0, 0)$, the origin.

Suppose we compare $f(x) = 2x^2$ to a new function $g(x)$ where

$$g(x) = 2(x - 3)^2$$

"Q3: a, h, k Sliders" in Quadratic Functions can help you visualize quadratic functions in the a-h-k form.

The function $g(x)$ tells us to subtract 3 from *x*, square the result, and then multiply by 2. The graph of this function is the same as the graph of the original function $f(x) = 2x^2$, except that it is shifted *to the right* by 3 units.

To remember the direction of the shift, compare the positions of the vertices of $f(x)$ and $g(x)$. Since $g(x) \geq 0$, then its vertex represents a minimum that occurs when $g(x) = 0$. Since $g(3) = 2(3 - 3)^2 = 2 \cdot 0 = 0$, the vertex for $g(x)$ is at $(3, 0)$. So the vertex has been shifted 3 units to the right, from $(0, 0)$ for $f(x)$ to $(3, 0)$ for $g(x)$. (See Figure 11.24.)

What if we compare the function $f(x) = 2x^2$ to a new function $j(x)$ where

$$j(x) = 2(x + 5)^2$$

The function j(x) tells us to add 5 to x, square the result, and multiply by 2. The graph of this function is the same as the graph of the original function $f(x) = 2x^2$, except

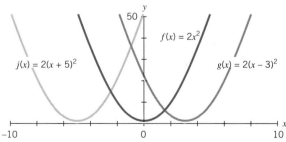

Figure 11.24   Graphs of $f(x) = 2x^2$, $g(x) = 2(x - 3)^2$, and $j(x) = 2(x + 5)^2$.

that it is shifted *to the left* by 5 units. The vertex of $j(x)$ is at $(-5, 0)$, or 5 units to the left of $(0, 0)$, the vertex of $f(x)$ (see Figure 11.24).

Figure 11.24 shows, on the same grid, the graphs of the functions

$$f(x) = 2x^2$$
$$g(x) = 2(x - 3)^2$$
$$= 2(x - (+3))^2 \qquad \text{graph of } f(x) \text{ shifted 3 units to the right}$$
$$j(x) = 2(x + 5)^2$$
$$= 2(x - (-5))^2 \qquad \text{graph of } f(x) \text{ shifted 5 units to the left}$$

In general, if we compare the function $f(x) = ax^2$ to a new function

$$g(x) = a(x - h)^2$$

the graph of $g(x)$ is the graph of $f(x)$ shifted horizontally h units. The shift will be to the right if h is positive and to the left if h is negative. In particular, the vertex of $g(x)$ at $(h, 0)$ is the vertex of $f(x)$ at $(0, 0)$ shifted h units.

Something to think about

How would the graph of $y = 2(x - 4)^3$ compare to the graph of $y = 2x^3$? What about the graph of $y = 2(x + 4)^3$?

Algebra Aerobics 11.6a

For each of the following problems, graph (a), (b), and (c) on the same grid. Compare the position of the vertex of (b) and of (c) to that of (a).

1. **a.** $y = x^2$ **c.** $y = (x - 2)^2$
    b. $y = (x + 3)^2$

2. **a.** $f(x) = 0.5x^2$ **c.** $j(x) = 0.5(x + 4)^2$
    b. $g(x) = 0.5(x - 1)^2$

3. **a.** $r = -2t^2$ **c.** $r = -2(t - 0.9)^2$
    b. $r = -2(t + 1.2)^2$

Something to think about

How do you think the graph of $y = 2(x - 4)^3 + 7$ compares to the graph of $y = 2x^3$? What about the graph of $y = 2(x - 4)^3 - 5$? Check your predictions by plotting the functions.

Shifting a parabola vertically k units   Suppose we wanted to shift the graph vertically, in the y direction. Recall from Section 11.1 that the constant term acts as an "elevator"; that is, in order to raise or lower a function we simply add a constant term. For example, the graph of the function $y = 2(x - 3)^2 + 10$ is the graph of $y = 2(x - 3)^2$ shifted up 10 units (Figure 11.25). The graph of $y = 2(x - 3)^2 - 5$ is the graph of $y = 2(x - 3)^2$ shifted down 5 units.

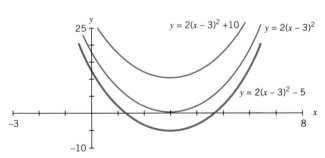

Figure 11.25   Three identically shaped parabolas, one 10 units above $y = 2(x - 3)^2$ and one 5 units below.

In general, to shift the function $f(x) = a(x - h)^2$ vertically k units, we add a constant term k. So the graph of the function

$$g(x) = a(x - h)^2 + k$$

is the same as the graph of $f(x)$ shifted vertically k units. The graph is shifted up if k is positive and down if k is negative. In particular, the vertex of $(h, 0)$ for $f(x)$ is vertically shifted k units to become (h, k), the vertex for $g(x)$.

Putting all this together we have:

For a graphic illustration of the shift from a-b-c to a-h-k form, see "Q7: From a-b-c to a-h-k Form" and "Q8: y = ax²
vs. y = a(x − h)² + k" in
Quadratic Functions.

A quadratic function in the *a-h-k form*

$$y = a(x - h)^2 + k$$

has a vertex at (h, k).

Its graph is the same as the graph of $y = ax^2$ that has been

shifted h units horizontally (to the right if $h > 0$, to the left if $h < 0$),

shifted k units vertically (up if $k > 0$, down if $k < 0$).

The quadratic function $g(z) = 11(z - 6)^2 - 2$ is in the *a-h-k* form, where $a = 11$, $h = 6$, and $k = -2$. Its vertex is at $(6, -2)$. Since the value of the coefficient a is 11, which is positive, the graph of $g(z)$ opens upward and the vertex represents a minimum.

Example 1

In the function $y = -3(x + 5)^2 + 10$, we have $a = -3$ and $k = 10$. We have to be careful in identifying the value for h. If we think of $x + 5$ as $x - (-5)$, then we have written the expression exactly in the $(x - h)$ format, and it is clearer that $h = -5$. The function then has a vertex at $(-5, 10)$. Since $a < 0$, the parabola opens downward and the vertex represents a maximum.

Example 2

Algebra Aerobics 11.6b

Graph the four parabolas on the same coordinate plane, and then compare the graphs of (b), (c), and (d) to that of (a).

1. **a.** $y = x^2$
    b. $y = (x - 2)^2$
    c. $y = (x - 2)^2 + 4$
    d. $y = (x - 2)^2 - 3$

2. **a.** $y = -x^2$
    b. $y = -(x + 3)^2$
    c. $y = -(x + 3)^2 - 1$
    d. $y = -(x + 3)^2 + 4$

The Relationship Between the *a-h-k* and the *a-b-c* Forms

A quadratic function in the *a-b-c* form

$$y = ax^2 + bx + c$$

may be written as a shifted parabola in the *a-h-k* form

$$y = a(x - h)^2 + k$$

by setting $h = -\dfrac{b}{2a}$ and $k = -\dfrac{b^2}{4a} + c$

The vertex of the graph is located at coordinates (h, k) and the graph is symmetrical about the vertical line $x = h$ passing through the vertex.

Converting from the a-h-k *to the* a-b-c *form of a quadratic*   Every function written in *a-h-k* form can be rewritten as a function in *a-b-c* form if we multiply out and group terms with the same power of *x*.

Example   3

Rewrite the quadratic $f(x) = 3(x + 7)^2 - 9$ in the *a-b-c* form.

SOLUTION

$f(x)$ is in the *a-h-k* form where $a = 3$, $h = -7$, and $k = -9$.

given	$f(x) = 3(x + 7)^2 - 9$
write out the factors	$= 3(x + 7)(x + 7) - 9$
multiply the factors	$= 3(x^2 + 14x + 49) - 9$
distribute the 3	$= 3x^2 + 42x + 147 - 9$
group the constant terms	$f(x) = 3x^2 + 42x + 138$

This function is in the *a-b-c* format with $a = 3$, $b = 42$, and $c = 138$.

Example   4

Find the equation of the parabola in Figure 11.26. Write it in *a-h-k* and *a-b-c* form.

SOLUTION

The vertex of the graph appears to be at $(2, 1)$ and the graph opens upward. Using this estimate of the vertex, we can substitute the coordinates of the vertex in the *a-h-k* form of the quadratic equation to get

$$y = a(x - 2)^2 + 1 \tag{1}$$

How can we find a value for a? If we can identify values for any other point (x, y) that lies on the parabola, we can substitute these values into Equation (1) to find a. The *y*-intercept, $(0, 3)$, is a convenient point to pick. Setting $x = 0$ and $y = 3$, we get

$$3 = a(0 - 2)^2 + 1$$
$$3 = 4a + 1$$
$$2 = 4a$$

so

$$a = 0.5$$

The equation in the *a-h-k* form is

$$y = 0.5(x - 2)^2 + 1$$

If we wanted it in the equivalent *a-b-c* form, we could square, multiply, and collect like terms to get

$$y = 0.5x^2 - 2x + 3$$

• • •

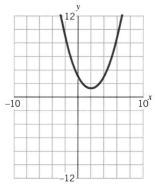

Figure 11.26 A mystery parabola.

Given a general quadratic in the *a-h-k* form we have:

$$f(x) = a(x - h)^2 + k$$

write out the factors $\quad = a(x - h)(x - h) + k$

multiply the factors $\quad = a(x^2 - 2hx + h^2) + k$

distribute the *a* $\quad = ax^2 - 2ahx + ah^2 + k$

group terms $\quad = ax^2 + (-2ah)x + (ah^2 + k)$

If we set $\qquad b = -2ah \qquad$ and $\qquad c = ah^2 + k$

then we have rewritten the function in the form $f(x) = ax^2 + bx + c$. So the *a-h-k* form is just a different way of writing a quadratic function. Notice that although we are using *h* and *k* in one version and *b* and *c* in the other, the coefficient *a* is the same in both forms.

*Two strategies for getting from the **a-b-c** to the **a-h-k** form of a quadratic*

Now suppose we want to go the other way. Suppose we are given a quadratic function, such as

$$g(x) = 3x^2 - 12x + 5$$

that is in the *a-b-c* form and we want to write it in the *a-h-k* form. We want to write

$$g(x) = a(x - h)^2 + k$$

for some values of *a*, *h*, and *k*.

Strategy 1: Using the Formula for the Vertex

Since the coefficient *a* is the same in both the *a-b-c* and the *a-h-k* form, we have $a = 3$. The coordinates of the vertex of a quadratic in the *a-b-c* form are given by $\left(-\dfrac{b}{2a}, -\dfrac{b^2}{4a} + c\right)$. For $g(x)$ we have $a = 3, b = -12$, and $c = 5$. So

$$-\frac{b}{2a} = -\frac{(-12)}{2 \cdot 3} = \frac{12}{6} = 2$$

and

$$-\frac{b^2}{4a} + c = -\frac{(-12)^2}{4 \cdot 3} + 5 = -\frac{144}{12} + 5 = -12 + 5 = -7$$

The vertex is at $(2, -7)$. In the a-h-k form the vertex is at (h, k) so we must have $h = 2$ and $k = -7$. Substituting for a, h, and k, we get

$$g(x) = 3(x - 2)^2 - 7$$

and the transformation is complete.

• • •

In general, to convert any quadratic function $f(x) = ax^2 + bx + c$ in the a-b-c form into an equivalent a-h-k form, the value of the coefficient a remains the same, $h = -\dfrac{b}{2a}$ and $k = -\dfrac{b^2}{4a} + c$, and therefore

$$f(x) = a\left(x + \frac{b}{2a}\right)^2 - \frac{b^2}{4a} + c$$

Strategy 2: "Completing the Square"   We can, of course, always use the formula to calculate the vertex directly and plug the values into the a-h-k format as we did above. But there is an alternate strategy, called *completing the square,* that is sometimes faster. We can convert the function $f(x) = x^2 - 4x + 9$ into a-h-k form using this method.

When a function is in a-h-k form, the term $(x - h)^2$ is a perfect square; that is, $(x - h)^2$ is the product of the expression $(x - h)$ times itself. So we examine separately the expression $x^2 - 4x$ and ask what constant term we would need to add to it in order to make it a perfect square. Since

$$(x - 2)(x - 2) = x^2 - 4x + 4,$$

the answer is 4. We can add 4 to $x^2 + 4x + 9$ if we also subtract 4, in order to preserve equality. So we have

given | $f(x) = x^2 - 4x + 9$
add and subtract 4 | $= x^2 - 4x + (4 - 4) + 9$
regroup terms | $= (x^2 - 4x + 4) - 4 + 9$
factor and simplify | $= (x - 2)^2 + 5$

We now have $f(x)$ in a-h-k form. The vertex is at $(2, 5)$.

Something to think about

By multiplying out and combining like terms, verify that

$$y = a\left(x + \frac{b}{2a}\right)^2 - \frac{b^2}{4a} + c$$

and

$$y = ax^2 + bx + c$$

are two forms of the same function.

Example   5

Convert the function $g(t) = -2t^2 + 12t - 23$ into a-h-k form.

SOLUTION

This function is more difficult to convert by completing the square, since a is not 1. We need first to factor out -2 *from the t terms only,* getting

$$g(t) = -2(t^2 - 6t) - 23 \tag{1}$$

It is the expression $(t^2 - 6t)$ for which we must complete the square. Since

$$t^2 - 6t + 9 = (t - 3)^2$$

we must add the constant term 9 *inside the parentheses* in Equation (1) in order to make the binomial a perfect square. Since everything inside the parentheses is multiplied by -2, we need to add 18 *outside* the parentheses in order to preserve equality.

So we have:

given	$g(t) = -2t^2 + 12t - 23$
factor out -2 from t terms	$= -2(t^2 - 6t) - 23$
add 9 inside parentheses and 18 outside parentheses	$= -2(t^2 - 6t + 9) + 18 - 23$
factor and simplify	$= -2(x - 3)^2 - 5$

We now have $f(x)$ in *a-h-k* form. The vertex is at $(3, -5)$.

Algebra Aerobics 11.6c

1. Convert the following functions into *a-h-k* form by completing the square.
   a. $f(x) = x^2 + 2x - 1$ c. $h(x) = -3x^2 - 12x$
   b. $j(z) = 4z^2 - 8z - 6$

2. Express each of the following functions in the form $y = ax^2 + bx + c$.
   a. $y = 2\left(x - \dfrac{1}{2}\right)^2 + 5$
   b. $y = -\dfrac{1}{3}(x + 2)^2 + 4$

3. Express each of the following functions in the form $y = a(x - h)^2 + k$.
   a. $y = x^2 + 6x + 7$ b. $y = 2x^2 + 4x - 11$

4. Find the coordinates of the vertex and the x- and y-intercepts, and graph the following functions.
   a. $y = x^2 + 8x + 11$ b. $y = 3x^2 + 4x - 2$

5. Find the coordinates of the vertex and the x- and y-intercepts, and graph the following functions.
   a. $y = 0.1(x + 5)^2 - 11$
   b. $y = -2(x - 1)^2 + 4$

11.7 WHY THE FORMULA FOR THE VERTEX AND THE QUADRATIC FORMULA WORK

Determining the Vertex by "Completing the Square"

We will show that, as stated in Section 11.2, the vertex of a quadratic function in the form $y = ax^2 + bx + c$ has coordinates

$$\left(-\frac{b}{2a}, -\frac{b^2}{4a} + c\right)$$

Given an arbitrary quadratic function $y = ax^2 + bx + c$, we can find the coordinates (h, k) of the vertex of the parabola by putting the function into *a-h-k* form $y = a(x - h)^2 + k$, using the strategy of completing the square introduced in Section 11.6.

As we did in Example 5 of Section 11.6, we start by factoring out the coefficient a from the x^2 and x terms. We get

$$y = a\left(x^2 + \frac{b}{a}x\right) + c \tag{1}$$

Now we need to find the right constant to add inside the parentheses in order to create a perfect square. Since

$$x^2 + \frac{b}{a}x + \frac{b^2}{4a^2} = \left(x + \frac{b}{2a}\right)\left(x + \frac{b}{2a}\right)$$

we need to add $\dfrac{b^2}{4a^2}$ *inside* the parentheses in Equation (1). Since all the terms in the

parentheses are multiplied by a, we must subtract $a\left(\dfrac{b^2}{4a^2}\right) = \dfrac{b^2}{4a}$ *outside* the paren-

theses. We have

given $y = ax^2 + bx + c$

factor out a $= a\left(x^2 + \dfrac{b}{a}x\right) + c$

add $\dfrac{b^2}{4a^2}$ inside and $-\dfrac{b^2}{4a}$
outside the parentheses $= a\left(x^2 + \dfrac{b}{a}x + \dfrac{b^2}{4a^2}\right) - \dfrac{b^2}{4a} + c$

factor $= a\left(x + \dfrac{b}{2a}\right)^2 - \dfrac{b^2}{4a} + c$

The final expression is in *a-h-k* form, with

$$h = -\frac{b}{2a} \qquad \text{and} \qquad k = -\frac{b^2}{4a} + c$$

So the vertex of any quadratic function in the form $y = ax^2 + bx + c$ is at $\left(-\dfrac{b}{2a}, -\dfrac{b^2}{4a} + c\right)$, which is the formula we stated in Section 11.2.

Deriving the Quadratic Formula

We will show that for any quadratic equation in the form $0 = ax^2 + bx + c$, the solutions or roots of the equation are given by

$$x = \frac{-b \pm \sqrt{b^2 - 4ac}}{2a}$$

We've just seen that any quadratic $f(x) = ax^2 + bx + c$ in *a-b-c* form can be translated into $f(x) = a(x - h)^2 + k$ in *a-h-k* form, where (h, k) are the coordinates of the vertex and $h = -\dfrac{b}{2a}$ and $k = -\dfrac{b^2}{4a} + c$. To find the zeros of a quadratic in the *a-h-k* form, we look for values of x that satisfy the equation

$$a(x - h)^2 + k = 0$$

First, we subtract k from both sides of the equation and divide by a to get

$$(x - h)^2 = -\frac{k}{a} \qquad\qquad (2)$$

Since the square of either a positive or a negative number is positive, then $(x - h)^2 \geq 0$. So Equation (2) can only be satisfied if $-k/a \geq 0$ or equivalently, if $k/a \leq 0$ (remember that multiplying through by a negative number reverses the inequality). Hence for Equation (2) to have any solutions, we must have k and a of opposite signs: either $k \geq 0$ and $a < 0$, or $k \leq 0$ and $a > 0$. The case $k/a = 0$ is somewhat special, and for now we shall suppose that $k/a < 0$.

There are then *two* values for $x - h$ that make Equation (2) true: a positive square root and a negative square root. We write

$$x - h = \pm \sqrt{-\frac{k}{a}}$$

Finally, we may write the two solutions as

$$x = h \pm \sqrt{-\frac{k}{a}} \tag{3}$$

Notice that the two solutions are symmetric about the axis of symmetry, the line $x = h$, which runs through the vertex. One solution is h plus $\sqrt{-\dfrac{k}{a}}$, and the other is h minus $\sqrt{-\dfrac{k}{a}}$. Since the whole graph is symmetric about the axis of symmetry, the positions of the zeros must be symmetric as well.

If the original quadratic function is given in the form $y = ax^2 + bx + c$, then we can substitute into Equation (3) the expressions we found previously for h and k in terms of a, b, and c. That is, substitute $h = -\dfrac{b}{2a}$ and $k = -\dfrac{b^2}{4a} + c$. The two roots are then

$$x = -\frac{b}{2a} \pm \sqrt{-\left(\frac{\dfrac{-b^2}{4a} + c}{a} \right)}$$

combine terms in the numerator

$$= -\frac{b}{2a} \pm \sqrt{-\left(\frac{\dfrac{-b^2 + 4ac}{4a}}{a} \right)}$$

simplify

$$= -\frac{b}{2a} \pm \sqrt{-\left(\frac{-b^2 + 4ac}{4a^2} \right)}$$

$$= -\frac{b}{2a} \pm \sqrt{\frac{b^2 - 4ac}{4a^2}}$$

and take the square root of $4a^2$

$$= -\frac{b}{2a} \pm \frac{\sqrt{b^2 - 4ac}}{2a}$$

combine terms

$$x = \frac{-b \pm \sqrt{b^2 - 4ac}}{2a}$$

We have arrived at the *quadratic formula*.

Something to think about

If $k/a = 0$, why does the quadratic formula still hold? In this special case, how many roots would the equation $0 = ax^2 + bx + c$ have?

11.8 GENERALIZING TO OTHER POLYNOMIALS

Quadratic functions are a member of the family of functions called polynomials. We introduced polynomial functions in Chapter 9 as the sum of one or more power functions that have positive integer powers. For example, if we add together the three power functions $y_1 = 16x^3$, $y_2 = -5x$, and $y_3 = 10$, we create the polynomial function $y = y_1 + y_2 + y_3$, where $y = 16x^3 - 5x + 10$. (Note, we can write y_3 as the power function $10x^0$.) We say that y is a polynomial function of *degree 3,* since the highest power of the independent variable x is 3.

Polynomials are simple functions in the sense that they can be created by adding, subtracting, or multiplying power functions; no divisions or root extractions are necessary. Because polynomials are the simplest functions that can exhibit a rich range of behavior, they are the "workhorse functions" widely used for modeling data and for approximating other functions.

A *polynomial function* $y = f(x)$ can be represented by an equation in the form

$$y = a_n x^n + a_{n-1} x^{n-1} + \ldots + a_1 x^1 + a_0$$

where each coefficient $a_n, a_{n-1} \ldots a_0$ is a constant and

n is a positive integer, called the *degree* of the polynomial (provided $a_n \neq 0$).

This form for a polynomial function looks pretty imposing, but think of what the notation means. The coefficients $a_n, a_{n-1}, a_{n-2} \ldots a_0$ are constants where a_0 is often called the *constant term.* The expressions $x^n, x^{n-1}, x^{n-2}, x^{n-3} \ldots x^1$ are powers of x where n is the degree of the polynomial.

For example, in the polynomial

$$y = 16x^3 - 5x + 10$$

or equivalently

$$y = 16x^3 + 0x^2 - 5x + 10$$

we have $a_3 = 16$, $a_2 = 0$, $a_1 = -5$, and $a_0 = 10$.

Example   1

Suppose you are making deposits of $2000 in a savings account each year, starting today. What annual interest must you earn if you want to have $10,000 in the account after 4 years? Assume the annual interest, compounded annually, is constant over these 4 years.[2]

SOLUTION

Let r be the annual interest rate. You want to find out what value of r gives you $10,000 after 4 years. If, for example, $r = 8\%$, then after one year, the $2,000 plus interest earned on the $2,000 equals

$$2000 + (.08)(2000) = 1.08(2000).$$

---

[2] Example modeled after a problem in Eric Connally, Deborah Hughes-Hallett, and Andrew M. Gleason, et. al., *Functions Modeling Change: A Preparation for Calculus* (Preliminary Edition), (John Wiley & Sons, New York, 1998) pp. 412–413.

At the end of 1 year your total includes the initial deposit, plus interest earned on that deposit, plus the second deposit.

amount after one year = (initial amount + interest on initial amount) + new deposit
$$= 1.08(2000) + 2000$$

Let $x = 1 + r$ be the annual multiplier or growth factor. If the account pays 8% interest, then $x = 1 + 0.08$. We can write the balance after 1 year in terms of x as

amount after one year = (initial amount)(growth factor) + new deposit
$$= 2000x + 2000$$

During the second year you will earn interest on the amount you had after 1 year. At the end of the second year you will also deposit another $2000. So the total amount after 2 years will be

amount after 2 years = (amount after 1 year)(growth factor) + new deposit
$$= (2000x + 2000)x + 2000$$
$$= 2000x^2 + 2000x + 2000$$

After 3 years, interest will again be earned and a fourth deposit made.

amount after 3 years = (amount after 2 years)(growth factor) + new deposit
$$= (2000x^2 + 2000x + 2000)x + 2000$$
$$= 2000x^3 + 2000x^2 + 2000x + 2000$$

A pattern is emerging. After a $2000 deposit is held for n years, it grows to $2000x^n$. At the end of 4 years, assuming you do not make an additional deposit, you will have

amount after 4 years = $2000x^4 + 2000x^3 + 2000x^2 + 2000x$

If you want to withdraw $10,000 at the end of 4 years instead of making another deposit, then you need to subtract $10,000 and your balance will be zero.

$$0 = 2000x^4 + 2000x^3 + 2000x^2 + 2000x - 10{,}000 \qquad (1)$$

If you can solve this equation for x, you can find out what interest rate r will give you $10,000 at the end of 4 years, since $x = 1 + r$. Dividing both sides of Equation (1) by 2000 you get

$$0 = x^4 + x^3 + x^2 + x - 5 \qquad (2)$$

A solution to Equation (2) will also be a zero for the function

$$f(x) = x^4 + x^3 + x^2 + x - 5$$

You can estimate any real zeros of $f(x)$ and hence solutions to Equation (2) by graphing the function (Figure 11.27) and then zooming in to estimate values for any x-intercepts (Figure 11.28).

There is only one positive x-intercept, at approximately (1.091, 0). Since $x = 1 + r$, if $x = 1.091$, then $r = x - 1 = 0.091$ in decimal form, or 9.1% when written as a percent. So you would need an interest rate of 9.1% in order to end up with $10,000 by investing $2000 each year for 4 years.

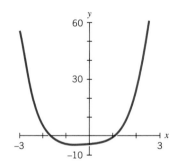

Figure 11.27   Graph of $f(x) = x^4 + x^3 + x^2 + x - 5$.

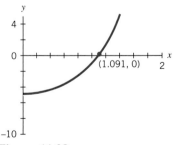

Figure 11.28   Zooming in on the x-intercept between 0 and 1.

Visualizing Polynomial Functions

What can we predict about the graph of a polynomial function from its equation? Examine the graphs of polynomials of different degrees in Figure 11.29. What can we observe from each of these pairs?

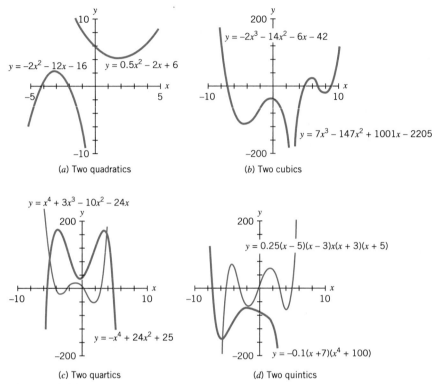

Figure 11.29   Graphs of pairs of polynomial functions
(*a*) of degree 2 (quadratics)
(*b*) of degree 3 (cubics)
(*c*) of degree 4 (quartics)
(*d*) of degree 5 (quintics)

1. The first thing we might notice is the number of times each graph "wiggles" or bends into a new direction. The quadratics bend once, the cubics seem to bend twice, the quartics three times, one quintic seems to bend four times, and the other quintic appears to bend twice. In general, *a polynomial function of degree n will bend at most n − 1 times.*

2. Second, we might notice the number of times each graph crosses the *x*-axis. Each quadratic crosses at most two times; the cubics each cross at most three times; the quartics cross at most four times; and the quintics cross at most five times. In general, *a polynomial function of degree n will cross the x-axis at most n times.*

3. Finally, imagine zooming way out on the graph, to look at it on a global scale. After all, the sections of the *x*-axis displayed in the four graphs of Figure 11.29 are really quite small, the largest containing values of *x* only between −10 and +10. Suppose we consider values of *x* between −1000 and +1000 or between −1,000,000 and +1,000,000. What will the graphs look like?

Figure 11.30 displays a graph of the two quintics in Figure 11.29(d). Here the displayed values of x extend between -1000 and $+1000$. The wiggles are no longer noticeable. The dominant feature to notice now is whether the two "arms" of the function extend indefinitely up or indefinitely down. For polynomial functions of odd degree, such as the quintics in Figure 11.30, the two "arms" extend in opposite directions, one up and one down. For polynomial functions of even degree, both "arms" extend in the same direction, either both up or both down. This is because given a polynomial

$$y = a_n x^n + a_{n-1} x^{n-1} + \ldots + a_1 x^1 + a_0$$

of degree n, for large values of x, the values of the leading term $a_n x^n$ will dominate the values of the other terms of smaller degree. In other words, for large values of x the function behaves a lot like the simple power function $y = a_n x^n$. *The degree of a polynomial determines its global shape.*

In summary,

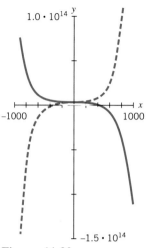

Figure 11.30 Zooming out on the graphs of the two quintics in Figure 11.29.

The degree n of a polynomial function determines

- that the graph will bend in the middle at most $n - 1$ times.
- that the graph will cross the x-axis at most n times.
- the global shape of the polynomial.

Finding the Intercepts of Polynomial Functions

Finding the y- (or vertical) intercept   The y-intercept is the point at which the graph of a function $y = f(x)$ crosses the y-axis. Given any general polynomial function of the form $y = f(x)$ where $y = a_n x^n + a_{n-1} x^{n-1} + \ldots + a_1 x^1 + a_0$, the y-intercept will occur when $x = 0$. The graph will cross the y-axis at $f(0) = a_n(0)^n + a_{n-1}(0)^{n-1} + \ldots + a_1(0)^1 + a_0 = a_0$, where a_0 is the constant term. Hence the coordinates of the y-intercept are $(0, a_0)$. We often shorten this to say that the y-intercept is a_0.

For example, the function $y = x^4 - 2x^2 - 3$ has a y-intercept at $(0, -3)$, or we say the y-intercept is -3 (Figure 11.31).

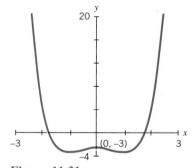

Figure 11.31   The graph of $y = x^4 - 2x^2 - 3$ has a y-intercept at -3.

The y-intercept is the point $(0, a_0)$ for any polynomial function of the form $y = f(x)$ where

$$y = a_n x^n + a_{n-1} x^{n-1} + \ldots + a_1 x^1 + a_0$$

(n is a positive integer and $a_n \neq 0$)

Finding the x- (or horizontal) intercepts   The x-intercepts are the points at which the graph of a function $y = f(x)$ crosses the x-axis. Since every point on the x-axis has a y-coordinate of 0, we often refer to the x-intercept only by its x-coordinate.

The values of x for which $f(x) = 0$ are called the *zeros* of the function. Just as with quadratics, not all the zeros may be real. Each real zero determines an x-intercept.

Given a polynomial function

$$f(x) = a_n x^n + a_{n-1} x^{n-1} + \ldots + a_1 x^1 + a_0$$

(where n is a positive integer and $a_n \neq 0$)

the *zeros* of the function are the values of x that make $f(x) = 0$.
Each real zero corresponds to an x-intercept.

For example, for the relatively simple function $f(x) = x^4 - 16$, we can set $f(x) = 0$ and solve the corresponding equation.

given	$f(x) = x^4 - 16$
set $f(x) = 0$	$0 = x^4 - 16$
add 16	$16 = x^4$
take 4th root	$x = \pm 2$

So both $+2$ and -2 are zeros of the function. Since they are both real numbers, they represent x-intercepts. So the graph of $f(x)$ crosses the x-axis at $(2, 0)$ and $(-2, 0)$.

The zeros correspond to solutions called the *roots* of the equation

$$0 = a_n x^n + a_{n-1} x^{n-1} + \ldots + a_1 x^1 + a_0$$

For example, we say that 2 and -2 are the *roots* of the equation $0 = x^4 - 16$.

The solutions of a polynomial equation

$$0 = a_n x^n + a_{n-1} x^{n-1} + \ldots + a_1 x^1 + a_0$$

(where n is a positive integer and $a_n \neq 0$)

are called *roots*.

Graphical strategies for estimating x-intercepts: zooming in
A graphing calculator or computer (preferably with a zoom function) is very useful in helping to identify the number of x-intercepts and their approximate values. For example, if we graph the function $y = x^4 - 2x^2 - 3$, we see that it crosses the x-axis twice, which means that it has two x-intercepts (Figure 11.32).

If we zoom in on the x-intercept to the right of the y-axis, we can obtain increasingly accurate, though still approximate, values for that intercept (Figures 11.33 and 11.34).

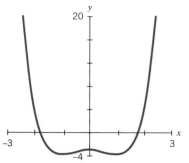

Figure 11.32 The graph of $y = x^4 - 2x^2 - 3$ shows two x-intercepts.

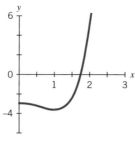

Figure 11.33 Zooming in on the x-intercept to the right tells us that its value is somewhere between 1.5 and 2.0.

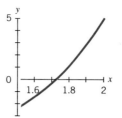

Figure 11.34 Zooming in again indicates that the x-intercept occurs between 1.7 and 1.8.

Algebra Aerobics 11.8a

Use a function graphing program (and its zoom feature) to estimate the number of x-intercepts and their approximate values for the following functions.

a. $y = 3x^3 - 2x^2 - 3$

b. $f(x) = x^2 + x + 3$

Using equations to find x-intercepts: factoring For polynomials of degree 4 or less, there are formulas that give us the exact values of the zeros (and hence the x-intercepts). We calculated x-intercepts for lines in Part I of this book and in this chapter we found the x-intercepts for polynomials of degree 2. Life isn't as easy when dealing with polynomial functions of higher degrees. The formulas for the zeros of third- and fourth-degree polynomials are extremely complicated. It has been proved that there are *no* general algebraic formulas for the zeros of polynomials of degree 5 or higher. We do know, however, that the zeros all lie within the complex number system. There are algebraic approximation methods that allow us to calculate values for the x-intercepts or real zeros of functions accurate to as many decimal places as we wish.

The basic strategy for finding the zeros of a function is always the same: set $f(x) = 0$ and try to solve the resulting equation for x.

Find the x-intercepts of the cubic function $f(x) = x^3 + x^2 - 2x$.

Example 2

SOLUTION

If a function can be written as a product of linear factors, then each factor corresponds to a real zero or x-intercept of the function. For example, the cubic function $f(x) = x^3 + x^2 - 2x$ can be factored into three linear terms.

given $f(x) = x^3 + x^2 - 2x$

if we factor out x, we get $= x(x^2 + x - 2)$

then factor the binomial $= x(x + 2)(x - 1)$

if we set $f(x) = 0$, we get $0 = x(x + 2)(x - 1)$

apply the zero product rule $x = 0$ or $x + 2 = 0$ or $x - 1 = 0$

solve each equation $x = 0$ or $x = -2$ or $x = 1$

The three factors x, $x + 2$, and $x - 1$ give us the corresponding real zeros, 0, −2, and 1, which we can see as x-intercepts in Figure 11.35.

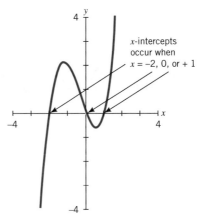

x-intercepts
occur when
$x = -2, 0,$ or $+1$

Figure 11.35   The x-intercepts of $f(x) = x^3 + x^2 - 2x$ occur when x is −2, 0, or 1.

Example   3

Find the zeros of the fifth-degree polynomial function
$$y = 0.25(x - 5)(x - 3)x(x + 3)(x + 5)$$

SOLUTION

given $y = 0.25(x - 5)(x - 3)x(x + 3)(x + 5)$

set $y = 0$ $0 = 0.25(x - 5)(x - 3)x(x + 3)(x + 5)$

apply zero product rule

$x - 5 = 0$ or $x - 3 = 0$ or $x = 0$ or $x + 3 = 0$ or $x + 5 = 0$

solve for x

$x = 5$ or $x = 3$ or $x = 0$ or $x = -3$ or $x = -5$

This polynomial has five real zeros and hence five x-intercepts that occur when x is 5, 3, 0, −3, or −5 (Figure 11.36).

Figure 11.36   Graph of $y = 0.25(x - 5)(x - 3)x(x + 3)(x + 5)$ with zeros or x-intercepts when $x = -5, -3, 0, 3, 5$.

Something to think about

In general, determining the factors of a quadratic expression, of a more general polynomial, or of a large integer is much more difficult than verifying that a given set of factors is correct; this fact is the foundation of modern "public-key" cryptographic techniques.

Algebra Aerobics 11.8b

Identify the degree of each polynomial and its y-intercept. Factor each of the following equations to find the x-intercepts. Graph each function to verify your work.

a. $y = 3x + 6$

b. $y = x^2 + 3x - 4$

c. $y = (x^2 + 2x - 15)(2x + 5)$

CHAPTER SUMMARY

A *quadratic function* or *polynomial function of degree 2* has the form $y = f(x)$ where

$$y = ax^2 + bx + c$$

(a, b, and c are constants and $a \neq 0$). The graph of a quadratic function is called a *parabola* and has the following properties.

It is in the shape of a symmetric "bowl."

It opens upward if $a > 0$ and downward if $a < 0$.

As $|a|$ (the absolute value or magnitude of a) increases, the parabola becomes narrower.

It maintains its basic shape but is shifted up if c is increased and down if c is decreased.

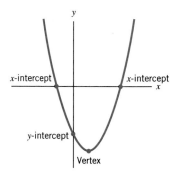

Graph of a quadratic with two x-intercepts.

It has a *vertex* with coordinates $\left(-\dfrac{b}{2a}, -\dfrac{b^2}{4a} + c \right)$.

If $a > 0$, the vertex represents a *minimum;* if $a < 0$, the vertex represents a *maximum.*

It has a *y*-intercept at the point with coordinates $(0, c)$ that is often abbreviated as just c.

It may have two, one, or no *x*-intercepts. If $(d, 0)$ is an *x*-intercept, we often abbreviate this and say the *x*-intercept is d.

To find the *x*-intercepts of a quadratic function $f(x) = ax^2 + bx + c$, we must solve or find the *roots* of the related equation $0 = ax^2 + bx + c$. The *quadratic formula* says that the roots of the equation are given by

$$x = \frac{-b \pm \sqrt{b^2 - 4ac}}{2a}$$

The roots may be real or imaginary. Imaginary numbers are in the form $a + bi$, where a and b are real numbers, $b \neq 0$, and $i = \sqrt{-1}$. Each real root corresponds to an *x*-intercept.

If there are one or two real roots, then the graph of $f(x)$ has one or two *x*-intercepts respectively. If the roots are not real, the graph has no *x*-intercepts.

Any value of x for which $f(x) = 0$ is called a *zero* of the function, f. Each *x*-intercept corresponds to a zero of a function.

There are three basic formats in which any quadratic function may be written: *a-b-c* form, factored form, and *a-h-k* form. A function $f(x)$ in *a-h-k* format is in the form

$$y = a(x - h)^2 + k$$

where (h, k) is the vertex.

A *polynomial function* $y = f(x)$ can be represented by an equation in the form

$$y = a_n x^n + a_{n-1} x^{n-1} + \ldots + a_1 x^1 + a_0$$

where each coefficient $a_n, a_{n-1} \ldots a_0$ is a constant and n is a positive integer, called the *degree* of the polynomial (provided $a_n \neq 0$).

The degree n of a polynomial function determines

that the graph will bend in the middle at most $n - 1$ times;

that the graph will cross the *x*-axis at most n times;

the global shape of the polynomial.

Exercises

1. **a.** For each of the following functions, evaluate $f(2)$ and $f(-2)$.

    i. $f(x) = x^2 - 5x - 2$ **iii.** $f(x) = -x^2 + 4x - 2$

    ii. $f(x) = 3x^2 - x$

    b. For each of the following functions, evaluate $g(1)$ and $g(-1)$.

    i. $g(x) = x^4 - 2x^3 + x$ **iii.** $g(x) = -2x^3 + 5x^2 - 4x + 3$

    ii. $g(x) = -x^3 + 2x^2 - 10$

2. For each set of functions, describe how the graphs are similar and how they are different.

    a. $f(x) = x^2$ $g(x) = 3x^2$ $h(x) = 0.5x^2$

    b. $f(x) = 3x^2 + 4$ $g(x) = -3x^2 + 4$ $h(x) = -3x^2 - 4$

    c. $f(x) = x^2$ $g(x) = 2x$ $h(x) = 2^x$

3. Match each of the following graphs with an equation. Explain your reasoning for each of your choices. (Note that the grid lines are 2 units apart.)

    $$f(x) = 2x^2 - 8x - 2 \qquad j(x) = -0.5x^2 - 2x + 3$$
    $$g(x) = 2x^2 - 8x + 3 \qquad k(x) = 2x - 5$$
    $$h(x) = 0.5x^2 - 2x + 3 \qquad k(x) = 0.5x^2 - 2x - 3$$
    $$i(x) = -2x^2 - x + 2$$

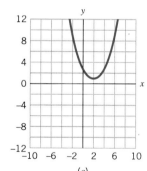

(a)

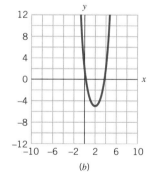

(b)

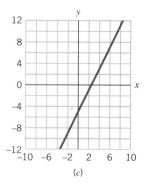

(c)

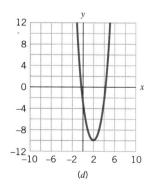

(d)

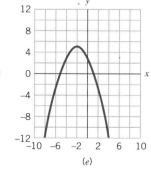

(e)

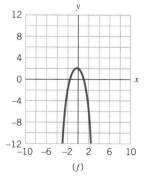

(f)

4. On the same graph, sketch the plots of the following functions and label each with its equation.

$$y = 2x^2 \qquad\qquad y = -2x^2 + 3$$
$$y = -2x^2 \qquad\qquad y = -2x^2 - 3$$

5. Construct the equations of two quadratic functions each with a y-intercept of 10, one with a graph that turns up and the other with a graph that turns down. Do a rough sketch for each function and identify how many horizontal intercepts each function has.

6. (Requires computer with course software.)
    Practicing skills with quadratics in the a-b-c form $y = ax^2 + bx + c$

    a. Open "Q2: Finding a, b, c" in *Quadratic Functions*. The software will pick a random quadratic function and display its graph. Predict the corresponding values of a, b, and c; graph your predicted quadratic; and then see whether your guess is correct. Repeat using integer and real values until you are consistently right.

    b. Open "Q4: Finding 3 points: a-b-c Form" in *Quadratic Functions*. The software will display a quadratic equation and ask you to click on any three points that would lie on the graph of the given equation. If you then click on "graph curve," the computer will graph the given equation as well as (if possible) a quadratic through your 3 points. Are the graphs the same? Continue the process until you are 100% accurate.

7. Write each of the following quadratic equations in function form (i.e., solve for y in terms of x). Find the vertex, and the y-intercept and x-intercepts, using any method. Finally, using these points, do a rough sketch of the quadratic function.

    a. $\qquad\qquad y + 12 = x(x + 1)$
    b. $2x^2 + 6x + 14.4 - 2y = 0$
    c. $\qquad\qquad y + x^2 - 5x = -6.25$
    d. $\qquad\qquad y - 8x = x^2 + 15$
    e. $\qquad\qquad y + 1 = (x - 2)(x + 5)$

8. For each part do a rough sketch of a graph of any function of the type

$$f(x) = ax^2 + bx + c$$

    a. where $a > 0$, $c > 0$, and the function has no real zeros
    b. where $a < 0$, $c > 0$, and the function has two real zeros
    c. where $a > 0$ and the function has one real zero

9. Solve the following equations using the quadratic formula. (Hint: Rewrite each equation so that one side of the equation is 0.)

    a. $6t^2 - 7t = 5$
    b. $3x(3x - 4) = -4$
    c. $(z + 1)(3z - 2) = 2z + 7$
    d. $(x + 2)(x + 4) = 1$

10. Use the discriminant to predict the number of x-intercepts for each function.

Then use the quadratic formula to find all the zeros. Identify the coordinates of any *x*-intercepts.

a. $y = 2x^2 + 3x - 5$ **c.** $f(x) = x^2 + 2x + 2$

b. $f(x) = -16 + 8x - x^2$

11. A baseball, hit straight up in the air, is at height

$$h = 4 + 50t - 16t^2$$

feet above the ground level at time *t* seconds after being hit. This formula is valid for $t \geq 0$, until the ball hits the ground.

a. What is the value for *h* when $t = 0$? What does this value represent in this context?

b. When does the baseball hit the ground?

c. When, if ever, is it 30 feet high? 90 feet high?

d. What is the maximum height that the baseball reaches? When does it reach that height?

12. The following function represents the relationship between time *t* and height *h* for objects thrown upward on the planet Pluto. For an initial velocity of 20 feet per second and an initial height above ground of 25 feet, we get

$$h = -2t^2 + 20t + 25$$

where *t* is the independent variable.

a. Find the coordinates of the point where the graph intersects the *h*-axis.

b. Find the coordinates of the vertex of the parabola.

c. At how many points will the graph cross the *t*-axis? If the graph crosses the *t*-axis, find the value of the horizontal intercept(s).

d. Sketch the graph. Label the axes.

e. Interpret the vertex in terms of time and height.

f. For what values of *t* does the mathematical model make sense?

13. At low speeds an automobile engine is not at its peak efficiency; efficiency initially rises with speed and then declines at the higher speeds. When efficiency is at its maximum, consumption rate of gas measured in gallons/hour is at a minimum. The gas consumption rate of a particular car can be modeled by the following equation, where *G* is gas consumption rate in gallons/hour and *M* is speed in miles/hour.

$$G = 0.00020 \, M^2 - 0.013M + 1.07$$

a. Construct a graph of gas consumption rate versus speed. Estimate the minimum gas consumption rate from your graph and the speed at which it occurs.

b. Using the equation for *G*, calculate the speed at which the gas consumption rate is at its minimum. What is the minimum gas consumption rate?

c. If you travel for 2 hours at peak efficiency, how much gas will you use and how far will you go?

d. If you travel at 60 miles/hour, what is your gas consumption rate? How long does it take to go the same distance that you calculated in (c)? (Recall that travel distance = speed × time traveled.) How much gas is required for the trip?

e.   Compare the answers for (c) and (d), which tell you how much gas is used for the same length trip at two different speeds. Is gas actually saved for the trip by traveling at the speed that gives minimum gas consumption rate?

f.   Using the function G, generate data for gas consumption rate measured in gallons/mile by computing the following table.

Speed of Car (miles/hour)	Measures of the Rate of Gas Consumption	
	(gallons/hour)	(gallons/hour) / (miles/hour) = gallons/mile
0		
10		
20		
30		
40		
50		
60		
70		
80		

Plot gallons/mile versus miles/hour. At what speed is gallons per mile at a minimum?

g.   Add a fourth column to the data table above. This time compute miles/gallon = (miles/hour)/(gallons/hour). Plot miles/gallon versus miles/hour. At what speed is miles/gallon at a maximum? This is the inverse of the preceding question; we are normally used to maximizing miles/gallon instead of minimizing gallons/mile. Does your answer make sense in terms of what you found for parts (b) and (f)?

14.   These quadratics are given in factored form. Find the x-intercepts directly, without using the quadratic formula. Will the vertex lie above or below the x-axis? Find the vertex and sketch the graph, labeling the x-intercepts.

a. $y = (x + 2)(x + 1)$

b. $y = 3(1 - 2x)(x + 3)$

15.   Solve each of the following by factoring and then sketch the graph, labeling the axes and horizontal intercepts.

a. $y = x^2 + 6x + 8$ **d.** $w = t^2 - 25$

b. $z = 3x^2 - 6x - 9$ **e.** $r = 4s^2 - 100$

c. $f(x) = x^2 - 3x - 10$ **f.** $g(x) = 3x^2 - x - 4$

16.   **a.**   Construct a quadratic function with roots $x = 1$ and $x = 2$.

b.   Is there more than one possible quadratic function for part (a)? Why?

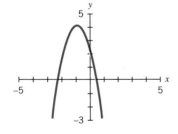

17.   Construct an equation for the accompanying parabola.

18.   For each part construct an equation that satisfies the given conditions.

a.   Has a constant rate of increase of $15,000/year

b.   Is a quadratic that opens up and has a vertex at $(1, -4)$

c.   Is a quadratic that opens down and has a single root

d. Is a quadratic with no real roots

e. Is a quadratic with roots at 1 and -3

19. (Requires computer with course software.) Practicing skills with quadratics in the *a-h-k* form $y = a(x - h)^2 + k$

Open "Q9: Finding 3 Points: *a-h-k* Form." The software will display a quadratic equation and ask you to click on any three points that lie on the graph. When you click on "graph curve," the computer will graph the given equation as well as the quadratic function that goes through your three points. Are the graphs the same? Continue the process until you are 100% accurate.

20. Identify the vertex and whether it represents a maximum or minimum and then sketch the graphs of each of the following functions:

a. $y = (x - 2)^2$

b. $y = \dfrac{1}{2}(x - 2)^2 + 3$

c. $y = -2(x + 1)^2 + 5$

d. $y = -0.4(x - 3)^2 - 1$

21. **a.** Find the equation of the parabola with a vertex of $(2, 4)$ that passes through the point $(1, 7)$.

b. Construct two different quadratic functions both with a vertex at $(2, -3)$ such that the graph of one function is concave up and the graph of the other function is concave down.

22. **a.** Write each of the following functions in both the *a-b-c* and *a-h-k* forms. In which form is it easier to find the intercepts? In which form is it easier to find the vertex?

$$y_1 = 2x^2 - 3x - 20 \qquad y_3 = 3x^2 + 6x + 3$$
$$y_2 = -2(x - 1)^2 - 3 \qquad y_4 = -(2x + 4)(x - 3)$$

b. Find the vertex, *x*- and *y*-intercepts and construct a graph for the functions in part (a). If you have access to a graphing calculator or function graphing program, check your work.

23. As we have seen, the distance that a freely falling object falls can be modeled by the quadratic function $d = 16t^2$, where t is measured in seconds and d in feet. There is a closely related function $v = 32t$ that gives the velocity, v, in feet per second at time, t, for the same freely falling body.

a. Fill in the missing values in the following table.

Time, t (seconds)	Distance, d (feet)	Velocity, v (feet/sec)
1		
1.5		
2		
		80
	56	

b. When $t = 3$, describe the associated values of d and v and what they tell you about the object at that time.

c. Sketch both functions, distance versus time and velocity versus time, on two different graphs but on the same page. Label the points from part (b) on the curves.

d. You are standing on a bridge looking down at a river below. How could you use a pebble to estimate how far you are above the water?

24. (Requires graphing calculator or computer.)

The following data show the average growth of the human embryo prior to birth.

Embryo Age (weeks)	Weight (grams)	Length (centimeters)
8	3	2.5
12	36	9
20	330	25
28	1000	35
36	2400	45
40	3200	50

Source: Reprinted with permission from Kimber et al., *Textbook of Anatomy and Physiology,* Prentice Hall, Upper Saddle River, N.J., 1955, "Embryo Age, Weight and Height," p. 785.

a. Plot weight versus age and find the closest quadratic model that you can to approximate the data.

b. According to your model, what would an average 32-week embryo weigh?

c. Comment on the domain for which your formula is reliable.

d. Plot length versus age; then construct a mathematical model for the length versus age of an embryo from 20 to 40 weeks.

e. Using your model compute the age at which an embryo would be 42.5 cm long.

25. A pilot has crashed in the Sahara Desert. She still has her maps, and she knows her position, but her radio is destroyed. Her only hope for rescue is to hike out to a highway that passes near her position. She needs to determine the closest point on the highway, and how far away it is.

a. The highway is a straight line, passing through a point 15 miles due north of her and another point 20 miles due east. Draw a sketch of the situation on graph paper, placing the pilot at the origin and labeling two points on the highway.

b. Construct an equation that represents the highway (using x for miles east, and y for miles north).

c. Now use the Pythagorean Theorem to represent the square of the distance, d, of the pilot to any point (x, y) on the highway.

d. Substitute the expression for y from part (b) into the equation from part (c) in order to write d^2 as a quadratic in x.

e. If we minimize d^2, we minimize the distance d. So let $D = d^2$, and write D as a quadratic function in x. Now find the minimum value for D.

f. What are the coordinates of the closest point on the highway, and what is the distance, d, to that point?

26. Parabolic mirrors have the useful property that there exists a special point called the *focus* lying on the axis of symmetry inside the parabolic curve. When any ray of light originating at the focus strikes the mirrored surface, it is reflected parallel to the axis of symmetry. Conversely, any rays coming into the mirror parallel to the axis of symmetry are reflected to the focus. If you have access to the course software, open the *Quadratic Functions* program and click on "Q10: Parabolic Reflector" for a demonstration. This property is used to focus incoming heat, light, sound, and television waves at a single point, or conversely to reflect in parallel lines any waves originating at the focus. Common applications are large parabolic dishes that concentrate waves received from TV transmitters, small light sources with parabolic reflectors (such as flashlights and car headlights) that cast a spot of light, and portable electric heaters backed by shiny metal in a parabolic shape to radiate heat. (See photo page 381.)

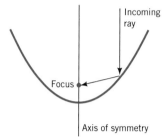

A parabolic reflector: any incoming ray parallel to the axis of symmetry will be reflected to the focus.

a. We know that the coordinates of the vertex are $\left(\dfrac{-b}{2a}, \dfrac{-b^2}{4a} + c\right)$. The focus is $\left|\dfrac{1}{4a}\right|$, the absolute value of $\dfrac{1}{4a}$, units above the vertex (if the graph opens up) or below the vertex (if the graph opens down). The distance between the focus and the vertex is called the *focal length*. Find an algebraic expression in terms of a, b, and c for the coordinates of the focus.

b. Calculate the coordinates of the focus and the focal length for the following parabolas.

   i. $y = x^2 + 3x + 2$ ii. $y = 2x^2 - 4x + 5$

c. Sketch a graph of each parabola in part (b) indicating the position of the focus.

d. Describe in words how the focal length varies depending on how narrow or wide the parabolic curve is.

27. (Assumes Exercise 26.) Suppose you are given a lightbulb 4 inches long with a spherical end having a 1-inch radius. Think of the filament as being located at the center of this sphere. You want to mount the lightbulb in a parabolic reflector so that the vertex is at the base of the bulb and filament is at the focus. The reflector is to extend 1 inch above the glass end of the bulb.

a. Sketch a picture of the lightbulb and the surrounding parabola.

b. What focal length is needed to get the filament at the focus of the parabola?

c. What is the equation of a parabola that gives this focal distance? (Hint: You may want to put the vertex at the origin.)

d. What is the diameter of the wide end of the reflector?

e. How much clearance to the side of the reflector will the bulb have at the widest part of the bulb?

28. (Assumes Exercise 26.) A designer proposes a parabolic satellite dish 5 feet wide across the top and 1 foot 3 inches deep.

a. What is the equation for the dish?

b. What is the focal length of the dish?

c. What is the diameter of the dish at the focus?

29. We dealt previously with systems of lines and ways to determine the coordinates of points where lines intersect. Once you know how to use the quadratic formula,

it's possible to determine where a line and a parabola, or two parabolas, intersect. As with two straight lines, at the point where the graphs of two functions intersect (*if* they intersect), they both have the same *x* value and the same *y* value.

a. Find the intersection of the parabola $y = 2.0x^2 - 3.0x + 5.1$ and the line $y = -4.3x + 10.0$.

b. Plot both functions, labeling any intersection point(s).

30. Open up the *FAM 1000 Census Graphs* course software file and in "F3: Regression with Multiple Subsets" examine mean personal wages versus age for all men, and then for all women. Note the linear regression formulas and the corresponding correlation coefficients for these two cases.

a. Given the shape of the data, it is also reasonable to consider quadratic models for wages (in dollars) versus age (in years) for women and men.

$$\text{women's wages} = -28.54(\text{age})^2 + 2626(\text{age}) - 36390$$
$$\text{men's wages} = -47.48(\text{age})^2 + 4587(\text{age}) - 67080$$

At what age does the quadratic model for women predict the same wages as the linear regression model? Calculate your answer and sketch a graph showing the answer.

b. Is there an age at which both quadratic models predict the same income for men and women? If so, what is it?

c. Comment on what domain might produce reliable values for linear and quadratic models in this case.

31. Complete the table below for the function $y = 3 - x - x^2$

x	y	Average Rate of Change = $\dfrac{\Delta y}{\Delta x}$ (over prior interval)	Rate of Change of the Average Rate of Change
-3	-3	N/A	N/A
-2	1	$\dfrac{1-(-3)}{-2-(-3)} = \dfrac{4}{1} = 4$	N/A
-1	3	$\dfrac{3-1}{-1-(-2)} = \dfrac{2}{1} = 2$	$\dfrac{2-4}{-1-(-2)} = \dfrac{-2}{1} = -2$
0	3	$\dfrac{3-3}{0-(-1)} = \dfrac{0}{1} = 0$	$\dfrac{0-2}{0-(-1)} =$
1	1		
2			
3			

Plot the average rate of change on the vertical axis, and *x* on the horizontal axis. What type of function does the graph represent? What is its slope (the rate of change of the average rate of change)?

32. Match each of the following with its graph.

    a. $y = 2x - 3$ **c.** $y = 3(2^x)$

    b. $y = 2 - x$ **d.** $y = (x^2 + 1)(x^2 - 4)$

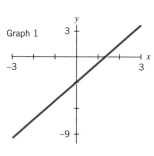

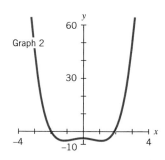

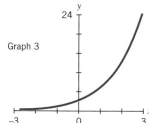

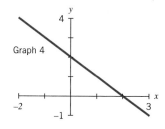

33. In each part, construct a polynomial function with the indicated characteristics.

    a. Crosses the x-axis at least three times

    b. Crosses the x-axis at -1, 3, and 10

    c. Has a y-intercept of 4 and degree of 3

    d. Has a y-intercept of -4 and degree of 5

34. The zeros of a polynomial may be real or imaginary. Real zeros correspond to x-intercepts on the graph of a function.

    a. If the degree of a polynomial is odd, then at least one of its zeros must be real. Explain why this is true.

    b. Sketch a polynomial function that has no real zeros and whose degree is

        i. 2   **ii.** 4

    c. Sketch a polynomial function of degree 3 that has exactly

        i. one real zero.   **ii.** three real zeros.

    d. Sketch a polynomial function of degree 4 that has exactly two real zeros.

35. Use a function graphing program (and its zoom feature) to estimate the number of x-intercepts and their approximate values for

    a. $y = 3x^3 - 2x^2 - 3$ **b.** $f(x) = x^2 + 5x + 3$

36. Calculate the x-intercepts of the following functions, then graph the functions to check your work.

    a. $y = 3x + 6$

    b. $y = x^2 + 3x - 4$

    c. $y = (x + 5)(x - 3)(2x + 5)$

37. Match each of the following with its graph.

a. $f(x) = x^2 + 3x + 1$ **d.** $f(x) = -\dfrac{1}{2}x^3 + x - 3$

b. $f(x) = -2x^2$ **e.** $f(x) = (x^2 + 1)(x^2 - 4)$

c. $f(x) = \dfrac{1}{3}x^3 + x - 3$ **f.** $f(x) = 3(2^x)$

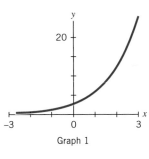

Graph 1

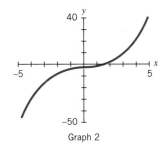

Graph 2

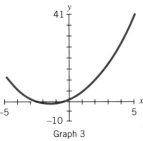

Graph 3

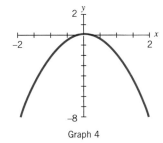

Graph 4

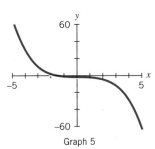

Graph 5

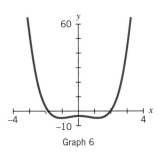

Graph 6

38. One method of graphing functions is called "addition of ordinates." For example, to graph $y = x + \dfrac{1}{x}$ using this method, we would first graph $y_1 = x$. On the same coordinate plane, we would then graph $y_2 = 1/x$. Then we would estimate the y-coordinates (called ordinates) for several selected x-coordinates by adding geometrically on the graph itself the values of y_1 and y_2 rather than by substituting numerically. This technique is often used when graphing the sum or difference of two different types of functions by hand, without the use of a calculator.

a. Use this technique to sketch the graph of the sum of the two functions graphed in the accompanying figure.

b. Use this technique to sketch the graph of $y = -x^2 + x^3$ for $-2 \le x \le 2$. Then use a graphing tool and compare.

c. Use this technique to graph $y = 2^x - x^2$ for $-2 \le x \le 2$.

EXPLORATION 11.1

Properties of Quadratic Functions

Objectives

- Part I: explore the effects of a, b, and c on the graph of quadratic equations in the *a-b-c* form: $y = ax^2 + bx + c$
- Part II: explore the effects of a, h and k on the graph of quadratic equations in the *a-h-k* form: $y = a(x - h)^2 + k$

Materials/Equipment

- graphing calculator or computer and several programs in *Quadratic Functions*
- graph paper

Procedure

Part I: Exploring Quadratics in the Form $y = ax^2 + bx + c$

We start by choosing values for the coefficients a, b, and c and graphing the resulting equations by hand. From these graphs we make predictions about the effects of a, b, and c on the graphs of the equations. Then you can check your predictions, using a graphing calculator or, if you have access to a computer, using one of the quadratic programs in the course software. Be sure to record your predictions and observations so you can share them with the class. Work in pairs and discuss your findings with your partner.

1. **Making Predictions**

   Start with the simplest case, where $b = 0$ and $c = 0$. The equation will have the form

   $$y = ax^2$$

   a. Make a data table and hand sketch the graph of $y = x^2$ using both positive and negative values for x. (Note that $a = 1$ in this example.) On the same graph sketch the graph of $y = 4x^2$ (here $a = 4$). Make a prediction of what the graph of $y = 2x^2$ will look like. Check your work and predictions with your partner.

   b. How would you describe your graphs? Are they symmetrical? When are the values for y increasing? Decreasing? Do the functions have a maximum or a minimum value? What is this value? Which graph shows y changing the fastest compared to x?

   c. Make a data table and sketch the graph of $y = -x^2$. Make predictions for what the graphs $y = -4x^2$ and $y = -2x^2$ will look like. Check your predictions with your partner.

   d. How would you describe these graphs? Are they symmetrical? When are the values for *y* increasing? Decreasing? Do the functions have a maximum or minimum value? What is this value? Which graph shows *y* changing the fastest compared to *x*? How do two graphs compare if one has a larger absolute value for *a*?

    e. Describe how changes in *a* affect the graph of the equation.

2. Testing your predictions

    Use a graphing calculator or a computer and "Q1: *a, b, c* Sliders" in the course software package *Quadratic Functions* to check your predictions.

    a. What effect does *a* have?

        Make predictions based on the graphs you constructed by hand.

- If using a graphing calculator, graph several equations where $b = 0, c = 0$, and the value of *a* is different for each equation. Do your predictions about the effect of *a* hold true? Now keep $b = 0$, but $c \neq 0$, and change the value of *a*. What is the effect? Do your predictions about the effect of *a* still hold true if both $b \neq 0$ and $c \neq 0$?

- If using the course software, open "Q1: *a, b, c* Sliders" in *Quadratic Functions* and first explore what happens when you set $b = 0$ and $c = 0$ and change the value of *a*. Dragging the line in the *a* column changes the value of *a*, which changes your equation and the corresponding graph on the screen. Did your predictions about *a* hold true? Now keep $b = 0$, but set $c \neq 0$, and change the value of *a*. What is the effect? Do your predictions about the effect of *a* still hold true if both $b \neq 0$ and $c \neq 0$?

        Summarize how changing *a* changes the graph. What happens to the graph if you change the sign of *a*? What happens if you keep the sign of *a* but increase the absolute value of *a*? Is there any point on the graph that doesn't change as you increase or decrease the size of *a*?

    b. What effect does *c* have? Using the method described above, explore how the coefficient *c* affects the graph.

    c. What effect does *b* have?

3. Class discussion

    Summarize your conclusions in your own words. Does the rest of the class agree?

Part I: Exploration-Linked Homework

1. Write a 60 second summary of your conclusions.

2. Open "Q2: Finding *a, b, c*" in *Quadratic Functions*. The software will pick a random quadratic function and display its graph. You predict the corresponding value of *a, b,* and *c* using the sliders, graph your predicted quadratic, and then see whether your guess is correct. Repeat, using integer and real values, until you are consistently right.

Part II: Exploring Quadratics in the Form $y = a(x - h)^2 + k$

Work with a partner, recording your results as you go. Again you'll start graphing by hand and then use technology to confirm your predictions.

1. Making Predictions

    a. Make a data table and sketch of the graph $y = 2x^2$ using both positive and negative values for *x*. On the same grid sketch the graphs of $y = 2(x - 1)^2$ and $y = 2(x + 1)^2$. Make a prediction about the graphs of $y = 2(x - 3)^2$ and $y = 2(x + 3)^2$. Check your work with your partner.

    b. What effect does replacing *x* with $(x - 1)$ have? What has happened to the vertex? The shape of the curve? What if you replace *x* with $(x + 1)$? With $(x - 3)$? With $(x + 3)$?

In general, how does the graph of $y = 2x^2$ compare with the graph of $y = 2(x - h)^2$ if h is positive? If h is negative? Do you and your partner agree?

c. Using what you know from Part I about the effect of adding a constant term to a quadratic, on a new graph sketch $y = 2x^2$, $y = 2x^2 + 4$, and $y = 2x^2 - 5$. In general, how does the graph of $y = 2x^2$ compare with the graph of $y = 2(x)^2 + k$ if k is positive? If k is negative? Check your predictions with your partner.

d. Now (without generating a data table) do a rough sketch of the graphs of the quadratic functions $y = 2(x - 1)^2 + 3$ and $y = 2(x + 1)^2 - 5$. Do you and your partner agree?

2. **Testing Your Predictions**

Use a graphing calculator or a computer and "Q3: a, h, k Sliders" in the course software package *Quadratic Functions* to test your predictions.

a. What effect does a have?

Choose $h = 0$ and $k = 0$. What is the form of the resulting function? What is the effect of changing the value of a? How does the role of a here compare to the role of a in equations of the form $y = ax^2 + bx + c$?

b. What effect does h have?

Make a prediction based on the graphs you have constructed by hand. Choose a nonzero value for a, set $k = 0$, and change the value of h.

- If using a graphing calculator, compare your graphs to the graph of the equation when $h = 0$.

Let $y_1 = ax^2$ $k = 0$ and $h = 0$

Use y_2, y_3, y_4, and so on, for functions where $k = 0$ and $h \neq 0$

- If using the course software, open "Q3: a, h, k Sliders," set $k = 0$, pick a value for a, and then "slide" h. What effect does changing the value of h have on the graph?

c. What effect does k have? Make a prediction. Then choose a nonzero value for a, set $h = 0$, and change k. Use a method similar to that described above in part (b) but this time change k.

d. What effect do h and k have together? What do you think will be the effect of changing both of them?

- If using a graphing calculator, choose a nonzero value for a. Now vary both h and k. Compare your graphs to the graph of the equation when $h = 0$ and $k = 0$.

Let $y_1 = ax^2$ $k = 0$ and $h = 0$

Use y_2, y_3, y_4, and so on, for functions where $k \neq 0$ and $h \neq 0$

- If using the software "Q3: a, h, k Sliders," choose a nonzero value for a. Now change both h and k. Can you predict the effects of changing a, h, and k? Can you describe the relationship among the four graphs?

3. **Class discussion**

Summarize your conclusions in your own words. Does the rest of the class agree?

Part II: Exploration-Linked Homework

Write a 60 second summary of your conclusions.

EXPLORATION 11.2

Getting to the Root of Things

Objective

- study several quadratic functions that describe physical phenomena

Materials/Equipment

- graphing calculator or function graphing program (optional)
- enclosed Function Profile Worksheets

Related Software

- "Q10: Parabolic Reflector" in *Quadratic Functions*

Procedure

Working in pairs, complete each of the following function profiles. Record your answers in your notebook.

FUNCTION PROFILE 1: $h = -4.84t^2 + 381.00$

This function describes the height, h, from the ground at time, t, for an object dropped from the top of the Empire State Building, h is in meters and t is in seconds.

GENERATING A QUICK GRAPH

Parabola opens up or down?

Coordinates of h-intercept

Coordinates of vertex

Make a quick sketch by hand.

DETERMINING THE HORIZONTAL INTERCEPTS

From your sketch, how many horizontal or t-intercepts do you expect to find?

Solve for t when $h = 0$.

List the coordinates of any t-intercepts.

If you have access to a graphing calculator or function graphing program, plot the function. Choose an appropriate range and domain that will show any intercepts. Use this plot to give estimates for the t-intercepts and check with your calculated values. Either print out and attach the graph with intercept estimates or write down these estimates.

INTERPRETING IN CONTEXT

How tall is the Empire State Building?

How far away from the ground is the object after

1 second? _____ 3 seconds? _____ 10 seconds? _____

Label these points by hand on your graph.

How many seconds pass before the object hits the ground? ____

For what values of t does the physical interpretation make sense in this problem?

FUNCTION PROFILE 2: $y = 16x^2 + 40x$

This function describes the distance traveled by an object that is thrown downward with an initial velocity of 40 ft/sec. y gives the distance in feet and x measures the time in seconds.

GENERATING A QUICK GRAPH

Parabola opens up or down?

Coordinates of y-intercept

Coordinates of vertex

Make a quick sketch by hand.

DETERMINING THE HORIZONTAL INTERCEPTS

From your sketch, how many x-intercepts do you expect to find?

Solve for x when $y = 0$.

List the coordinates of any x-intercepts.

If you have access to a graphing calculator or function graphing program, plot the function. Choose an appropriate range and domain that will show any intercepts. Use this plot to give estimates for the x-intercepts and check with your calculated values. Either print out and attach the graph and x-intercept estimates, or record these estimates.

INTERPRETING IN CONTEXT

How many feet has the object traveled in

0 seconds? ____ 2 seconds? ____ 8 seconds? ____

Label these points by hand on your graph.

For what domain does the function have physical meaning in this problem?

FUNCTION PROFILE 3: $r = s^2 - 14s + 49$

The graph of this function describes a cross section of a parabolic reflector, a mirror whose shape you can think of as being generated by rotating a parabola about the vertical line of symmetry that runs through its vertex. This cross section is a vertical slice that contains the axis of symmetry of the parabola.

GENERATING A QUICK GRAPH

Parabola opens up or down?

Coordinates of r-intercept

Coordinates of vertex

Make a quick sketch of the graph by hand.

DETERMINING THE HORIZONTAL INTERCEPTS

From your sketch, how many s-intercepts do you expect to find?

Solve for s when $r = 0$.

List the coordinates of any s-intercepts.

If you have access to a graphing calculator or function graphing program, plot the function. Choose an appropriate range and domain that will show any intercepts. Use it to give estimates for the s-intercepts and check with your calculated values. Either print out and attach the graph and s-intercept estimates, or write down these estimates.

INTERPRETING IN CONTEXT

Parabolic mirrors have the unusual property that there exists a special point inside the parabolic shape called the focus. When any ray of light originating from the focus strikes the mirrored surface, it is reflected parallel to the axis of the reflector. Conversely, any rays coming into the mirror parallel to the axis are reflected to the focus. Open up the *Quadratic Functions* and click on "Q10: Parabolic Reflector" for a demonstration. These properties are used in parabolic mirrors for telescopes and in parabolic radio antennae.

The coordinates for the focus for the general parabola $y = ax^2 + bx + c$ are given by $\left(\dfrac{-b}{2a}, \dfrac{1 - b^2}{4a} + c\right)$. Find the coordinates of the focus in the function example above. Show your work. Plot the point on your graph.

Generate a sketch of what would happen to at least two rays of light emitted from this focus that hit the parabolic surface.

EXPLORATION 11.3

Constructing Under Constraints

Objectives

- construct polynomial functions that represent physical phenomena
- explore the effects of various constraints on constructing geometric figures

Materials

- graphing calculator or function graphing program (optional)
- graph paper

Procedure

Class Discussion

Assume you wish to enclose a rectangular region and are constrained by a fixed perimeter of 24 meters. The classic problem is a farmer who has a fixed amount of fencing material with which to enclose a rectangular grazing area. Find the dimensions of the rectangle that will contain the greatest area.

1. Solving by collecting and graphing data.

   Fill in the data table below. Assume that the perimeter of the rectangle must always equal 24 meters. You may want to use graph paper to sketch some of the rectangles.

Dimension of One Side of Rectangle (in meters), x	Dimension of Other Side of Rectangle (in meters)	Area of Rectangle (in square meters), A
0	12	0
1	11	11
2		
3		
4		
5		
6		
7		
8		
9		
10		
11		

   a. From the table, what appears to be the maximum value for the area?

   b. Plot your data by hand, using x as the independent variable and A as the dependent variable. From the shape of the graph, what kind of function might A be?

   c. If you have a graphing calculator or function graphing program, enter the values for x and A, plot the data, and generate a best fit function. Use your function to determine the dimensions of the rectangle with the greatest area.

2. Solving by constructing equations.

Let

x = dimension of one side of the rectangle

Then

? = dimension of other side of the rectangle (expressed in terms of x)
Hint: Remember that the perimeter is fixed at 24 meters.

Now write the area, A, as a function of x.

$$A = ?$$

Graph this function (using a function graphing program if you have one). Find the coordinates of the vertex. Does the vertex represent a maximum or minimum? What do the coordinates of the vertex represent in this situation? How does this function compare to the best fit function you generated in part (1)?

3. Analysis of results

- Of all the possible rectangles with a perimeter of 24 meters, what are the dimensions of the rectangle with the greatest area?

- If the perimeter of the rectangle were held constant at 200 meters, what would your equation be? How could you find the dimensions of the rectangle with the greatest area?

- If the perimeter of the rectangle were held constant at P meters, what would your equation be?

On Your Own or in a Small Group

Choose one of the following problems to solve. Use the strategy of constructing equations and then graphing the equations to solve each problem.

1. A box has a base of 10 feet by 20 feet, and a height of h feet. If we decrease each side of the base by h feet, for what value of h will the volume be maximized?

2. Find the dimensions of a cylinder with the maximum volume if the surface area (including two bases) is held constant at 80 sq. meters.

3. A gardener wants to grow carrots along the side of her house. In order to protect the carrots from wild rabbits, the plot must be enclosed by a wire fence. The gardener wants to use 16 feet of fence material left over from a previous project. Assuming that she constructs a rectangular plot, using the side of her house as one edge, what is the area of the largest plot she can construct?

Useful Formulas

Volume of rectangular solid = area of base · height

Volume of cylinder = area of base · height

$= \pi r^2 \cdot$ height

Surface area of cylinder = (circumference of cylinder · height) + (2 · area of base)

$= 2\pi r h + 2\pi r^2$

(where the height of the cylinder is perpendicular to the base and the cylinder has two bases)

Logarithmic Links: Logarithmic, Exponential and Power Functions

Overview

How does the noise level of a rock band compare to that of a typical conversation? How can we determine the doubling time for a population of *E. coli* bacteria? How can we tell whether an exponential or a power function would be the better model of the relationship between heartbeat and body mass for animals? To answer these questions we return to logarithms.

In Chapters 7 and 8 we considered logarithms as numbers. In this chapter we consider the *logarithmic function* and its intimate link with the exponential function. We introduce the *natural logarithm* that has as its base Euler's number *e*, a constant of nature. We examine strategies for using logarithmic scales to help find the best fit function for a data set.

In the first Exploration we study the functional properties of common and natural logarithms. In the second Exploration we revisit the U.S. population data and look for ways to find an appropriate function model.

After reading this chapter you should be able to:

- use logarithms to solve exponential equations
- recognize and use the constant *e*
- understand and apply the basic properties of common and natural logarithms
- construct an exponential model for continuous compounding
- use semi-log and log-log plots to find a mathematical model for a data set

12.1 USING LOGARITHMS TO SOLVE EXPONENTIAL EQUATIONS

Estimating Solutions to Exponential Equations

We saw in Chapter 8 that a simple model for the growth of *E. coli* bacteria is given by the exponential equation

$$\text{number of } E.\ coli \text{ bacteria} = (\text{initial number of bacteria}) \cdot 2^{\text{time period}}$$

If we let N = number of bacteria and t = time (in 20-minute periods), and the initial number of bacteria is 100, we can rewrite this function as

$$N = 100 \cdot 2^t$$

Given a value for t, we can use this equation to find a corresponding value for N.

We could also ask, given a value for N, can we find a corresponding value for t? For example, at what time t will the value of N be 1000?

We can estimate a value for t that makes $N = 1000$ by using data tables or graphs. In Table 12.1 when $t = 3$, $N = 800$. When $t = 4$, $N = 1600$. Since N is steadily increasing, N equals 1000 for a value of t that is somewhere between 3 and 4.

We can also estimate the value for t when $N = 1000$ by looking at a graph of the function (Figure 12.1). By locating the position on the vertical axis where $N = 1000$, we can move over horizontally to find the corresponding point on the function graph. By moving from this point vertically down to the t-axis, we can estimate the t value for this point. The value for t appears to be approximately 3.3, so the coordinates of the point are roughly (3.3, 1000).

Table 12.1	
Values for $N = 100 \cdot 2^t$	
t	N
0	100
1	200
2	400
3	800
4	1,600
5	3,200
6	6,400
7	12,800
8	25,600
9	51,200
10	102,400

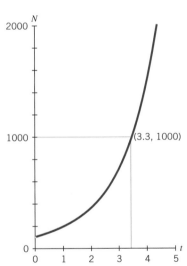

Figure 12.1 Estimating a value for t when $N = 1000$ on a graph of $N = 100 \cdot 2^t$.

An alternate stategy is to set $N = 1000$ and solve the equation for the corresponding value for t.

Start with the equation $N = 100 \cdot 2^t$

set $N = 1000$	$1000 = 100 \cdot 2^t$
divide both sides by 100	$10 = 2^t$

Then we are left with the problem of finding a solution to the equation

$$10 = 2^t$$

This equation is unlike any other we have solved. We can estimate the value for t that satisfies the equation. Since $2^3 = 8$ and $2^4 = 16$, then $2^3 < 10 < 2^4$. So $2^3 < 2^t < 2^4$ and therefore t is between 3 and 4, which agrees with our previous estimates.

Using these strategies we can find approximate solutions to the equation $10 = 2^t$. Strategies for finding exact solutions to such equations require the use of logarithms. Hence we will study the properties of logarithms in order to solve exponential equations for the value of the exponent.

`Algebra Aerobics 12.1a

1. Given the equation $M = 250 \cdot 3^t$, find values for M when $t = 0, 1, 2,$ and 3.

2. Use Table 12.1 to estimate the value of t when N is

   a. 2000   **b.** 50,000

3. Use the following graph of $y = 3^x$ to estimate the solution to each of the following equations.

   a. $3^x = 7$

   b. $3^x = 0.5$

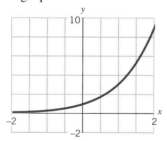

Graph of $y = 3^x$.

4. Use the accompanying table to estimate the solution to each equation.

   a. $5^x = 73$

   b. $5^x = 0.36$

x	5^x
-2	0.04
-1	0.2
0	1
1	5
2	25
3	125
4	625

Properties of Logarithms

Recall that

The *logarithm base 10 of x* is the power of 10 you need to produce x.

$$\log_{10} x = c \qquad \text{means that} \qquad 10^c = x.$$

Logarithms base 10 are called *common logarithms* and $\log_{10} x$ is usually abbreviated $\log x$.

$$10^5 = 100,000 \qquad \text{is equivalent to saying that} \qquad \log 100,000 = 5.$$

$$10^{-3} = 0.001 \qquad \text{is equivalent to saying that} \qquad \log 0.001 = -3$$

Using a calculator we can solve equations such as

$$10^x = 80$$

We can rewrite it in equivalent form $\qquad \log 80 = x$

use a calculator to compute the log $\qquad 1.903 \approx x$

But to solve equations such as $10 = 2^t$ that involve exponential expressions where the base is not 10, we need to know more about logarithms. As the following tables suggest, the properties of logarithms follow directly from the definition of logarithms and from the properties of exponents.

Properties of Exponents

If a is any positive real number and p and q are any real numbers, then

1. $a^p \cdot a^q = a^{p+q}$
2. $a^p/a^q = a^{p-q}$
3. $(a^p)^q = a^{p \cdot q}$
4. $a^0 = 1$

Corresponding Properties of Logarithms

If A and B are positive real numbers and p is any real number, then

1. $\log (A \cdot B) = \log A + \log B$
2. $\log (A/B) = \log A - \log B$
3. $\log A^p = p \log A$
4. $\log 1 = 0$ (since $10^0 = 1$)

In addition

5. $\log (10^x) = x$
6. $10^{\log x} = x$

Finding the common logarithm of a number means finding its exponent when the number is written as a power of 10. So when you see "logarithm" think "exponent," and the properties of logarithms make sense.

As we learned in Chapter 7, the log of 0 or a negative number is not defined. But when we take the log of a number, we can get 0 or a negative value. For example, $\log 1 = 0$ and $\log 0.1 = -1$.

We will list a rationale for each of the properties of logarithms and prove Properties 1, 5 and 6. We leave the other proofs as exercises.

Something to think about

Try expressing in words each of the other properties of logarithms in terms of exponents.

PROPERTY 1 $\qquad\qquad \log (A \cdot B) = \log A + \log B$

RATIONALE

Property 1 of exponents says that when we multiply two terms with the same base, we keep the base and *add* the exponents; that is, $a^p \cdot a^q = a^{(p+q)}$.

Property 1 of logs says that if you write $A \cdot B$, A and B each as powers of 10, then the exponent of $A \cdot B$ is the sum of the exponents of A and B.

PROOF

If we let log $A = x$, then	$10^x = A$
and if log $B = y$, then	$10^y = B$
we have two equal products	$A \cdot B = 10^x \cdot 10^y$
by laws of exponents	$= 10^{x+y}$
Now rewrite using logs	$\log (A \cdot B) = x + y$
substitute log A for x and log B for y	$\log (A \cdot B) = \log A + \log B$

and we arrive at our desired result.

Use Properties (1) and (5) to verify that $\log (10^2 \cdot 10^3) = \log 10^2 + \log 10^3$.

Example   1

SOLUTION

On the left side $\log (10^2 \cdot 10^3) = \log (10^{2+3}) = \log 10^5 = 5$.

On the right side $\log 10^2 + \log 10^3 = 2 + 3 = 5$

Since both sides equal 5, the left and the right sides of our original equation are equal.

Verify that $\log (10^5 \cdot 10^{-7}) = \log 10^5 + \log 10^{-7}$.

Example   2

SOLUTION

On the left side $\log (10^5 \cdot 10^{-7}) = \log (10^{5-7}) = \log 10^{-2} = -2$

On the right side $\log 10^5 + \log 10^{-7} = 5 - 7 = -2$

Since both sides equal -2, then $\log (10^5 \cdot 10^{-7}) = \log 10^5 + \log 10^{-7}$.

PROPERTY 2 $\log (A/B) = \log A - \log B$

RATIONALE

Property 2 of logs says that when we divide terms with the same base, we keep the base and *subtract* exponents, i.e., $a^p/a^q = a^{p-q}$. Property 2 of logs says that if we write A and B each as powers of 10, then the exponent of A/B equals the exponent of A *minus* the exponent of B.

Verify that $\log (10^2/10^3) = \log 10^2 - \log 10^3$.

Example   3

SOLUTION

On the left side we have $\log (10^2/10^3) = \log (10^{2-3}) = \log (10^{-1}) = -1$

and on the right side $\log 10^2 - \log 10^3 = 2 - 3 = -1$

Example   4

Verify that $\log(10^5/10^{-7}) = \log 10^5 - \log 10^{-7}$.

SOLUTION

$$\log(10^5/10^{-7}) = \log(10^{5-(-7)} = \log(10^{12}) = 12$$

and

$$\log 10^5 - \log 10^{-7} = 5 - (-7) = 12$$

PROPERTY 3 $\log A^p = p \log A$

Something to think about

Why is $\log \sqrt{AB}$ halfway between $\log A$ and $\log B$?

RATIONALE

Since

$$\log A^2 = \log(A \cdot A) = \log A + \log A = 2 \log A$$
$$\log A^3 = \log(A \cdot A \cdot A) = \log A + \log A + \log A = 3 \log A$$
$$\log A^4 = \log(A \cdot A \cdot A \cdot A) = 4 \log A$$

it seems reasonable to expect that in general

$$\log A^p = p \log A$$

Example   5

Simplify each expression, and if possible evaluate with a calculator.

$$\log 10^3 = 3 \log 10 = 3 \cdot 1 = 3$$
$$\log 2^3 = 3 \log 2 \approx 3 \cdot 0.301 \approx 0.903$$
$$\log \sqrt{3} = \log 3^{(1/2)} = (1/2) \log 3 \approx (1/2) \cdot 0.477 \approx 0.239$$
$$\log x^{-1} = (-1) \cdot \log x = -\log x$$
$$\log 0.01^a = a \log 0.01 = a \cdot (-2) = -2a$$

PROPERTY 4 $\log 1 = 0$

RATIONALE

Since $10^0 = 1$ by definition, the equivalent statement using logarithms is $\log 1 = 0$.

PROPERTIES 5 AND 6

$$\log(10^x) = x \qquad \text{and} \qquad 10^{\log x} = x$$

RATIONALE FOR $\log(10^x) = x$

Finding the logarithm of a number base 10 involves writing the number as 10 to a power, and then identifying the exponent. Since 10^x is already written as 10 to the power x, then $\log(10^x) = x$.

PROOF

By Property 3 $\log 10^x = x \log 10$

evaluate $\log 10$ $= x(1)$

$$\log 10^x = x$$

RATIONALE FOR $\qquad\qquad 10^{\log x} = x$

By definition, $\log x$ is the number such that when 10 is raised to that power the result is x.

PROOF

Let $\qquad\qquad\qquad\qquad\qquad\qquad y = \log x$

rewrite using definition of logarithm $\qquad 10^y = x$

substitute $\log x$ for y $\qquad\qquad\qquad 10^{\log x} = x$

Use the properties of logarithms to write the following expression as the sum or differences of several logs.

Example 6

$$\log\left(\frac{x(y-1)^2}{\sqrt{z}}\right)$$

SOLUTION

By Properties 1 and 2 $\qquad \log\left(\dfrac{x(y-1)^2}{\sqrt{z}}\right) = \log x + \log(y-1)^2 - \log z^{1/2}$

by Property 3 $\qquad\qquad\qquad\qquad\quad = \log x + 2\log(y-1) - \dfrac{1}{2}\log z$

We call this process *expanding the expression.*

Use the properties of logarithms to write the following expression as a single logarithm.

Example 7

$$2\log x - \log(x-1)$$

SOLUTION

By Property 3 $\qquad 2\log x - \log(x-1) = \log x^2 - \log(x-1)$

by Property 2 $\qquad\qquad\qquad\qquad\quad = \log\left(\dfrac{x^2}{x-1}\right)$

We call this process *contracting the expression.*

Common Error

Probably the most common error in using logarithms stem from confusion over the division property. For example,

$$\log 10 - \log 2 = \log\left(\frac{10}{2}\right)$$

but

$$\log 10 - \log 2 \neq \frac{\log 10}{\log 2}$$

Example 8

Solve for x.

$$1 = \log x + \log (x + 3)$$

SOLUTION

Use Property 1	$1 = \log [x (x + 3)]$
rewrite using definition of log	$10^1 = x (x + 3)$
multiply and subtract 10	$0 = x^2 + 3x - 10$
factor	$0 = (x + 5)(x - 2)$
solve	$x = -5 \quad$ or $\quad x = 2$

Since logarithms are not defined for negative numbers, then $\log (-5)$ is not defined. Hence we must discard $x = -5$ as a solution. So $x = 2$ is the only solution.

Algebra Aerobics 12.1b

1. Using only the properties of exponents and the definition of logarithm, verify that
   a. $\log (10^5/10^7) = \log 10^5 - \log 10^7$
   b. $\log [10^5 \cdot (10^7)^3] = \log (10^5) + 3 \log (10^7)$

2. Expand, using the properties of logarithms.

   $$\log \sqrt{\frac{2x - 1}{x + 1}}$$

3. Contract, expressing your answer as a single logarithm.

   $$\frac{1}{3} [\log x - \log (x + 1)]$$

4. Solve for x.
   a. $\log (x + 6) = \log (x + 20) - \log (x + 2)$
   b. $\log (x + 12) - \log (2x - 5) = 2$

5. Show that $\log 10^3 - \log 10^2 \neq \dfrac{\log 10^3}{\log 10^2}$

Answering Our Original Question: Using Logarithms To Solve Exponential Equations

Remember the question in Section 12.1 that started this chapter? We wanted to find out how long it would take 100 *E. coli* bacteria to become 1000. To find an exact solution, we needed to solve the equation $1000 = 100 \cdot 2^t$, or by dividing by 100, the equivalent equation, $10 = 2^t$. Now we have the necessary tools to solve such equations.

given	$10 = 2^t$
take the logarithm of each side	$\log 10 = \log 2^t$
use Property 3 of logs	$\log 10 = t \log 2$
divide both sides by log 2	$\dfrac{\log 10}{\log 2} = t$
or	$t = \dfrac{\log 10}{\log 2}$ time periods

When $N = 1000$, t is log 10/log 2. We have found a strategy using logarithms that produces an *exact* mathematical answer.

It's hard to judge the size of the number log 10/log 2, but we can obtain as accurate an estimate as needed by using a calculator. Since log 10 = 1 and log 2 ≈ 0.301, then

$$t = \frac{\log 10}{\log 2}$$

$$\approx \frac{1}{0.301}$$

$$\approx 3.32 \text{ time periods}$$

which is consistent with our previous estimates of a value between 3 and 4, approximately equal to 3.3.

Since each time period, t, represents 20 minutes, then 3.32 time periods represents $3.32 \cdot (20 \text{ minutes}) = 66.4$ minutes. So in the model, the bacteria would increase from the initial number of 100 to 1000 in a little over 66 minutes.

How long would it take for 100 bacteria to increase to 2000 bacteria?

Example   9

SOLUTION

Given that

$$N = 100 \cdot 2^t$$

where N = number of bacteria, and t = time (in 20-minute periods), then we must set $N = 2000$ and solve the equation

$$2000 = 100 \cdot 2^t$$

divide both sides by 100 $20 = 2^t$

take the log of both sides $\log 20 = \log 2^t$

use Property 3 of logs $\log 20 = t \log 2$

divide by log 2 $\dfrac{\log 20}{\log 2} = t$

evaluate using a calculator $\dfrac{1.301}{0.301} \approx t$

simplify and switch sides to get $t \approx 4.32$ time periods

So when $N = 2000$, $t = 4.32$ time periods. Since each time period is 20 minutes, it takes $4.32 \cdot (20 \text{ minutes}) = 86.4$ minutes, or a little over 86 minutes for the bacteria to increase from 100 to 2000.

We could also arrive at the same answer if we noted that since 2 is the base (or the growth factor) of the exponential function $N = 100 \cdot 2^t$, then the number of bacteria doubles during each time period t. Each time period is 20 minutes. Since it takes about 66 minutes or 3.3 time periods for 100 bacteria to grow to 1000, it will take one additional time period of 20 minutes for the bacteria to double from 1000 to 2000. Hence the total time to grow from 100 to 2000 bacteria will be $3.3 + 1 = 4.3$ time periods or approximately $66 + 20 = 86$ minutes.

Example   10

As we saw in Chapter 8, the equation $P = 250(1.05)^n$ gives the value of $250 invested at 5% interest (compounded annually) for n years. How many years does it take for the initial $250 investment to double to $500?

SOLUTION

a. Estimating the answer

If $R = 5\%$ per year, then the rule of 70 (discussed in Section 8.4) estimates the doubling time as

$$70/R = 70/5 = 14 \text{ years}$$

b. Calculating a more precise answer.

We can set $P = 500$ and solve the equation

$$500 = 250(1.05)^n$$

divide both sides by 250	$2 = (1.05)^n$
take the log of both sides	$\log 2 = \log (1.05)^n$
use Property 3 of logs	$\log 2 = n \log 1.05$
divide by log 1.05	$\log 2 / \log 1.05 = n$
evaluate with a calculator	$0.301/0.021 \approx n$
simplify and switch sides	$n \approx 14.3 \text{ years}$

So the estimate of 14 years using the rule of 70 was pretty close.

Example   11

In Chapter 8 we used the function $f(t) = 100 \left(\dfrac{1}{2}\right)^{\left(\frac{t}{28}\right)}$ to measure the remaining amount of radioactive material as 100 milligrams (mg) of strontium-90 decayed over time t (in years). How many years would it take for there to be only 10 mg of strontium-90 left?

SOLUTION
We set $f(t) = 10$ and solve the equation

$$10 = 100 \cdot \left(\dfrac{1}{2}\right)^{\left(\frac{t}{28}\right)}$$

divide both sides by 100

$$0.1 = \left(\dfrac{1}{2}\right)^{\left(\frac{t}{28}\right)}$$

take the log of both sides

$$\log 0.1 = \log\left(\left(\dfrac{1}{2}\right)^{\left(\frac{t}{28}\right)}\right)$$

use Property 3 of logs

$$\log 0.1 = \left(\dfrac{t}{28}\right)\log\left(\dfrac{1}{2}\right)$$

divide by log (1/2)

$$\dfrac{\log 0.1}{\log (1/2)} = \dfrac{t}{28}$$

note that $\log 0.1 = \log 10^{-1} = -1$
and $1/2 = 0.5$

$$\dfrac{-1}{\log (0.5)} = \dfrac{t}{28}$$

evaluate using a calculator

$$\frac{-1}{-0.301} \approx \frac{t}{28}$$

simplify

$$3.32 \approx \frac{t}{28}$$

multiply by 28 and switch sides

$$t \approx 93 \text{ years}$$

So it takes almost a century for 100 mg of strontium-90 to decay to 10 mg.

Algebra Aerobics 12.1c

1. Solve for *t* in the equation

$$60 = 10 \cdot 2^t$$

2. Using the model $N = 100 \cdot 2^t$ for bacteria growth where *t* is measured in 20 minute time periods, how long will it take for the bacteria count to reach 7000? To reach 12,000?

3. First use the rule of 70 to estimate how long it would take $1000 invested at 6% compounded annually to double to $2000. Then use logs to find a more precise answer.

4. Use the equation in Example 11. How long will it take for 100 milligrams of strontium-90 to decay to 1 milligram?

12.2 BASE *e*

A Brief Introduction to *e*

We saw in Chapter 7 that we can construct expressions using any positive number *a* as a base and any real number as an exponent. Similarly, we can construct logarithms using any positive number *a* as a base. There are two standard bases used with logarithms. Base 10 may seem most logical when you are first learning about logarithms. But in most scientific applications, another base, a special number called *e* (named after Euler, a Swiss mathematician), turns out to be a much more natural choice. The value of *e* is approximately 2.71828. (You can use 2.72 as an estimate for *e* in most calculations. Your calculator probably has an e^x key for more accurate computations.) The number *e* is irrational; it cannot be written as the quotient of two integers or as a repeating decimal. Like π, *e* is a fundamental mathematical constant.

> The irrational number *e* is a fundamental mathematical constant whose value is approximately 2.71828.

First we'll discuss why *e* is important, and then we'll see how any exponential function can be written using *e*.

Continuous Compounding

The number *e* arises naturally in cases of continuous growth at a specified rate. For example, suppose we invest $100 in a bank account that pays interest of 6% per year. To compute the amount of money we have at the end of 1 year, we must also know how often the interest is credited to our account, that is, how often it is *compounded*.

The simplest case is the type we have examined in Chapter 8, where the interest is compounded once per year. We earn interest of 6% on our principal, so at the end of 1 year we have

$$\$100 + 0.06 \cdot (\$100) = \$100 \cdot (1 + 0.06)$$
$$= \$100 \cdot (1.06)$$
$$= \$106$$

Now suppose that the interest is compounded twice a year. Since the rate is 6% *per year,* the interest computed at the end of each 6-month period is 6%/2 or 3%. So at the end of the first half year, we have $100 · (1 + 0.03) = $100 · (1.03) = $103. At the end of the second half year, we earn 3% interest on our new balance of $103. So at the end of one year our $100 has become:

$$\$103 \cdot (1.03) = (\$100 \cdot 1.03) \cdot (1.03)$$
$$= \$100 \cdot (1.03)^2$$
$$= \$106.09$$

We earn 9 cents more when interest is credited twice per year than when it is credited once per year; the difference is a result of interest earned during the second half year on the $3 in interest credited at the end of the first half year. In other words, we're starting to earn interest on interest. To earn the same amount with only annual compounding, we would need an interest rate of 6.09%. We call 6% the *nominal interest rate* and 6.09% the *effective interest rate* or *effective annual yield.*[1] This distinction was not useful before, because when interest is compounded only once a year the nominal and effective rates are the same.

Next, suppose that interest is compounded quarterly, or four times per year. In each quarter, we receive one-quarter of 6%, or 1.5% interest. Each quarter, our investment is multiplied by $1 + 0.015 = 1.015$ and, after the first quarter, we earn interest on the interest we have already received. At the end of 1 year we have received interest four times, so our initial $100 investment has become:

$$\$100 \cdot (1.015)^4 \approx \$106.14$$

In this case, the effective interest rate (or effective annual yield) is about 6.14%.

We may imagine dividing the year into smaller and smaller time intervals and computing the interest earned at the end of 1 year. The effective interest rate will be slightly more each time (Table 12.2).

In general, if we calculate the interest on $100 *n* times a year when the nominal interest rate is 6%, we get

$$\$100 \left(1 + \frac{0.06}{n}\right)^n$$

---

[1] Banks call the nominal interest rate the annual percentage rate or APR. When describing accounts, banks are required by law to list both the nominal (the APR) and the effective interest rates. The effective interest rate is the one that tells you how much interest you will actually earn each year.

Table 12.2
Investing $100 for One Year at a Nominal Interest Rate of 6%

Number of Times Interest Computed During the Year	Value of $100 at End of One Year ($)				Effective Annual Interest Rate (%)
1	$100(1 + 0.06) =$	$100(1.06) =$		106.00	6.00
2	$100(1 + 0.06/2)^2 =$	$100(1.03)^2 \approx$	$100(1.0609) \approx$	106.09	6.09
4	$100(1 + 0.06/4)^4 =$	$100(1.015)^4 \approx$	$100(1.0614) \approx$	106.14	6.14
6	$100(1 + 0.06/6)^6 =$	$100(1.010)^6 \approx$	$100(1.0615) \approx$	106.15	6.15
12	$100(1 + 0.06/12)^{12} =$	$100(1.005)^{12} \approx$	$100(1.0617) \approx$	106.17	6.17
24	$100(1 + 0.06/24)^{24} =$	$100(1.0025)^{24} \approx$	$100(1.0618) \approx$	106.18	6.18
⋮					
n	$100(1 + 0.06/n)^n$				

We may imagine increasing the number of periods, n, without limits, so that interest is computed every week, every day, every hour, every second, and so on. The surprising thing is that the term by which 100 gets multiplied, namely

$$\left(1 + \frac{0.06}{n}\right)^n$$

does not get arbitrarily large. Examine Table 12.3.

Table 12.3
Value of $(1 + 0.06/n)^n$ as *n* Increases

Compounding Period	*n* (No. of Compoundings per Year)		App. value of $(1 + 0.06/n)^n$
Once a day		365	1.0618313
Once an hour	$365 \cdot 24 =$	8,760	1.0618363
Once a minute	$365 \cdot 24 \cdot 60 =$	525,600	1.0618365
Once a second	$365 \cdot 24 \cdot 60 \cdot 60 =$	31,536,000	1.0618365

Where does e fit in? As n, the number of compounding periods per year, increases, the value of $(1 + 0.06/n)^n$ approaches $1.0618365 \approx e^{0.06}$. You can confirm this on your calculator. As n gets arbitrarily large, we can think of the compounding occurring at each instant. We call this *continuous compounding*.

So if we invest $100 at 6% continuously compounded, at the end of 1 year we will have

$$\$100 \cdot e^{0.06} \approx \$100 \cdot 1.0618365 = \$106.18365$$

An annual interest rate of 6% *compounded continuously* is equivalent to an annual interest rate of 6.18365% *compounded once a year*. Six percent is the nominal interest rate, and 6.18365% is the effective interest rate, the one that tells you exactly how much money you will make after one year.

What if we invest $100 for x years at a nominal interest rate of 6%? If the interest is compounded n times a year, the annual growth factor is $(1 + 0.06/n)^n$; that is, every year the $100 is multiplied by $(1 + 0.06/n)^n$. After x years the $100 is multiplied by $(1 + 0.06/n)^n$ a total of x times or equivalently multiplied by $[(1 + 0.06/n)^n]^x = (1 + 0.06/n)^{nx}$. So $100 will be worth

$$\$100(1 + 0.06/n)^{nx}$$

If the interest is compounded continuously, the annual growth factor is $e^{0.06}$; that is, every year the $100 is multiplied by $e^{0.06}$. After x years the $100 is multiplied by $e^{0.06}$ a total of x times or equivalently multiplied by $(e^{0.06})^x = e^{0.06x}$. So $100 will be worth

$$\$100e^{0.06x}$$

In summary, if we wish to invest our $100 over x years at a nominal interest rate of 6%, then we will have:

$100(1 + 0.06/n)^{nx}$ if the interest is compounded n times a year

$100e^{0.06x}$ if the interest is compounded continuously

Generalizing Our Results

If we invest P dollars at an annual interest rate of r (in decimal form) compounded n times a year, then at the end of 1 year we will have

$$P\left(1 + \frac{r}{n}\right)^n$$

At the end of x years we have

$$P\left(1 + \frac{r}{n}\right)^{nx}$$

where r is the nominal interest rate.

> The value of P dollars invested at an annual interest rate r (expressed in decimal form) compounded n times a year for x years equals
>
> $$y = P\left(1 + \frac{r}{n}\right)^{nx}$$
>
> where r is the *nominal interest rate*.

Just as $\left(1 + \dfrac{0.06}{n}\right)^n$ approaches $e^{0.06}$ as n gets very large, $\left(1 + \dfrac{r}{n}\right)^n$ approaches e^r.

> $\left(1 + \dfrac{r}{n}\right)^n$ approaches e^r as n increases toward infinity.

Note that in the special case where $r = 1$, as n gets arbitrarily large $\left(1 + \dfrac{1}{n}\right)^n$ approaches $e^1 = e$.

$$\left(1 + \frac{1}{n}\right)^n \text{ approaches } e \text{ as } n \text{ increases toward infinity.}$$

So if P dollars are invested at r percent (in decimal form) compounded continuously, then after x years we have

$$P \cdot (e^r)^x = P \cdot e^{rx}$$

In summary,

> The value of P dollars invested at an annual interest rate r (expressed in decimal form) compounded continuously for x years equals
>
> $$y = P \cdot e^{rx}$$
>
> where r is the *nominal interest rate*.

If you have $250 to invest, and you are quoted a nominal interest rate of 4%, construct the equations that will tell you how much money you will have if the interest is compounded once a year, quarterly, once a month, or continuously. In each case calculate the value after 10 years.

Example   1

SOLUTION

Table 12.4

Investing $250 at a Nominal Interest Rate of 4% for Different Compounding Intervals

Number of Compoundings per Year	Value After x Years ($)		Approximate Value (in $) when $x = 10$ Years
1	$250 \cdot (1 + 0.04)^x$	$= 250 \cdot (1.04)^x$	370.06
4	$250 \cdot (1 + 0.04/4)^{4x}$	$= 250 \cdot (1.01)^{4x}$	372.22
12	$250 \cdot (1 + 0.04/12)^{12x}$	$\approx 250 \cdot (1.00333)^{12x}$	372.56
Continuous	$250 \cdot e^{0.04x}$	$= 250 \cdot e^{0.04x}$	372.96

If the nominal interest rate is 7% compounded continuously, what is the effective interest rate?

Example   2

SOLUTION

The nominal interest rate of 7% is compounded continuously, so the equation $y = Pe^{0.07x}$ describes the amount y that an initial investment P is worth after x years. Using a calculator we find that $e^{0.07} \approx 1.073$. The equation could be rewritten as $y = P(1.073)^x$. So the effective interest rate is about 7.3%.

Example 3

You have a choice between two bank accounts. One is a passbook account in which you receive simple interest of 5% per year, compounded once per year. The other is a 1-year certificate of deposit, which pays interest at the rate of 4.8% per year, compounded continuously. Which account is the better deal?

SOLUTION

Since the interest on the passbook account is compounded once a year, the nominal and effective interest rates are both 5%. The equation $y = P(1.05)^x$ can be used to describe the amount y that the initial investment P is worth after x years.

The 1-year certificate of deposit has a nominal interest rate of 4.8%. Since this rate is compounded continuously, the equation $y = Pe^{0.048x}$ describes the amount y that the initial investment P is worth after x years. Since $e^{0.048} \approx 1.049$, the equation can also be written as $y = P(1.049)^x$, and the effective interest rate is 4.9%. So the passbook account is a better deal.[2]

Algebra Aerobics 12.2

1. Find the amount accumulated after 1 year on an investment of $1000 at 8.5% compounded

   a. annually   **b.** quarterly   **c.** continuously

2. Find the effective interest rate if the nominal interest rate is given below and interest is compounded continuously.

   a. 4%   **b.** 12.5%   **c.** 18%

3. The value for e is often defined as the number that $(1 + 1/n)^n$ approaches as n gets arbitrarily large. Use your calculator to fill in the table below. Use your exponent key (x^y or y^x) to evaluate the last column. Is your value consistent with the approximate value for e of 2.71828 given in the text?

n	$1/n$	$1 + (1/n)$	$[1 + (1/n)]^n$
1			
100	0.01	$1 + 0.01 = 1.01$	$(1.01)^{100} \approx 2.7048138$
1,000			
1,000,000			
1,000,000,000			

12.3 THE NATURAL LOGARITHM

The *common logarithm* uses 10 as a base. The *natural logarithm* uses e as a base and is written $\ln x$ rather than $\log_e x$. Scientific calculators have a key that computes $\ln x$.

---

[2] You need to be aware, when opening an account on which the interest is compounded a finite number of times during the year, of whether or not you lose interest if you withdraw the money early. For example, frequently in accounts that are compounded quarterly, if you withdraw your money before the quarter is up, you will not receive any interest for that quarter.

The *logarithm base e* of x is the power of e you need to produce x. Logarithms base e are called *natural logarithms* and are written as $\ln x$.

$$\ln x = c \qquad \text{means that} \qquad e^c = x$$

The properties for natural logarithms are similar to the properties for common logarithms.

Properties of Common Logarithms

If A and B are positive real numbers and p is any real number, then

1. $\log (A \cdot B) = \log A + \log B$
2. $\log (A/B) = \log A - \log B$
3. $\log A^p = p \log A$
4. $\log 1 = 0$ (since $10^0 = 1$)
5. $\log (10^x) = x$
6. $10^{\log x} = x$

Properties of Natural Logarithms

If A and B are positive real numbers and p is any real number, then

1. $\ln (A \cdot B) = \ln A + \ln B$
2. $\ln (A/B) = \ln A - \ln B$
3. $\ln A^p = p \ln A$
4. $\ln 1 = 0$ (since $e^0 = 1$)
5. $\ln (e^x) = x$
6. $e^{\ln x} = x$

Like the common logarithm, $\ln A$ is not defined when $A \le 0$.

Property 5 of natural logarithms says that $\ln (e^x) = x$, since x is the power of e needed to produce e^x. Applying this property we have

$$\ln e = \ln e^1 = 1$$
$$\ln e^4 = 4$$

Example 1

Property 6 says that $e^{\ln x} = x$, since by definition $\ln x$ is the power of e needed to produce x. As a result we have

$$e^{(\ln x^2)} = x^2$$
$$e^{3 \ln (x)} = e^{(\ln x^3)} = x^3$$

If the effective annual interest rate on an account compounded continuously is 0.0521, estimate the nominal interest rate.

Example 2

SOLUTION

If 0.0521 is the effective interest rate, then the equation $y = P(1.0521)^x$ represents the value, y, of P dollars after x years. To find the nominal interest rate that is continuously compounded, we must write y in the form Pe^{rx} or equivalently $P(e^r)^x$. So the expression e^r must equal 1.0521. We need to solve for r in the equation

$$1.0521 = e^r$$

take ln of both sides	$\ln 1.0521 = \ln e^r$
use a calculator	$0.0508 \approx \ln e^r$
use Property 5	$0.0508 \approx r$

So the nominal interest rate is approximately 5.08%. In other words, compounding annually once at 5.21% is equivalent to compounding continuously at 5.08%.

Example 3

Expand, using the laws of logarithms,

$$\ln \sqrt{\frac{x+3}{x-2}}$$

SOLUTION

$$\ln \sqrt{\frac{x+3}{x-2}} = \ln \left(\frac{x+3}{x-2} \right)^{1/2} = \frac{1}{2} \ln \left(\frac{x+3}{x-2} \right) = \frac{1}{2} [\ln (x+3) - \ln (x-2)]$$

Example 4

Contract, expressing the answer as a single logarithm,

$$\frac{1}{3} \ln (x-1) + \frac{1}{3} \ln (x+1)$$

SOLUTION

$$\frac{1}{3} \ln (x-1) + \frac{1}{3} \ln (x+1) = \frac{1}{3} [\ln (x-1) + \ln (x+1)]$$

$$= \frac{1}{3} \ln [(x-1)(x+1)]$$

$$= \ln [(x-1)(x+1)]^{(1/3)}$$

$$= \ln (x^2 - 1)^{(1/3)} \quad \text{or} \quad \ln \sqrt[3]{x^2 - 1}$$

Example 5

Solve for t

$$10 = e^t$$

SOLUTION

Given	$10 = e^t$
take ln of both sides	$\ln 10 = \ln e^t$
use Property 5	$\ln 10 = t$
evaluate and switch sides	$t \approx 2.303$

Algebra Aerobics 12.3

1. Evaluate without a calculator.

   a. $\ln e^2$ d. $\ln \dfrac{1}{e^2}$

   b. $\ln \dfrac{1}{e}$ e. $\ln 1$

   c. $\ln \sqrt{e}$

2. Expand $\ln \dfrac{\sqrt{x+2}}{x(x-1)}$

3. Contract, expressing your answer as a single logarithm.
   $$\ln x - 2\ln(2x-1)$$

4. Find the nominal rate on an investment compounded continuously if the effective rate is 6.4%.

5. Determine how long it takes for $10,000 to grow to $50,000 at 7.8% compounded continuously.

6. Solve the following equations for x.

   a. $e^{x+1} = 10$

   b. $e^{x-2} = 0.5$

12.4 LOGARITHMIC FUNCTIONS

For any $x > 0$, both the common and natural logarithms define unique numbers $\log x$ and $\ln x$, respectively, so we can define two functions

$$y = \log x \qquad \text{and} \qquad y = \ln x$$

What will the graphs look like? We know something about the graphs since:

Exploration 12.1 and course software "E8: Logarithmic Sliders" in Exponential & Log Functions *will help you understand the properties of logarithmic functions.*

If $x > 1$,	$\log x$ and $\ln x$ are both positive.
If $x = 1$,	$\log 1 = 0$ and $\ln 1 = 0$.
If $0 < x < 1$,	$\log x$ and $\ln x$ are both negative.
If $x \le 0$,	neither logarithm is defined.

Both graphs lie to the right of the y-axis, since they are only defined for $x > 0$. They both cross the x-axis at $(1, 0)$ and lie above the x-axis to the right of the x-intercept. Between the origin and $(1, 0)$, the graphs lie below the x-axis.

We can use a calculator to construct a table of values and then sketch the graphs of both functions. See Table 12.5 and Figure 12.2 on the next page.

The graphs of common and natural logarithmic functions share a distinctive shape. Both are asymptotic to the y-axis; that is, as x gets closer and closer to 0, the graphs of $\log x$ and $\ln x$ approach negative infinity. Both graphs come very close to, but never touch, the y-axis. The graphs meet once at the common x-intercept $(1, 0)$. Both functions grow large very slowly when $x > 1$. As $x \to +\infty$, both $\log x$ and $\ln x \to +\infty$. The domain for both functions is all positive real numbers, since the logarithms of 0 and negative numbers are not defined. The range for both is all real numbers.

Table 12.5		
Evaluating $\log x$ and $\ln x$		
x	$y = \log x$	$y = \ln x$
0.001	−3.000	−6.908
0.01	−2.000	−4.605
0.1	−1.000	−2.303
1	0.000	0.000
2	0.301	0.693
3	0.477	1.099
4	0.602	1.386
5	0.699	1.609
6	0.778	1.792
7	0.845	1.946
8	0.903	2.079
9	0.954	2.197
10	1.000	2.303

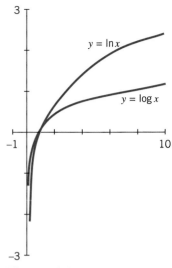

Figure 12.2 Graphs of $y = \log x$ and $y = \ln x$.

Measuring Acidity: The pH Scale

Chemists use the pH scale to measure acidity.[3] The pH is given by the equation

$$pH = -\log [H^+]$$

where $[H^+]$ designates the concentration of hydrogen ions. Chemists use the symbol H^+ for hydrogen ions and the brackets [ ] mean "the concentration of."

$[H^+]$ is measured in molar units, M. Typical concentrations range from $10^{-15}\,M$ to $10\,M$. Table 12.6 and Figure 12.3 show a set of values and the graph for pH.

Table 12.6	
Calculating pH Values	
$[H^+]$ (in molar units M)	pH
10^{-15}	15.000
10^{-10}	10.000
1	0.000
5	−0.699
10	−1.000

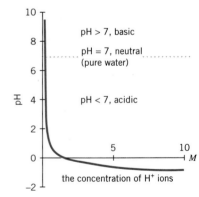

Figure 12.3 Graph of the pH function.

The graph is the standard logarithmic graph flipped over the horizontal axis, because of the negative sign in front of the log.

[3] For more details about the pH scale see Exercise 39 in Chapter 7.

Note that multiplying $[H^+]$ by 10 (increasing $[H^+]$ by 1 order of magnitude), *decreases* the pH by 1. For example, when $[H^+] = 1$, then pH $= -\log 1 = 0$. When $[H^+] = 10$, then pH $= -\log 10 = -1$.

Pure water has a pH of 7 and is considered neutral. A substance with a pH > 7 is called *basic* (or *alkaline*). One with a pH < 7 is called *acidic*. The lower the pH, the higher the acidity and the higher the concentration of hydrogen ions.

Vinegar has a pH of 3. Compare its acidity to that of pure water.

Example 1

SOLUTION
Since vinegar's pH is < 7, it is acidic and has a higher concentration of H^+ than pure water. Its pH is 4 *less* than pure water's, so vinegar is 10^4 or 10,000 times more acidic!

Ammonia has a pH of 10. Compare its H^+ ion concentration to pure water's.

Example 2

SOLUTION
Since ammonia's pH is > 7, it is basic (or alkaline). Its pH is 3 *more* than pure water's, so its H^+ concentration (or acidity) is less, only $10^{-3} = 1/10^3$ or 1/1000 that of water.

• • •

The rain in many parts of the world is becoming increasingly acidic. The burning of fossil fuels (such as coal and oil) by power plants and automobile exhausts releases gaseous impurities into the air. The impurities contain oxides of sulfur and nitrogen that combine with moisture in the air to form drops of dilute sulfuric and nitric acids. An acid, by definition, releases hydrogen ions in water. High concentrations of hydrogen ions damage plants and water resources (such as the lakes of New England and Sweden) and erode structures (such as the outdoor fountain statues in Rome) by removing oxygen molecules. Some experts feel that the acid rain dilemma may be one of the greatest environmental problems facing the world in the near future.

Measuring Noise: The Decibel Scale

The decibel scale was designed to reflect human perception of sounds.[4] When it is very quiet it is easy to notice a small increase in sound intensity. The same increase in intensity in a noisy environment would not be noticed; it would take a much bigger change to be detected by humans. The same is true for light. If a 50-watt light bulb is replaced with a 100-watt light bulb, it is easy to notice the difference in brightness. But if you replace 500 watts with 550 watts, it would be very hard to distinguish the 50-watt difference. So the decibel scale, like the pH scale for acidity or the Richter scale for earthquakes, is logarithmic; that is, it measures order of magnitude changes.

Noise levels are measured in units called *decibels,* abbreviated dB. The name is in honor of the inventor of the telephone, Alexander Graham Bell. If we designate by I_0 the intensity of a sound at the threshold of human hearing (10^{-16} watts/cm^2), and we

---

[4] Two scientists, Weber and Fechner, studied the psychological response to intensity changes in stimuli. Their discovery that the perceived change was proportional to the logarithm of the intensity change of the stimulus is called the Weber-Fechner stimulus law.

let I represent the intensity of an arbitrary sound (measured in watts/cm^2), then the noise level N of that sound measured in decibels is defined to be

$$N = 10 \log \left(\frac{I}{I_0} \right)$$

The expression $\left(\dfrac{I}{I_0} \right)$ gives the *relative intensity* of a sound compared to the reference value of I_0. For example, if $\dfrac{I}{I_0} = 100$, then the noise level, N, is equal to

$$N = 10 \log (100) = 10 \log (10^2) = 10(2) = 20 \text{ dB.}$$

Table 12.7 shows relative intensities, the corresponding noise levels (in decibels), and how people perceive that noise level. Note how much the relative intensity (the ratio I/I_0) of a sound source must increase for people to discern differences. Each time we *add* 10 units on the decibel scale, we *multiply* the relative intensity by 10 (or one order of magnitude).

Something to think about

A noise emission statute enforced by the Massachusetts Registry of Motor Vehicles requires that the noise level of a motorcycle not exceed 82–86 decibels. Jay McMahon, chairman of the Modified Motorcycle Association says; "It's just another way for the Registry to stick it to motorcycle operators. No one complains when a Maserati or a tractor trailer drives by." Does the statute seem fair to you? Why?

Table 12.7

How Decibel Levels Are Perceived

Relative Intensity, I/I_0	Decibels (dB)	Average Perception
1	0	Threshold of hearing
10	10	Sound-proof room, very faint
100	20	Whisper, rustle of leaves
1,000	30	Quiet conversation, faint
10,000	40	Quiet home, private office
100,000	50	Average conversation, moderate
1,000,000	60	Noisy home, average office
10,000,000	70	Average radio, average factory, loud
100,000,000	80	Noisy office, average street noise
1,000,000,000	90	Loud truck, police whistle, very loud
10,000,000,000	100	Loud street noise, noisy factory
100,000,000,000	110	Elevated train, deafening
1,000,000,000,000	120	Thunder of artillery, nearby jackhammer
10,000,000,000,000	130	Threshold of pain, ears hurt

Example   3

What is the decibel level of a typical rock band playing with an intensity of 10^{-5} watts/cm^2? How much more intense is the sound of the band than an average conversation?

SOLUTION

Given $I_0 = 10^{-16}$ watts/cm^2, and letting $I = 10^{-5}$ watts/cm^2 and N represent

the decibel level, by definition $N = 10 \log \left(\dfrac{I}{I_0} \right)$

substitute for I and I_0	$= 10 \log (10^{-5}/10^{-16})$
Property 2 of exponents	$= 10 \log (10^{11})$
Property 5 of logs	$= 10 \cdot 11$
	$= 110$ decibels

So the noise level of a typical rock band is about 110 decibels.

According to Table 12.7, an average conversation measures about 50 decibels. So the noise level of the rock band is 60 decibels higher. Each increment of 10 decibels corresponds to a one order of magnitude increase in intensity. So the sound of a rock band is about six orders of magnitude or $10^6 = 1,000,000$ times more intense than an average conversation.

Example 4

What's wrong with the following statement?

"A jet airplane landing at the local airport makes 120 decibels of noise. If we allow three jets to land at the same time, there will be 360 decibels of noise pollution."

SOLUTION

There will certainly be three times as much sound intensity, but would we perceive it that way? According to Table 12.7, 120 decibels corresponds to a relative intensity of 10^{12}. Three times that relative intensity would equal $3 \cdot 10^{12}$. So the corresponding decibel level would be

	$N = 10 \log (3 \cdot 10^{12})$
	$= 10 (\log 3 + \log (10^{12}))$
use a calculator	$\approx 10(0.477 + 12)$
	$\approx 10(12.477)$
multiply and round off	≈ 125 decibels

So three jets landing will produce a decibel level of 125, not 360. We would only perceive a slight increase in the noise level.

Algebra Aerobics 12.4

1. How would the graphs of $y = \log x^2$ and $y = 2 \log x$ compare?

2. Draw a rough hand sketch of the graph $y = -\ln x$. Compare it to the graph of $y = \ln x$.

3. A typical pH value for rain or snow in the northeastern United States is about 4. Is this basic or acidic? What is the corresponding hydrogen concentration? How does this compare to the hydrogen concentration of pure water?

4. What is the decibel level of a sound whose intensity is $1.5 \cdot 10^{-12}$ watts/cm^2?

5. If the intensity of a sound increases by a factor of 100, what is the increase in the decibel level? What if the intensity is increased by a factor of 10,000,000?

12.5 WRITING EXPONENTIAL FUNCTIONS USING BASE e

Translating from Base a to Base e

We can use logarithms to show that any exponential function can be rewritten using e as a base.

Consider the bacterial growth we described with the equation $N = 100 \cdot 2^t$. The bacteria don't all double at the same time, precisely at the beginning of each time period t. A continuous growth pattern is much more likely. If we are to rewrite this equation to reflect continuous compounding, we want to use base e. To do that, we need to write 2 as e^r for some r.

If $\qquad\qquad\qquad\qquad 2 = e^r$

take ln of both sides $\qquad \ln 2 = \ln e^r$

use Property 5 $\qquad\qquad \ln 2 = r$

By substituting $e^r = e^{\ln 2}$ for 2, we can rewrite

$$N = 100 \cdot 2^t \qquad \text{as} \qquad N = 100\,(e^{\ln 2})^t = 100\,e^{(\ln 2)t}$$

Since $\ln 2 \approx 0.693$, we also have $N \approx 100 e^{0.693t}$.

In general, if $\qquad\qquad\qquad\qquad y = Ca^x$

is an exponential function (where $a > 0$), we can always find an r such that $a = e^r$. By substituting for a we can rewrite y as

$$y = C(e^r)^x = Ce^{rx}$$

We can then write e^r in terms of $\ln a$.

Given $\qquad\qquad\qquad\qquad\qquad\qquad e^r = a$

take ln of both sides of the equation $\qquad \ln e^r = \ln a$

use Property 5 of ln $\qquad\qquad\qquad\qquad r = \ln a$

See "E4: $y = Ce^{rx}$ Sliders" and "E5: Comparing $y = Ca^x$ to $y = Ce^{rx}$" in Exponential & Log Functions.

So $y = Ca^x$ can be rewritten as $y = Ce^{rx} = Ce^{(\ln a)x}$. If $y = Ca^x$ and $a > 1$, the function represents exponential growth. For exponential growth, $\ln a = r$ is greater than 0. When $0 < a < 1$, the function represents decay. For exponential decay, $\ln a = r$ is less than 0.

Given a function $y = Ce^{rx}$, in a financial setting r is called the nominal interest rate. In general applications, r is called the *instantaneous growth rate*.

For any number $a > 0$, the function $y = C \cdot a^x$ may be rewritten $y = C \cdot e^{rx}$, where $r = \ln a$. We call r the *instantaneous growth rate*.

If $a > 1$, then $r > 0$ and the function represents growth.

If $0 < a < 1$, then $r < 0$ and the function represents decay.

If $a = 1$, then $r = 0$ and the function is constant.

Writing an exponential function in the form $y = Ca^x$ or $y = Ce^{rx}$ (where $r = \ln a$) is a matter of emphasis, since the graphs and functional values are identical. When we use $y = Ca^x$, we implicitly think of the growth taking place at discrete points in time, whereas the form $y = Ce^{rx}$ emphasizes the notion of continuous growth. For example, $y = 100(1.05)^x$ could be interpreted as giving the value of \$100 invested for x years at 5% compounded annually, whereas its equivalent form

$$y = 100(e^{\ln 1.05})^x \approx 100e^{0.049x}$$

suggests that the money is invested at 4.9% compounded continuously.

In Chapter 8 we saw that the function $S = 100\left(\dfrac{1}{2}\right)^{\left(\frac{t}{28}\right)}$ measures the amount of radioactive strontium-90 remaining as 100 milligrams (mg) decay over time t (in years). Rewrite the function using base e.

Example 2

SOLUTION

We must rewrite 1/2 (or 0.5) as a power of e; that is, we must solve the equation

$$0.5 = e^k$$

take ln of both sides $\ln 0.5 = \ln e^k$

apply Property 5 of ln $\ln 0.5 = k$

evaluate ln 0.5 $-0.693 \approx k$

We get $0.5 \approx e^{-0.693}$

Substituting into the original function, we get

$$S = 100\,(0.5)^{(t/28)}$$
$$\approx 100\,(e^{-0.693})^{(t/28)}$$
$$\approx 100\,e^{(-0.693t/28)}$$
$$\approx 100\,e^{-0.0248t}$$

Note that since the base 1/2 (or 0.5) is less than 1, the function represents decay. When the function is rewritten using base e, the value of -0.0248 for r, the instantaneous growth rate, is negative.

Use a continuous compounding model to describe the growth of Medicare expenditures.

Example 3

SOLUTION

In Chapter 8 we found a best fit function for Medicare expenditures to be

$$y = 8.7 \cdot 1.14^x$$

where x = years since 1970 and y = Medicare expenditures in billions of dollars. This implies a growth rate of 14% compounded annually. To describe the same growth in terms of continuous compounding we need to rewrite 1.14 as a power of e. Hence we need to solve

$$1.14 = e^r$$

Take ln of both sides	$\ln 1.14 = \ln e^r$
use Property 5 of ln	$\ln 1.14 = r$
evaluate ln 1.14	$0.131 \approx r$
substitute for r	$1.14 \approx e^{0.131}$

Substituting into the original function gives

$$y \approx 8.7 \cdot e^{0.131x}$$

So the instantaneous growth rate is 13.1%. Note that since the original base, 1.14, is greater than 1 and represents exponential growth, then the instantaneous growth rate is positive.

Example   4

We now have the tools to prove the rule of 70 introduced in Chapter 8. Recall the rule said that if a quantity is growing (or decaying) at R percent per time period, then the time it takes the quantity to double (or halve) is approximately 70/R time periods. For example, if a quantity increases by a rate, R, of 7% each month, the doubling time is about 70/7 = 10 months.

PROOF

Let f be an exponential function of the form

$$f(t) = Ce^{rt}$$

Then C is the initial quantity, r is the nominal growth rate, and t represents time. Recall that r is in decimal form, and R is the equivalent amount expressed as a percentage. Let r represent exponential growth, so $r > 0$. Since the doubling time for an exponential function is constant, we need only calculate the time for any given quantity to double. In particular, we can determine the time it takes for the initial amount C (at time $t = 0$) to become twice as large; that is, we can calculate the value for t such that

$$f(t) = 2C$$

set	$Ce^{rt} = 2C$	
divide by C	$e^{rt} = 2$	
take ln of both sides	$\ln e^{rt} = \ln 2$	
evaluate, and use ln Property 5	$rt \approx 0.693$	(1)

$R = 100r$, so $r = R/100$. We round 0.693 up to 0.70.

Substitute in Equation (1)	$(R/100) \cdot t \approx 0.70$
multiply both sides by 100	$R \cdot t \approx 70$
divide by R	$t \approx 70/R$

So the time t it takes for the initial amount to double is approximately 70/R, which is what the rule of 70 claims.

We leave the similar proof about half-lives, where an $r < 0$ represents decay, to the exercises.

Determining the Equation of an Exponential Function Through Two Points

Question   The amount of carbon dioxide (CO_2) in the atmosphere has been increasing steadily since at least 1860, primarily because of the increased burning of fossil fuels such as coal, oil, and gasoline. In 1890 the CO_2 concentration (in parts per million by volume) was roughly 290. This means that out of, say, one million cubic feet of air, 290 cubic feet were CO_2. By 1970 the concentration had risen to about 320. Many scientists describe the increase as exponential and believe that the CO_2 concentration will continue to rise at the same pace for the next 100 to 200 years. Construct a model of CO_2 concentration, starting from 1860, that describes the increase in CO_2 concentration over time.

Solution   If we let our independent variable x = number of years since 1860, then 1890 corresponds to $x = 30$ and 1970 corresponds to $x = 110$. Let Q denote our dependent variable, the CO_2 concentration. Then when $x = 30$, $Q = 290$, and when $x = 110$, $Q = 320$. Our goal then is to construct an exponential function through the two points $(30, 290)$ and $(110, 320)$. The function will be of the basic form

$$Q = Ce^{rx}$$

where Q is the CO_2 concentration x years after 1860, and C is the initial CO_2 concentration in 1860.

Since the points $(30, 290)$ and $(110, 320)$ must satisfy our equation, substituting for x and Q we have

$$290 = Ce^{30r} \qquad \text{and} \qquad 320 = Ce^{110r}.$$

In order to solve for r, we start by dividing Ce^{110r} by Ce^{30r}.

$$\frac{Ce^{110r}}{Ce^{30r}} = \frac{320}{290}$$

We can simplify the left-hand side:

cancel the C's $\dfrac{Ce^{110r}}{Ce^{30r}} = \dfrac{e^{110r}}{e^{30r}}$

use Property 2 of exponents $= e^{110r-30r}$

$= e^{80r}$

Simplifying the right-hand side gives us:

$$320/290 \approx 1.1034$$

Setting the simplified expressions for the left-hand side and the right-hand side equal, we have

$$e^{80r} \approx 1.1034$$

take ln of both sides $\ln e^{80r} \approx \ln 1.1034$

use Property 5 of ln $80r \approx 0.09844$

divide by 80 $r \approx 0.00123$

With that value for the exponent r, our function model will be of the form $Q = Ce^{0.00123x}$. We need to find a value for the coefficient C, the initial amount of CO_2 concentration in 1860.

Substitute one of the two points in our equation for Q to solve for C.

substitute the point (30, 290) $\qquad 290 = Ce^{(0.00123)(30)}$

multiply $\qquad\qquad\qquad\qquad\quad = Ce^{0.0369}$

use a calculator to evaluate $e^{0.0369}$ $\quad 290 \approx C(1.037)$

divide by 1.037, switch sides $\qquad\quad C \approx 280$

So the concentration of CO_2 in 1860 is about 280 parts per million by volume.
The desired function is then

$$Q \approx 280e^{0.00123x}$$

To double check that this is a reasonable answer, verify that the other point (110, 320) satisfies the equation; that is, using your calculator, evaluate $280e^{0.00123x}$ when $x = 110$ to verify that you get approximately 320.

Algebra Aerobics 12.5

1. Identify each of the following exponential functions as representing growth or decay.

    a. $M = Ne^{-0.029t}$ **c.** $Q = 375e^{0.055t}$

    b. $K = 100(0.87)^r$

2. Rewrite each of the following as continuous-growth models using base e.

    a. $y = 1000(1.062)^t$ **b.** $y = 50(0.985)^t$

3. Iodine-131 is a radioactive substance that decays exponentially. If the amount of iodine-131 left after 8 days is 2.40 micrograms, and after 20 days is 0.88 micrograms, construct a function using e to describe its decay. How many micrograms of iodine-131 were there initially, and what was the instantaneous decay rate?

12.6 USING LOGARITHMIC SCALES TO FIND THE BEST FUNCTION MODEL

The course software "E11: Semi-log Plots of $y = Ca^x$," in Exponential & Log Functions *and "P2: Log-log Plots of Power Functions" in* Power Functions *can help you visualize the ideas in this section.*

Throughout this course, we have examined several different families of functions including linear, exponential, logarithmic, and power. In this section we address the question: "How can we determine a reasonable functional model for a given set of data?" The simple answer is: "Look for a way to plot the data so it appears as a straight line."

We look for straight line representations not because the world is intrinsically linear, but because straight lines are easy for humans to recognize and manipulate. In Part I, we considered linear functions, whose graphs are straight lines when plotted on a "standard" plot, with linear scales on both axes. In Chapter 8, we saw how the graph of an exponential function appeared as a straight line when plotted on a *semi-log plot* using a logarithmic scale on the vertical axis and a standard linear scale on the horizontal axis. The graph of a power function appears as a straight line when plotted using a logarithmic scale on both axes. We call such a graph a *log-log plot.*

Let's examine the following functions:

$$y = 3 + 2x \qquad \text{linear function}$$
$$y = 3 \cdot (2^x) \qquad \text{exponential function}$$
$$y = 3x^2 \qquad \text{power function}$$

In Figure 12.4 we plot the three functions using a linear scale on both the x- and y-axes. Only the linear function appears as a straight line; the power and exponential functions curve steeply upwards.

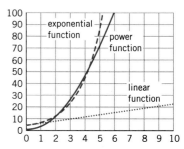

Figure 12.4 Graph of a linear, a power, and an exponential function on a "standard" plot, with linear scales on both axes.

Semi-log Plots

On the semi-log plot in Figure 12.5, we see that the exponential function now is a straight line, and the power and linear functions curve downward. From Chapter 7, we know that moving a fixed distance on a logarithmic scale corresponds to *multiplying* by a constant factor. In exponential growth, the dependent variable is multiplied by the growth factor each time 1 unit is added to the independent variable. Thus, as we saw in Chapter 8, exponential growth will always appear as a straight line on a *semi-log* plot, where the independent variable is plotted on a linear axis, and the dependent variable on a logarithmic axis. This is one of the easiest and most reliable ways to recognize exponential growth in a data set.

A logarithmic scale has the added advantage of allowing us to display clearly a wide range of values. In Figure 12.4 the vertical axis only goes to 100 units, whereas in Figure 12.5, the vertical axis extends to 1,000 units.

The properties of logarithms help us understand why exponential functions appear as straight lines on a semi-log plot. Let's examine our exponential function

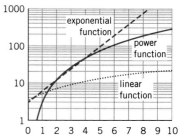

Figure 12.5 Graph of a linear, a power, and an exponential function on a semi-log plot, with a logarithmic scale on the vertical and linear scale on the horizontal axis.

$$y = 3 \cdot (2^x) \qquad (1)$$

take the logarithm of both sides	$\log y = \log\,[3 \cdot (2^x)]$	
use Property 1 of logs	$= \log 3 + \log(2^x)$	
use Property 3 of logs	$\log y = \log 3 + x\,(\log 2)$	(2)

Equation (2) is an equivalent form of Equation (1). If we use a calculator to evaluate log 3 and log 2 we can rewrite Equation (2) as

$$\log y = 0.48 + 0.30x$$

Representing log y as Y, we can rewrite this equation as

$$Y = 0.48 + 0.30x \qquad (3)$$

It's easier then to see that Y (or log y) depends linearly on x. On a standard linear plot, the graph of the points (x, Y) [or equivalently $(x, \log y)$] that satisfy Equation (3) is a straight line. Its slope is 0.30 or log 2, the logarithm of the growth factor of the original exponential function, Equation (1). (See Figure 12.6.)

Figure 12.6 also shows the direct translation between plotting y on a logarithmic scale versus plotting log y on a linear scale. Many graphing calculators do not have the ability to switch between using linear and logarithmic scales, so one can in effect change the scale on the vertical axis from linear to logarithmic, by plotting log y in-

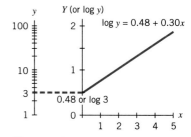

Figure 12.6 Comparing y on a logarithmic scale to log y on a linear scale.

stead of y. Notice that on the log y axis, the units are now evenly spaced, and we can use the standard strategies for finding the slope and vertical intercept.

Using Figure 12.6 we can interpret the slope and the vertical intercept. When $x = 0$, the vertical intercept on the Y (or log y) axis is 0.48, which equals log 3. This is the value of Y in the equation $Y = 0.48 + 0.30x$ when $x = 0$. On the y-axis the vertical intercept is 3, which is the value of y in the original equation $y = 3 \cdot (2^x)$ when $x = 0$.

The slope of the line $Y = 0.48 + 0.30x$ is 0.30 or log 2. Thinking in terms of the x and Y axes, each time we add 1 unit to x, we must add log 2 units to Y (or log y) in order to stay on the line.

If we think in terms of the x- and y-axes, adding 1 unit to x means multiplying y by 2 in order to return to the line. So adding 1 to x translates to *adding* log 2 to log y in Equation (2) or *multiplying* y by 2 in Equation (1) where $y = 3 \cdot 2^x$. These translations make sense if we remember that logarithms are exponents.

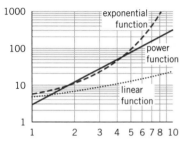

Figure 12.7   Graph of a linear, a power, and an exponential function on a log-log plot.

Log-log Plots

On the log-log plot in Figure 12.7, the power function now appears as a straight line.

Again we can use the properties of logarithms to understand why power functions appear as straight lines on a log-log plot. Let's analyze our power function

$$y = 3x^2 \tag{3}$$

take the logarithm of both sides $\log y = \log (3x^2)$

use Property 1 of logs $= \log 3 + \log x^2$

use Property 3 of logs $\log y = \log 3 + 2 \log x \tag{4}$

Equations (3) and (4) are equivalent ways of saying the same thing. If we evaluate log 3, we get $\log y = 0.48 + 2 \log x$. By letting $X = \log x$ and $Y = \log y$, we can rewrite the equation as

$$Y = 0.48 + 2X \tag{5}$$

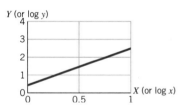

Figure 12.8   The graph of $Y = 0.48 + 2X$.

This helps us see that Y depends linearly on X, or equivalently, that log y depends linearly on log x. If we plot Y vs. X (or equivalently log y vs. log x) the slope, 2, of the straight line described by Equation (5) and graphed in Figure 12.8 is the exponent of the original power function (3).

Since the slope is 2, each time X increases by 1 unit, Y increases by 2 units. Equivalently, when log x increases by 1 unit, log y increases by 2 units. Each time we increase log x by 1, we increase x by a factor of 10. In Equation (3) each time x increases by a factor of 10, y increases by a factor of 10^2.

The linear function looks almost like a straight line on our log-log plot in Figure 12.7. This is perhaps not surprising, since if the constant term were 0, the linear function would be a power function $y = mx$, where the power of x is 1.

Conclusion

We can tell the difference among the graphs of these three types of functions by plotting them on different types of axes. To summarize:

The graph of a linear function appears as a straight line on a standard plot.

The graph of an exponential function appears as a straight line on a semi-log plot. The slope of the line is the logarithm of the growth factor.

The graph of a power function appears as a straight line on a log-log plot. The slope of the line is the exponent of the power function.

Something to think about

On what type of plot would a logarithmic function appear as a straight line?

Algebra Aerobics 12.6a

1. Identify the type of function for the following:

$$y = 4x^3, \quad y = 3x + 4, \quad \text{and} \quad y = 4 \cdot (3^x).$$

2. Which of the functions in Problem (1) would have a straight line graph on a standard linear plot? On a semi-log plot? On a log-log plot?

3. If possible, use technology to check your previous answers. Remember that if you cannot switch from a lin-ear to a logarithmic scale on an axis, then plot the log of the number instead. For example, plotting $(x, \log y)$ will produce a graph equivalent to a semi-log plot. Plotting $(\log x, \log y)$ will produce a graph equivalent to a log-log plot.

4. For each straight line graph in Problem (3), use the original equation in Problem (1) to predict the slope of the line.

Using Semi-log and Log-log Plots to Investigate Data

Accumulated Federal Debt

We can now use our knowledge of semi-log and log-log plots to help find the type of function that best describes a set of data. Table 12.8 shows the accumulated debt of the federal government since 1970. Figure 12.9 is a log-log plot of the data and Figure 12.10 is a semi-log plot of the data.

The points on the semi-log plot show a more linear pattern than the points on the log-log plot. Thus we expect that an exponential function would provide a better fit to

Example   1

See Excel or graph link file
FEDDEBT

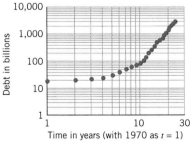

Figure 12.9   Log-log plot of federal debt using 1970 as year 1.

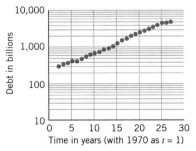

Figure 12.10   Semi-log plot of federal debt using 1970 as year 1.

Something to think about

Why is there no 0 on an axis with a logarithmic scale?

the data than a power function. Using what we know about logarithms we can find the equation of the exponential function that best describes the data. One strategy is to transform the data by calculating the logarithm of the debt, D, or log D, as shown in Table 12.9. As in the graphs in Figure 12.9 and 12.10, we set 1970 as year 1. In Chapter 8, we saw that it is useful to reference a base year when using time series. Since we need to take log(t) for the log-log plot in Figure 12.9, we cannot start with $t = 0$ since log 0 is undefined.

Table 12.8		Table 12.9		
Accumulated Gross Federal Debt		**Time (t) in Years 1970 = Year 1**	**Debt (D) Billions of $**	**Log (D)**
Year	**Billions of $**			
1970	361	1	361	2.56
1971	408	2	408	2.61
1972	436	3	436	2.64
1973	466	4	466	2.67
1974	484	5	484	2.68
1975	541	6	541	2.73
1976	629	7	629	2.80
1977	706	8	706	2.85
1978	777	9	777	2.89
1979	829	10	829	2.92
1980	909	11	909	2.96
1981	994	12	994	3.00
1982	1137	13	1137	3.06
1983	1371	14	1371	3.14
1984	1564	15	1564	3.19
1985	1817	16	1817	3.26
1986	2120	17	2120	3.33
1987	2346	18	2346	3.37
1988	2601	19	2601	3.42
1989	2868	20	2868	3.46
1990	3206	21	3206	3.51
1991	3599	22	3599	3.56
1992	4003	23	4003	3.60
1993	4410	24	4410	3.64
1994	4644	25	4644	3.67
1995 (est.)	4961	26	4961	3.70

A function graphing program provides the following equation for a best fit regression line of the transformed data in Figure 12.11. The numbers are rounded to the nearest hundredth.

$$\log D = 2.47 + 0.05t \tag{1}$$

To solve for D, we need to "undo" our original transformation of the data. We know that $10^{\log D} = D$. So if we use each side of Equation (1) as an exponent for the base 10, we get:

$$10^{\log D} = 10^{(2.47 + 0.05t)}$$

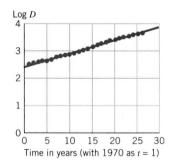

Figure 12.11 Transformed debt data: log(debt) versus time.

substitute D for $10^{\log D}$	$D = 10^{(2.47+0.05t)}$	
use Property (1) and (3) of exponents	$= (10^{2.47})(10^{0.05})^t$	
evaluate $10^{0.05}$	$= (10^{2.47})1.12^t$	
evaluate $10^{2.47}$	$D \approx (295) \cdot 1.12^t$	(2)

Equation (2) is equivalent to Equation (1) and expresses D in terms of t. Hence, an exponential model for the accumulated federal debt is $D = (295) \cdot 1.12^t$. Our model tells us that the annual growth factor is 1.12 and thus there has been an annual growth rate in the accumulated federal debt of approximately 12% per year since 1970.

Analyzing weight and height data

In 1938, Katherine Simmons and T. Wingate Todd measured the average weight and height of children in Ohio between the ages of 3 months and 13 years. Table 12.10 shows their data.

Let's examine the relationship between height and weight. Figure 12.12 shows height vs. weight for boys and for girls together on a linear scale. The filled circles are data for boys, the open squares for girls. Table 12.10 shows that in general baby girls are smaller than baby boys, and teenage girls are larger than teenage boys. But by superimposing the height/weight data for boys and girls in Figure 12.12, we see that at the same weight the heights of boys and girls are roughly the same.

Example   2

See Excel or graph link file CHILDSTAT.

Table 12.10

Measuring Children

Age (yrs)	Weight in kg		Height in cm	
	Boys	**Girls**	**Boys**	**Girls**
1/4	6.5	5.9	61.3	59.3
1	10.8	9.9	76.1	74.2
2	13.2	12.5	87.4	86.2
3	15.2	14.7	96.2	95.5
4	17.4	16.8	103.9	103.2
5	19.6	19.2	110.9	110.3
6	22.0	22.0	117.2	117.4
7	24.8	24.5	123.9	123.2
8	28.2	27.9	130.1	129.3
9	31.5	32.1	136.0	135.7
10	35.6	35.2	141.4	140.8
11	39.2	39.5	146.5	147.8
12	42.0	46.6	151.1	155.3
13	46.6	52.0	156.7	159.9

Source: Data adapted from D'Arcy W. Thompson, *On Growth and Form* (New York: Dover Paperback, 1992) p. 105.

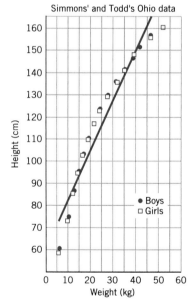

Figure 12.12   Standard linear plot of Simmons and Todd's height/weight data for boys and girls.

The line in Figure 12.12 shows the linear model that best approximates the combined data. The line does not describe the data very well, and it is not reasonable to consider that height is a linear function of weight for growing children. Other models may describe the data better.

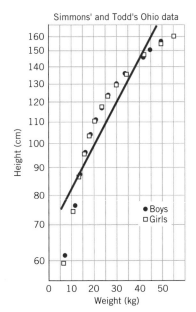

Figure 12.13 Semi-log plot of Simmons and Todd's height/weight data for boys and girls.

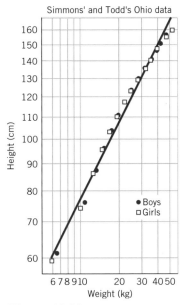

Figure 12.14 Log-log plot of Simmons and Todd's height/weight data.

Figure 12.13 shows the same data, this time using a linear scale for weight, and a logarithmic scale for height. The line shows the best model of the form

$$\log H = b + mW \tag{3}$$

where H is the height (in cm) and W is the weight (in kg). As we saw in Example 1, this equation is equivalent to the exponential model

$$H = Ca^W \qquad \text{where } C = 10^b \text{ and } a = 10^m$$

On this semi-log plot, the exponential model appears as a straight line. Clearly, it does not fit the data well, and it is not reasonable to suppose that height is an exponential function of weight.

Figure 12.14 again shows the same data, but this time using logarithmic scales on both axes. The line shows the best model of the form

$$\log H = b + m \log W \tag{4}$$

If we use each side of the equation as an exponent for the base 10, we get

$$10^{\log H} = 10^{(b + m \log W)}$$

use Property 5 of logs	$H = 10^{(b + m \log W)}$
Property 1 of exponents	$= 10^b \cdot 10^{m \log W}$
Property 3 of logs	$= 10^b \cdot 10^{\log W^m}$
Property 5 of logs	$H = 10^b \cdot W^m$

If we let $C = 10^b$ we get the following power model which is equivalent to Equation (4).

$$H = C \cdot W^m$$

The slope of the regression line in Figure 12.14 is approximately 0.48 which is equal to m. Thus

$$H = C \cdot W^{0.48}$$

In this case, the line is a reasonably good approximation to the data, and certainly the best fit of the three alternatives. It is also plausible to argue that height is a power-law function of weight for growing children. Approximating 0.48 by 0.5 since the data are fairly rough, we may argue that

$$H \propto W^{1/2} \qquad \text{or} \qquad W \propto H^2.$$

where the symbol $\propto$ means "is directly proportional to."

Let's think about whether this is a reasonable exponent. Suppose children grew *self-similarly;* that is, they kept the same shape as they grew from the age of 3 months to 13 years. Then, as we have discussed in Section 9.1 their volume, and hence their weight, would be proportional to the cube of their height:

$$W = kH^3 \qquad \text{(self-similar growth).}$$

But of course children do not grow self-similarly; they become proportionately more slender as they grow from babies to young adults. (See Figure 12.15.) Their weight therefore grows less rapidly than would be predicted by self-similar growth, and we expect an exponent less than three for weight as a function of height.

On the other hand, suppose that children's bodies grew no wider as their height increased. Since weight is proportional to height times cross-sectional area, with constant cross-sectional area we would expect weight to increase linearly with height. So $W \propto H^1$. But certainly, children do become wider as they grow taller, so we expect the exponent of H to be greater than 1. Therefore the experimentally determined exponent of close to 2 seems reasonable.

There is also a fourth possibility for plotting our height-weight data: we may use a linear scale for the height axis, and a logarithmic scale for the weight axis. On such a plot, a straight line will represent a model in which weight is an exponential function of height.

If we do this plot, the data fall very close to a straight line. Based only on the data, we might argue that an exponential function is a reasonable model. However, it is very difficult to construct a plausible physical explanation of such a model, and it would not help us understand how children grow. It may be, of course, that this relationship reveals some unsuspected physical or biological law. Perhaps you, the reader, will be the one to find some previously unsuspected explanation.

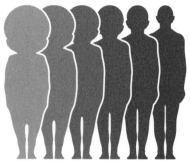

0.42 0.75 2.75 6.75 12.75 25.75
Years

Figure 12.15   The change in human body shape with increasing age.

Source: Thomas A. McMahon and John T. Bonner, *On Size and Life* (New York: Scientific American Books, 1983), p. 32.

Algebra Aerobics 12.6b

1. Interpret the slopes of 1.2 and 1.0 on the accompanying graph in terms of arm length and body height.

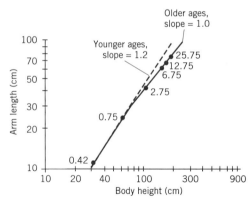

Source: Thomas A. McMahon and John T. Bonner, *On Size and Life* (New York: Scientific American Books, 1983), p. 32.

2. For each of the following graphs examine the scales on the axes. Then decide whether a linear, exponential, or power function would be the most appropriate model for the data.

a.

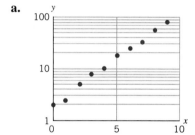

b.

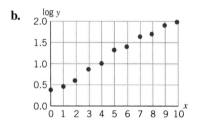

c.

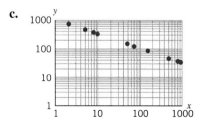

12.7 ALLOMETRY: THE EFFECT OF SCALE

Allometry is the study of how the relationships among different physical attributes of an object change with scale. In biology, allometric studies naturally focus on relationships in living organisms. Biologists look for general "laws" that can describe the relationship for instance between height and weight (as in Example 2 in 12.6) or surface area and volume (as in Chapter 9) as the organism size increases. The relationships may be within one species or across species. These laws are characteristically power functions. Sometimes we can predict the exponent by simple reasoning about physical properties and verify our prediction by examining the data; other times we can measure the exponent from data but cannot yet give a simple explanation for the observed value.

Surface Area vs. Body Mass

In Chapter 9 we made the argument that larger animals have relatively less surface area than smaller ones. Let's see if we can describe this relationship as a power law and then look at some real data to see if our conclusion was reasonable.

Since all animals have roughly the same mass per unit volume, their mass should be directly proportional to their volume. We will substitute mass for volume in our discussion since we can determine mass easily by weighing an animal. Since volume is measured in cubical units of some length, if animals of different sizes have roughly the same shape, then we would expect mass, M, to be proportional to the cube of the length, L, of the animal,

$$M \propto L^3, \tag{1}$$

that is, $M = k_1 \cdot L^3$ for some constant k_1.

Also, we would expect surface area, S, to be proportional to the square of the length,

$$S \propto L^2, \tag{2}$$

that is, $M = k_2 L^2$ for some constant k_2.

By taking the cube root and square root respectively in relationships (1) and (2), we may rewrite them in the form

$$M^{(1/3)} \propto L \qquad \text{and} \qquad S^{(1/2)} \propto L$$

Since $M^{(1/3)} \propto L$ and $L \propto S^{(1/2)}$, we can now eliminate L, and combine these two statements into the prediction that

$$S^{(1/2)} \propto M^{(1/3)}$$

By squaring both sides we have the equivalent relationship

$$S \propto M^{(2/3)}$$

It is useful to eliminate length as a measure of the size of an animal, since it is a little more ambiguous than mass. Should a tail, for instance, be included in the length measurement?

This prediction, that surface area is directly proportional to the 2/3 power of body mass, can be tested experimentally. Figure 12.16 shows data collected for a wide range of mammals, from mice with a body mass on the order of 1 gram to elephants, whose body mass is more than a million grams (1 ton). Here surface area, S, measured in cm^2, is plotted on the vertical axis and body mass, M, in grams, is on the horizontal. The scales on both axes are logarithmic.

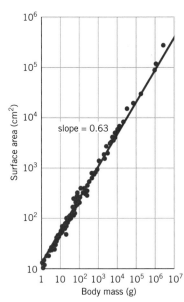

Figure 12.16 A log-log plot of body surface vs. body mass for a wide range of mammals.

Source: Thomas A. McMahon and John T. Bonner, *On Size and Life* (New York: Scientific American Books, 1983) p. 130.

The relationship looks quite linear. Since this is a log-log plot, that implies that S is directly proportional to M^p for some power p. The best fit line has a slope of approximately 2/3, which implies that

$$S \propto M^{(2/3)}$$

confirming our simple prediction of a 2/3 power law. We have approximate verification of the fact that mammals of different body masses have roughly the same proportion of surface area to the 2/3 power of body mass. This power function is just another way of describing our finding in Section 9.1.

To see that the slope of the line corresponds to the exponent 2/3, notice that at the lower left corner it passes through the point whose coordinates are $(1, 10)$. In the middle of the graph, it passes through the point whose coordinates are $(10^3, 10^3)$, and near the upper right corner it passes through the point with coordinates $(10^6, 10^5)$. That is, every time the horizontal coordinate increases by a factor of 10^3, the vertical coordinate increases by a factor of 10^2. The line is described by the equation

$$S = 10\, M^{(2/3)} \qquad (3)$$

where area S is measured in cm^2, and mass M in grams.

Metabolic Rate vs. Body Mass

Another example of scaling among animals of widely different sizes is metabolic rate as a function of body mass. This relationship is very important in understanding the mechanisms of energy production in biology. Figure 12.17 shows a log-log plot of metabolic heat production, H, in kilocalories per day versus body mass, M, in kilograms for a range of land mammals.

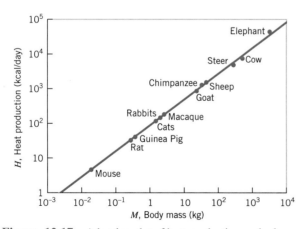

Source: Thomas A. McMahon and John T. Bonner, *On Size and Life* (New York: Scientific American Books, 1983) p. 65.

Figure 12.17   A log-log plot of heat production vs. body mass.

The graph again looks quite linear, so we would expect H to be directly proportional to a power of M. The slope of this line, and hence the exponent of M, is $3/4 = 0.75$. As body mass increases by 4 factors of 10 say from $10^{(-1)}$ kg to 10^3 kg, metabolic rate increases by 3 factors of 10 from 10 kcal/day to 10^4 kcal/day. The relationship is given approximately by

$$H = 151\, M^{(3/4)} \qquad (4)$$

where M is mass in kilograms, and H is metabolic heat production in kilocalories (thousands of calories) per day.

This scaling relationship is called "Kleiber's law," after the American veterinary scientist who first observed it in 1932. It has been verified by many series of subsequent measurements, though its cause is not fully understood.

• • •

Animals have evolved biological modifications partially to avoid the consequences of scaling laws. For example, our argument for the 2/3 power law of surface area versus body mass was based on animals' keeping roughly the same shape as their size increases. But elephants have relatively thicker legs than gazelles in order to provide the extra strength needed to support their much larger weight. Such changes in shape are very important biologically but are too subtle to be seen in the overall trend of the data we have shown. Despite these adaptations, inexorable scaling laws give an upper limit to the size of land animals, since eventually the animal's weight would become too heavy to be supported by its body. That's why large animals, such as whales, must live in the ocean, and why giant creatures in science fiction movies couldn't exist in real life.

Allometric laws provide an overview of the effects of scale, valid over several orders of magnitude. They help us compare important traits of elephants and mice, or of children and adults, without getting lost in the details. They help us to understand the limitations imposed by living in three dimensions.

Algebra Aerobics 12.7

1. What is the approximate surface area for a human being whose mass is 70 kg (7×10^4 g)? First estimate the answer by locating the proper point on the graph (Figure 12.16), then compute it using Equation (3). How do your answers compare? Translate your answers into lb. and inches.

2. The following graph shows the relationship between heart rate (in beats per minute) and body mass (in kilograms) for mammals ranging over many orders of magnitude in size.

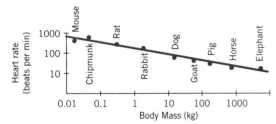

Source: Robert E. Ricklefs, *Ecology,* (San Francisco: W.H. Freeman, 1990) p. 66.

  a. Would a linear, power, or exponential function be the best model for the relationship between heart rate and body mass?

  b. The slope of the best fit line shown on the graph is -0.23. Construct the basic form of the function model.

  c. Interpret the slope of -0.23 in terms of heart rate and body mass.

3. Using both the graph (Figure 12.17) and Equation (4), find the rate of heat production does this model predict for a 70 kg human being? Considering that each kilocalorie of heat production requires consumption of one food calorie, does this seem about right?

4. Weight is proportional to length cubed, but the ability to support the weight—as measured by the cross-sectional area of the bones—is proportional to length squared. Explain why Godzilla or King Kong could only exist in movies.

We come now to the end of our study of different families of functions. We have used these functions to help us describe the world around us. As one naturalist wrote

It seldom happens, outside of the exact sciences, that we comprehend the mathematical aspect of a phenomenon enough to "define" (by formulae and constants) the curve which illustrates it. But, failing such thorough comprehension, we can at least speak of the "trend" of our curves and put into words the character and the course of the phenomena they indicate. . . . When the curve becomes, or approximates to, a mathematical one, the types are few to which it is likely to belong. A straight line, a parabola, or hyperbola, an exponential or a logarithmic curve . . . , a sine-curve or sinusoid . . . ; suffice for a wide range of phenomena.

[D'Arcy Wentworth Thompson, *On Growth and Form,* (New York: Dover Paperback, 1992) p. 38.]

For you, we hope this is the beginning of many more explorations in mathematics.

For further explorations, see the collection "Exploring on your own" on the CD or check our web site at: http://www.wiley.com/college/math/mathem/kimeclark

CHAPTER SUMMARY

Logarithms

The *logarithm base 10 of x* is the power of 10 you need to produce *x*: that is,

$$\log_{10} x = c \text{ means that } 10^c = x.$$

Logarithms base 10 are called *common logarithms* and $\log_{10} x$ is usually abbreviated as log *x*. The properties of logarithms follow directly from the definition of logarithms and from the properties of exponents.

Properties of exponents	Corresponding properties of logarithms
If *a* is any positive real number and *p* and *q* are any real numbers, then **1.** $a^p \cdot a^q = a^{(p+q)}$ **2.** $a^p/a^q = a^{(p-q)}$ **3.** $(a^p)^q = a^{(p \cdot q)}$ **4.** $a^0 = 1$	If *A* and *B* are positive real numbers and *p* is any real number, then **1.** $\log (A \cdot B) = \log A + \log B$ **2.** $\log (A/B) = \log A - \log B$ **3.** $\log A^p = p \log A$ **4.** $\log 1 = 0$ (since $10^0 = 1$) In addition **5.** $\log (10^x) = x$ **6.** $10^{\log x} = x$

The log of 0 or a negative number is not defined. But when we take the log of a number, we can get 0 or a negative value.

While we can estimate solutions to equations of the form $10 = 2^t$, where a variable occurs in the exponent, we can use logarithms to find an exact answer. For example,

given	$10 = 2^t$
take logarithm of both sides	$\log 10 = \log (2^t)$
apply Property 3 of logs	$\log 10 = t \log 2$
divide by $\log 2$	$t = \dfrac{\log 10}{\log 2}$

Using a calculator, we can obtain as accurate an estimate as needed. We have

$$t \approx \frac{1}{0.301} \approx 3.32$$

Base e: Natural Logarithms

The irrational number e is a fundamental mathematical constant whose value is approximately 2.71828.

The value of P dollars invested at an annual interest rate r (in decimal form) compounded n times a year for x years equals

$$y = P (1 + r/n)^{nx}$$

where r is the *nominal interest rate*. We can solve for x given a particular value of y by taking the log of both sides. The actual amount of interest earned per year is called the *effective interest rate*.

The value of $(1 + 1/n)^n$ approaches e as n approaches infinity.

The number e is used to describe *continuous compounding*. For example, the value of P dollars invested at an annual interest rate r (expressed in decimal form) compounded continuously for x years can be calculated using the formula

$$y = P \cdot e^{rx}$$

where r is the nominal interest rate.

We define the *logarithm base e* or the *natural logarithm* of x as the power of e you need to produce x. The natural log of x is written as $\ln x$. The properties of natural logarithms correspond directly to those of common logarithms.

Properties of natural logarithms

If A and B are positive real numbers and p is any real number, then

1. $\ln (A \cdot B) = \ln A + \ln B$
2. $\ln (A/B) = \ln A - \ln B$
3. $\ln A^p = p \ln A$
4. $\ln 1 = 0$ (since $e^0 = 1$)
5. $\ln (e^x) = x$
6. $e^{\ln x} = x$

As with common logarithms, $\ln (x)$ is not defined if $x \leq 0$.

Logarithmic Functions

For any $x > 0$, both the common and natural logarithms define unique numbers $\log x$ and $\ln x$ respectively. So we can define two functions

$$y = \log x \qquad \text{and} \qquad y = \ln x \qquad \text{(where } x > 0 \text{ for both functions).}$$

The graphs of these functions have similar shapes. Both are defined only when $x > 0$, are increasing everywhere and are asymptotic to the y-axis.

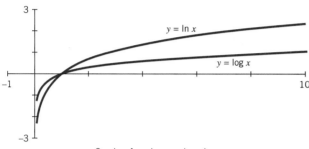

Graphs of $y = \log x$ and $y = \ln x$

For any number $a > 0$, the function $y = C \cdot a^x$ may be rewritten $y = C \cdot e^{rx}$, where $r = \ln a$. We call r the *instantaneous growth rate*.

If $a > 1$, then $r > 0$ and the function represents growth.
If $0 < a < 1$, then $r < 0$ and the function represents decay.
If $a = 1$, then $r = 0$ and the function is constant, $y = C$.

Semi-log and Log-log Graphs

In order to find the best function to model a relationship between two variables, we look for a way to plot the data so it appears as a straight line.

- The graph of a linear function appears as a straight line on a standard plot that has linear scales on both axes.
- The graph of an exponential function appears as a straight line on a semi-log plot that has a linear scale on the horizontal axis and a logarithmic scale on the vertical axis. The slope is the logarithm of the growth factor.
- The graph of a power function appears as a straight line on a log-log plot that has logarithmic scales on both axes. The slope is the exponent of the power function.

Allometry is the study of how the relationships among different physical attributes of an object change with scale. These relationships can often be described by power functions.

Exercises

1. Use the following table from Chapter 8 to estimate the number of years it would take $100 to become $300 at an interest rate compounded annually of
   a. 3%   **b.** 5%   **c.** 7%

	Compound Interest over 40 Years		
Years	Value of $100 at 3% ($)	Value of $100 at 5% ($)	Value of $100 at 7% ($)
0	100	100	100
10	134	163	197
20	181	265	387
30	243	432	761
40	326	704	1,497

Remaining drug dosage in milligrams.

2. The accompanying graph shows the concentration of a drug in the human body as the initial amount of 100 mg dissipates over time. Estimate when the concentration becomes
   a. 60 mg   **b.** 40 mg   **c.** 20 mg

3. Determine x if we know that log x equals
   a. -3 **d.** 0
   b. 6   **e.** 1
   c. 1/3   **f.** -1

4. Prove Property 2 of logarithms:
$$\log (A/B) = \log A - \log B \quad (A \text{ and } B \text{ both } > 0)$$

5. Given that log $5 \approx 0.699$, without using a calculator determine the value of
   a. log (25)   **b.** log (1/25)   **c.** log 10^{25} **d.** log 0.0025
   Be sure to show your work. Check your answers with a calculator.

6. Expand using the properties of logarithms.
   a. $\log (x^2 y^3 \sqrt{z} - 1)$ **b.** $\log \dfrac{A}{\sqrt[3]{BC}}$

7. Contract using the properties of logarithms and express your answer as a single logarithm.
   a. $3 \log K - 2 \log (K + 3)$
   b. $-\log m + 5 \log (3 + n)$

8. Solve for x.
   a. $2^x = 7$ **d.** $\log (x + 3) + \log 5 = 2$
   b. $(\sqrt{3})^{x+1} = 9^{2x-1}$ **e.** $\log x + \log (x + 1) = 1$
   c. $12(1.5)^{x+1} = 13$ **f.** $\ln (x - 1) = 2$

9. Prove Property 3 of common logarithms: $\log A^p = p \log A$ (where $A > 0$).

10. Returning to Exercise 1, now *calculate* the number of years it would take $100 to become $300 at an interest rate compounded annually at
    a. 3% b. 5% c. 7%

11. If the amount of drug remaining in the body after t hours is given by $f(t) = 100\left(\dfrac{1}{\sqrt{2}}\right)^t$ (graphed in Exercise 2), then calculate
    a. the number of hours it would take for the initial 100 mg to become
       i. 60 mg ii. 40 mg iii. 20 mg
    b. the half-life of the drug.

12. In Chapter 8 we saw that the function $N = N_0 \cdot 1.5^t$ described the actual number N of *E. coli* bacteria after t time periods (of 20 minutes each) starting with an initial bacteria count of N_0.
    a. What is the doubling time?
    b. How long would it take for there to be 10 times the original amount of bacteria?

13. Assume $10,000 is invested at a nominal interest rate of 8.5%. Write the equations that give the value of the money after n years and determine the effective interest rate if the interest is compounded
    a. annually
    b. semiannually
    c. quarterly
    d. continuously

14. Assume you invest $2000 at 3.5% compounded continuously.
    a. Construct an equation that describes the value of your investment at year t.
    b. How much will $2000 be worth after 1 year? 5 years? 10 years?
    c. How long will it take $2000 to double to $4000?
    d. How long will it take P dollars to double to $2P$ dollars?

15. The half-life of uranium-238 is about 5 billion years. Assume you start with 10 grams of U-238 that decays continuously.
    a. Construct an equation to describe the amount of U-238 remaining after x billion years.
    b. How long would it take for 10 grams of U-238 to become 1 gram?

16. You want to invest money for your newborn child so that she will have $50,000 for college on her 18th birthday. Determine how much you should invest if the best annual rate that you can get on a secure investment is
    a. 6.5% compounded annually
    b. 9% compounded quarterly
    c. 7.9% compounded continuously

17. Determine the doubling time for money invested at the rate of 12% compounded
    a. annually
    b. quarterly
    c. continuously

18. a. Phosphorus-32 is used to mark cells in biological experiments. If P-32 has a continuous daily decay rate of 0.0485, what is its half-life?

b. Phosphorus-32 can be quite dangerous to work with if the experimenter fails to use the proper shields, its high-energy radiation extends out to 610 cm. Because disposal of radioactive wastes is increasingly difficult and expensive, laboratories often store the waste until it is within acceptable radioactive levels for disposal with nonradioactive trash. For instance, the rule of thumb for the laboratories of a large east coast university and medical center is that any waste containing radioactive material with a half-life under 65 days must be stored for 10 half-lives before disposal with the nonradioactive trash.

    i. For how many days would P-32 have to be stored?

    ii. What percentage of the original P-32 would be left at that time?

19. Write an equivalent equation in exponential form.

  a. $n = \log 35$ **b.** $\ln (N/N_0) = -kt$

20. Write an equivalent equation in logarithmic form.

  a. $N = 10^{-t/c}$ **b.** $I = I_0 \cdot e^{-k/x}$

21. Prove that $\ln (A \cdot B) = \ln A + \ln B$ where A and B are positive real numbers.

22. Find the nominal interest rate, if a bank advertises that the effective interest rate on an account compounded continuously is

  a. 3.43% on a checking account   **b.** 4.6% on a NOW account

23. Assume $f(t) = Ce^{rt}$ is an exponential decay function (so $r < 0$). Prove the rule of 70 for halving times; that is, if a quantity is decreasing at $R\%$ per time period t, then the number of time periods it takes for the quantity to halve is approximately $70/R$. (Hint: $R = 100r$.)

24. Expand

  a. $\ln \left(\dfrac{x}{y\sqrt{2}} \right)^2$ **b.** $\ln \left(\dfrac{K^2 L}{M + 1} \right)$

25. Contract, expressing your answer as a single logarithm.

  a. $3 \ln R - (1/2) \ln P$ **b.** $\ln N - 2 \ln N_0$

26. The barometric pressure, p, in millimeters of mercury at height h in kilometers above sea level is given by the equation $p = 760e^{-0.128h}$. At what height is the barometric pressure 200 mm?

27. The stellar magnitude M of a star is defined as $\dfrac{-5}{2} \log \left(\dfrac{B}{B_0} \right)$ where B is the brightness of the star, and B_0 is a constant.

  a. If you plotted B on the horizontal and M on the vertical axis, where would the graph cross the B axis?

  b. Without calculating any other coordinates, draw a rough sketch of the graph of M. What is the domain? On your graph, indicate about where Rigel (with an average magnitude of -7.1) and Sirius (with an average magnitude of $+1.45$) would be positioned.

  c. As the brightness B increases, does the magnitude M increase or decrease? Is a sixth magnitude star brighter or dimmer than a first magnitude star?

d. If the brightness of a star is increased by a factor of 5, by what factor does the magnitude increase or decrease?

28. Logarithms can be constructed using any positive number as a base.

$$\log_a x = y \qquad \text{means that} \qquad a^y = x$$

x	$y = \log_3 x$
1/9	
1/3	
1	
3	
9	
27	

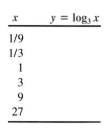

a. Complete the table in the margin and sketch the graph of $y = \log_3 x$

b. Now make a small table and sketch the graph of $y = \log_4 x$. (Hint: To simplify computations, try using powers of 4 for values of x.)

29. (Adapted from Hugh D. Young's *University Physics,* Vol. 1. (Reading, Mass.: Addison-Wesley, 1992.) p. 591.)
If you listen to a 120 decibel sound for about 10 minutes, your threshold of hearing will typically shift from 0 dB up to 28 dB for a while. If you are exposed to a 92-dB sound for 10 years, your threshold of hearing will be permanently shifted to 28 dB. What intensities correspond to 28 and 92 dB? (See p. 458.)

30. In all the sound problems so far, we did not take into account the distance between the sound source and the listener. In fact, in Chapter 9 we saw that sound intensity obeys an inverse square law. The intensity is inversely proportional to the square of the distance from the sound source. So, $I = k/r^2$, where I is intensity, r is the distance from the sound source, and k is a constant.

Suppose that you are sitting a distance R from the TV, where its sound intensity is I_1. Now you move to a seat twice as far from the TV, a distance $2R$ away, where the sound intensity is I_2.

a. What is the relationship between I_1 and I_2?

b. What is the relationship between the decibel levels associated with I_1 and I_2?

31. If there are a number of different sounds being produced simultaneously, the resulting intensity is the sum of the individual intensities. How many decibels louder is the sound of quintuplets crying than the sound of one baby crying? (See p. 458.)

32. The number of neutrons in a nuclear reactor can be predicted from the equation $n = n_0 e^{t/T}$ where $n =$ number of neutrons at time t (in seconds), $n_0 =$ the number of neutrons at time $t = 0$, and $T =$ the reactor period, the doubling time of the neutrons (in seconds). When $t = 2$ seconds, $n = 11$, and when $t = 22$ seconds, $n = 30$. Find the initial number of neutrons, n_0, and the reactor period, T, both rounded to the nearest whole number.

33. Biologists believe that, in the deep sea, species density decreases exponentially with the depth. The graph in the margin shows data collected in the North Atlantic. Sketch an exponential decay function through the data. Then identify two points on your curve, and generate an equation for your function.

34. Identify each of the following functions as representing growth or decay.
    a. $Q = Ne^{-0.029t}$ **b.** $h(r) = 100(0.87)^r$ **c.** $f(t) = 375e^{0.055t}$

35. Rewrite each of the following functions using base e.
    a. $N = 10(1.045)^t$ **b.** $Q = 5 \cdot 10^{-7} \cdot (0.072)^A$

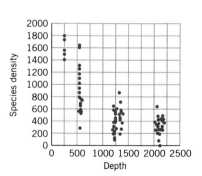

Source: Data collected by Ron Etter, UMass/Boston Biology Department.

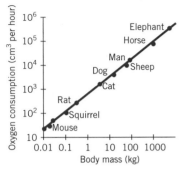

Source: Robert E. Ricklefs, *Ecology,* (San Francisco: W.H. Freeman, 1990) p. 67.

36. The graph in the margin shows oxygen consumption vs. body mass in mammals.

    a. Would a power or exponential function be the best model of the relationship between oxygen consumption and body mass?

    b. The slope of the best fit line shown on the graph is approximately 3/4. Construct the basic form of the functional model that you chose in part (a).

    c. Interpret the slope in terms of oxygen consumption and body mass. In particular, by how much does oxygen consumption increase when body mass increases by a factor of 10? By a factor of 10^4?

37. **a.** The figure below on the left shows the relationship between population density and length of an organism. The slope of the line is -2.25. Express the relationship between population density and length in terms of direct proportionality.

    b. The figure below on the right shows the relationship between population density and body mass of mammals. The slope of the line is -0.75. Express the relationship between population density and body mass in terms of direct proportionality.

    c. Are your statements in (a) and (b) consistent with the fact that body mass is directly proportional to length^3? Hint: calculate length^3 to the -0.75 power.

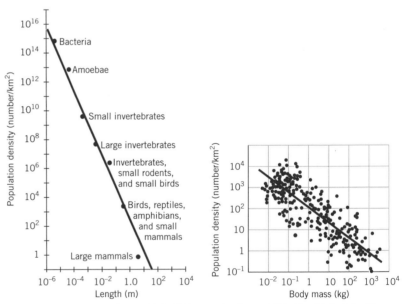

Source: Thomas A. McMahon and John T. Bonner, *On Size and Life* (New York: Scientific American Books, 1983) p. 228.

38. (optional use of technology)

    The data on page 483 give the typical mass for some birds and their eggs:

    a. Calculate the third column in the data table to see if the ratio of egg mass to adult bird mass is the same for all of the birds. Write a sentence that describes what you discover.

Species	Adult Bird Mass, grams	Egg Mass, grams	Egg/Adult Ratio
ostrich	113,380.0	1,700.0	0.015
goose	4,536.0	165.4	
duck	3,629.0	94.5	
pheasant	1,020.0	34.0	
pigeon	283.0	14.0	
hummingbird	3.6	0.6	

Source: Size, Function and Life History by William A. Calder III, (Boston: Harvard University Press, 1984.)

b. Graph egg mass (vertical axis) and adult bird mass (horizontal axis) using each of the following types of plots:

   i.   standard linear

  ii.   semi-log

 iii.   log-log

c. Examine the three graphs you made and determine which looks closest to linear. Find a linear equation in the form $Y = mX + b$ to model the line. Remember that any value, such as the vertical intercept, that you read off a log scale has been converted to the log of the number by the scale. So for example, on a log-log plot, $Y = b + mX$ is actually of the form

$$\log y = b + m (\log x),$$

where $Y = \log y$ and $X = \log x$.

d. Once you have a linear model, transform it to a form that gives egg mass as a function of adult bird mass.

e. What egg mass does your formula predict for a 12.7 kilogram turkey?

f. A giant hummingbird (Patagona gigas) lays an egg of 2 gram mass. What size does your formula predict for the mass of an adult bird?

39. Use the accompanying graphs to answer the following questions:

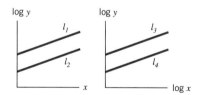

a. Assume l_1 and l_2 are straight lines that are parallel. In each case what type of equation would describe y in terms of x? How are the equations corresponding to l_1 and l_2 similar? How are they different?

b. Assume l_3 and l_4 are also parallel straight lines. For each case, what type of equation would describe y in terms of x? How are the equations corresponding to l_3 and l_4 similar? How are they different?

40. Find the equation of the line in each of the following graphs. Rewrite each equation, expressing y in terms of x.

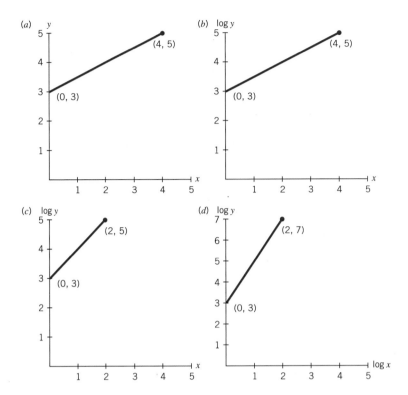

EXPLORATION 12.1

Properties of Logarithmic Functions

Objective

- explore the effects of a and c on the graph of the logarithmic functions

    $$y = c \log (ax) \qquad \text{and} \qquad y = c \ln (ax) \qquad \text{where } a > 0 \quad \text{and} \quad x > 0$$

Materials/Equipment

- graphing calculator or computer with "E8: Logarithmic Sliders" in *Exponential & Log Functions* in course software or a function graphing program
- graph paper

Procedure

Making Predictions

1. *The effect of a on the equation y $= \log (ax)$ where a > 0 and x > 0.*

    a. Why do we need to restrict a and x to positive values?

    b. Using the properties of logarithms, write the expression $\log (ax)$ as the sum of two logs. Discuss with your partner what effect you expect a to have on the graph. Jot down your predictions in your notebook. Now let $y = \log (2x)$, and $y = \log (3x)$ (where $a = 2$ and 3, respectively). If $\log 2 \approx 0.302$ and $\log 3 \approx 0.477$, complete the data table below and, by hand, sketch the three graphs on the same grid. Do your results confirm your predictions? What do you expect to happen to the graph of $y = \log (ax)$ if larger and larger positive values are substituted for a?

    Evaluating $y = \log (ax)$ when $a = 1, 2,$ and 3

x	$y = \log x$	$y = \log (2x)$	$y = \log (3x)$
0.001	−3.000		
0.010	−2.000		
0.100	−1.000		
1.000	0.000		
5.000	0.699		
10.000	1.000		

    c. Discuss with your partner how you think the graphs of $y = \log x$ and $y = \log (ax)$ will compare if $0 < a < 1$. Jot down your predictions. Complete the following small data table and sketch the graphs of $y = \log x$ and $y = \log (x/10)$ on the same grid. Do your predictions and your graph agree? Predict what would happen to the graph of $y = \log (ax)$, if smaller and smaller positive values were substituted for a.

Evaluating $y = \log(ax)$ when $a = 1$ and $1/10$

x	$y = \log x$	$y = \log(x/10)$
0.001	−3.000	
0.010	−2.000	
0.100	−1.000	
1.000	0.000	
5.000	0.699	
10.000	1.000	

d. How do you think your findings for $y = \log(ax)$ relate to the function $y = \ln(ax)$?

2. *The effect of c on the equation $y = c \log x$*

   a. When $c > 0$

   Talk over with your partner your predictions for what will happen as c increases. Fill in the table below and on the same grid do a quick sketch of the functions $y = \log x$, $y = 2 \log x$, and $y = 3 \log x$ (where $c = 1$, 2, and 3, respectively). Were your predictions correct? What do you expect to happen to the graph of $y = c \log x$ as you substitute larger and larger positive values for c?

Evaluating $y = c \log x$ when $c = 1$, 2, and 3

x	$y = \log x$	$y = 2 \log x$	$y = 3 \log x$
0.001	−3.000		
0.010	−2.000		
0.100	−1.000		
1.000	0.000		
5.000	0.699		
10.000	1.000		

   b. When $c < 0$

   Why can c be negative, when a had to remain positive? How do you think the graphs of $y = \log x$ and $y = c \log x$ will compare if $c < 0$? Write down your prediction. Complete the small data table below and sketch the graphs of $y = -\log x$ and $y = -2 \log x$ on the same grid. Check your results and your predictions with your partner. What will happen to the graph of $y = c \log x$ if you substitute a negative value for c. What happens if c remains negative, but $|c|$ gets larger and larger (for example, $c = -10, -150, -5000$, etc.)?

Evaluating $y = c \log x$ when $c = -1$ and -2

x	$y = -\log x$	$y = -2 \log x$
0.001		
0.010		
0.100		
1.000		
5.000		
10.000		

c. How do your findings on $y = c \log x$ relate to the function $y = c \ln x$?

3. *Generalizing your results*

Talk over with your partner the effect of varying both a and c on the general functions $y = c \log (ax)$ and $y = c \ln (ax)$. Try predicting the shape of the graph of such functions as $y = 3 \log (2x)$ or $y = -2 \log (x/10)$. Have each partner construct a small table, and graph the results of one such function. Compare your findings.

Testing Your Predictions

Test your predictions either by using "E8: Logarithmic Sliders" in *Exponential & Log Functions,* or creating your own graphs with a graphing calculator or a function graphing program.

a. Try changing the value for a in different functions of the form $y = \log (ax)$ or $y = \ln (ax)$. Be sure to try values of $a > 1$ and values such that $0 < a < 1$.

b. Try changing the value for c in functions of the form $y = c \log x$ and $y = c \ln x$. Let c assume both positive and negative values.

Summarizing Your Results

Write a 60 second summary describing the effect of varying a and c in functions of the form $y = c \log (ax)$ and $y = c \ln (ax)$.

Exploration-Linked Homework

1. Use your knowledge of logarithmic functions to predict the shape of the graphs of the following:

The decibel scale: given by the function $N = 10 \log \left(\dfrac{I}{I_0} \right)$ where I is the intensity of a sound and I_0 is the intensity of sound at the threshold of human hearing.

Stellar magnitude: given by the function $M = \dfrac{-5}{2} \log \left(\dfrac{B}{B_0} \right)$ where B is the brightness of a star and B_0 is the brightness of our sun.

Use technology to check your predictions.

2. Explore the effect of changing the base: that is, the effect of changing b in $y = c \log_b (ax)$.

EXPLORATION 12.2

Revisiting the U.S. Population Data

Objectives

- represent and interpret the U.S. population data on semi-log and log-log plots
- construct an equation or set of equations to model the U.S. population data

Materials/Equipment

- function graphing program, spreadsheet or graphing calculator
- U.S. population data files (called USPOP) in Excel or graph link form on CD-ROM

Procedure

1. **a.** Load the following data on the U.S. population from Chapter 3 into your graphing calculator, spreadsheet, or function graphing program (using USPOP file).

Population of the U.S. 1790 to 1995	
Year	(in millions)
1790	3.9
1800	5.3
1810	7.2
1820	9.6
1830	12.9
1840	17.1
1850	23.2
1860	31.4
1870	39.8
1880	50.2
1890	63.0
1900	76.0
1910	92.0
1920	105.7
1930	122.8
1940	131.7
1950	151.3
1960	179.3
1970	203.3
1980	226.5
1990	248.7
1992	255.4
1994	260.7
1995	263.0

   b. Generate a third column for time, t, that references t to a base year. We suggest referencing 1790 as $t = 10$ (you can think of 10 as the number of years since 1780). You want to avoid using 0 as initial value for t since log (0) is undefined, and we will be constructing a log-log plot.

2. **If using a graphing calculator**

   Instructions are in the Graphing Calculator Workbook on the CD-ROM.

   Transform the data to log (time) and log (population). (See examples 1 and 2 in Sec. 12.6). Use your third column values for time where time's referenced to a base year. Plot log (time) on the horizontal axis and log (population) on the vertical axis. Now, plot time on the horizontal axis and log (population) on the vertical axis. Interpret your graphs. Which graph shows a more linear pattern? What type of function could be used to model the data?

   Find and interpret a best fit linear function for the graph which shows a more linear pattern.

   If using a spreadsheet or function graphing program

   Construct semi-log and log-log plots of the data, graphing time on the horizontal axis and population on the vertical axis. Interpret your graphs. Which graph shows a more linear pattern? Would a linear, power, or exponential function best model the relationship between population and time?

   Transform the data using logarithms as described above for graphing calculators. Graph the transformed data and find a best fit linear function. (See Examples 1 and 2 in Sec. 12.6).

3. Using your best fit linear function, construct and interpret an equation of the population, P, in terms of time, t.

4. Examine your graphs and decide if a piecewise linear function might be more appropriate for this set of data. If so, construct a set of equations for this function. Use this function to find and interpret a non-linear piecewise function for the population, P, in terms of time, t.

Summarizing Your Results

Construct a 60 second summary of your results. Include at least one graph and equation in the summary. You may want to include information from Sections 3.1 and 3.2 in the text.

Anthology of Readings

Contents

Science Musings

True Nature of Math Remains, in Sum, a Mystery

Chet Raymo

If you are afflicted with math anxiety, you may not like what I'm about to say.

Our most certain knowledge of the world is mathematical. Non-mathematical knowledge is held suspect by scientists. Dependence upon mathematics as the arbiter of truth defines the modern, Western way of knowing. For better or worse, it is the source of our power, our wealth and our physical well-being.

There it is. Take it or leave it. Math offers our best crack at figuring out what the world is all about.

And the big mystery is—no one knows what mathematics is, or why it works so unreasonably well.

Not even mathematicians.

Or the scientists who use mathematics as their language of discovery.

There are, I suppose, three possible explanations:

1. Mathematics is an invention of the human mind, like the English language or the game of chess, that has proven particularly useful for expressing patterns we observe in nature.
2. Nature is itself mathematical. We learn mathematics by observing the way the world works.
3. Mathematics exists as a kind of Platonic ideal, beyond and outside of nature. All reality, including the human mind, participates in some mysterious way in this preexisting order.

All of which is a wordy way of saying:

1. We invent math.
2. We discover math.
3. We are math.

> **Math offers our best crack at figuring out what the world is all about.**

John D. Barrow, professor of astronomy at the University of Sussex in Britain, takes up the mysterious effectiveness of mathematics in his recent book, "Pi in the Sky: Counting, Thinking and Being." He takes us through the history of mathematics, from counting on fingers to the high-blown abstractions of theoretical physicists who use mathematics to investigate the Big Bang moment of creation.

He's after the answers: What is math? Why does it work? Why is science up to its ears in numbers—nay, to the very tip-top of its

Chet Raymo is a professor of physics at Stonehill College and the author of several books on science.

head? Barrow's answer: Uhhh . . . well . . . I guess . . . maybe . . . dunno.

The mystery remains.

And that, says Barrow, is the point.

His claim: "We have found that at the roots of the scientific image of the world lies a mathematical foundation that is itself ultimately religious."

I think he's right.

And if I were a theologian, that is exactly where I would start constructing a concept of God that is relevant to our time—with mathematics.

At first blush, the idea sounds preposterous, even blasphemous. But wait. Isn't mathematics the most effective medium of exchange between the human mind and the cosmos? Through math we have come to know the Big Bang, the universe of galaxies, the unfolding infinities of space and time. We have plunged into the heart of matter and explored mysteries of life and mind. Nothing else has taken us to there. Math, only math.

And we haven't the foggiest idea why it works.

By and large, theologians still offer us the God of Michelangelo's Sistine Chapel ceiling—a gray-bearded version of ourselves ensconced in some paltry heaven up in the sky. "Oh, don't be silly," they'll protest, "the gray-bearded man is just a metaphor for something we cannot fully express." Exactly. So why stick with a medieval metaphor that opens a gulf between science and religion, between knowing and feeling? Why not adopt a metaphor that embraces the full richness of the cosmos, a metaphor that links our minds to all that exists?

Namely, mathematics.

The human mind has evolved in response to the world in which we live, in the same way as did our senses of sight, sound, and smell. If we are mathematical creatures, it is because the world is in some deeply mysterious sense mathematical. Call it, if you will, the mind of God. The phrase is metaphorical, but so is every other definition of God. The mathematical metaphor links us profoundly, deeply, to all of creation in a way that is consistent with the spirit and the substance of modern science.

> **And maybe, just maybe, we are math incarnate—to the core of our souls.**

That's where I'd start, if I were a theologian.

But I'm not a theologian. I don't even have a religion. I just know a ripping good mystery when I see one.

Of the three possible explanations for the mysterious effectiveness of mathematics, maybe all are true. Maybe we invent math. Maybe we discover it. And maybe, just maybe, we are math incarnate—to the core of our souls. Maybe, just maybe, mathematics is our glimpse of the eternal, omnipresent, creative foundation of the world. Why else does it work so astonishingly well?

Barrow may have got it exactly right: "Our ability to create and apprehend mathematical structures in the world is . . . a consequence of our oneness with the world. We are the children as well as the mothers of invention."

U.S. Government Definitions of Census Terms

by Anthony Roman (1994), Center for Survey Research, University of Massachusetts, Boston.

Housing Unit

The government defines *housing unit* as a "house, an apartment, or a single room occupied as separate living quarters." The exact definition becomes more complex when one considers what constitutes separate living quarters. Perhaps the easiest way to define a housing unit is to describe what it is not. The following are not housing units and therefore the people living in them do not make up households: 1) most units in rooming or boarding houses where people share kitchen facilities, 2) units in transient hotels or motels, 3) college dormitories, 4) bunk houses, 5) group quarters living arrangements such as military housing, convents, prisons, and mental institutions, and 6) units within other housing units which do not have direct access from a hallway or outside. Most other living arrangements should come under the definition of a housing unit and therefore a household.

Household & Family

A *household* includes all persons occupying a housing unit. A *family* consists of all people in a household who are related by blood, marriage, or adoption. By strict definition, a family needs to have a minimum of two people, a householder and at least one other person. A householder is defined as the person in whose name the housing unit is owned or rented. If more than one such person exists, or if none exist, than any adult household member can be designated as a householder.

A household may contain no families. This, in fact, is not uncommon. Although they may consider themselves one, a group of single unrelated people or an unmarried couple living together are not considered a family. A household may also contain more than one family, although this is rare. In most cases in which multiple families share a single housing unit, there is at least one member of one family who is related in some way to a member of another family. By definition then this group becomes one large family. For example, if a husband and wife rent a house and the wife's cousin and her two children come to live with them, this constitutes one large family since all members are related by blood, marriage, or adoption.

Size of Household

The *size of a household* is the number of persons who are residing in the household at the time of interview and who do not usually live elsewhere. A visitor staying temporarily at someone's house is not part of the household. A person who is away on vacation or in a hospital is a member of the household if he or she usually lives there. The status of college students may be the hardest to determine. If they usually live away from the household, they are not part of the household.

Household vs. Family Income

Household income is the sum of all incomes earned by all who live in the household. If a husband and wife each earn $30,000, their child earns $5,000, and an unrelated boarder living in an extra bedroom earns $15,000, then the household income is $80,000. All four members of the household are considered to be living in a household with an income of $80,000, even though all household members may not have access to that entire amount.

For the purposes of income reporting, *family income* is considered to be the sum of all incomes earned by an individual and all other family members. In the previous example, the husband, wife, and child would all be considered as part of a family whose income is $65,000. Although by strict definition, a family of one cannot exist, for the purposes of the distribution of total household income, the boarder would be considered a family of one whose income is $15,000.

Employment Status

A person's *employment status* falls into one of three distinct categories: employed, unemployed, or not in the labor force. An employed person is anyone at least 16 years old who worked <u>last week</u> for pay or for his or her own family's business, regardless of the number of hours worked. Persons with a job but who did not work last week due to illness, vacation, or other reasons, are also considered employed. An unemployed person is one who did not work at all last week, but who was available to accept a job and has looked for work during the last 4 weeks. A person is not in the labor force if he or she did not work at all last week and either hasn't looked for work during the last four weeks or did not want a job. Retired persons, housewives, and students are the most common examples of persons not in the labor force. A special class of persons called "discouraged workers" has recently been added to those who are not in the labor force. Discouraged workers are people who did not work last week, and have been out of work for a long time. They may have looked for work in the past, but have not looked in the last four weeks.

What constitutes "looking for work?" Interviewing for a job, sending resumes to companies, or answering newspaper ads are all considered looking for work. Reading the newspaper "help wanted" ads without following up is not considered looking for work.

Race

The Census Bureau classifies people into one of five distinct *racial categories*: 1) White, 2) Black, 3) American Indian, Eskimo or Aleut, 4) Asian or Pacific Islander, and 5) Other. Included among whites are those who claim their race is Canadian, German, Near Eastern Arab, or Polish among many others. Included among blacks are those who claim their race is Jamaican, Haitian, or Nigerian among many others. Included among Asian and Pacific Islanders are people who claim their race is Chinese, Filipino, Vietnamese, Hawaiian or Samoan among many others. Included among the "other" race category, are people who consider themselves to be multiracial and many people of Hispanic origin.

Ethnicity

An individual's *ethnicity* refers to what the person considers to be his or her origin. A person may consider him or herself Polish, Irish, African, Hispanic, English or one of many other ethnic origins. Origin means different things to different people. For example, origin can be interpreted as a person's ancestry, nationality group, lineage, or country of birth of the individual, his or her parents or ancestors. A person can have dual ethnicity, but may identify more closely with one and therefore consider themselves to be of that ethnicity.

Ethnicity is distinct from race. People of a given ethnicity may be of any race. A person can be black and English, or white and Greek. A person of Hispanic ethnicity may also consider her or himself to be black, white or neither of the two. The Census Bureau asks only one specific question on ethnicity which is, "Do you consider yourself to be of Hispanic origin?" This question is asked of all respondents along with questions of race. This produces a more accurate count of people of Hispanic origin than asking the single question, "Are you white, black, Hispanic, or something else?" Many people of Hispanic origin will choose white or black to this single question.

There is no single list of ethnic categories now in use. Many people use race, Hispanic origin, and a question pertaining to origin to infer ethnicity. This way Poles can either be separated from Czechs or combined as Eastern Europeans.

Wages vs. Total Income

Wages are defined as the amount of money earned from jobs or services performed. Total income includes money received from all sources. Total income includes wages as well as interest on bank accounts, stock dividends, royalties, retirement distributions, inheritances and government subsidies such as welfare payments.

The Median Isn't the Message

Stephen Jay Gould

My life has recently intersected, in a most personal way, two of Mark Twain's famous quips. One I shall defer to the end of this essay. The other (sometimes attributed to Disraeli), identifies three species of mendacity, each worse than the one before—lies, damned lies, and statistics.

Consider the standard example of stretching truth with numbers—a case quite relevant to my story. Statistics recognizes different measures of an "average," or central tendency. The *mean* is our usual concept of an overall average—add up the items and divide them by the number of sharers (100 candy bars collected for five kids next Halloween will yield 20 for each in a just world). The *median*, a different measure of central tendency, is the halfway point. If I line up five kids by height, the median child is shorter than two and taller than the other two (who might have trouble getting their mean share of the candy). A politician in power might say with pride, "The mean income of our citizens is $15,000 per year." The leader of the opposition might retort, "But half our citizens make less than $10,000 per year." Both are right, but neither cites a statistic with impassive objectivity. The first invokes a mean, the second a median. (Means are higher than medians in such cases because one millionaire may outweigh hundreds of poor people in setting a mean; but he can balance only one mendicant in calculating a median).

The larger issue that creates a common distrust or contempt for statistics is more troubling. Many people make an unfortunate and invalid separation between heart and mind, or feeling and intellect. In some contemporary traditions, abetted by attitudes stereotypically centered upon Southern California, feelings are exalted as more "real" and the only proper basis for action—if it feels good, do it—while intellect gets short shrift as a hang-up of outmoded elitism. Statistics, in this absurd dichotomy, often become the symbol of the enemy. As Hilaire Belloc wrote, "Statistics are the triumph of the quantitative method, and the quantitative method is the victory of sterility and death."

This is a personal story of statistics, properly interpreted, as profoundly nurturant and life-giving. It declares holy war on the downgrading of intellect by telling a small story about the utility of dry, academic knowledge about science. Heart and head are focal points of one body, one personality.

> In 1982, I learned I was suffering from a rare and serious cancer. After surgery, I asked my doctor what the best technical literature on the cancer was. She told me, with a touch of diplomacy, that there was nothing really worth reading. I soon realized why she had offered that humane advice: my cancer is incurable, with a median mortality of eight months after discovery.

Stephen Jay Gould teaches biology, geology, and the history of science at Harvard University.

In July 1982, I learned that I was suffering from abdominal mesothelioma, a rare and serious cancer usually associated with exposure to asbestos. When I revived after surgery, I asked my first question of my doctor and chemotherapist: "What is the best technical literature about mesothelioma?" She replied, with a touch of diplomacy (the only departure she has ever made from direct frankness), that the medical literature contained nothing really worth reading.

Of course, trying to keep an intellectual away from literature works about as well as recommending chastity to *Homo sapiens,* the sexiest primate of all. As soon as I could walk, I made a beeline for Harvard's Countway medical library and punched mesothelioma into the computer's bibliographic search program. An hour later, surrounded by the latest literature on abdominal mesothelioma, I realized with a gulp why my doctor had offered that humane advice. The literature couldn't have been more brutally clear: mesothelioma is incurable, with a median mortality of only eight months after discovery. I sat stunned for about fifteen minutes, then smiled and said to myself: so that's why they didn't give me anything to read. Then my mind started to work again, thank goodness.

If a little learning could ever be a dangerous thing, I had encountered a classic example. Attitude clearly matters in fighting cancer. We don't know why (from my old-style materialistic perspective, I suspect that mental states feed back upon the immune system). But match people with the same cancer for age, class, health, socioeconomic status, and, in general, those with positive attitudes, with a strong will and purpose for living, with commitment to struggle, with an active response to aiding their own treatment and not just a passive acceptance of anything doctors say, tend to live longer. A few months later I asked Sir Peter Medawar, my personal scientific guru and a Nobelist in immunology, what the best prescription for success against cancer might be. "A sanguine personality," he replied. Fortunately (since one can't reconstruct oneself at short notice and for a definite purpose), I am, if anything, even-tempered and confident in just this manner.

Hence the dilemma for humane doctors: since attitude matters

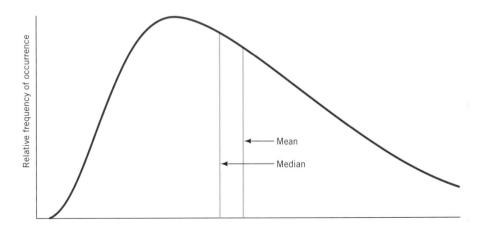

so critically, should such a sombre conclusion be advertised, especially since few people have sufficient understanding of statistics to evaluate what the statements really mean? From years of experience with the small-scale evolution of Bahamian land snails treated quantitatively, I have developed this technical knowledge—and I am convinced that it played a major role in saving my life. Knowledge is indeed power, in Bacon's proverb.

The problem may be briefly stated: What does "median mortality of eight months" signify in our vernacular? I suspect that most people, without training in statistics, would read such a statement as "I will probably be dead in eight months"—the very conclusion that must be avoided, since it isn't so, and since attitude matters so much.

I was not, of course, overjoyed, but I didn't read the statement in this vernacular way either. My technical training enjoined a different perspective on "eight months median mortality." The point is a subtle one, but profound—for it embodies the distinctive way of thinking in my own field of evolutionary biology and natural history.

We still carry the historical baggage of a Platonic heritage that seeks sharp essences and definite boundaries. (Thus we hope to find an unambiguous "beginning of life" or "definition of death," although nature often comes to us as irreducible continua.) This Platonic heritage, with its emphasis on clear distinctions and separated immutable entities, leads us to view statistical measures of central tendency wrongly, indeed opposite to the appropriate interpretation in our actual world of variation, shadings, and continua. In short, we view means and medians as the hard "realities," and the variation that permits their calculation as a set of transient and imperfect measurements of this hidden essence. If the median is the reality and variation around the median just a device for its calculation, the "I will probably be dead in eight months" may pass as a reasonable interpretation.

But all evolutionary biologists know that variation itself is nature's only irreducible essence. Variation is the hard reality, not a set of imperfect measures for a central tendency. Means and medians are the abstractions. Therefore, I looked at the mesothelioma statistics quite differently—and not only because I am an optimist who tends to see the doughnut instead of the hole, but primarily because I know that variation itself is the reality. I had to place myself amidst the variation.

When I learned about the eight-month median, my first intellectual reaction was: fine, half the people will live longer; now what are my chances of being in that half. I read for a furious and nervous hour and concluded, with relief: damned good. I possessed every one of the characteristics conferring a probability of longer life: I was young; my disease had been recognized in a relatively early stage; I would receive the nation's best medical treatment; I had the world to live for; I knew how to read the data properly and not despair.

Another technical point then added even more solace. I immediately recognized that the distribution of variation about the eight-month median would almost surely be what statisticians call "right skewed." (In a symmetrical distribution, the profile of variation to the left of the central tendency is a mirror image of variation to the right. In skewed distributions, variation to one side of the central tendency is more stretched out—left skewed if extended to the left, right skewed if stretched out to the right.) The distribution of variation had to be right skewed, I reasoned. After all, the left of the distribution contains an irrevocable lower boundary of zero (since mesothelioma can only be identified at death or before). Thus there isn't much room for the distribution's lower (or left) half—it must be scrunched up between zero and eight months. But the upper (or right) half can extend out for years and years, even if nobody ultimately survives. The distribution must be right skewed, and I needed to know how long the extended tail ran—for I had already concluded that my favorable profile made me a good candidate for that part of the curve.

The distribution was, indeed, strongly right skewed, with a long tail (however small) that extended for several years above the eight month median. I saw no reason why I shouldn't be in that small tail, and I breathed a very long sigh of relief. My technical knowledge had helped. I had read the graph correctly. I had asked the right question and found the answers. I had obtained, in all probability, that most precious of all possible gifts in the circumstances—substantial time. I didn't have to stop and immediately follow Isaiah's injunction to Hezekiah—set thine house in order: for thou shalt die, and not live. I would have time to think, to plan, and to fight.

One final point about statistical distributions. They apply only to a prescribed set of circumstances—in this case to survival with mesothelioma under conventional modes of treatment. If circumstances change, the distribution may alter. I was placed on an experimental protocol of treatment and, if fortune holds, will be in the first cohort of a new distribution with high median and a right tail extending to death by natural causes at advanced old age.

It has become, in my view, a bit too trendy to regard the acceptance of death as something tantamount to intrinsic dignity. Of course I agree with the preacher of Ecclesiastes that there is a time to love and a time to die—and when my skein runs out I hope to face the end calmly and in my own way. For most situations, however, I prefer the more martial view that death is the ultimate enemy—and I find nothing reproachable in those who rage mightily against the dying of the light.

The swords of battle are numerous, and none more effective than humor. My death was announced at a meeting of my colleagues in Scotland, and I almost experienced the delicious pleasure of reading my obituary penned by one of my best friends (the so-and-so got suspicious and checked; he too is a statistician, and didn't expect to find me so far out on the left tail). Still, the incident provided my first good laugh after the diagnosis. Just think, I almost got to repeat Mark Twain's most famous line of all: the reports of my death are greatly exaggerated.

CHANCE News[*]

Prepared by J. Laurie Snell, with help from William Peterson, Fuxing Hou and Ma. Katrina Munoz Dy, as part of the CHANCE Course Project supported by the National Science Foundation.

Please send comments and suggestions for articles to:

jlsnell@dartmouth.edu

Back issues of Chance News and other materials for teaching a CHANCE course are available from the Chance Web Data Base.

http://www.geom.umn.edu/locate/chance

===

" Data, data everywhere, but not a thought to think "
    Jesse Shera's paraphrase of Coleridge.

===

We found this quote in John Paulos' new book *A Mathematician Reads the Newspaper*.

FROM OUR READERS

Jerry Johnson sent us the following excerpts from a discussion on a journalism listserve group.

I teach statistics at the University of Texas at Arlington. Two weeks ago I read in the science section of a local paper an article defining the difference between the median, the mean, and the average. Everything was fine until he defined the mean as the average of the largest and smallest numbers in a set of data. I have always used and taught that the (arithmetic) mean is the same as the average of the numbers. When I talked to him about this, he indicated that this was the definition given in an Associated Press list of definitions. Is this the definition used by journalists?

Thanks for everyone who answered my question concerning the mean. I have contacted the Associated Press and hope to change their definition. One of the problems seems to come from dictionaries that define the mean as midway between extremes. I contacted Merriam-Webster and got one editor there to agree that midway between extremes is in a philosophical sense and not a mathematical sense. Webster's New World Dictionary has a more specific definition as "a middle or intermediate position as to place, time, quantity, kind, value,..."

After discussion with Norm Goldstein, Director of APN Special Projects, the Associated Press Style Book will be modified to indicate that the calculation of the mean is identical to that of the average.

[*] Chance News *is an electronic magazine. For instructions to subscribe to this magazine, send an E-mail to* jlsnell@dartmouth.edu

The New York Times Op Ed Wednesday May 29, 1996

A Fragmented War on Cancer

By Hamilton Jordan

ATLANTA

It has been 25 years since President Richard Nixon declared war on cancer. Having had two different cancers, I am a survivor of that war and a grateful beneficiary. My first, an aggressive lymphoma, was treated with an experimental therapy developed at the National Cancer Institute. Ten years later, my prostate cancer was detected early by the simple P.S.A. blood test, a diagnostic tool supported by Federal grants.

But I am also a symbol of the limited success of that war. The treatments I received were merely updated versions of the methods used 25 years ago. A powerful cocktail of chemicals killed my lymphoma while ravaging my body. A surgeon, using an elegant procedure with no permanent side effects, cut out my prostate.

Scientists are still looking for both the "magic bullet" that kills only

Hamilton Jordan, who was President Jimmy Carter's chief of staff, is a board member of Capcure, a nonprofit organization that finances prostate cancer research.

cancer cells and the genetic switch that turns off random cancer growth or prevents genetic flaws from causing cancer. While significant progress has been made, twice as many people will be diagnosed with cancer this year as in 1971, and twice as many will die. One in three women and one in two men will have cancer in their lifetimes. The raw data suggest we are on the verge of an epidemic. What happened to the "war"?

Groups compete for a shrinking pie.

● Our rhetoric exceeded our commitment. Dr. Donald Coffey, a cancer researcher, says we promised a war but financed only a few skirmishes. The Federal budget expresses our national priorities: The Federal Aviation Administration, for example, will spend $8.92 billion to make air travel safe. The chances of dying in an airline accident are one in two million. But the National Cancer Institute will spend only $2.2 billion — one-tenth of a cent of every Federal tax dollar — to find a cure for the disease that kills

more Americans in a month than have died in all commercial aviation accidents in our history.

● We created expectations not based on scientific reality. Our political leaders failed to appreciate the simple reality that there is not just one cancer but more than 100 cancers that have all defied a single solution. At the same time, the numerous organizations representing those different cancers have fought among themselves for bigger slices of a shrinking pie instead of forging a consensus on behalf of a larger pie.

● Huge successes with some cancers have been offset by rises in others. In addition, mortality from other diseases has declined, leaving an aging, cancer-prone population.

With adequate financing, breakthroughs in cancer prevention and treatment are likely over the next decade. Yet promising research that would have been automatically financed a decade ago is rejected today because of belt-tightening, discouraging brilliant young investigators from entering cancer research in the first place.

Is one-tenth of one cent enough to find a cure for a disease that will strike 40 percent of Americans? You will not think so when cancer strikes you or your loved ones.   □

The New York Times Editorials/ Letters, Saturday June 1, 1996

Promote Cancer Treatment, Not Cancer Phobia

To the Editor:

In "A Fragmented War on Cancer" (Op-Ed, May 29), Hamilton Jordan says, "The raw data suggest we are on the verge of an epidemic."

His assertions that cancer will be diagnosed in twice as many people and that twice as many will die of it this year than in 1971 are misleading.

More people will be diagnosed with and die of cancer this year than in 1971 because the population is now much larger. Moreover, cancer is much more common for older

age groups than younger age groups, and the proportion of Americans who are in older age groups is larger this year than in 1971.

After you take into account population size and the number of people in different age groups, recent national cancer data suggest a slight downward trend in overall cancer mortality during the current decade.

A reported increase in the diagnosis of cancer usually reflects changes in diagnostic and reporting practices rather than an actual increase in cancer incidents.

It was also misleading for Mr. Jordan to present as evidence of an emerging cancer epidemic the often quoted

figures, "One in three women and one in two men will have cancer in their lifetimes." Such high lifetime risks figures actually mean many people will live long enough lives to be likely to develop cancer. These figures will mask much lower risk of people of any given age developing cancer in the next 10, 20 or 30 years.

While I support Mr. Jordan's continued efforts to promote progress in cancer prevention and treatment, I encourage him to be more careful lest he unwittingly promote cancer phobia.

William M. London Dir. of Public Health, American Council on Science and Health New York, May 29, 1996

Presentation of Mathematical Papers at Joint Mathematics Meetings and Mathfests

Robert M. Fossum
Secretary
American Mathematical Society

Kenneth A. Ross
Associate Secretary
Mathematical Association of America

VISUAL AIDS AT JOINT MATHEMATICS MEETINGS

Preparation of Transparencies for the Overhead Projector

The most frequent complaint heard at meetings is that transparencies are unreadable.

Although transparencies for the overhead projector can be written on while the talk is in progress, you are strongly encouraged to prepare them in advance and test them out with a projector. Writing as you lecture is not a good idea. For one thing, it invites recording too many details. In addition, if you make an error, you will either fail to notice it or else will cross it out and correct it. Either way the audience is distracted.

Transparency copy may be prepared by word processing on a computer or by hand. Typewritten copy is not recommended as the characters are too small to read. To prepare transparencies in advance you are advised as follows:

Preparation of text by hand If you write by hand, use either a #2 lead pencil or a thick felt pen (such as Flair # 844-01 or Pentel Permanent Point Bullet Marker MM50) on sheets of ordinary white 8 1/2″ × 11″ paper.

Never write in script; it is too hard to read. If you prepare a transparency by hand, you should print. (This does not mean all caps.)

Prepare all copies on a hard-surfaced desk or table, not on a cloth-covered table, blotter, or other resilient surface.

There should be no smudges, erasures, or corrections on a prepared transparency.

General instructions

1. Leave at least a 1″ margin on all four sides of the text.
2. Use characters not less than 1/2″ in height. If you use word processing software, use bold type, 14-point or larger. Please note that the projected

image is distorted in such a way that the upper part of each page is considerably larger than the lower part. To balance the image, make the characters on the bottom of the sheet larger and farther apart than those at the top.

3. Use no more than 12 lines per sheet, and leave ample space between lines.

4. Limit each transparency to one topic. Complicated problems may, however, be simplified in presentation by the use of overlays, which consist of several acetate sheets hinged together like the pages of a book. A complex image can be built up from simple components added to the picture, one at a time, by turning the pages. This mode of presentation can be very effective but calls for careful preparation.

Please Note: The speaker who needs to illustrate while lecturing will be provided with a supply of blank transparency sheets and an overhead projector pen. Use only black, blue, green, or red overhead projector pens. Do not use pink, yellow, orange, or any pastel colors.

Use of the Overhead Projector

Overhead projection equipment is relatively easy to use; however, a speaker unfamiliar with the overhead projector should practice with it prior to the lecture in order to become familiar with its features and feel comfortable with it during the lecture.

For maximum effectiveness in the use of the projector, please note the following suggestions:

1. Keep your shoulder out of the way! If it is lighted by the projector, it is blocking the screen.

2. Avoid distracting the audience by continually turning around to look at the screen. Be sure, however, to glance at the screen when placing a new transparency on the projector to ensure its proper placement.

3. Keep in mind that the projector's lamp can be turned on or off to direct the audience's attention to the speaker or to the screen as desired.

Additional Guidelines for Speakers

1. Speak loudly enough to be heard in all areas of the room.

2. Practice your speech, timing yourself to ensure that important points are not rushed and ample time is left for a summary.

3. Define key terms briefly.

Slopes

by Meg Hickey

As children we learn about slopes as something hard to walk up and fun to slide down. Because humans find it much easier to walk on flat surfaces, we are physically aware of the increase in effort required to climb hilly or sloped ground. Some sloped ground has recreational possibilities, such as skiing or sledding, or dramatic interest in a marathon, such as Heartbreak Hill in the Boston Marathon. Though we experience slopes in different ways, we all know that once you have expended energy to go up them, they offer the potential to roll or slide down, and, the steeper the slope, the faster the movement downwards.

There are a lot of practical uses of slopes that involve their potential for moving people or things downwards, by the force of gravity upon them. In buildings, sloped roofs are designed to shed water and snow to the ground. Streets are sloped from the middle down to the gutters so that rain can run along the gutters into the catch basins. Planted areas and paved parking areas need to be sloped at least 1 foot down for every 100 feet for the surface to drain well. Sloped pipes also use gravity to move drain water and sewage down into the main sanitary sewer in the street. A slope as small as 1/8″ down for every foot of horizontal length is sufficient for plumbing pipes to drain well: note that a slope of 1/8″ per foot is equivalent to 1/8″ per 12″ or 1 unit down for every 96 horizontal units, roughly the same as for surface drainage.

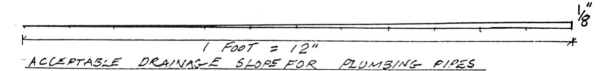

1 FOOT = 12″

ACCEPTABLE DRAINAGE SLOPE FOR PLUMBING PIPES

There are a number of ways to specify how steep a slope is:

With roofs slope is often referred to as "pitch" and might be specified as 5:12 which means 5 feet up for every 12 horizontal feet. Sometimes carpenters refer to slopes as "rise over run" meaning the ratio of vertical to horizontal. Slopes can be laid out with a carpenter's framing square.

When a slope is specified as a percentage, such as 40%, then the vertical rise is 40 feet for a horizontal run of 100 feet. You can use any other units like meters or yards and you will get the same slope.

Meg Hickey is a mechanical engineer and architect who teaches Architectural Engineering at The Massachusetts College of Art.

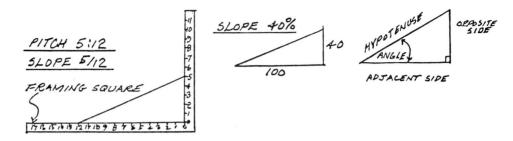

Slopes can also be specified by the angle at which they rise from the horizontal. The angle can be found by trigonometry very easily if you know the rise and run of a slope, since the *tangent* of an angle is the rise/run. On a triangle showing the rise, run, and angle of a slope, the tangent is sometimes called the "opposite side over the adjacent side" referring to the sides of the triangle relative to the angle.

The angle of a slope is the *arctangent* of the slope ratio. Many calculators have an ATAN function which gives angles from tangents; if your calculator doesn't have it, you can look it up in math books with trigonometric tables.

For a triangle with 40 rise and 100 run, the tangent is:

$$40/100 = 0.40$$
$$ATAN(0.40) = 20.8°$$

What is the slope, from the horizontal, of this italic type?

Roof plan A is the view looking down on the roof of the Victorian house shown (both on page 507). The arrows show the down direction of drainage, and the dotted path shows the path of a raindrop starting on the top of the tower and rolling down to the ground. Notice the small section of ridged roof behind the chimney: it is called a "cricket"; if it was not there the water and snow would drain into the back of the chimney and cause a leakage problem.

Now look at roof plan B (page 507) which shows three proposed additions to an L shaped building. You should be able to identify three potential leakage problems with this plan. Where will the snow build up?

If an earth slope is too steep it may cause erosion. Roots of trees and shrubs can help to prevent erosion of planted banks, but if the slope is greater than 50%, retaining walls are required to stop the earth from sliding. Because slope is critical to ground water drainage and building and planting potential, architects start with a contour map of a site, usually available from the U.S. Geological Survey map service. From these you can draw a sideways cut-through view, called a "section" showing the slope of the land.

One of the obvious problems in placing a building on sloped land is that you need to avoid having all the surface water runoff running into your cellar—as in Section A. What is done is to slope the ground away from the house, all the way around it, as in Section B.

If you look at the contour map of Amesbury on page 508 you can compute the slope of the ski runs on the northeast side of

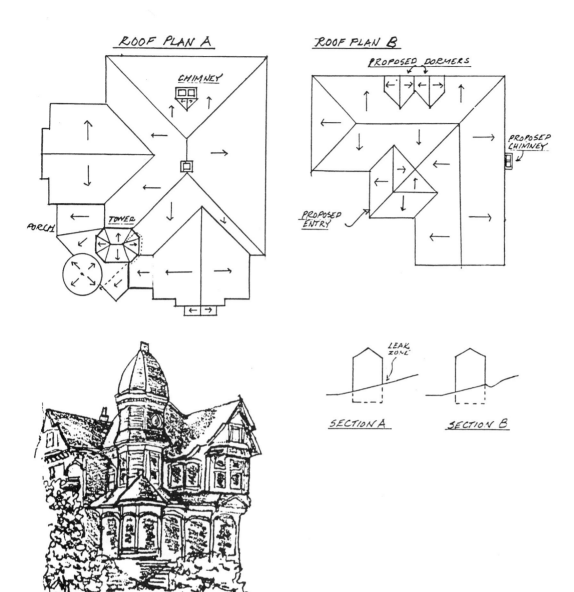

ROOF PLAN A

ROOF PLAN B

CHIMNEY

PROPOSED DORMERS

PROPOSED CHIMNEY

PORCH

TOWER

PROPOSED ENTRY

LEAK ZONE

SECTION A

SECTION B

Powwow Hill. The scale is one centimeter represents 200 meters, and the contour lines are 3 meter differences in height. The shorter run starts on the 69m. contour line and ends down at the 27m. line for a rise of $69 - 27 = 45$m.; it covers about 1 centimeter on the map, so the horizontal distance is 200m. The average slope then is $45/200 = .225$ or 22.5%.

Disregarding the slight inconvenience of ending up in the lake, calculate the slope % on the opposite, southwest side of the hill, from the 60 meter contour line to the base.

After the Vietnam war many young veterans confined to wheelchairs successfully lobbied for much needed improvements in the architectural standards for wheelchair access to public buildings and housing. As a result many intersections now have curb cuts where a section of curb has been replaced with a sloped ramp from

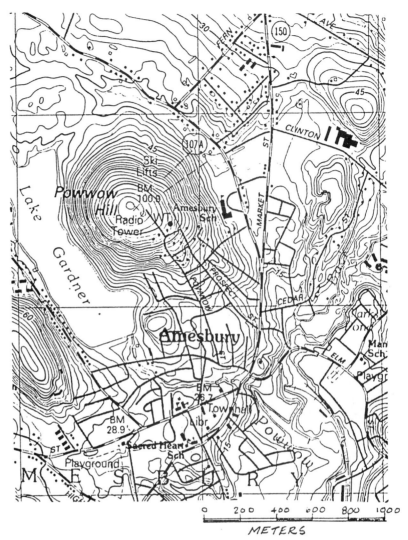

From Massachusetts Highway Department & the U.S. Geological Survey

sidewalk to street area, and the steepness of ramps for rising into buildings has been changed from 1 foot of rise for every 10 feet of horizontal length to 1 foot of rise for every 12 feet of ramp length. This gentler slope makes it easier for people with less strength to feel more in control of the downwards roll of the chair, and more able to roll the chair up the ramp. Curb cuts, because they are shorter, are allowed to be steeper but should be kept under 15%. If wheelchair ramps are specified 1:12 and the front door is 3 feet above the ground, then the ramp must be 12 × 3' or 36 feet long.

Ramps that rise around a central space have been used in two famous buildings: the Guggenheim art museum in New York City has a spiralling ramped gallery 5 stories tall, and the Boston Aquarium has a series of ramps winding around a huge fish tank about 3 stories tall. The museum was designed by Frank Lloyd Wright, and the aquarium by the Cambridge Seven Associates. Both of these places are worth a visit.

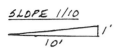

SLOPE 1/10

SLOPE 1/12

GUGGENHEIM MUSEUM

Another practical use of slopes is in making 3D objects that have angled surfaces rather than just horizontal and vertical planes. People who sew clothing are familiar with producing sloped surfaces in skirts or trousers where the fabric must be sloped in from the larger hip to the smaller waist. For example, if a 40″ hip must be reduced to a 30″ waist, 10″ must be gradually removed over the 7″ distance from hip to waist. If the entire 10″ was removed only at the side seams, the garment would fit badly over the butt and stomach. These parts can be fitted with sloped "darts," which taper the fabric in gently to the waist. Darts are made by folding the fabric and stitching at an angle down to the edge of the fold. The tapering resulting from a dart is twice the width of the top of the dart.

For the skirt shown: if the front and back are each 20″ wide with a 1″ taper on each side from the hip to the top, there remains 18″ at the top of each piece; this is still too big since half of the 30″ waist is 15″. The front and back pieces each have to be reduced from 18″ to 15″ with darts. If the front has 2 darts, one each side, each dart can take up half of the 3″ difference required, or 1.5″. The top of the dart when folded is 1.5″/2 = .75″.

Incidentally, the old-fashioned word for a basic clothing pattern is "sloper," deriving from the old English "slupan"—to slip. This

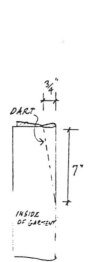

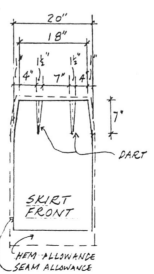

word is not only the root of clothing to slip over your body, but you can see how words like slip and slope are related to the same root.

Another example of slopes in clothing is the high heel. If cowboy boots are sold with a 2″ heel in both men's and women's sizes, ranging from a women's size 5, 7.5″ long, to a men's size 13, 12.5″ long, you can calculate the difference in slope for each case.

The small size has a slope of 2/7.5 = .29 = 29%; the large size has a slope of 2/12.5 = .16 = 16%.

What angles are the 2 slopes?

What size heel would a woman with a size 5 foot have to buy in order to have a slope of 16%?

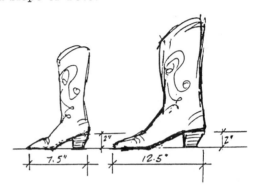

All of the examples of slopes cited so far have been physical, and measurable in space as one dimension divided by another, in effect, a ratio. The word "rate," which comes from the same root as "ratio," is often used for slopes on graphs. We talk about "birth rate" (16.2 live births per 1000 population in 1991 in the U.S.), "tax rate" (5 cents per dollar = 5%), or "rate of travel" (miles/hr or mph).

Any 2 related quantities which can be graphed, such as U.S. population versus time, or position of mercury in a thermometer versus temperature, can produce a plotted line which has one or more slopes along it. If the line is not straight you can find the slope for short segments.

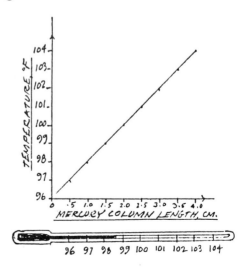

For example, if you set your car trip meter to 0, then read the mileage every 5 minutes for a 25 mile trip from the center of one town to the suburbs of another, then plot distance versus time, you can find the average speed during any 5 minute interval, or you can find the overall average speed of the whole trip.

minutes	miles
0	0
5	1.0
10	2.5
15	5.0
20	8.0
25	12.0
30	17.0
35	21.5
40	25.0

The average speed for the whole trip is:

25 miles/40 minutes = 5/8 mile per minute

or, to get it in mph, multiply by 60 minutes per hour:

60 min/hr × 5/8 mile/min = 37.5 mph

You know that at some times you were travelling faster or slower than this average. You can get a more accurate picture of the variations in speed by looking at individual time segments.

During the first 5 minutes 1 mile was covered. The average speed is 1 mile per 5 minutes, or 1/5 mile per minute: this can be read as the slope of the first line segment. At the same rate for an hour, since there are 60 minutes in an hour, you would go 60 × 1/5 mile = 12 miles. The speed is then 12 miles per hour (mph).

During the next segment 2.5 − 1 = 1.5 miles was covered in 5 minutes. The speed was:

(1.5/5) miles/min × 60 min/hr = 18 mph.

Calculate the average mph for the segment from 25 to 30 minutes.

Obviously, if you took mileage data at smaller time intervals, you could get a more accurate reading of the speed as it changes. The branch of mathematics called differential calculus teaches how to find slopes for infinitesimal segments of a graphed line or curve.

Once you know the slope of a line you have a ratio you can apply to similar right angled triangles of the same slope but different size as shown on page 512. The law of Pythagoras, A × A + B × B = C × C, and the proportional rules of similar triangles, A/a = B/b = C/c, can be used to find unknown dimensions. Here is one worked out example, and one for you to try:

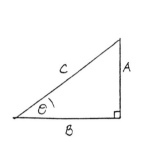

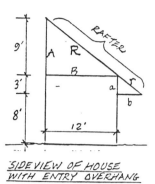

SIDE VIEW OF HOUSE WITH ENTRY OVERHANG

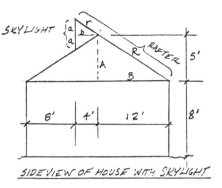

SIDE VIEW OF HOUSE WITH SKYLIGHT

For the house with entry overhang shown above, how long is the total rafter, R + r?

Using Pythagoras: A × A + B × B = R × R
9 × 9 + 12 × 12 = R × R = 81 + 144 = 225
$R = \sqrt{225} = 15$

Using similar triangles, because triangles A B R and a b r have the same slope:

a/A = r/R

3/9 = r/15

15·(3/9) = r = 5

So: Total Rafter = R + r = 15 + 5 = 20

For the house section with skylight, how long is the total rafter, R + r, and what is the height of the skylight, a + a?

References

Architectural Graphic Standards, Ramsey & Sleeper, John Wiley & Sons, N.Y.

Vogue Pattern Book, Vogue Pattern Service.

Webster's Unabridged Dictionary, Merriam-Webster Inc.

United States Coastal & Geodetic Survey.

Frank Lloyd Wright, Solomon R. Guggenheim Museum.

Daughters of Painted Ladies, Pomada, Larsen, Keister; E.P. Dutton, N.Y.

Sketches by Meg Hickey, building illustrations by Myrna Kustin and Juanita Jones.

Statistical Portrait—
University of Massachusetts/Boston

Table 5

SAT Scores of New Freshmen by College/Program, Ten-Year Trend
(Excluding the DSP Program, Learning Disabled and Foreign Students)

		1986	1987	1988	1989	1990	1991	1992	1993	1994	1995
College of	SATVerbal	453	453	464	464	449	447	432	431	433	430
Arts &	SATMath	468	475	479	488	471	473	477	464	483	474
Sciences	Combined	921	928	943	952	920	920	909	895	916	904
	[N]	[429]	[444]	[330]	[299]	[294]	[262]	[240]	[256]	[211]	[251]
College of	SATVerbal	448	453	449	409	436	428	423	418	412	413
Management	SATMath	512	529	513	507	520	502	518	500	490	515
	Combined	960	982	962	916	956	930	941	918	902	928
	[N]	[108]	[93]	[68]	[54]	[47]	[36]	[32]	[52]	[31]	[32]
College of	SATVerbal	467	431	442	475	NA	NA	NA	NA	NA	NA
Education	SATMath	437	457	445	485	NA	NA	NA	NA	NA	NA
	Combined	904	888	887	960	NA	NA	NA	NA	NA	NA
	[N]	[15]	[26]	[19]	[15]						
Human	SATVerbal	436	425	432	385	453	363	397	399	413	416
Performance &	SATMath	439	467	476	426	442	407	407	453	489	463
Fitness	Combined	875	892	908	811	895	770	804	852	902	879
	[N]	[15]	[11]	[20]	[18]	[6]	[6]	[15]	[12]	[7]	[12]
College of	SATVerbal	441	419	444	454	439	488	457	420	387	408
Nursing	SATMath	480	459	478	457	443	522	502	466	537	521
	Combined	921	878	922	911	882	1010	959	886	924	929
	[N]	[13]	[15]	[11]	[9]	[7]	[6]	[17]	[13]	[3]	[8]

Table 1

Undergraduate Admissions Summary: Fall 1987 - 1995

FRESHMEN

	1987	1988	1989	1990	1991	1992	1993	1994	1995
Applied	3349	3274	3272	2775	2487	2378	2494	2356	2439
Decision Ready	2925	2932	2804	2420	2181	2035	2237	1963	2110
Admitted	1877	1601	1648	1516	1388	1366	1496	1320	1363
Admit Rate	64.2%	54.6%	58.8%	62.6%	63.6%	67.1%	66.9%	67.2%	64.6%
Enrolled	1033	818	823	751	745	734	800	662	691
Admitted but Deferred	94	41	53	68	70	53	34	41	23
Yield Rate	55.0%	51.1%	49.9%	49.5%	53.7%	53.7%	53.5%	50.2%	50.7%
Not Admitted:									
Denied	1048	1331	1156	904	793	669	741	643	747
Incomplete Applications	424	342	468	355	306	343	257	393	329

TRANSFERS

	1987	1988	1989	1990	1991	1992	1993	1994	1995
Applied	3468	3683	4118	3653	3189	3375	3382	2875	2761
Decision Ready	3037	3012	3398	3149	2808	2912	3018	2480	2374
Admitted	2574	2224	2623	2610	2455	2555	2741	2249	2081
Admit Rate	84.8%	73.8%	77.2%	82.9%	87.4%	87.7%	90.8%	90.7%	87.7%
Enrolled	1668	1319	1619	1552	1442	1505	1666	1397	1225
Admitted but Deferred	239	170	182	218	166	170	110	93	79
Yield Rate	64.8%	59.3%	61.7%	59.5%	58.7%	58.9%	60.8%	62.1%	58.9%
Not Admitted:									
Denied	463	788	775	539	353	357	277	231	293
Incomplete Applications	431	671	720	504	381	463	364	395	387

TOTAL UNDERGRADUATES

	1987	1988	1989	1990	1991	1992	1993	1994	1995
Applied	6817	6957	7390	6428	5676	5753	5876	5231	5200
Decision Ready	5962	5944	6202	5569	4989	4947	5255	4443	4484
Admitted	4451	3825	4271	4126	3843	3921	4237	3569	3444
Admit Rate	74.7%	64.4%	68.9%	74.1%	77.0%	79.3%	80.6%	80.3%	76.8%
Enrolled	2701	2137	2442	2303	2187	2239	2466	2059	1916
Admitted but Deferred	333	211	235	286	236	223	144	134	102
Yield Rate	60.7%	55.9%	57.2%	55.8%	56.9%	57.1%	58.2%	57.7%	55.6%
Not Admitted:									
Denied	1511	2119	1931	1443	1146	1026	1018	874	1040
Incomplete Applications	855	1013	1188	859	687	806	621	788	716

Office of Institutional Research, UMass Boston: Fall 1995

Table 13

Enrollment Trends at UMass Boston (State Funded Enrollment), Fall Semesters 1986-1995

	1986	1987	1988	1989	1990	1991	1992	1993	1994	1995
Total Enrollment	12919	13574	12451	12584	12478	11606	11775	12136	12142	11602
Full-Time Enrollment (Hct)	6969	7448	7007	6964	7002	6556	6561	6657	6532	6064
FTE Enrollment	8983	9526	8921	8921	8863	8300	8439	8607	8552	8095
Matriculated Undergraduate	9065	9615	9283	9514	9216	8589	8693	8972	8556	8007
% Full-Time	66.8%	68.0%	67.9%	66.6%	68.0%	67.9%	66.9%	65.1%	66.2%	64.6%
Matriculated Graduate	1010	1474	1678	1756	1802	1890	1897	1958	2035	2258
% Full-Time	33.4%	30.9%	27.0%	24.0%	26.4%	27.4%	30.6%	32.9%	32.5%	30.8%
Non-degree Students	2844	2485	1490	1314	1460	1127	1185	1206	1551	1337
% Undergraduate	82.3%	79.9%	75.0%	77.5%	76.7%	76.5%	68.9%	69.3%	67.0%	74.0%
Total Undergraduate										
HCT Enrollment	11406	11601	10399	10532	10336	9451	9509	9808	9595	8997
FTE Enrollment	8005	8252	7605	7613	7489	6891	6933	7035	6849	6312
% Female	54.6%	55.8%	56.9%	56.5%	55.6%	55.0%	53.5%	53.0%	53.3%	53.7%
Mean Age	27	27	27	27	27	28	28	28	28	29
Median Age	24	24	24	24	25	25	25	25	25	25
Total Graduate										
HCT Enrollment	1513	1973	2052	2052	2142	2155	2266	2328	2547	2605
FTE Enrollment	978	1274	1316	1308	1374	1409	1506	1572	1703	1783
% Female	58.6%	60.4%	62.6%	62.0%	64.0%	64.7%	64.2%	63.6%	63.3%	62.3%
Mean Age	34	34	34	35	34	35	34	34	34	35
Median Age	33	33	32	33	32	33	32	32	32	32

Office of Institutional Research, UMass Boston: Fall 1995

Table 27

Degrees Conferred by Level, Academic Year 1984-85 to 1994-94

	AY 83-84	AY 84-85	AY 85-86	AY 86-87	AY 87-88	AY 88-89	AY 89-90	AY 90-91	AY 91-92	AY 92-93	AY 93-94	AY 94-95
UMass Boston												
Certificate	24	24	52	38	31	53	41	44	59	62	82	55
Bachelor's	1010	1251	1149	1232	1319	1401	1535	1630	1636	1579	1579	1467
Graduate Certificates										8	25	23
Master's	89	97	140	192	252	358	421	427	474	508	502	478
CAGS	6	8	11	14	12	25	19	26	32	30	31	22
Doctorate					1	2	1	2	2	3	4	8
Total	1129	1380	1352	1476	1615	1839	2017	2129	2203	2190	2223	2053
Public Institutions												
Bachelor's	12574	12769	12849	13599	13211	13714	14235	14728	14678	n.a.	n.a.	n.a.
Statewide												
Bachelor's	40135	39780	39779	40748	40308	41613	42825	43520	44487	n.a.	n.a.	n.a.

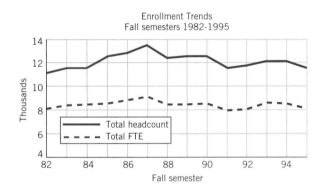

Enrollment Trends
Fall semesters 1982-1995

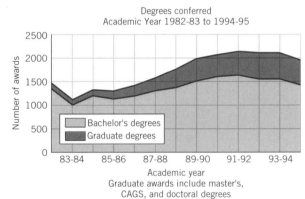

Degrees conferred
Academic Year 1982-83 to 1994-95

Graduate awards include master's,
CAGS, and doctoral degrees

Statistical Portrait—
The University of Southern Mississippi*

The University of Southern Mississippi

10-YEAR TREND:
ENROLLMENT BY ETHNIC GROUP

Hattiesburg Campus
Fall 1987-1996

FALL SEMESTER	CAUCASIAN		ASIAN		BLACK		AMERICAN INDIAN		HISPANIC		TOTAL
1987	8,716	82.2%	283	2.7%	1,506	14.2%	25	0.2%	72	0.7%	10,602
1988	9,125	82.6%	270	2.4%	1,575	14.3%	27	0.2%	54	0.5%	11,051
1989	9,576	83.0%	248	2.1%	1,650	14.3%	21	0.2%	49	0.4%	11,544
1990	9,872	82.9%	260	2.2%	1,702	14.3%	20	0.2%	58	0.5%	11,912
1991	10,146	82.2%	307	2.5%	1,811	14.7%	24	0.2%	60	0.5%	12,348
1992	9,468	81.1%	329	2.8%	1,791	15.3%	26	0.2%	66	0.6%	11,680
1993	9,187	80.0%	311	2.7%	1,874	16.3%	33	0.3%	82	0.7%	11,487
1994	9,202	79.4%	306	2.6%	1,953	16.9%	23	0.2%	103	0.9%	11,587
1995	9,454	78.0%	291	2.4%	2,218	18.3%	38	0.3%	112	0.9%	12,113
1996	9,626	77.0%	325	2.6%	2,385	19.1%	33	0.3%	128	1.0%	12,497

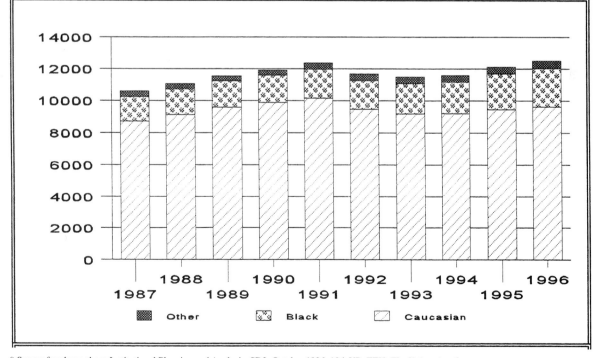

* Source for above chart: Institutional Planning and Analysis, SD3, October 1996. 10th YR_ETH. *The University of Southern Mississippi Fact Book 1996–1997.*

The University of Southern Mississippi
10-YEAR TREND:
ENROLLMENT BY GENDER

Hattiesburg Campus
Fall 1987-Fall 1996

FALL SEMESTER	MEN		WOMEN		TOTAL
1987	4,781	45.1%	5,821	54.9%	10,602
1988	4,951	44.8%	6,100	55.2%	11,051
1989	5,009	43.4%	6,535	56.6%	11,544
1990	5,098	42.8%	6,814	57.2%	11,912
1991	5,309	43.0%	7,039	57.0%	12,348
1992	5,055	43.3%	6,625	56.7%	11,680
1993	5,023	43.7%	6,464	56.3%	11,487
1994	5041	43.5%	6546	56.5%	11,587
1995	5192	42.9%	6921	57.1%	12,113
1996	5215	41.7%	7282	58.3%	12,497

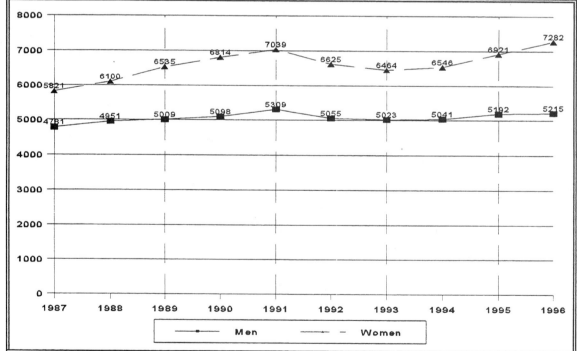

SOURCE: Institutional Planning and Analysis, SD1
DATE: March 17, 1997

10YR_GND

The University of Southern Mississippi

RETENTION OF FIRST-TIME ENTERING FRESHMEN
(All First-Time Freshmen)

Hattiesburg Campus
Fall 1983-1995

	ALL FIRST-TIME FRESHMEN	AVERAGE ACT	CUMULATIVE GRADUATION AND CONTINUATION RATES							
			AFTER 1ST YEAR	AFTER 2ND YEAR	AFTER 4TH YEAR		AFTER 6TH YEAR		AFTER 8TH YEAR	
			CONTINUED	CONTINUED	GRADUATED	CONTINUED	GRADUATED	CONTINUED	GRADUATED	CONTINUED
1983	1,179	19.7	72.3%	61.7%	21.3%	29.3%	37.9%	9.7%	42.7%	5.0%
1984	1,151	19.8	72.7%	59.3%	18.1%	31.8%	35.6%	10.6%	40.7%	6.0%
1985	1,093	19.7	71.6%	57.9%	20.8%	29.4%	38.0%	10.9%	43.5%	6.6%
1986	1,015	20.1	72.0%	60.6%	18.9%	31.4%	35.0%	13.1%	42.3%	6.3%
1987	1,033	20.3	75.7%	65.0%	19.9%	34.7%	39.9%	12.7%	47.0%	6.5%
1988	1,183	20.0	74.0%	61.5%	19.5%	34.9%	39.5%	10.0%	44.2%	5.2%
1989	1,204	20.4	76.5%	63.0%	22.2%	32.1%	41.2%	11.8%	47.3%	6.3%
1990	1,084	21.6	78.9%	63.1%	21.7%	32.2%	39.5%	13.1%	46.2%	6.9%
1991	1,124	21.8	71.3%	58.4%	16.9%	33.3%	35.4%	12.0%		
1992	1,004	21.8	71.9%	55.6%	17.7%	31.4%				
1993	1,017	21.7	72.9%	61.5%						
1994	1,063	21.7	73.9%	60.5%						
1995	1,235	21.5	72.7%							
1996										
1997										

SOURCE: Institutional Planning and Analysis, Retention Report
DATE: November, 1996

RETEN.ALL

The University of Southern Mississippi

10-YEAR TREND:
ENROLLMENT BY CLASSIFICATION

Hattiesburg Campus
Fall 1987-Fall 1996

FALL SEMESTER	FRESHMEN	SOPHOMORE	JUNIOR	SENIOR	GRADUATE	TOTAL	% CHANGE
1987	1,786	1,458	2,250	3,487	1,621	10,602	-4.74%
1988	1,931	1,563	2,461	3,424	1,672	11,051	4.24%
1989	1,950	1,659	2,571	3,597	1,767	11,544	4.46%
1990	1,773	1,722	2,725	3,898	1,794	11,912	3.19%
1991	1,818	1,605	2,886	4,148	1,891	12,348	3.66%
1992	1,550	1,470	2,522	4,299	1,839	11,680	-5.41%
1993	1,580	1,385	2,468	4,172	1,882	11,487	-1.65%
1994	1,685	1,405	2,410	3,956	2,131	11,587	0.87%
1995	1,797	1,472	2,575	4,024	2,245	12,113	4.54%
1996	1,843	1,551	2,611	4,225	2,267	12,497	3.17%

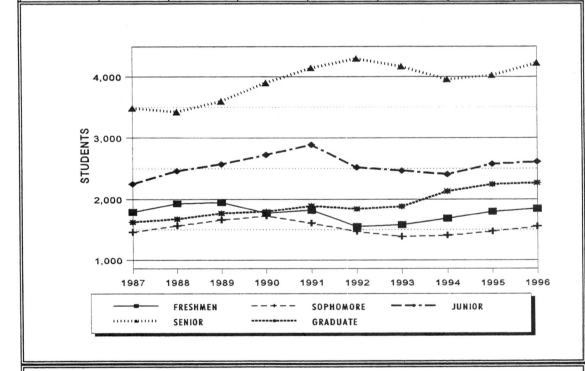

SOURCE: Institutional Planning and Analysis
DATE: March 17, 1997

10YR_CLS

The University of Southern Mississippi

DEGREES AWARDED:
5-YEAR PROFILE BY TYPE OF DEGREE

Fiscal Year 1991-92 through 1995-96

	1991-92	1992-93	1993-94	1994-95	1995-96
UNDERGRADUATE DEGREES	2,425	2,581	2,383	2,274	2,068
Bachelor of Arts	129	150	176	193	227
Bachelor of Fine Arts	18	21	33	31	28
Bachelor of Music	16	23	16	17	17
Bachelor of Music Education	18	16	15	19	15
Bachelor of Science	1,709	1,708	1,570	1,513	1,266
BS in Business Administration	419	480	394	326	314
Bachelor of Science in Nursing	98	153	138	145	173
Bachelor of Social Work	18	30	41	30	28
MASTER'S DEGREES	714	776	744	704	775
Master of Arts	41	53	36	39	45
Master of Art Education	2	9	4	2	0
MA in the Teaching of Languages	0	5	10	17	29
Master of Business Administration	81	56	51	28	35
Master of Education	247	239	191	216	232
Master of Fine Arts	5	2	6	4	6
Master of Library Science	19	37	41	29	38
Master of Music	6	7	8	11	3
Master of Music Education	6	17	12	11	13
Master of Professional Accountancy	11	11	13	20	16
Master of Public Health	0	8	9	13	22
Master of Science	227	275	282	257	244
Master of Science in Nursing	5	7	38	20	41
Master of Social Work	64	50	43	37	51
SPECIALIST DEGREES	19	22	9	14	26
DOCTORAL DEGREES	95	93	95	103	127
Doctor of Education	18	10	11	10	11
Doctor of Music Education	1	0	1	1	0
Doctor of Musical Arts	1	4	3	3	4
Doctor of Philosophy	75	79	80	89	112
TOTAL DEGREES AWARDED	3,253	3,472	3,231	3,095	2,996

SOURCE: Institutional Planning and Analysis, Degrees Awarded by Major, STU0510
DATE: March 18, 1997

DEG_TYPE

The University of Southern Mississippi

DEGREES AWARDED:
5-YEAR PROFILE BY COLLEGE AND TYPE OF DEGREE

Fiscal Year 1991-92 through 1995-96

	1991-92	1992-93	1993-94	1994-95	1995-96
THE ARTS	56	67	69	72	67
Bachelor of Arts	4	7	5	5	7
Bachelor of Fine Arts	18	21	33	31	28
Bachelor of Music	16	23	16	17	17
Bachelor of Music Education	18	16	15	19	15
BUSINESS ADMINISTRATION	526	586	459	360	323
BS in Business Administration	419	480	394	326	314
Bachelor of Science	107	106	65	34	9
EDUCATION AND PSYCHOLOGY	647	638	579	533	478
Bachelor of Science	630	617	564	510	455
Bachelor of Arts	17	21	15	23	23
HEALTH AND HUMAN SCIENCES	353	429	423	437	439
Bachelor of Science	237	246	244	262	238
Bachelor of Science in Nursing	98	153	138	145	173
Bachelor of Science in Social Work	18	30	41	30	28
LIBERAL ARTS	490	532	529	521	477
Bachelor of Science	384	411	373	356	281
Bachelor of Arts	106	121	156	165	196
SCIENCE AND TECHNOLOGY	353	329	324	351	284
Bachelor of Science	351	328	324	351	283
Bachelor of Arts	2	1	0	0	1

SOURCE: Institutional Planning and Analysis, Degrees Awarded by Major, STU0510   Page 1 of 2
DATE: March 18, 1997   DEG_COLL

The University of Southern Mississippi

DEGREES AWARDED:
5-YEAR PROFILE BY COLLEGE AND TYPE OF DEGREE

Fiscal Year 1991-92 through 1995-96

	1991-92	1992-93	1993-94	1994-95	1995-96
THE GRADUATE SCHOOL	**828**	**891**	**848**	**821**	**928**
Master of Arts	41	53	36	39	45
Master of Art Education	2	9	4	2	0
MA in the Teaching of Languages	0	5	10	17	29
Master of Business Administration	81	56	51	28	35
Master of Education	247	239	191	216	232
Master of Fine Arts	5	2	6	4	6
Master of Library Science	19	37	41	29	38
Master of Music	6	7	8	11	3
Master of Music Education	6	17	12	11	13
Master of Professional Accountancy	11	11	13	20	16
Master of Public Health	0	8	9	13	22
Master of Science	227	275	282	257	244
Master of Science in Nursing	5	7	38	20	41
Master of Social Work	64	50	43	37	51
Specialist in Education	19	22	9	14	26
Specialist in English	0	0	0	0	0
Specialist in History	0	0	0	0	0
Doctor of Education	18	10	11	10	11
Doctor of Music Education	1	0	1	1	0
Doctor of Musical Arts	1	4	3	3	4
Doctor of Philosophy	75	79	80	89	112
TOTAL DEGREES AWARDED	**3,253**	**3,472**	**3,231**	**3,095**	**2,996**

SOURCE: Insitutional Planning and Analysis, Degrees Awarded by Major, STU0510
DATE: March 18, 1997

Page 2 of 2
DEG_COLL

Quantitative Examples

Edited by John G. Truxal

VERBAL SAT SCORES

One of the major arguments in federal education circles concerns whether to publish national test results on a state-by-state or even district-by-district basis. The National Assessment of Educational Progress (NAEP) has historically released data only on a regional basis so particular states or districts are not criticized. The SAT scores for 1989 are available by state as shown below.

SAT Scores   1989

Mean SAT Verbal Score

1.	Iowa	512	27.	Washington	448
2.	N. Dakota	500	28.	N. Hampshire	447
3.	Utah	499	29.	Alaska	443
4.	S. Dakota	498	30.	Oregon	443
5.	Kansas	495	31.	Nevada	439
6.	Nebraska	487	32.	Connecticut	435
7.	Tennessee	486	33.	Delaware	435
8.	New Mexico	483	34.	Maryland	434
9.	Alabama	482	35.	Maine	431
10.	Oklahoma	479	36.	Massachusetts	432
11.	Wisconsin	477	37.	Vermont	435
12.	Kentucky	477	38.	Virginia	430
13.	Minnesota	477	39.	Rhode Island	429
14.	Louisiana	473	40.	New Jersey	423
15.	Mississippi	472	41.	Pennsylvania	423
16.	Missouri	471	42.	California	422
17.	Arkansas	471	43.	Florida	420
18.	Montana	469	44.	New York	419
19.	Idaho	465	45.	Texas	415
20.	Wyoming	462	46.	Indiana	412
21.	Illinois	462	47.	D.C.	407
22.	Colorado	458	48.	Hawaii	406
23.	Michigan	458	49.	Georgia	402
24.	Arizona	452	50.	South Carolina	399
25.	Ohio	451	51.	North Carolina	397
26.	W. Virginia	448			

Before we jump to the conclusion that Iowa is the optimum place to educate children and North Carolina is the worst, we might also want to consider the following data.

Percent Taking SAT

1.	Iowa	5		27.	Washington	39
2.	N. Dakota	5		28.	N. Hampshire	66
3.	Utah	5		29.	Alaska	42
4.	S. Dakota	6		30.	Oregon	63
5.	Kansas	10		31.	Nevada	23
6.	Nebraska	10		32.	Connecticut	75
7.	Tennessee	12		33.	Delaware	60
8.	New Mexico	11		34.	Maryland	60
9.	Alabama	8		35.	Maine	59
10.	Oklahoma	8		36.	Massachusetts	72
11.	Wisconsin	12		37.	Vermont	63
12.	Kentucky	10		38.	Virginia	59
13.	Minnesota	15		39.	Rhode Island	63
14.	Louisiana	9		40.	New Jersey	67
15.	Mississippi	4		41.	Pennsylvania	63
16.	Missouri	13		42.	California	44
17.	Arkansas	7		43.	Florida	47
18.	Montana	20		44.	New York	69
19.	Idaho	17		45.	Texas	43
20.	Wyoming	14		46.	Indiana	55
21.	Illinois	17		47.	D.C.	67
22.	Colorado	29		48.	Hawaii	52
23.	Michigan	12		49.	Georgia	59
24.	Arizona	23		50.	South Carolina	55
25.	Ohio	23		51.	North Carolina	57
26.	W. Virginia	15				

The relation between these two data sets is displayed clearly in the following graph.

Mean SAT
Verbal Score

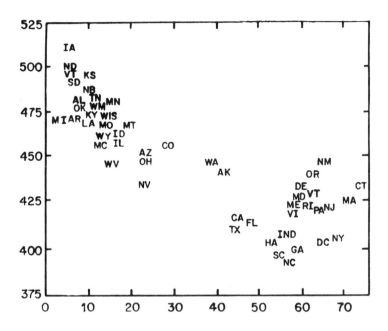

Percent of Seniors Taking the SAT
Source: The College Board

When we look at test scores from *past* years we can clearly see that the Mean SAT Average goes down as the percentage of students taking the test rises. For example, partial data from two times four years apart show changes in average Math scores:

	Old		New	
	%	Average	%	Average
Montana	9	547	20	523
Arizona	11	512	23	500
Colorado	17	520	29	508

Who Collects Data and Why?

By Bob Martin

GOVERNMENT AGENCIES

The most politically charged, and therefore the most discussed, databases have in the past been those maintained by the Federal government: e.g., those of the Census Bureau, the Social Security Administration, all branches of the military, the FBI and other security agencies, and myriad lesser-known Federal agencies which inevitably compile extensive databases of use to someone.

Less known to the public, but equally important, are various obscure computer files established by the 50 state governments and thousands of local government entities. Unlike the well known Federal databases, many state and local data files are in the public domain—available for free and for a fee to any interested party. The compilation and sale of such files is a growing business for database companies.

Important state and local data archives—often computerized and readily accessible—include:

- Public registries of births, deaths, marriages and divorces, real estate property sales and assessments, and so on.
- State department of motor vehicles listings of automobile ownership and accidents.
- Local tax records (such as property tax payments), lists of residents and voter registration lists.

BANKS AND INSURANCE FIRMS

Commercial banks, insurance companies of all kinds, and health care insurers in particular all assemble massive amounts of data on the finances, health status, and other personal business of nearly all Americans. While this data is not public information as such, it is very often available for a fee to outside parties.

Even the most confidential kinds of information are often available with little difficulty to those who know how to gain access: for example, certain health insurance firms will provide employers with detailed monthly reports on the health care expenditures and diagnoses of their employees (e.g., John Jones saw a psychiatrist four times for depression last month).

Bob Martin is the administrator for a large health care organization.

Other commonly used criteria include what are called the "recency, frequency and amount" of each list member's purchases during a certain period. For example, you could purchase a mailing list whose members had each bought at least $50 worth of books on at least three different occasions during the past six months.

Various databases firms specialize in compiling and continuously updating lists based on certain public sources. For example, one firm tracks every house sold in the nation, using local property tax and courthouse records as raw material. The firm compiles not only the buyer's and seller's identity but the sales price, location, and other data. This data is valuable to many other firms—for example, to carpet and furniture retailers for whom recent homebuyers are prime customers.

HOW IS THIS DATA USED?

As indicated above, all of this data is compiled, sold, and used for many commercial purposes. The most obvious of these purposes is direct marketing—especially the sale of retail consumer goods by mail and by telephone.

As computer memories become ever-larger and the speed and complexity of computer capabilities grow, the cross-tabulation and synthesis and the many types of database mentioned above is becoming remarkably sophisticated.

One example of such sophistication is the Claritas Corporation, a database and marketing firm which has analyzed the U.S. population on the basis of zip code and sub-area classifications. Each small geographical zone—down to the neighborhood level—is given one of about 40 tag names, such as "Shotguns and Pick-up trucks," "Brains and Money," and so on. These categories are based not only on income, but other variables such as occupation, family size, property values in the area, spending habits, and so on. Claritas continuously updates and sells this data not only to direct marketers but to many types of businesses deciding where to initiate new ventures.

Besides direct marketing, other uses of data include the following:

- Deciding where to locate new stores, malls, distribution hubs, office buildings, residential developments and other enterprises.

U.S. Government Data Collection

by Anthony Roman, 1994,
Center for Survey Research,
University of Massachusetts/Boston

The primary data collection arm of the Federal government is the Bureau of the Census. Although most people are familiar with the Bureau of the Census through the decennial Census, many people are unaware that its data collection activities continue throughout each and every year. The U. S. Government routinely collects information about many aspects of American life, both personal and corporate, through surveys.

One major aspect of the Census Bureau's work involves economic surveys of American businesses. The Bureau tracks manufacturing firms, importing and exporting firms, wholesale and retail trade businesses and virtually any kind of business. Among the data collected from these surveys are the types and amounts of products being manufactured, sold, and exported. The number of workers being employed within each type of industry is also estimated. Economic forecasters and legislators rely on these data when setting policies designed to keep the U.S. economy stable. On many days, the business section of U.S. daily newspapers will contain articles describing how a recent government report on "manufactured durable goods" or "exports" or some other economic variable, affected the stock market or had direct or indirect impact on our economic situation.

The Bureau of the Census collects many other types of data through its demographic surveys conducted with randomly selected U.S. residents. The National Health Interview Survey collects information about American health and the proliferation of various types of diseases. National crime trends are examined using data from the National Crime Survey which measures victimization and exposures to crime. The Consumer Expenditures Surveys collect information about U.S. residents' purchasing habits, from houses and cars to pickles and bread. The Survey of Income and Program Participation examines the components of Americans' income and to what extent people are or are not availing themselves of government programs. This survey, for example, is used to estimate the number of families receiving food stamps and the number who are eligible to receive them but don't.

The largest, ongoing demographic survey is the Current Population Survey (CPS). The CPS collects data from approximately 60,000 households each month in order to estimate employment, unemployment, and other characteristics of the general labor force. Each month, network news broadcasts, local news broadcasts, and newspapers will detail how the national or statewide unemployment rate went up or down the previous month. An increase of 1 tenth of 1 percent in the national unemployment rate may mean that about 60,000 more Americans were unemployed last month as compared to the month before. Needless to say, the numbers are carefully watched by legislators, policy makers, and business leaders.

The origin of the Current Population Survey corresponds to the origin of government data collection in general. During the Great Depression of the 1930's, it became painfully clear that government leaders lacked sufficient information to develop strategies for dealing with the economic plight of U.S. citizens. How many Americans were out of work? Was the South in worse shape than the Midwest? Were farmers in worse shape than laborers? Was Ohio better off than Oklahoma? Which industries were strong and which were about to fail? Leaders had no answers to any of these questions and no centralized database had been kept for this type of information. The Enumerative Check Census of 1937 was the first attempt by government to conduct a nationwide probability survey and measure unemployment. The experience led to a great deal of research on methods of conducting large-scale probability surveys and methods of estimating unemployment and other important national statistics. The Sample Survey of Unemployment began in March 1940 and has been conducted monthly ever since. The Census Bureau took over conducting the survey in August 1942 which became known as the Current Population Survey (CPS).

Since its inception as a national survey of approximately 8,000 households, the CPS has grown in size and complexity. It has become useful as a tool for the equitable distribution of federal funds and is used by researchers to study many attributes related to employment and unemployment in the U.S.

In general, the data collection conducted by the Census Bureau for the Federal government has provided a wealth of information for legislators and researchers to help understand the effects of national policies and the relationships among earnings, industry and job classification, and a host of demographic characteristics.

Current Population Reports   Special Studies Series P23-189

POPULATION

PROFILE
OF THE UNITED STATES
1995

Issued July 1995

U.S. Department of Commerce
Ronald H. Brown, Secretary
David J. Barram, Deputy Secretary

Economics and Statistics Administration
Everett M. Ehrlich, Under Secretary
for Economic Affairs

BUREAU OF THE CENSUS
Martha Farnsworth Riche, Director

Educational Attainment

ANDREA ADAMS

The Nation's educational level has risen dramatically in the past 50 years.

Since the Bureau of the Census first collected data on educational attainment in the 1940 census, educational attainment among the American people has risen substantially. In 1940, one-fourth (24.5 percent) of all persons 25 years old and over had completed high school (or more education), and 1 in 20 (4.6 percent) had completed 4 or more years of college. By 1993, over four-fifths (80.2 percent) had completed 4 years of high school or more, and over one-fifth (21.9 percent) had completed 4 or more years of college.

The increase in educational attainment over the past half century is primarily due to the higher educational attainment of young adults, combined with the attrition of older adults who typically had less formal education. For example, the proportion of persons 25 to 29 years old who were high school graduates rose from 38.1 percent in 1940 to 86.7 percent in 1993, while for persons 65 years old and over, it increased from 13.1 to 60.3 percent.

There is no difference in the educational attainment of young men and women.

Differences in educational attainment between men and women have historically been attributed to differences in attainment at the college level. In 1940, the percentages of men and women 25 to 29 years old completing 4 or more years of college were close to equal, but at a very low level (6.9 percent compared with 4.9 percent). Between 1940 and 1970, both sexes increased their college attainment, but men's gains were significantly greater. The college completion rates for men and women 25 to 29 years old in 1970 were 20.0 and 12.9 percent,

respectively. Since 1970, however, the college gains of young adult women have outstripped those of young adult men, until by 1993, there was no statistical difference in the proportions of men and women 25 to 29 years old with 4 or more years of college (23.4 and 23.9 percent, respectively).

Educational attainment levels continue to rise for race and Hispanic groups.

Blacks have made substantial progress in narrowing the educational attainment gap relative to Whites. In 1940, only 7.7 percent of Blacks 25 years old and over had completed high school, compared with 26.1 percent of Whites. In 1965, the corresponding figures were 27.2 and 51.3 percent, respectively. By 1993, 70.4 percent of Blacks 25 years old and over had completed high school, compared with 81.5 percent of Whites. Hence, the

Percent of Persons Who Have Completed High School or College: Selected Years 1940 to 1993

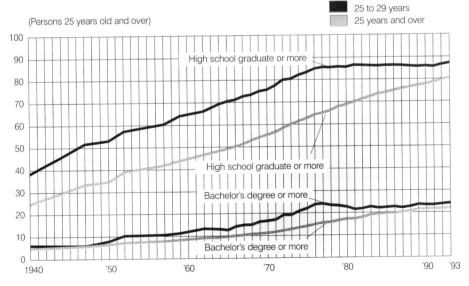

(Persons 25 years old and over)

25 to 29 years
25 years and over

difference between the Black and White rates was smaller in 1993 than in the earlier years.

Among persons 25 to 29 years old in 1940, only 10.6 percent of Black and Other-races men had completed 4 years of high school, compared with 38.9 percent of White men. By 1993, there was no statistical difference in the proportions of Black men and White men who had completed high school: 85.0 and 86.0 percent, respectively. Similar gains were made by young Black women but they remained different from White women in 1993, when 80.9 percent of Black women 25 to

29 years old had completed 4 years of high school, compared with 88.5 percent of White women. In 1940, the proportions were 13.8 percent of Black and Other-races women and 43.4 percent of White women.

Although the proportion of Blacks 25 years old and over who have completed college has increased since 1940, it is about one-half the proportion of their White counterparts (12.2 percent compared with 22.6 percent in 1993). Among young adults 25 to 29 years old in 1993, Blacks were more than half as likely as Whites (13.2 percent compared with 24.7 percent)

to have completed 4 or more years of college. Data for persons of Hispanic origin[1] have not been collected for as long a period as for race groups, but the patterns also indicate some improvement in educational attainment over time. Among Hispanics 25 years old and over in 1993, 53.1 percent had completed high school, up from 36.5 percent in 1974. Completion of college stood at 9.0 percent, a significant increase from the level of 5.5 percent in 1974.

[1]Persons of Hispanic origin may be of any race. These data do not include the population of Puerto Rico.

For Further Information

See: Current Population Reports, Series P20-476, *Educational Attainment in the United States: March 1993 and 1992.*

Contact: Rosalind R. Bruno
or Andrea Adams
Education and
Social Stratification
Branch
301-457-2464

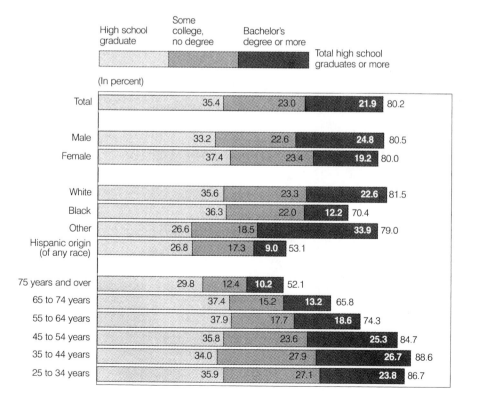

Educational Attainment of Persons 25 Years Old and Over, by Sex, Race, Hispanic Origin, and Age: March 1993

	High school graduate	Some college, no degree	Bachelor's degree or more	Total high school graduates or more
Total	35.4	23.0	21.9	80.2
Male	33.2	22.6	24.8	80.5
Female	37.4	23.4	19.2	80.0
White	35.6	23.3	22.6	81.5
Black	36.3	22.0	12.2	70.4
Other	26.6	18.5	33.9	79.0
Hispanic origin (of any race)	26.8	17.3	9.0	53.1
75 years and over	29.8	12.4	10.2	52.1
65 to 74 years	37.4	15.2	13.2	65.8
55 to 64 years	37.9	17.7	18.6	74.3
45 to 54 years	35.8	23.6	25.3	84.7
35 to 44 years	34.0	27.9	26.7	88.6
25 to 34 years	35.9	27.1	23.8	86.7

(In percent)

Money Income

WILFRED T. MASUMURA

Household income declined between 1989 and 1993.

Real median household income (in 1993 dollars) in the United States fell 7.0 percent from $33,585 in 1989 to $31,241 in 1993.[1] Most recently, real median household income fell 1.0 percent between 1992 and 1993. This continuing decline occurred during and after the 1990-91 recession.

Household income varied by household composition.

In 1993, the median income of married-couple households was $43,129. The median income was much less for households with a female householder, no husband present ($18,545), and for nonfamily households, mainly one-person households ($18,880).[2]

Households with the oldest householders and the

---

[1]Changes in "real" income refer to comparisons after adjusting for inflation based on changes in the Consumer Price Index.

[2]The difference between $18,880 and $18,545 is not statistically significant.

youngest householders had the lowest median incomes in 1993. Households with householders 65 years old and over had a median income of only $17,751. Somewhat higher was the median income of households with householders 15 to 24 years old, $19,340. In contrast, households with householders 45 to 54 years old had the highest median income, $46,207.

Household income varied by race and ethnic origin.

In 1993, Asian and Pacific Islander households had the highest median income ($38,347); whereas, Black households had the lowest ($19,532). The 1993 median income was $32,960 for White households and $22,886 for Hispanic households.[3]

---

[3]Persons of Hispanic origin may be of any race. These data do not include the population of Puerto Rico. Due to the small number of American Indian, Eskimo, and Aleut households, a median income figure for them would be statistically unreliable.

Household income varied by the number of earners in a household.

Households with no earners had the lowest median income in 1993 ($11,807); whereas, households with two or more earners had the highest median income ($49,430). Households with one earner had a median income of $25,560.

Household income varied by the householder's job status.

In 1993, the median household income of householders who were employed year-round, full-time was $44,834. Among householders who were part-time workers, the median household income was $21,608. Householders who did not work had a median household income of only $14,787.

Median earnings of year-round, full-time workers varied by gender.

In 1993, the median earnings of year-round, full-time workers was $30,407 for men and $21,747 for women. The

---

Median Household Income, 1989, 1992, and 1993

(In 1993 dollars)

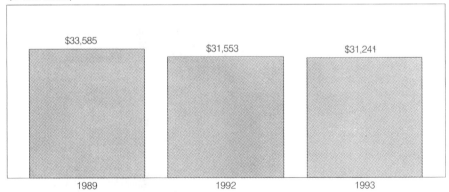

$33,585 — 1989
$31,553 — 1992
$31,241 — 1993

Percent change: 1992-93 = 1.0 percent decline
1989-93 = 7.0 percent decline

female-to-male earnings ratio for year-round, full-time workers was 0.72, comparable to the all-time high reached in 1990.[4]

Median earnings of year-round, full-time workers varied by occupation.

In 1993, among male year-round, full-time workers, the median earnings was $42,722 for executives and managers; $32,327 for sales workers; and $27,653 for precision production, craft, and repair workers. For women, the figures were $28,876, $18,743, and $21,357, respectively.

Median earnings of year-round, full-time workers varied by educational attainment.

In 1993, the median earnings of male year-round, full-time workers 25 years old and over with a college degree was $45,987. In comparison, the median earnings for those

[4]The earnings data and female-to-male earnings ratio for 1989 and 1990 were modified based on the inclusion of data on members of the Armed Forces.

with only a high school diploma was $26,820, and the median earnings for those with some high school education but no diploma was $21,402. For female year-round, full-time workers, the comparable figures were $32,291, $19,168, and $14,700, respectively.

The distribution of income has become somewhat more unequal over time.

The household income distribution changed over the past 25 years. In 1993, those at the bottom 20 percent of the income distribution received less of the Nation's income than previously, while those at the top 20 percent received more.

In 1968, the poorest 20 percent of households received 4.2 percent of the aggregate household income. By 1993, their share declined to just 3.6 percent. In contrast, the highest 20 percent of households received 42.8 percent of the aggregate household income in 1968. By 1993, their share had increased to 48.2 percent.

Those in the middle of the income distribution also received proportionally less of the Nation's income in 1993 than previously. The middle 60 percent of households received 53.0 percent of the aggregate household income in 1968. By 1993, their share had declined to 48.2 percent.

For Further Information

See: Current Population Reports, Series P60-188, *Income, Poverty, and Valuation of Noncash Benefits: 1993.*

Contact: Wilfred T. Masumura
Income Statistics Branch
301-763-8576

Share of Aggregate Household Income, by Quintile: 1968 to 1993

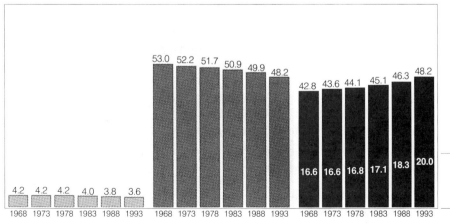

Lowest 20 percent — 1968: 4.2, 1973: 4.2, 1978: 4.2, 1983: 4.0, 1988: 3.8, 1993: 3.6

Middle 60 percent — 1968: 53.0, 1973: 52.2, 1978: 51.7, 1983: 50.9, 1988: 49.9, 1993: 48.2

Highest 20 percent — 1968: 42.8, 1973: 43.6, 1978: 44.1, 1983: 45.1, 1988: 46.3, 1993: 48.2; (top 5 percent) 1968: 16.6, 1973: 16.6, 1978: 16.8, 1983: 17.1, 1988: 18.3, 1993: 20.0

Wealth, Income, and Poverty

By Mark Maier

Economists differentiate between wealth, a stock of value such as a house or stock, and income, a flow of value over time, such as a salary or interest payment. Data on wealth and income typically come from different sources, each with its own set of problems. There is relatively little information on the distribution of wealth in the United States. One reason for the scarcity is difficulty of measurement. Because many items of wealth have not been sold recently, they do not have a readily identifiable price. A second reason for the lack of data on wealth is its ownership by a relatively small group who are not eager to share information about how much wealth they own. As the controversies below illustrate, this second factor severely restricts how much we know about wealth. Unlike wealth, incomes leave a "paper trail" of readily measured dollar amounts. As a result, income is well documented in a variety of government and private-sector sources. But, as this chapter demonstrates, there is still serious disagreement about fundamental findings, including the trend in well-being for the typical family, and the extent of poverty in the United States.

DATA SOURCES

Wealth

Federal Reserve Survey   The most comprehensive survey on wealth holdings is conducted by the U.S. Federal Reserve, a quasi-independent government body that acts as the country's central bank. At three-year intervals during the 1980s, the Fed conducted its own independent survey of several thousand households, asking 100 pages of questions about each family's financial status.

> *Data Sample: The 1986 Survey of Consumer Finances found that 19.3 percent of households owned stocks, with a median value of $6,000, and an average value of more than $80,000.*

Survey of Income and Program Participation   The second major U.S. wealth survey is conducted by the Census Bureau. These data were collected on a regular basis only between 1850 and 1890, and then not again until 1984 in the Survey of Income and Program Participation (SIPP).

From Mark Maier, *The Data Game: Controversies in Social Science Statistics,* 1991.

Where the Numbers Come from

Organizations	Data Sources	Key Publications
Board of Governors U.S. Federal Reserve System	Survey of Consumer Finances	*Federal Reserve Bulletin*
Bureau of the Census U.S. Department of Commerce	Survey of Income and Program Participation; U.S. Census of Population; Current Population Survey	*Census of Population; Current Population Reports* (Series P–60, P–70)
Bureau of Labor Statistics U.S. Department of Labor	Establishment Survey	*Employment and Earnings; Current Wage Developments*
Internal Revenue Service U.S. Department of the Treasury	Tax returns	*Statistics of Income*
Summary data in *Statistical Abstract of the U.S.*		

Data Sample: In the 1984 survey, equity in homes comprised 59.7 percent of the wealth of households with incomes less than $10,000, but only 30.1 percent of the wealth of households with incomes over $48,000.

Indirect Estimates from Tax Records Historical data on wealth distribution are calculated by two major methods. The "estate-multiplier" method pioneered by Robert J. Lampman of the National Bureau for Economic Research uses Internal Revenue Service records for the 1 percent of estates subject to taxes (only holdings greater than $600,000 in 1987). A related method called "income capitalization" works backwards from tax reports of rent, dividends, and interest to estimate wealth from which these incomes are derived.

Data Sample: Lampman measured a decline in the share of wealth held by the top 1 percent from 36 percent in 1929 to 26 percent in 1956.

Direct Counts The extraordinarily rich usually are well-known individuals about whom information can be obtained from public sources. Based on stock-ownership records, media coverage, and independent investigation, *Forbes* and *Fortune* magazines each estimate wealth holdings for a select number of these individuals.

Income

U.S. Census Every decennial Census of Population since 1940 has included questions about income, providing researchers with a tremendously detailed data source, but one that is available only at ten-year intervals.

> *Data Sample: In the 1980 U.S. Census, McAllen-Edinburg-Mission, Texas, had 35.2 percent of the population below the poverty line, the highest rate for all U.S. metropolitan areas.*

Current Population Survey   By far the most frequently cited source of income data is the Census Bureau's Current Population Survey (CPS). Covering over 60,000 households, the CPS is the largest survey taken in between census years.

> *Data Sample: According to the Current Population Survey, Manchester-Nashua, New Hampshire, had the fastest increasing per capita income between 1979 and 1983 of all U.S. metropolitan areas.*

Bureau Labor Statistics' Establishment Survey   Data collected in the BLS Establishment Survey are often quite accurate because the survey is based on employer records, rather than on respondent recall. But the Establishment Survey measures earnings for individual jobs, not the total income for a worker, who may have more than one employer, or for a family with several separate sources of income.

> *Data Sample: In the March 1985 Establishment Survey, the lowest-paying manufacturing industry was "Children's dresses and blouses," with average weekly earnings of $170.16 for nonsupervisory production workers.*

Panel Study of Income Dynamics   For comparisons over time, or longitudinal studies, researchers frequently turn to a private survey, the University of Michigan's Panel Study of Income Dynamics (PSID). Begun with 5,000 households in 1968, PSID followed 7,000 households in 1986, including many of the original sample, as well as those split off when children married or couples separated.

CONTROVERSIES

Are the Rich Getting Richer?

Are the rich getting richer? The answer is "we don't know." Attempts during the 1980s to measure wealth distribution generated considerable controversy, but ended in continued uncertainty whether wealth ownership is becoming more or less concentrated. Nonetheless, this research failure illustrates the problem of surveying wealth, or any other variable so unequally distributed that even large surveys are unlikely to include those who own a significant part of the total wealth.

In 1983 the U.S. Federal Reserve changed its usual format for measuring wealth by adding 438 individuals already known to be wealthy. In theory, this commonly used method of "enriching" the data sample should have increased our knowledge about wealth holdings. Indeed, the survey indicated that the share of wealth going to the top 0.5 percent increased to 35 percent in 1983 from 25 percent in 1962. But these startling results were not destined for much public notice. As usual, the Federal Reserve survey was published in the *Federal Reserve Bulletin,* from which the numbers made their way into academic studies and textbooks.

Then, in 1986, the Joint Economic Committee of Congress released its own interpretation of the Federal Reserve's study using catchy labels—"super rich," "very rich," "rich," and "everyone else"—to underscore the vast holdings of the wealthy that apparently increased since 1962. After a barrage of media reports on these results, the Reagan administration asked the Federal Reserve to reexamine the survey. Suddenly an error appeared: one of the 438 wealthy individuals was not as rich as he or she reported. Removing this individual caused the estimate for the share owned by the super rich to fall by almost 10 percent, wiping out the apparent increase since 1962.

The accuracy of the one data point is difficult to determine. A 1986 follow-up survey showing this person's wealth to be only $2.3 million led the Fed to conclude that the reported 1983 wealth of $200 million was a coding error. Critics responded that the individual, known to own Texas oil and gas wells, indeed may have suffered a financial setback reported between 1983 and 1986. Whatever the actual circumstances of this one Texas magnate, the dispute shows how difficult it is to measure wealth. A basic research principle is to include a survey sample that is large enough that errors for a single individual do not affect overall findings. Even the relatively large Federal Reserve survey—less than 4,000 households—was still too small to accurately measure changes in wealth holdings because so few individuals owned so much.

What Is Wealth?

For the less-than-extremely wealthy, Federal Reserve Board data are relatively accurate. Overall they measure total average net worth in 1986 at over $145,000 per household, while the median net worth was nearly $44,000 (see Box 8.1 on the difference between average and median). For black and Hispanic households, median net worth was only $11,000 in 1986, about one-fifth of median net worth for white households.

Most household wealth is in housing, an average value of $80,000 in 1986 for those who owned a home. For some research purposes, it is best to exclude ownership of homes and other consumer durable goods in order to focus on ownership of wealth that is a source of economic power such as stocks and bonds. This "financial wealth" is even more unequally distributed than total

---

Box 8.1.   Mean, Median, and Mode

The *median* is the data point for which one-half the observations are above and one-half below; the *mean* is the average, or in this case, total income divided by number in the sample; the *mode* is the data point with the most observations, a statistic rarely used in income analysis. Median income is the most common starting point for research on the "typical" household. A relatively small number of families with very high incomes skews the mean at about 15 percent higher than the median. Thus, mean annual family income in the United States was about $24,000 in 1980, while the median income was $21,000. The major drawback to using the median is that it is unchanged by redistribution of income above and below it. For example, when the poor become poorer and the rich become richer, it is possible for the median to stay the same.

---

wealth; the top 10 percent of households owned 86 percent of financial wealth in 1983. Some researchers argue that we should move in the other direction, expanding the definition of wealth to include the value of pensions and social security. Although most people do not consider these benefits to be wealth, they are similar to bonds and other arrangements that promise a source of future income. Former chief economic adviser in the Reagan administration, Martin Feldstein, advocated such an approach, by which measured inequality in U.S. wealth holdings would be reduced by nearly 30 percent.

Researchers need to choose carefully between these different measures of wealth. For analysis of the distribution of resources, total wealth measures are appropriate, perhaps also including the value of retirement funds. But if the research goal is to understand how wealth affects power, then it might be proper to use a narrower definition that excludes housing and retirement benefits.

Who Is the Richest of Them All?

In 1987, the wealthiest individuals in the *Fortune* and *Forbes* surveys were as follows. Both *Forbes* and *Fortune* openly discuss problems in finding out about wealth; every year an individual will catapult to near the top when it is discovered that he or she is in fact quite wealthy. A previously uncounted Mars Candy Company benefactor, billionaire heiress Jacqueline Mars Vogel was belatedly "discovered" and added to the *Forbes* list in 1987. Lester Crown, owner of General Dynamics stock inherited from his father, as well as real estate and sports teams, weighed in at $5.7 billion in the 1987 *Fortune* list, but only $2.1 billion according to *Forbes.*

One source of such discrepancies is the lack of information on privately held wealth. The Mars Company, for example, is one of the small number of large firms still owned entirely by a few individuals. Unlike publicly held corporations, there is no stock price to

Fortune (worldwide)	*Forbes* (U.S. only)
Wealth (in billions)	
1. Sultan Hassanal Bolkiah $25.0	1. Sam Walton $8.5
2. King Fahd $20.0	2. John Werner Kluge $3
3. Samuel and Donald Newhouse $7.5	3. Henry Ross Perot $2.9

estimate the value of the company. Similarly, *Forbes* and *Fortune* must estimate the value of property that has not been bought or sold in recent years. Finally, family trusts complicate wealth measurement by dispersing fortunes among various descendants. As a result, *Forbes* publishes a separate wealth list for total family holdings, led by the DuPonts, Gettys, and Rockefellers.

Are We Better Off?

Is the typical American better off than twenty years ago? This simple question has provoked a broad spectrum of answers depending on the data source used. Between 1970 and 1987, Americans were somewhere between 10 percent *worse* off measured by average weekly earnings and more than 30 percent *better* off measured by income per capita. At stake in the interpretation of these data is the direction of U.S. economic policy. During the 1980s, supporters of the economic policies of presidents Reagan and Bush pointed to the optimistic income data, while critics emphasized data that showed an income decline.

Per Capita Income   The rosiest view of economic progress during the 1980s, indicating 20 percent growth, is based on per capita personal income, a statistic from the U.S. Commerce Department national income accounts. Per capita income is simply the total of all wages, interest, rents, and other incomes divided by the number of people in the country. Although this number is often cited as evidence of success in the Reagan-Bush economic program, critics charge that the statistic is misleading because the number of wage earners has increased relative to the number of dependents. Or to make the same point with a story: Assume a household of four people in 1980 with one person working outside the home earning $20,000 had a per capita income of $5,000. If in 1990, the original worker's income falls to $18,000, but a second family member now works outside the home, and earns $10,000, then per capita income increases to $7,000 ([$18,000 + $10,000] ÷ 4). In other words, total income increased, but only because more people were working, each of whom earned less.

Earnings   What is the trend in earnings of the typical worker? This question changes the original question from one of *income* from all sources including rents, interests, and dividends, to one of *earnings,* including only wages and salaries. (Sometimes earnings is defined

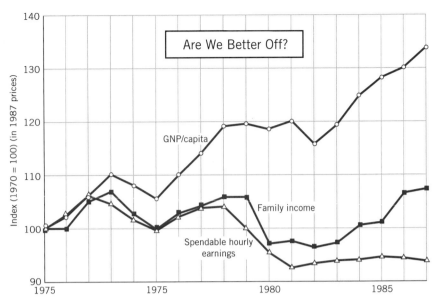

Figure 8.1.

also to include self-employment income and farm income.) For many years the Bureau of Labor Statistics Establishment Survey measured "spendable weekly earnings for the typical family of four with one full time worker and three dependents." According to this measure, this statistically "average" family was better off in 1959 than in 1981! In 1982, the Reagan administration suspended publication of the series on the grounds that it was inflammatory and inaccurate.

Even critics of the administration agreed that this measure used outdated assumptions about tax rates and the composition of the workforce. But, even though correcting for these deficiencies was a relatively simple matter, no official replacement data were introduced. Consequently, University of Michigan economist Thomas E. Weisskopf attempted to revive the series on his own, using spendable *hourly* earnings data in order to avoid the problems with the discredited *weekly* earnings data. According to Weisskopf, this new statistical series still shows steady deterioration of earnings after 1972, falling by more than 13 percent by 1987. Thus, it appears that pay levels for U.S. workers indeed stagnated during the 1970s and 1980s (see Figure 8.1).

Family Income   One way to combine the effect of additional workers with the effect of lower earnings is to look at the trend in income for household units. In fact, the most widely cited statistic on income is featured in the annual Current Population Survey publication, "Money and Income of Households, Families and Persons." According to this source, income was nearly constant during most of the 1970s and 1980s at about $29,000 in 1987 buying power, and improved to only $30,853 in 1987. This statistic is often quoted, but usually without reference to serious shortcomings in the data.

One problem is that the U.S. Census carefully defines a "family" as two or more related individuals living together, thus leaving out more than 30 million individuals in 1990 who lived in nonfamily households, either as single-person households or with unrelated individuals. Median income for families is more than double the median income for nonfamily individuals, so looking only at families overstates the typical income of U.S. households. On the other hand, in recent years nonfamily household incomes increased faster than the average, so research only on families underestimates improvement in overall well-being.

A second problem with family income data is that they do not take into account the decline in family size from 3.58 in 1970 to 3.28 in 1980. According to one measure of family well-being, these smaller families are 20 percent better off because there are fewer children to support. However, other researchers object that these families might have fewer children precisely because of stagnating incomes. In this view, families are worse off because they could not afford as many children as they might have liked.

Because of these problems, researchers need to be careful in using the family income data. Overall, the limited improvement in the median family income level was a worrisome trend during the 1970s and 1980s. But used by itself it does not tell the whole story about changing living standards.

Good Jobs, Bad Jobs

Average well-being, whether measured by per capita income, earnings, or family income, is based on a single statistic that may hide important changes in the *distribution* of income. For example, it is possible for average income to remain unchanged at the same time that it is possible for the rich to become richer and the poor to become poorer. In fact, according to some researchers precisely this trend was under way during the 1980s, an assertion that provoked an often bitter debate about the actual distribution of pay for newly created jobs in the U.S. economy.

Box 8.2.   The Economic Consequences of Divorce

According to Lenore Weitzman's 1985 book, *The Divorce Revolution,* women suffer a 73 percent loss in economic status after divorce. Criticism published subsequently in the journal *Demography* suggested that this widely cited statistic is quite inaccurate. Weitzman's sample consisted of only 228 men and women who became divorced in Los Angeles County in 1977. Data from the far larger Panel Study of Income Dynamics measured an average of about 30 percent loss in economic status for women in the first year after divorce. The critics conclude that the economic well-being of men and women indeed diverges substantially in the years after divorce, but not nearly to the extent publicized by Weitzman's findings.

This controversy is a lesson for researchers about the potential pitfalls in using income data. Popular magazine articles titled "A Surge in Inequality," "The Shrinking Middle Class," "Low-Pay Jobs: The Big Lie," and "Chicken Little Income Statistics" confused readers with seemingly convincing statistics for opposite conclusions. Less frequently reported were the reasons for these differences. When the political controversy ended, it turned out that we had learned quite a bit about trends in U.S. income distribution.

It is well documented that pay levels decline during economic bad times, sending individuals down from middle to lower income levels. The debate about the disappearing middle class occurred just after the 1981–82 recession, so results depended critically on how those years were treated. For example, economists Barry Bluestone and Bennett Harrison received nationwide attention for their study showing that nearly one-half of net new jobs created between 1979 and 1985 paid less than $7,400 (in 1986 buying power). But researchers at the Bureau of Labor Statistics attributed the trend to changing economic conditions, in particular the 1981–82 economic downturn.

A second rejoinder to Bluestone and Harrison came from Brookings Institution economist Robert Lawrence, who attributed most of the growth in low-wage jobs to the sudden appearance of baby-boom youth in the labor market. As these workers gain in age and experience, Lawrence believed income distribution would move closer to its previous norm. Finally, other researchers attributed Bluestone and Harrison's results to incorrect use of inflation adjustments. The relationship between inflation measures and the measurement of income distribution is discussed in chapter 11. To summarize, by using a different inflation index, American Enterprise Institute researchers Marvin Kosters and Murray Ross measured a *decline* in low-wage jobs for the same time period that Bluestone and Harrison measured their much-publicized increase in low-wage jobs.

Each of these researchers made quite different policy recommendations. Bluestone and Harrison advocated government-sponsored industrial policy to increase the number of high-paying jobs. Lawrence argued against industrial policy, calling instead for retraining and relocation assistance for low-paid, young workers. Kosters and Ross opposed all such government intervention in favor of continued reliance on private markets. The jury is still out on which policies will best help the U.S. economy. Nonetheless, it is possible to assess what we have learned about income distribution in this debate. First, there is widespread agreement that a shift has occurred toward low-pay jobs for some groups. Even Kosters and Ross, who gave the most optimistic appraisal of the U.S. economy, measured an increase in the proportion of low-paid jobs for full-time, year-round workers. In addition, the number of part-timers (who obviously earn less) has increased, although there is debate about how to measure the effect of the increase in part-time work on overall pay levels. Overall, there has been an increase in the proportion of low-income individuals, but the impact on households or families is a more complex picture because, as noted above, households may have higher earnings at the same time that each individual earns less.

Statistics for Every Theory?   At this point it may be tempting to question the usefulness of income statistics that tell such conflicting stories. If policy advisers from every political persuasion can find ample evidence to support any position, what good are income statistics? The answer is that each income statistic measures a slightly different concept; they must be used together, with careful distinction about what is being measured. For example, per capita income provides an indication of well-being, but it tells us nothing about the distribution of that well-being, which must be measured by other statistics. The confusing statistics on earnings and family income showed that major changes were occurring in the labor force and the family in addition to an underlying slowdown in the growth of income. Finally, the good jobs—bad jobs debate illustrated what can go wrong when researchers and the popular media reporting the findings do not pay careful attention to the limitations inherent in the data. Research by Bluestone and Harrison faced considerable criticism when it was shown that alternative assumptions caused a reversal in their findings. The lesson for researchers is that it is necessary to anticipate objections by proving the robustness of results under different assumptions. Indeed, subsequent research confirmed Bluestone and Harrison's conclusion that low-pay jobs increased, but in a politically acrimonious climate opposed to their interventionist policy recommendations, these later findings received little attention.

What Is Poverty?

Research on poverty typically confronts a perplexing question: how to define poverty? The easy solution is to rely on the official U.S. Census Bureau poverty line, the most commonly used in social science research. But, as demonstrated by its origins, the Census Bureau poverty line is quite arbitrary, and therefore inadequate for some purposes.

The Official Poverty Line   During the early 1960s, Molly Orshansky, a Social Security Administration staff economist, was asked to develop a definition of poverty for the War on Poverty. By her own admission, this "Orshansky" poverty line was a compromise between scientific justification and political expediency. She began with a U.S. Agriculture Department estimate for a nutritionally sound diet, and then estimated the total poverty budget based on the proportion of food expenditures in a typical total family budget. Orshansky was unhappy with these assumptions, in particular the food budget, which was intended as a stop-gap grocery list but required a sophistication of purchasing and food preparation that was not likely to be practiced by poor families. But Orshansky was under pressure from the administration to calculate a poverty line at about $3,000 for a family of four, low enough so that the War on Poverty could reasonably be expected to help all those designated as poor. She had more leeway in defining poverty for other-size families, so she set the poverty line on slightly more favorable terms for smaller-size households and families with more than four people.

Not surprisingly, these highly arbitrary poverty lines have been criticized on many grounds. But researchers disagree whether this measure under- or overestimates the actual rate of poverty.

The Poverty Line Is Too Low   The official U.S. poverty line is an absolute standard, adjusted only for inflation. (In 1990 there was renewed debate about how to adjust for inflation, including a Census Bureau proposal to use a new inflation adjustment that would substantially reduce the official poverty line.) Over the years since 1965 the standard of living in the United States has increased, so that on a *relative* basis, the poverty line has fallen. In 1965, it was just under one-half the U.S. median income; by 1986 the official poverty line was less than one-third of the median income. If the poverty line measured poverty at a constant *relative* rate, then measured poverty would have increased between 1965 and 1983, instead of falling, as it did in the official absolute standard.

The Poverty Line Is Too High   Others have criticized the Orshansky poverty line for overestimating the poverty line. For example, economist Rose Friedman points out that poor families spend a higher proportion of their income on food than the ratio used by Orshansky. Friedman recalculates Orshansky's poverty level with new ratios, estimating U.S. poverty at one-half its official level.

A second correction that reduces the apparent level of poverty is to include the value of noncash government programs such as food stamps, school meals, housing subsidies, and medical care. One such effort measured government programs at their market value, that is, what it would cost the poor to buy services comparable to those provided by the government. By this estimate, the poverty rate falls by about one-third, which prompted President Reagan's chief economic adviser Martin Anderson to conclude that the War on Poverty had been "won." But many economists question the relevance of market values for adjusting the poverty rate. In particular, medical care is so expensive in the private sector that the market value of Medicare and Medicaid is sufficient by itself to lift—in theory—almost all the elderly poor out of poverty.

Implications   The official "Orshansky" poverty line has withstood the test of time. It makes sense for many research projects because conclusions can be compared readily with other studies that also use this poverty line. Nonetheless, researchers should be aware of biases in the official number. Most notably, the absolute standard means that poverty today is measured by the same living standards used in 1965 (again, corrected only for inflation). The arbitrary assumptions made by Orshansky may be less of a problem. *Any* poverty line will be arbitrary, and the problems with the official one in part cancel out one another: Friedman's complaint about the food budget ratio is offset by the low food budget originally used, and the failure to include noncash benefits for the poor is offset by tax breaks and other government assistance available to the nonpoor.

One alternative research strategy is to adopt more than one poverty measure, including both absolute and relative standards. By using several poverty lines, it is possible to demonstrate the constancy, or robustness, of results under alternative assumptions. For example, the issue of how to count noncash benefits can be resolved by looking at alternative poverty measures published by the U.S. Census Bureau. Replacing the questionable market valuation for noncash benefits with more realistic assumptions reduces by about one-half the number of the poor who are raised above the poverty line by their noncash benefits.

Do the Poor Stay Poor?

The official poverty rate is based on a snapshot view. If our research interest is to find out if the same households are poor in different years, then we need longitudinal studies, that is, surveys that reinterview respondents over a long period of time. Surprising data from the University of Michigan's Panel Study of Income Dynamics (PSID) and the Census Bureau's Survey of Income and Program Participation (SIPP) show remarkable movement in and out of poverty. The PSID found that only 2.6 percent of the population was poor in eight out of ten years between 1969 and 1978, while over 24 percent were poor for at least one year. Similarly, in SIPP, only 6 percent of the population was poor in *every* month of 1984, but 26 percent were poor for at least one month.

For some conservative policy advisers, these data prove that poverty is less serious than CPS data indicate. In this view, most poor people require little government assistance because their poverty is temporary. Moreover, the small number who stay poor have become dependent on welfare programs and would gain by being forced to find regular employment. On the other hand, liberal social scientists often emphasize the surprisingly large proportion of the population experiencing temporary poverty. In this view, the welfare system is a much-needed security cushion for many families who are precariously close to the poverty line. Along similar lines, sociologist Mary Jo Bane used longitudinal data to show that most poverty occurred because of income or job changes, not because of family composition changes such as divorce or abandonment. Bane concludes: "the problem of poverty should be addressed by devoting attention to employment, wages and the development of skills necessary for productive participation in the labor force rather than hand-wringing about the decline of the family." This policy recommendation has not been tested, but it shows how one can use dynamic data available in the PSID or SIPP to go beyond the debate about the "actual" low or high rate of poverty.

SUMMARY

Without improved data on wealth, we will not understand its distribution in the United States. At present, however, the barriers to better data appear insurmountable. The uneven distribution of wealth means that even extremely large surveys will tell us nothing about the very wealthy; it is unlikely that any of the *Forbes* 400 will be interviewed in the extremely large-sample Current Population Survey, and even more improbable that they will be included in the Federal Reserve survey. For example, the enriched sample of 438 known-to-be-wealthy households in the 1983 Federal Reserve survey included none of the wealthiest Americans identified by business magazines.

Additional data on income also would be helpful, for example, replacing the spendable weekly earnings series and increasing funding to the Survey of Income and Program Participation. But in general, simply increasing the quantity of data will not be sufficient. As economist Isabel Sawhill concludes, "We are swamped with facts about people's incomes and about the number and composition of people who inhabit the lower tail, but we don't know very much about the process that generates these results." In other words, we need more data about income dynamics, that is, when and how income changes over time, and we need data about why incomes differ for different jobs.

Controversies about the distribution of income and the rate of poverty both demonstrate the need to test different assumptions in research projects. This kind of sensitivity analysis can preempt criticism about what may appear to be arbitrary choices such as years chosen for study, inflation indexes, or poverty lines. Such care can help researchers reach an audience beyond the already convinced.

Finally, despite the apparent existence of contradictory statistics, researchers have been able to reach important conclusions about income distribution and poverty. In other words, although it is possible to "lie with statistics," bending the numbers to agree with one's prejudice, it is also possible to use income statistics correctly if one understands the limitations of the underlying data.

CASE STUDY QUESTIONS

1. Even though the Census Bureau spends considerable effort training its survey takers to gather accurate data, respondents may nonetheless misrepresent their income. One study comparing Current Population Survey

(CPS) data with tax returns showed that actual self-employment income was 24 percent higher than reported in the survey, actual government transfer programs other than social security were 42 percent higher, and actual property income such as interest, rent, and dividends was 135 percent higher. How might these possible errors affect research on wealth and income distribution?

2. Current Population Survey data on income exclude capital gains, that is, income derived from the profitable sale of homes, stocks and bonds, or other investments. What effect does this exclusion have on studies of the distribution of income?

3. A 1987 *Ebony* magazine article celebrated a more than 50 percent increase between 1980 and 1986 in the number of black families with incomes between $25,000 and $50,000. By contrast, there was only an 18 percent increase in white families within this income bracket. How might these statistics exaggerate the improvement of income for blacks relative to income for whites?

4. The "feminization of poverty" has received much attention based on statistics showing that over one-half of all poor families were headed by women during the 1980s, up from only 25 percent in 1959. But research on the dynamics of poverty by Mary Jo Bane shows that for blacks, most poor female-headed households were formed from households that were *already* poor. Based on this data, why do some policy advisers recommend employment programs instead of increased child support enforcement as the best way to reduce poverty rates?

5. Many studies of income focus exclusively on family incomes. What biases are introduced by leaving out nonfamily households when we study the well-being of the "typical" household?

6. The gap between men's and women's pay has received much attention in recent years. Usually the gap was measured by the ratio of women's pay to men's pay, which was approximately 69 percent in 1987. But there are several potential problems with this measure.

   a. Most researchers use *weekly* earnings measured in the Current Population Survey. However *annual* earnings that include second jobs measure a slightly larger pay gap. Why?

   b. Most researchers compare pay for only full-time, year-round workers. Why would the pay gap appear larger if all workers were studied? What are the reasons for studying only full-time, year-round workers? What would be the reasons for studying all workers?

   c. The measured pay gap between men and women closed by 7 percentage points between 1979 and 1987. But during this period men's median pay fell by more than 7 percent. How does this additional statistic affect assessment of women's progress in achieving equality?

Education and Income

Statistical Portrait of the Nation

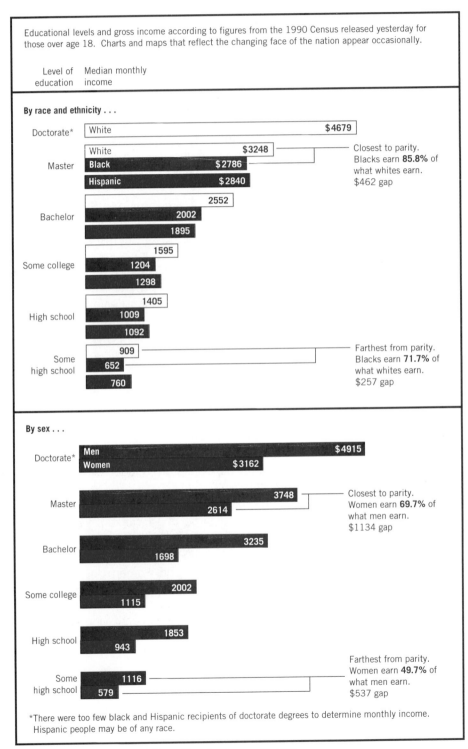

Educational levels and gross income according to figures from the 1990 Census released yesterday for those over age 18. Charts and maps that reflect the changing face of the nation appear occasionally.

Level of education / Median monthly income

By race and ethnicity . . .

Doctorate*   White $4679

Master
- White $3248
- Black $2786
- Hispanic $2840

Closest to parity. Blacks earn **85.8%** of what whites earn. $462 gap

Bachelor
- 2552
- 2002
- 1895

Some college
- 1595
- 1204
- 1298

High school
- 1405
- 1009
- 1092

Some high school
- 909
- 652
- 760

Farthest from parity. Blacks earn **71.7%** of what whites earn. $257 gap

By sex . . .

Doctorate*
- Men $4915
- Women $3162

Master
- 3748
- 2614

Closest to parity. Women earn **69.7%** of what men earn. $1134 gap

Bachelor
- 3235
- 1698

Some college
- 2002
- 1115

High school
- 1853
- 943

Some high school
- 1116
- 579

Farthest from parity. Women earn **49.7%** of what men earn. $537 gap

*There were too few black and Hispanic recipients of doctorate degrees to determine monthly income. Hispanic people may be of any race.

Graph taken from The New York Times, Thursday, January 28, 1993.

Linear Regression Summary

A linear regression line is an attempt to approximates the trend of a set of data points that looks as though it is headed roughly along a straight path. If you want to know what's behind the scenes in calculating the linear regression line on the computer screen, here it is:

STEP 1
Start with a set of data points, x and y coordinates.

STEP 2
Find the average point $\bar{x}$, $\bar{y}$; (called "x bar" and "y bar", these symbols mean "average x" and "average y"). To do this add all the x values and divide by the number of points; then do the same for the y values.

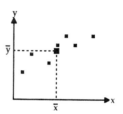

STEP 3
Find the vertical rise, Δy, and the horizontal run, Δx, for the line connecting each point to the average point.
On the data table this is $\Delta x = x - \bar{x}$ and $\Delta y = y - \bar{y}$ for each point.

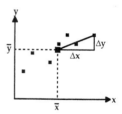

STEP 4
Notice that if we are trying to find a slope for the average line, the lines connecting the average point to points farther away are likely to be closer to what we need. If we are computing an average slope we need to give greater weight in proportion to the distance from the average point.

To give more importance to the slopes from the points farther away from the average point, $\bar{x}$, $\bar{y}$, we multiply each Δx and each Δy by the Δx for that point.
Now you have $\Delta x \cdot \Delta x$ and $\Delta x \cdot \Delta y$ for each point.
(Note that you can also multiply each Δx and each Δy by the Δy instead, this gives a different y weighted line, not used in this course.)

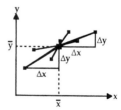

STEP 5
Now sum the $\Delta x \cdot \Delta y$ for all the points; this is $\Sigma \Delta x \cdot \Delta y$.
Then sum the $\Delta x \cdot \Delta x$ for all the points; this is $\Sigma \Delta x \cdot \Delta x$.
(Σ stands for "the sum of")

The x weighted average slope for all the points is $m = \dfrac{\Sigma \Delta x \cdot \Delta y}{\Sigma \Delta x \cdot \Delta x}$

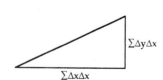

STEP 6
Find "b" by using $\bar{y} = m\bar{x} + b$ so $b = \bar{y} - m\bar{x}$

Now we know m and b we can use them to write the linear regression line equation y = mx + b.

Draw the line y = mx + b through the average point, $\bar{x}$, $\bar{y}$, using slope m and y-intercept b.

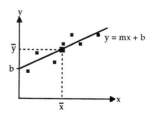

Linear Regression Calculation Example

7 babies between the age of 0 and 24 months are measured for length in inches. If we call the age x and the length y the data points are:

age, x	length, y
1	15
3	18
6	20
10	25
14	27
18	30
23	33

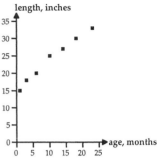

STEP 1: Plot points.

STEP 2: Calculate average point. (Round to 2 decimal places.)

$$\text{average } x = \bar{x} = \frac{1+3+6+10+14+18+23}{7} = 75/7 = 10.71$$

$$\text{average } y = \bar{y} = \frac{15+18+20+25+27+30+33}{7} = 168/7 = 24.00$$

x	y	$\Delta x = x - \bar{x}$	$\Delta y = y - \bar{y}$	$\Delta x \bullet \Delta x$	$\Delta y \bullet \Delta y$	$\Delta x \bullet \Delta y$
1	15	-9.71	-9	94.28	81	87.39
3	18	-7.71	-6	59.44	36	46.26
6	20	-4.71	-4	22.18	16	18.84
10	25	-0.71	1	0.50	1	-0.71
14	27	3.29	3	10.82	9	9.87
18	30	7.29	6	53.14	36	43.74
23	33	12.29	9	151.04	91	110.61
Sum 75	168			391.40	270	316.00

STEP 3: Calculate $\Delta x = x - \bar{x}$ and $\Delta y \; y - \bar{y}$

STEP 4: Calculate $\Delta x \bullet \Delta x$ and $\Delta y \bullet \Delta y$ and $\Delta x \bullet \Delta y$

STEP 5: Calculate (for an x weighted line)

$$m = \frac{\sum \Delta x \cdot \Delta y}{\sum \Delta x \cdot \Delta x} = \frac{316.00}{391.40} = 0.81$$

(For a y weighted line use $m = \frac{\sum \Delta x \cdot \Delta y}{\sum \Delta y \cdot \Delta y}$)

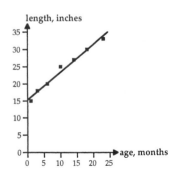

STEP 6: Calculate $b = \bar{y} - m\bar{x} = 24 - 0.81 \bullet 10.71 = 15.32$

So the linear regression line (x weighted) equation is:

$$y = 0.81 \, x + 15.32$$

The Correlation Coefficient

Karen Callaghan, Ph.D
The University of Massachusetts-Boston

A correlation coefficient measures the strength and direction of a linear association between two variables. It ranges from -1 to $+1$. The closer the absolute value is to 1, the stronger the relationship. A correlation of zero indicates that there is no linear relationship between the variables. The coefficient can be either negative or positive. The scatterplots below indicate two linear associations of the same strength but opposite directions.

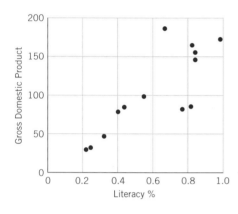 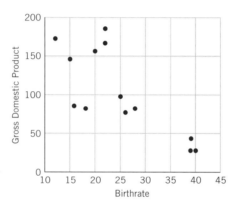

The graph on the left represents the relationship between literacy % and gross domestic product. As literacy % goes up, the gross domestic product goes up, so the correlation coefficient is positive. The graph on the right represents the relationship between birthrate and gross domestic product. As birthrates go up, the gross domestic product goes down, so the coefficient is negative.

The scatterplots on the next page will give you an intuitive grasp of the correlation coefficient, labelled as cc. Figure (a) represents a perfect correlation of $+1$, all the points fall on a perfectly straight line with a positive slope, an unlikely occurrence in any social science data set. Figure (b) represents a strong correlation where the behavior of one variable is similar, but not identical to the behavior of the other variable (e.g., like miles traveled versus amount of gasoline used).

In Figure (c), a correlation of .50 represents a moderately strong positive relationship. The relationship in Figure (d) is weak, so the coefficient is only .25.

The scatterplot in Figure (e) looks like a shotgun blast so the correlation is zero; the x and y variables are not linearly related.

You might be surprised to learn that the coefficient in Figure (f) is also zero. This is because the correlation coefficient measures a linear association, while the relationship in Figure (f) is curvilinear.

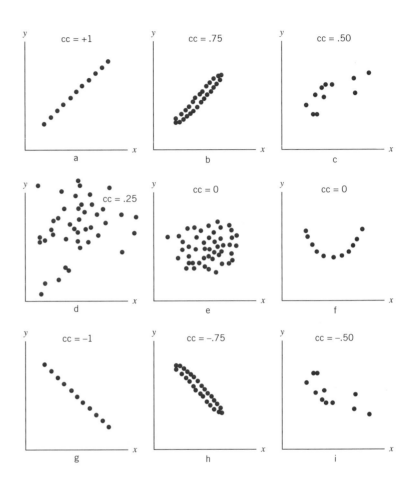

Figures (g), (h), (i) are mirror images of Figures (a), (b), (c). All the correlation coefficients are negative. Figure (g) represents a perfect correlation of −1, all the points fall on a straight line with a negative slope, an unlikely occurence in any social science data set. Figure (h) represents a strong negative correlation. In Figure (i) the correlation is moderately strong.

COMPUTING THE CORRELATION COEFFICIENT (CC.)

While a perfect correlation is easy to decipher, it is difficult to guess the coefficient of weaker correlations. That is why Karl Pearson developed a precise mathematical measure of correlation known as Pearson's r, which is called cc. throughout the text. For those who are interested in knowing how the correlation coefficients are actually calculated, the steps are outlined below.

$$\text{Correlation Coefficient} = \frac{\Sigma(x - \overline{x}) \cdot (y - \overline{y})}{\sqrt{\Sigma(x - \overline{x})^2 \cdot (y - \overline{y})^2}}$$

The symbol Σ is "sigma," the Greek letter S, which is a mathematical shorthand meaning "sum up." So if x takes on values 2, 3, and 6, then: $\Sigma x = 2 + 3 + 6 = 11$.

Literacy % and Life Expectancy for Six Selected Nations

	(x) Literacy %	(y) Life Expectancy
	.29	42
	.92	77
	.52	58
	.55	47
	.40	48
	.66	55
	3.34	327

STEP 1:
Start with a set of data, x and y points. Each data point is kept in a separate row.

STEP 2:
Find $\overline{x}$, the mean of x; and $\overline{y}$, the mean of y. To do this add the values of x and divide by the number of points; then do the same for y.

$\overline{x} = \Sigma x/n = 3.34/6 = .56$
$\overline{y} = \Sigma y/n = 327/6 = 54.50$

STEP 3:
Subtract $\overline{x}$ from each value of x; subtract $\overline{y}$ from each value of y to get a new table of rows.

$x - \overline{x}$	$y - \overline{y}$
−.27	−12.50
.36	22.50
−.04	3.50
−.01	−7.50
−.16	−6.50
.50	.10

STEP 4:
Take the products of each row in step 3 and sum them up.

Products

−.27 × −12.50 =	3.38
.36 × 22.50 =	8.10
−.04 × 3.50 =	−0.14
−.01 × −7.50 =	0.07
−.16 × −6.50 =	1.04
.10 × .50 =	0.05

Sum of products = 12.50

STEP 5:

Take each x value in step 3, square it and sum all the points; then do the same for y

	x squared	y squared
$-.27^2 = .07$	$-12.50^2 = 156.25$
$.36^2 = .13$	$22.50^2 = 506.25$
$-.04^2 = .00$	$3.50^2 = 12.25$
$-.01^2 = .00$	$-7.50^2 = 56.25$
$-.16^2 = .03$	$-6.50^2 = 42.25$
$.10^2 = \underline{.01}$	$.50^2 = \underline{.25}$
Sums of Squares $= .24$	$= 773.50$

STEP 6:

Take the square root of the product of the sums of the squares in step 5.

$$\sqrt{.24 \cdot 773.50} = 13.62$$

STEP 7:

Divide the sum in step 4 by the value in step 6 to calculate the correlation coefficient.

$$cc. = \frac{12.50}{13.62} = .92$$

So the correlation coefficient is .92 which says that literacy percent and life expectancy are strongly related.

Suppose we reverse the x and y axis. Now "x," the independent variable, is life expectancy and "y," the dependent variable, is literacy percent. You do not need to retrace steps 1 through 6 again. The correlation is still .92 since the formula does not make a distinction between the x and y variables. You cannot test whether literacy percent affects life expectancy, or life expectancy affects literacy percent. You can only test whether these two variables vary together in a linear relationship.

An example of a negative correlation.

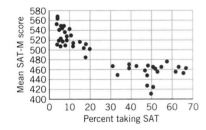

News reports sometimes rate the quality of educational systems by comparing the mean of scores of seniors in each state on college entrance exams. (See discussion p.121 and reading "Verbal SAT Scores.") This method is misleading because the percent of high school seniors who take any particular college entrance test varies greatly from state to state. The figure to the side shows a scatterplot of the mean score on the Scholastic Aptitude Test (SAT) mathematics examination for high school seniors in each state compared with the percent of graduates in each state who took the test.

The negative association between the two variables is evident: SAT scores tend to be lower in states where the percent of students who take the test is higher. Only 3% of the seniors in the three highest scoring states took the SAT.

Performing Arts—The Economic Dilemma

by William J. Baumol and William G. Bowen

A Study of Problems Common to Theater, Opera, Music, and Dance

APPENDIX X–1

Number of Performances and Audience Size

While the typical length of season is currently increasing, primarily in response to the demands of the performers, it is by no means obvious that the increased number of performances will bring with it a corresponding increase in the size of the total audience. There is some reason to expect that when more performances are provided, more tickets will be sold. If most of the orchestras that are lengthening their seasons are also developing new types of concerts, new series and different types of orchestral service, these may attract new persons to attend or they may lead others to attend more often. The new performances will probably take place on nights of the week or at times of the year which are convenient for some persons who would otherwise not have been able to come. But, other things being equal, we may well suspect that more frequent performance will serve, in part, just to spread the concert audience more thinly.

Since, as we saw in Chapter VIII, cost per concert does decrease up to a point as the number of concerts goes up, it is important to see whether the added concerts can be expected to bring an additional audience of average size along with them. Obviously, if attendance per concert declines sufficiently as the number of concerts grows, economies of scale will not help finances.

We compared average attendance per concert with the number of concerts per season for two of the major orchestras for which we had obtained statistically significant cost functions (see the appendix to Chapter VIII). Appendix Figure X–A shows what we found for the same orchestra whose unit cost curve is depicted in Figure VIII–4. It is evident from the diagram that average attendance does fall somewhat with the number of concerts.[1] Furthermore, the decline shown in the diagram is probably a considerable understatement of the rate of decrease involved in the underlying relationship. Many of the historical increases in number of con-

---

[1] Actually the decline is small in terms of the historical relationship shown in Appendix Figure X–A. An increase of 1 per cent in number of concerts yields at 1964 attendance levels a decrease of about 0.3 per cent in attendance per concert, a loss of about 8 persons per concert. With an average of slightly more than 100 concerts per season, this would reduce attendance by a total of about 800 admissions. Since, however, the new concert would on the average be attended by about 2,500 persons, it would bring in a net gain of some 1,700 paid admissions.

APPENDIX FIGURE X–A

RELATION BETWEEN ATTENDANCE PER CONCERT AND
NUMBER OF CONCERTS, FOR A MAJOR ORCHESTRA

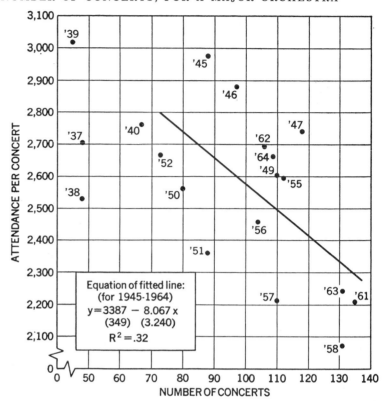

certs occurred only in response to popular demand when the potential audience was already available. If the increased number of performances had been undertaken haphazardly, without regard for autonomous demand changes, and if the new concert had been required, as it were, to hunt up its own audience, one may surmise that the resulting decrease in attendance per concert would have been far more marked. But as more orchestras strive for year-round operation, the number of concerts per season is bound to increase, whether or not they are accompanied by a considerable decline in attendance per concert.

While a larger supply of performances does increase the *total* audience, it undoubtedly yields diminishing returns in terms of audience per concert, and so it is not always an unmixed financial blessing.

His Stats Can Oust a Senator or Price a Bordeaux

By Peter Aseltine
Staff Writer

Princeton-Metro The Times
Sunday May 8, 1994

PRINCETON BOROUGH—Orley Ashenfelter's days are numbered.

Behind the tragedy of discrimination and the bumbling of bureaucrats, the Princeton econometrician sees numbers.

Behind the bouquet of a fine wine—more numbers.

Ashenfelter's ability to see the telling patterns of the world and interpret them with numbers has led people to seek the Princeton University professor's expertise for a wide variety of very practical reasons.

Last month a federal judge in Philadelphia relied partly on Ashenfelter's analysis of voting patterns to give a Pennsylvania Senate seat to a losing Republican candidate. The judge found the Democratic winner had used fraud in soliciting absentee votes in the election in Philadelphia's Second District.

The case raised a difficult question, Ashenfelter says, because while the fraud justified unseating the Democrat, William Stinson, it was unclear whether the proper remedy was to seat the Republican, Bruce Marks, or hold a new election.

Marks had outpolled Stinson on the voting machines, 19,691 to 19,127. But in absentee ballots, Stinson received 1,391 votes to Marks' 366, winning the November 1993 special election by 461 votes.

U.S. District Judge Clarence C. Newcomer found that Stinson workers had improperly influenced absentee voters, mostly in Latino and African-American neighborhoods, in some cases marking the voters' ballots or forging their names.

In February, Newcomer threw out the absentee votes and ordered that Marks be seated. The following month, however, a federal appellate court ruled that while the judge was not wrong to unseat Stinson, he should not have seated Marks without first analyzing whether the fraud actually had swayed the election.

"It was a tricky business for the judge, because the issue really was whether it was reasonable to assume that Marks had actually won that election," Ashenfelter said. "The circuit court wanted some evidence that, in the absence of the fraud on the voters, there would have been a presumption that Marks could have been expected to get at least enough of the absentee votes so that he wouldn't be overwhelmed."

A Princeton economics professor uses statistics to uncover the mysteries of human behavior—from voting patterns that changed the shape of Pennsylvania's Senate, to the price of fine French wines.

Ashenfelter, a 51-year-old economics professor, was appointed as an expert for the new hearing by Newcomer, who had taken one of the statistics seminars that Ashenfelter regularly teaches for federal judges. Two other statistics experts testified, one hired by Marks and one hired by a group of voters opposed to him.

Ashenfelter has testified in court about a dozen times, mostly in discrimination cases, which often rely on statistical analysis.

What Ashenfelter did in this case was quite simple, he says. He used a standard statistical technique called regression analysis to compare machine votes to absentee votes for each of the 22 senatorial elections held in Philadelphia in the last decade. What he found was that the difference between Democratic and Republican votes on the machines generally is a good indicator of the difference in absentee votes.

Ashenfelter used a scatter diagram to show that when the difference between Democratic and Republican machine votes is plotted against the difference in absentee votes, points representing the elections tend to fall around a line representing an ideal correlation between the machine and absentee tallies.

The graph shows that the 1993 election falls well outside the pattern of typical elections. Ashenfelter's analysis would predict a 133-vote advantage for Marks in absentee votes, given his 564-vote margin on the machines. Instead, Stinson had a 1,025-vote advantage.

Ashenfelter calculated that the probability was less than 1 percent that the deviation of 1,158 votes between his predicted result and the actual result was simply due to random changes in voting patterns. He calculated that the probability that Stinson could have received enough absentee votes to win was about 6 percent.

Ashenfelter said surveys of absentee voters conducted by The Philadelphia Inquirer and the Republicans provided information that was extrapolated in court to produce estimates of fraudulent votes that were surprisingly close to his deviation figure of 1,158.

Ashenfelter said there was an amusing moment in court when a lawyer for the Democratic-controlled board of elections tried to suggest that Ashenfelter's analysis would not establish public confidence in a decision to seat Marks. The circuit court had called in its opinion for "evidence and an analysis . . . worthy of the confidence of the electorate."

"He had asked me in my deposition whether I considered myself an expert on establishing the public confidence, and I had said, 'No,'" Ashenfelter said. "So in cross-examination, he got up, and he said, in the usual lawyer way, 'Isn't it true, Dr. Ashenfelter, that you are no more of an expert on what would establish the public confidence than I am?' He had been having his way with me, so at that point I just thought, if a guy asks you a question like that, you've got to let him have it. So I said, 'Well, I may be more of an expert than you are.' Well, the courtroom just cracked up."

Newcomer ultimately issued a new order to seat Marks, based on the testimony of the experts. The appeals court upheld the decision and Marks was sworn in on April 28, returning control of the Pennsylvania Senate to the Republicans.

Ashenfelter is modest about his role in the important case. He seems more excited about the prospect of using his court analysis as a teaching tool. He said many other professors have asked for it.

"It's very simple as these things go," he said. "The reason people want to use it for teaching is because it's hard to come up with such a simple example that's so telling for an actual problem."

Ashenfelter spent only a week developing his analysis for the election case. He currently is working on a much bigger project for the state of New York, evaluating the state's Wicks Law, which requires that state construction projects be done without a prime contractor. By forcing the state to act as prime contractor, the law has probably cost New York taxpayers a billion dollars, Ashenfelter said.

"Expecting bureaucrats to manage a construction project is a pretty expensive proposition," he said.

Ashenfelter's greatest claim to fame, however, is a regression analysis that predicts the prices of Bordeaux wine vintages by charting rainfall and temperature during each growing season. His analysis has proved quite accurate and he publishes his predictions twice each year in a newsletter called "Liquid Assets."

The newsletter has nearly a thousand subscribers, including some wine-industry insiders in Bordeaux. But Ashenfelter said the industry still refuses to use the information in setting prices.

"Most people in the industry refuse to believe this," he said. "They will not change prices to reflect quality, and they have never had to do it in the past. So the whole system is built on this bizarre sucker operation. The wine writers—I guess they're the ones who go along with it. It's a mystery to me."

Meanwhile, however, Ashenfelter is storing away some underpriced vintages. His days may be numbered in one sense, but he's looking forward to enjoying the fruits of his numerical labors for years to come.

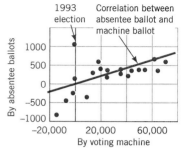

Princeton Professor Orley Ashenfelter used a scaller diagram to show that when the difference between Democratic and Republican machine votes is plotted against the difference in absentee votes for senatorial elections held in Philadelphia in the last decade, points representing the elections tend to fall around a line representing an ideal correlation between the machine and absentee tallies. The graph shows that the 1993 election falls well outside the shaded area where 95% of the results would be expected to fall. The probability that the deviation in the 1993 results was simply due to random changes in voting patterns is less than 1 percent, Ashenfelter calculated.

THE NEW YORK TIMES, MONDAY, FEBRUARY 3, 1992

North Dakota, Math Country

By Daniel Patrick Moynihan

WASHINGTON

In his State of the Union Message, the President reaffirmed his commitment to making our country "the world leader in education," adding that to do so, "We must revolutionize America's schools."

He didn't say how. But he asked for help. *And help is at hand!*

I have discovered the formula. In a flash of insight. Like Leo Szilard in London waiting to cross on a green light, thinking up nuclear fission, or James Powell thinking up magnetic levitation while waiting for a toll ticket at the Bronx-Whitestone Bridge.

I am a little old for that sort of thing, but still it happened. I was allotted two minutes in a gathering of Democrats last week to explain, yet again, that there is simply no significant connection between school expenditure and pupil achievement.

I had a chart. Utah at the bottom, under $3,000 average expenditure per pupil per year. New York at the top, over $6,000. Now what of average scores on the national eighth-grade exam in 1990? Well, No. 1 was North Dakota, eighth from the bottom in spending; No. 9 was Idaho, third from the bottom. (As for Utah, the test was never given there but the state has the highest high school graduation rate.) New York ranked 20th.

Uh huh, nodded the audience. Same old stuff. Then it came to me. "Fellow countrymen!" I exclaimed. "If you would improve your state's math scores, move your state closer to the Canadian border!"

The whole room got it!

On to the Congressional Budget Office! Please, I asked, get me the correlation between math scores and distance of state capitals from the

Daniel Patrick Moynihan is Democratic Senator from New York.

Canadian border. Back came the answer. A negative 0.522 — which may be the strongest correlation known to education, and which means that the further a capital is from the border, the lower its test score. By contrast, the correlation between expenditures and math tests is a paltry 0.203.

Not coincidentally, these findings may provide the first empirical support for the "theory of climates" of the 16th-century French philoso-

Border Watch

Average score on national eighth grade math exam, 1990, correlated with distance from Canada.

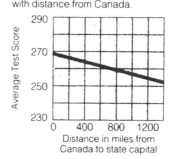

Source: Congressional Budget Office

The New York Times

pher Jean Bodin, who much influenced Montesquieu, so important to U.S. constitutional history. Bodin found "northerners unequaled in wars and industry, and southerners ... unequaled in the contemplative sciences, but the habitants of the median region ... particularly fit ... for the blossoming of arts and letters." That France should occupy the median region is perhaps no accident.

Even so, the indicated course for American education is obvious. Disadvantaged states should establish summer capitals in the Thousand Islands of the St. Lawrence River, which happens to include New York State territory bordering on French-speaking Quebec.

Furthermore, those long summer evenings in St. Lawrence skiffs might be given over in part to reading the literature on educational achievement that begins with the 1966 report on Equality of Educational Opportunity, known as the Coleman Report.

More spending on schools doesn't improve scores.

This study, commissioned under the Civil Rights Act of 1964, determined that after a point there is precious little association between school resources and school achievement. The resources that matter are those the student brings to the school, including community traditions that value education. Or don't.

The President was right enough in stating that "the major cause of problems of the cities is the dissolution of the family." But it is not just slum families: affluent families, too — the self-destructive poor and the self-indulgent affluent.

The plain fact is that in a rush of findings in the 1960's, social science disabused us of most of what we used to "know" about social issues. Generally speaking, social policy has not been able to accept this. In this climate, conservatives have fared best because of a general inclination to leave things alone anyway. (Last year's 1990 State of the Union goal — that Americans will be "first in the world in math and science achievement" by the year 2000 — is pure fantasy. But the Administration knows it, and knows also that it will be long out of office by the time the country finds out.)

Liberalism, by contrast, is reduced to a mantra. That's fine if you are a fakir; if a child, not so good.   □

How a Flat Tax Would Work, For You and for Them

By David Cay Johnston

Ideas and Trends *The New York Times*
Sunday, January 21, 1996

The big issue of the Presidential campaign in these first weeks of 1996 has been how Americans should tax themselves. And most of the talk is about a flat tax on incomes, with a large deduction for individuals so that lower-income people would pay no income tax.

There are about as many flat tax variants floating around as there are would-be occupants of the White House. But all derive from "The Flat Tax" (Hoover Institution Press, 1995), by Robert Hall and Alvin Rabushka, both economists at the Hoover Institution at Stanford University, who devised the concept in 1981 and have been refining it ever since.

The two economists say their plan, with a 19 percent flat tax rate, would spur investment and growth by taxing income only once, eliminating taxes on capital gains and taxing individuals only on wages, salaries and pensions. It would allow no deduction for mortgage interest and charitable contributions. Interest and dividends would not be taxed directly, but would be affected by a business tax, also at 19 percent. (Businesses would no longer be able to deduct the cost of fringe benefits, except pensions, for their employees, however.)

So how would a flat tax affect the typical family? And how would it affect the taxes of the Presidential candidates who have pushed the issue to center stage?

The New York Times recalculated the taxes that some of the major candidates would have owed in 1994 had the Hall-Rabushka flat tax been in effect, consulting with the Republican candidates' advisers and making minor adjustments to eliminate anomalies in the complex tax returns that are irrelevant to a flat tax.

The two candidates most closely identified with the flat tax—Steve Forbes, the magazine owner, and Patrick J. Buchanan, the television commentator—would not make their tax returns available. Mr. Forbes said he would never disclose his returns; Mr. Buchanan said he would make his public if he wins the Republican nomination.

President Clinton and three prominent Republican candidates—Senator Bob Dole and Phil Gramm and Lamar Alexander, the for-

Sources: Tax Foundation (average family's return).
Those candidates that provided tax returns were consulted on redefining 1994 income into categories appropriate to the flat tax form.

mer Education Secretary—made available copies of their 1994 Federal income tax returns. (Mr. Clinton opposes the flat tax, as does Mr. Alexander. Mr. Dole has said he would go along with a flat tax. Mr. Gramm favors a 16 percent flat tax.)

All four candidates would enjoy considerable savings in personal income taxes under the Hall-Rabushka flat tax. A comparison also requires calculating the effect of the Hall-Rabushka business tax on business income, but even with both taxes, the taxes owed by each of the four candidates would decline—although probably none would save as much as Mr. Forbes, a fervent flat-tax advocate who is one of the wealthiest men in the United States.

Senator Dole and his wife, who had $607,000 in income in 1994, would save about $49,000; the Clintons whose income was $267,000, would save about $21,000.

Senator Gramm, who with his wife reported $305,000 in income, would save about $40,000; his aides emphasized that under his own proposal, the Senator's taxes would decline even more sharply.

Mr. Alexander's savings would be about $184,000 on an income of about $1 million. The savings would presumably be significantly greater for Mr. Forbes, who has an income of $1.4 million as chief executive officer of Forbes Inc. and a net worth that Fortune magazine last week estimated at $439 million.

The picture is not so bright for the Smiths, a hypothetical family of four used by the Tax Foundation, a conservative research group, in its analyses of the various flat-tax plans. The Smiths, who have $50,000 in total income, nearly all from wages, would see no change in their individual tax bill. And if interest rates decline as predicted, the Smiths, who collected $1,425 in interest in 1994, would come out a few dollars behind.

Let's begin by admitting one thing that may be central to any thought we have about beginnings. Whether it is the mythic concept of divine power or the scientific concept of sheer vastness, we end up with something we cannot conceive.

We can put matters into words, numbers, expressions, formulas, but does that help? And if you cannot grasp it—

Imagine*

James E. Gunn[†]

IMAGINE ALL THE MATTER IN EXISTENCE
   GATHERED TOGETHER
IN THE CENTER OF THE UNIVERSE—
ALL THE METEORS
COMETS
MOONS
PLANETS
STARS
NEBULAS
THE MYRIAD GALAXIES
ALL COMPACTED INTO ONE GIANT PRIMORDIAL
   ATOM,
ONE MONOBLOC, MASSIVE BEYOND
   COMPREHENSION,
DENSE BEYOND BELIEF . . .
THINK OF WHITE DWARFS, THINK OF NEUTRON
   STARS,
THEN MULTIPLY BY INFINITY. . . .
THE CENTER OF THE UNIVERSE? THE UNIVERSE
   ITSELF.
NO LIGHT, NO ENERGY COULD LEAVE, NONE
   ENTER.
PERHAPS TWO UNIVERSES, ONE WITHIN THE
   GIANT EGG WITH EVERYTHING
AND ONE OUTSIDE WITH ALL NOTHING . . .
DISTINCT, UNTOUCHABLE . . .
IMAGINE!
ARE YOU IMAGINING?

ALL THE MATTER BROUGHT TOGETHER,
THE UNIVERSE A SINGLE, INCREDIBLE
   MONOBLOC,
SEETHING WITH INCOMPREHENSIBLE FORCES
   AND POTENTIALS
FOR COUNTLESS EONS OR FOR INSTANTS
(WHO MEASURES TIME IN SUCH A UNIVERSE?),
AND THEN . . .
BANG!
EXPLOSION!
BEYOND EXPLOSION!
TEARING APART THE MONOBLOC, THE
   PRIMORDIAL ATOM,
THE GIANT EGG HATCHING WITH FIRE AND FURY,
CREATING
THE GALAXIES

THE NEBULAS
THE SUNS
THE PLANETS
THE MOONS
THE COMETS
THE METEORS
SENDING THEM HURTLING IN ALL DIRECTIONS
   INTO SPACE
CREATING SPACE
CREATING THE EXPANDING UNIVERSE
CREATING EVERYTHING . . .
IMAGINE!

YOU CAN'T IMAGINE?
WELL, THEN, IMAGINE A UNIVERSE POPULATED
   WITH STARS AND
GALAXIES, A UNIVERSE FOREVER EXPANDING,
UNLIMITED,
WHERE GALAXIES FLEE FROM EACH OTHER,
THE MOST DISTANT SO RAPIDLY
THAT THEY REACH THE SPEED OF LIGHT
AND DISAPPEAR FROM OUR UNIVERSE
AND WE FROM THEIRS . . .
IMAGINE MATTER BEING CREATED
   CONTINUOUSLY,
A HYDROGEN ATOM POPPING INTO EXISTENCE
HERE AND THERE,
HERE
    AND
     THERE,
PERHAPS ONE ATOM OF HYDROGEN A YEAR
WITHIN A SPACE THE SIZE OF THE HOUSTON
   ASTRODOME,
AND OUT OF THESE ATOMS,
PULLED TOGETHER BY THE UNIVERSAL FORCE
   OF GRAVITATION,
NEW SUNS ARE BORN,
NEW GALAXIES TO REPLACE THOSE FLED
   BEYOND OUR PERCEPTION,
THE EXPANDING UNIVERSE
WITHOUT END,
WITHOUT BEGINNING . . .

YOU CAN'T IMAGINE?
WELL, PERHAPS IT IS ALL A FANTASY . . .

---

* Excerpted from *The Listeners*.

[†]James Gunn is a well known science fiction writer who teaches English at The University of Kansas.

The University of Kansas

Department of English

Feb. 2, 1995

Myrna Kustin
c/o CALC Lab
Massachusetts College of Art
621 Huntington Avenue
Boston, MA 02115-5882

Dear Myrna Kustin,

I am returning the signed permission form you sent to Martin Greenberg, who reprinted the piece of verse called "Imagine" in Creations . . . .

It may be of interest to you and the readers of Explorations in Algebra that "Imagine" was part of a novel titled The Listeners, published by Scribner's in 1972 and reprinted several times since then, that concerned a century-long project to pick up communications from extraterrestrials, a scientific project that now goes under the name of SETI. In the context of the novel, "Imagine" was the invention of the computer used to compile and analyze data from civilizations and languages on Earth as well as signals from the stars, and in this process the computer itself becomes self-aware and begins putting together information in various creative ways that its human programmers are not aware of. One of those creative ways is poetry, and "Imagine" is the culmination of that process; the poem has not title in the novel but is merely another part of information and musing upon it that is recorded in the interchapter sections that I called "Computer Run."

Sincerely,

James Gunn

Powers of Ten

A book about the relative size of things in the universe
and the effect of adding another zero

Philip and Phylis Morrison
and
The Office of Charles and Ray Eames
based on the film Powers of Ten
by The Office of Charles and Ray Eames

POWERS OF 10: HOW TO WRITE NUMBERS LARGE AND SMALL

This book uses a notation based on counting how many times 10 must be multiplied by itself to reach an intended number: For example, 10×10 equals 10^2, or 100; and $10 \times 10 \times 10$ equals 10^3, or 1000. Multiplying a number by itself produces a *power* of that number: 10^3 is read out loud as "ten to the third power," and is another way to say one thousand. In this case, there is no great advantage, but it is much easier and clearer to write or say 10^{14} than 100,000,000,000,000 or one hundred trillion. After 10^{14}, we even run low on names. The number written above in smaller type—the 14 in the last example—is called an *exponent,* and the powers notation is often called *exponential notation.*

It is not hard to grasp the positive powers of ten—10^4, 10^7, 10^{19}—and how they work; but the negative powers—10^{-2} or 10^{-3}—are another matter. If the exponent tells how many times the 10 is to be self-multiplied, what can an exponent of -5 (negative five) mean? The system requires a negative exponent to signal division by 10 a certain number of times: 10^{-1} equals 1 divided by 10, or 0.1 (one-tenth); 10^{-2} equals 0.1 divided by 10, or 0.01 (one-hundredth). Because *adding* 1 to the exponent easily works out to be the equivalent of multiplying by 10, it is self-consistent that subtracting 1 there works out to a division by 10. It is all a matter of placing zeros. Adding another terminal zero is simply to multiply by 10: 100×10 equals 1,000. Putting another zero next after the decimal point is to *divide* by 10: $0.01 \div 10$ equals 0.001. The powers notation makes these operations even clearer.

But what of 10^0? That seems a strange number. However, notice it is equal to 10^1 (10) divided by 10 (or to 10^{-1} multiplied by 10). Although surprising, it is at least logical that 10^0 should be equal to 1.

Because you can make any power of ten ten times larger by adding 1 to its exponent ($10^4 \times 10 = 10^5$), it follows that to multi-

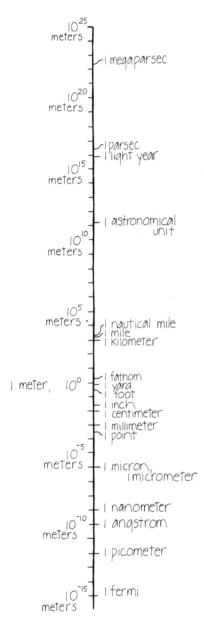

ply by 100 you add 2 to the exponent: $10^3 \times 100 = 10^5$ or $1000 \times 100 = 100,000$. In general, you can multiply one power of ten by another simply by adding their exponents: $10^6 \times 10^3 = 10^9$. Subtracting the exponents is the equivalent of division: $10^7 \div 10^5 = 10^2$.

All numbers, not only numbers that are exact powers of ten, like 100 or 10,000, can be written with the help of exponential notation. The number 4000 is 4×10^3; 186,000 is 1.86×10^5. This convenient scheme is referred to as scientific notation.

All of this can be extended to basic multipliers other than ten: $2^4 = 2 \times 2 \times 2 \times 2$ (the fourth power of two); $12^2 = 12 \times 12$, and $8^{-1} =$ one-eighth. (But note that $2^0 = 1$, $12^0 = 1$, and $8^0 = 1$.)

Logarithms arise from extensions of this scheme.

The symbol $\sim$ is mathematicians' shorthand for "approximately" or "about."

UNITS OF LENGTH

Grow you own food and build your own house, and no formal units of measurement much interest you; such is the general rule of thumb. But commerce has implied agreement on units of measurement. The legal yard has long been displayed for the use of Londoners, and the meter is still open to public comparison on a wall of a Paris building.

The system we call metric is the work of the savants of Revolutionary Paris in the 1790s. Even their determination to celebrate both novelty and reason met limits: Our modern second, minute, and hour remain resolutely nondecimal. That was no oversight—the metric day of ten hours, each of a hundred minutes with a hundred seconds to the minute, was formally adopted. But the scheme met fierce resistance. About the only costly mechanism every middle-class family then proudly owned was a clock or watch, not to be rendered at once useless by any mere claim of consistency! Practice won out over theory.

In much the same way, people who today frequently use units in a particular context are not always persuaded to sacrifice appropriateness to consistency. We list here a few nonstandard units of linear measurement that retain their utility even in these metric days, some even within the sciences.

Cosmic Distances

Parsec The word is a coinage from *parallax* of one *second*. The *parsec* is in common use among astronomers because it hints at the surveyor's basic technique of measuring stellar distance by using triangulation. The standard parallax is the apparent shift in direction of a distant object at six-month intervals as the observer moves with the orbiting earth. It is defined so that the radius of the earth's orbit seen from a distance of one parsec spans an angle of one second of arc. The nearest known star to the sun is more than one parsec away.

Light-year   This graspable interstellar unit rests on the relationship between cosmic distance and light travel over time.

The speed of light in space is 3.00×10^8 meters per second; in one year light thus moves 9.46×10^{15} meters, which is usually rounded off to 10^{16} meters, especially since only a few cosmic distances are so well known that the roundoff is any real loss of accuracy.

Astronomical unit   The mean distance between sun and earth is a fine baseline for surveying the solar system; it is a typical length among orbits. 1 AU = 1.50×10^{11} meters. *Note:* 1 parsec = 3.26 light-years = 206,300 AU. The interstellar and the solar-system scales plainly differ; intergalactic distances run to megaparsecs.

Terrestrial Lengths

Miles, leagues, etc.   These are units suited for earthbound travel, for distances at sea, or for road distances between cities. Nobody ever measured cloth by the mile, or train rides by the parsec.

Yards, feet, meters   These rest on human scale, in folklore the length of some good king's arm. They suit well the sizes of rooms, people, trucks, boats. Textiles are yard goods. The meter was defined more universally, but clearly it was meant to supplant the yard and the foot. It was related to the size of the earth: One quadrant of the earth's circumference was defined as exactly 10^7 meters, or 10^4 kilometers. In 1981 the meter is defined with great precision in terms of the wavelength of a specific atomic spectral line. It is "1,650,763.73 wavelengths in vacuum of the radiation corresponding to the transitions between the levels $2p_{10}$ and $5d_5$ of the krypton-86 atom."

Inches, centimeters, etc.   The same king's thumb? Human-scale units intended for the smaller artifacts of the hand: paper sizes, furniture, hats, or pies.

Line, millimeter, point   Small units for fine work are relatively modern. The seventeenth-century French and English line was a couple of millimeters, and the printer's point measure is about 0.35 mm. Film, watches, and the like are commonly sized by millimeters. The pioneer microscopist Antony van Leeuwenhoek used sand grains as his length comparison, coarse and fine: He counted one hundred of the fine grains to the common inch of his place and time. Smaller measurement units are generally part of modern science, and thus usually metric.

Atomic Distances

Angstroms, fermis, etc.   Once atoms became the topic of meaningful measurement, new small units of length naturally came into specialized use. The Swedish physicist Anders Ångström a century ago pioneered wavelength measurements of the solar spectrum. He expressed his results in terms of a length unit just 10^{-10} meters long. It has remained in widespread informal use bearing his name: convenient because atoms measure a few angstroms (Å) across. The impulse for such useful jargon words is by no means ended; nuclear particles are often measured in fermis, after the Italian physicist Enrico Fermi. 1 fermi equals 10^{-15} meters.

Angles and Time

Angles are measured, especially in astronomy, by a nonmetric system that goes back to Babylon. A circle is 360 degrees; 1 degree = 60 minutes of arc; 1 minute = 60 arc-seconds. An arc-second is roughly the smallest angle that the image of a star occupies, smeared as it is by atmospheric motion. This page, viewed from about twenty-five miles away, would appear about one arc-second across.

Time measurement shares the cuneiform usage of powers of sixty. Note that a year of 365.25 days of 24 hours, each of 60 minutes with 60 seconds apiece, amounts to about 3.16×10^7 seconds.

What seest thou else
In the dark backward and abysm of time?
Wm. Shakespeare
The Tempest

The Cosmic Calendar*

Carl Sagan

The world is very old, and human beings are very young. Significant events in our personal lives are measured in years or less; our lifetimes in decades; our family genealogies in centuries; and all of recorded history in millennia. But we have been preceded by an awesome vista of time, extending for prodigious periods into the past, about which we know little—both because there are no written records and because we have real difficulty in grasping the immensity of the intervals involved.

Yet we are able to date events in the remote past. Geological stratification and radioactive dating provide information on archaeological, palenotological and geological events; and astrophysical theory provides data on the ages of planetary surfaces, stars, and the Milky Way Galaxy, as well as an estimate of the time that has elapsed since that extraordinary event called the Big Bang—an explosion that involved all of the matter and energy in the present universe. The Big Bang may be the beginning of the universe, or it may be a discontinuity in which information about the earlier history of the universe was destroyed. But it is certainly the earliest event about which we have any record.

The most instructive way I know to express this cosmic chronology is to imagine the fifteen-billion-year lifetime of the universe (or at least its present incarnation since the Big Bang) compressed into the span of a single year. Then every billion years of Earth history would correspond to about twenty-four days of our cosmic year, and one second of that year to 475 real revolutions of the Earth about the sun. On the next two pages I present the cosmic chronology in three forms: a list of some representative pre-December dates; a calendar for the month of December; and a closer look at the late evening of New Year's Eve. On this scale, the events of our history books—even books that make significant efforts to deprovincialize the present—are so compressed that it is

* From: Carl Sagan, *The Dragons of Eden: Speculations on the Evolution of Human Intelligence,* Ballantine Books, NY, 1977.

Pre-December Dates

Big Bang	January 1
Origin of the Milky Way Galaxy	May 1
Origin of the solar system	September 9
Formation of the Earth	September 14
Origin of life on Earth	~September 25
Formation of the oldest rocks known on Earth	October 2
Date of oldest fossils (bacteria and blue-green algae)	October 9
Invention of sex (by microorganisms)	~November 1
Oldest fossil photosynthetic plants	November 12
Eukaryotes (first cells with nuclei) flourish	November 15

~ = approximately

Cosmic Calendar
DECEMBER

Sunday	Monday	Tuesday	Wednesday	Thursday	Friday	Saturday
	1 Significant oxygen atmosphere begins to develop on Earth.	2	3	4	5 Extensive vulcanism and channel formation on Mars.	6
7	8	9	10	11	12	13
14	15	16 First worms.	17 Precambrian ends. Paleozoic Era and Cambrian Period begin. Invertebrates flourish.	18 First oceanic plankton. Trilobites flourish.	19 Ordovician Period. First fish, first vertebrates.	20 Silurian Period. First vascular plants. Plants begin colonization of land.
21 Devonian Period begins. First insects. Animals begin colonization of land.	22 First amphibians. First winged insects.	23 Carboniferous Period. First trees. First reptiles.	24 Permian Period begins. First dinosaurs.	25 Paleozoic Era ends. Mesozoic Era begins.	26 Triassic Period. First mammals.	27 Jurassic Period. First birds.
28 Cretaceous Period. First flowers. Dinosaurs become extinct.	29 Mesozoic Era ends. Cenozoic Era and Tertiary Period begin. First cetaceans. First primates.	30 Early evolution of frontal lobes in the brains of primates. First hominids. Giant mammals flourish.	31 End of the Pliocene Period. Quatenary (Pleistocene and Holocene) Period. First humans.			

December 31

Origin of *Proconsul* and *Ramapithecus,* probable ancestors of apes and men	~1:30 P.M.
First humans	~10:30 P.M.
Widespread use of stone tools	11:00 P.M.
Domestication of fire by Peking man	11:46 P.M.
Beginning of most recent glacial period	11:56 P.M.
Seafarers settle Australia	11:58 P.M.
Extensive cave painting in Europe	11:59 P.M.
Invention of agriculture	11:59:20 P.M.
Neolithic civilization; first cities	11:59:35 P.M.
First dynasties in Sumer, Ebla and Egypt; development of astronomy	11:59:50 P.M.
Invention of the alphabet; Akkadian Empire	11:59:51 P.M.
Hammurabic legal codes in Babylon; Middle Kingdom in Egypt	11:59:52 P.M.
Bronze metallurgy; Mycenaean culture; Trojan War; Olmec culture; invention of the compass	11:59:53 P.M.
Iron metallurgy; First Assyrian Empire; Kingdom of Israel; founding of Carthage by Phoenicia	11:59:54 P.M.
Asokan India; Ch'in Dynasty China; Periclean Athens; birth of Buddha	11:59:55 P.M.
Euclidean geometry; Archimedean physics; Ptolemaic astronomy; Roman Empire; birth of Christ	11:59:56 P.M.
Zero and decimals invented in Indian arithmetic; Rome falls; Moslem conquests	11:59:57 P.M.
Mayan civilization; Sung Dynasty China; Byzantine empire; Mongol invasion; Crusades	11:59:58 P.M.
Renaissance in Europe; voyages of discovery from Europe and from Ming Dynasty China; emergence of the experimental method in science	11:59:59 P.M.
Widespread development of science and technology; emergence of a global culture; acquisition of the means for self-destruction of the human species; first steps in spacecraft planetary exploration and the search for extraterrestrial intelligence	Now: The first second of New Year's Day

necessary to give a second-by-second recounting of the last seconds of the cosmic year. Even then, we find events listed as contemporary that we have been taught to consider as widely separated in time. In the history of life, an equally rich tapestry must have been woven in other periods—for example, between 10:02 and 10:03 on the morning of April 6th or September 16th. But we have detailed records only for the very end of the cosmic year.

The chronology corresponds to the best evidence now

available. But some of it is rather shaky. No one would be astounded if, for example, it turns out that plants colonized the land in the Ordovician rather than the Silurian Period; or that segmented worms appeared earlier in the Precambrian Period than indicated. Also, in the chronology of the last ten seconds of the cosmic year, it was obviously impossible for me to include all significant events; I hope I may be excused for not having explicitly mentioned advances in art, music and literature or the historically significant American, French, Russian and Chinese revolutions.

The construction of such tables and calendars is inevitably humbling. It is disconcerting to find that in such a cosmic year the Earth does not condense out of interstellar matter until early September; dinosaurs emerge on Christmas Eve; flowers arise on December 28th; and men and women originate at 10:30 P.M. on New Year's Eve. All of recorded history occupies the last ten seconds of December 31; and the time from the waning of the Middle Ages to the present occupies little more than one second. But because I have arranged it that way, the first cosmic year has just ended. And despite the insignificance of the instant we have so far occupied in cosmic time, it is clear that what happens on and near Earth at the beginning of the second cosmic year will depend very much on the scientific wisdom and the distinctly human sensitivity of mankind.

Earthquake Magnitude Determination

BOX 15.1 In 1935, Dr. Charles F. Richter of the California Institute of Technology introduced to seismology a scale which made it possible to compare earthquakes throughout the world. The scale is now known as the **Richter magnitude scale.** This standardized measure of earthquakes is based on the logarithm of the maximum amplitude of seismic waves recorded by a Wood-Anderson seismograph; plus a factor which takes into account the weakening of the waves as they spread away from the earthquake focus. Hence, nearly the same value of magnitude can be determined by seismologists throughout California in a matter of minutes. Although Dr. Richter used a Wood-Anderson seismograph to develop the earthquake magnitude scale, it is possible to empirically relate the recordings produced by other types of seismometers to this scale (fig. 1).

Figure 2 illustrates how a seismologist would determine the magnitude of a particular earthquake from a Wood-Anderson seismogram. Two measurements are required: (1) maximum trace amplitude in millimeters (point "A") and (2) difference between arrival times of the primary "P" and secondary "S" seismic waves (point "B"). The values of these two measurements are then found on each side of the nomograph and connected by a line. That point where the line crosses the middle column is the value of the Richter magnitude of the earthquake, which in this example is magnitude 5.

Because magnitude is based on a logarithmic scale, a unit change in magnitude corresponds to a change in trace amplitude by a factor of ten as shown by the dashed line on the nomograph. Note that a magnitude 3 earthquake is defined on the Richter magnitude scale as an earthquake that will cause a peak amplitude of 1 mm to be recorded on a Wood-Anderson seismograph at an epicentral distance of 100 km (shown by dotted line on nomograph).

Source: R. Sherburne, *California Geology,* July, 1977.

The Richter Scale

To determine the magnitude of an earthquake, connect on the chart

A the maximum amplitude recorded by a standard seismometer, and

B the distance of that seismometer from the epicenter of the earthquake (or the difference in times of arrival of the P and S waves) by a straight line, which crosses the center scale at the magnitude.

- - - A unit change in magnitude corresponds to a decrease in seismogram amplitude by a factor of ten

········ Definition of a magnitude 3 earthquake

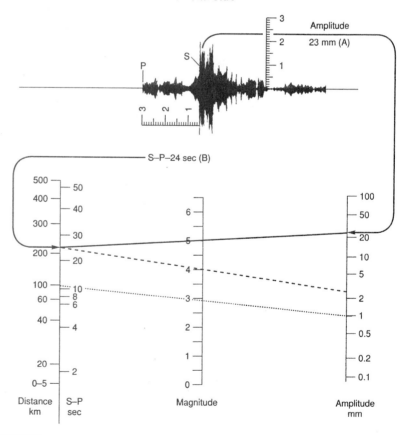

Figure 2

This diagram illustrates how seismologists determine earthquake magnitude using a Wood-Anderson seismograph recording and a magnitude determination chart.

From Robert N. Waallen, *Introduction to Physical Geography,* 1992.

Size and Shape

Stephen Jay Gould

Who could believe an ant in theory?
A giraffe in blueprint?
Ten thousand doctors of what's possible
Could reason half the jungle out of being.

John Ciardi's lines reflect a belief that the exuberant diversity of life will forever frustrate our arrogant claims to omniscience. Yet, however much we celebrate diversity and revel in the peculiarities of animals, we must also acknowledge a striking "lawfulness" in the basic design of organisms. This regularity is most strongly evident in the correlation of size and shape.

Animals are physical objects. They are shaped to their advantage by natural selection. Consequently, they must assume forms best adapted to their size. The relative strength of many fundamental forces (gravity, for example) varies with size in a regular way, and animals respond by systematically altering their shapes.

The geometry of space itself is the major reason for correlations between size and shape. *Simply by growing larger,* any object will suffer continual decrease in relative surface area when its shape remains unchanged. This decrease occurs because volume increases as the cube of length (length × length × length), while surface increases only as the square (length × length): in other words, volume grows more rapidly than surface.

Why is this important to animals? Many functions that depend upon surfaces must serve the entire volume of the body. Digested food passes to the body through surfaces; oxygen is absorbed through surfaces in respiration; the strength of a leg bone depends upon the area of its cross section, but the legs must hold up a body increasing in weight by the cube of its length. Galileo first recognized this principle in his *Discorsi* of 1638, the masterpiece he wrote while under house arrest by the Inquisition. He argued that the bone of a large animal must thicken disproportionately to provide the same relative strength as the slender bone of a small creature.

One solution to decreasing surface has been particularly important in the progressive evolution of large and complex organisms: the development of internal organs. The lung is, essentially, a richly convoluted bag of surface area for the exchange of gases; the circulatory system distributes material to an internal space that cannot be reached by direct diffusion from the external surface of large organisms; the villi of our small intestine increase the surface area

From Stephen Jay Gould, *Ever Since Darwin: Reflections in Natural History,* 1985.

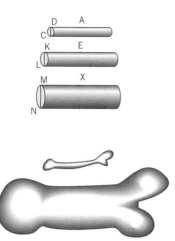

An approximation of Galileo's original illustration of the relationship between size and shape. To maintain the same strength, large cylinders must be relatively thicker than small ones. For exactly the same reason, large animals have relatively thick leg bones.

available for absorption of food (small mammals neither have nor need them).

Some simpler animals have never evolved internal organs; if they become large, they must alter their entire shape in ways so drastic that plasticity for further evolutionary change is sacrificed to extreme specialization. Thus, a tapeworm may be 20 feet long, but its thickness cannot exceed a fraction of an inch because food and oxygen must penetrate directly from the external surface to all parts of the body.

Other animals are constrained to remain small. Insects breathe through invaginations of their external surface. Oxygen must pass through these surfaces to reach the entire volume of the body. Since these invaginations must be more numerous and convoluted in larger bodies, they impose a limit upon insect size: at the size of even a small mammal, an insect would be "all invagination" and have no room for internal parts.

We are prisoners of the perceptions of our size, and rarely recognize how different the world must appear to small animals. Since our relative surface area is so small at our large size, we are ruled by gravitational forces acting upon our weight. But gravity is negligible to very small animals with high surface to volume ratios; they live in a world dominated by surface forces and judge the pleasures and dangers of their surroundings in ways foreign to our experience.

An insect performs no miracle in walking up a wall or upon the surface of a pond; the small gravitational force pulling it down or under is easily counteracted by surface adhesion. Throw an insect off the roof and it floats gently down as frictional forces acting upon its surface overcome the weak influence of gravity.

The relative weakness of gravitational forces also permits a mode of growth that large animals could not maintain. Insects have an external skeleton and can only grow by discarding it and secreting a new one to accommodate the enlarged body. For a period between shedding and regrowth, the body must remain soft. A large mammal without any supporting structures would collapse to a formless mass under the influence of gravitational forces; a small insect can maintain its cohesion (related lobsters and crabs can grow much larger because they pass their "soft" stage in the nearly weightless buoyancy of water). We have here another reason for the small size of insects.

The creators of horror and science-fiction movies seem to have no inkling of the relationship between size and shape. These "expanders of the possible" cannot break free from the prejudices of their perceptions. The small people of *Dr. Cyclops*, *The Bride of Frankenstein*, *The Incredible Shrinking Man*, and *Fantastic Voyage* behave just like their counterparts of normal dimensions. They fall off cliffs or down stairs with resounding thuds; they wield weapons and swim with olympic agility. The large insects of films too numerous to name continue to walk up walls or fly even at dinosaurian dimensions. When the kindly entomologist of *Them* discovered that the giant queen ants had left for their nuptial flight, he quickly calculated this simple ratio: a normal ant is a fraction of an inch long and can fly hundreds of feet; these ants are many feet long and must be able to fly as much as 1,000 miles. Why, they could be as far away as Los Angeles! (Where, indeed, they were, lurking in the sewers.) But the ability to fly depends upon the surface area of wings, while the weight that must be borne aloft increases as the cube of length. We may be sure that even if the giant ants had somehow circumvented the problems of breathing and growth by molting, their sheer bulk would have grounded them permanently.

Other essential features of organisms change even more rapidly with increasing size than the ratio of surface to volume. Kinetic energy, in some situations, increases as length raised to the fifth power. If a child half your height falls down, its head will hit with not half, but only 1/32 the energy of yours in a similar fall. A child is protected more by its size than by a "soft" head. In return, we are protected from the physical force of its tantrums, for the child can strike with, not half, but only 1/32 of the energy we can muster. I have long had a special sympathy for the poor dwarfs who suffer under the whip of cruel Alberich in Wagner's *Das Rheingold*. At their diminutive size, they haven't a chance of extracting, with mining picks, the precious minerals that Alberich demands, despite the industrious and incessant leitmotif of their futile attempt.[1]

This simple principle of differential scaling with increasing size

[1] A friend has since pointed out that Alberich, a rather small man himself, would only wield the whip with a fraction of the force we could exert—so things might not have been quite so bad for his underlings.

may well be the most important determinant of organic shape. J. B. S. Haldane once wrote that "comparative anatomy is largely the story of the struggle to increase surface in proportion to volume." Yet its generality extends beyond life, for the geometry of space constrains ships, buildings, and machines, as well as animals.

Medieval churches present a good testing ground for the effects of size and shape, for they were built in an enormous range of sizes before the invention of steel girders, internal lighting, and air conditioning permitted modern architects to challenge the laws of size. The small, twelfth-century parish church of Little Tey, Essex, England, is a broad, simple rectangular building with a semicircular apse. Light reaches the interior through windows in the outer walls. If we were to build a cathedral simply by enlarging this design, then the area of outer walls and windows would increase as length squared, while the volume that light must reach would increase as length cubed. In other words, the area of the windows would increase far more slowly than the volume that requires illumination. Candles have limitations; the inside of such a cathedral would have been darker than the deed of Judas. Medieval churches, like tapeworms, lack internal systems and must alter their shape to produce more external surface as they are made larger. In addition, large churches had to be relatively narrow because ceilings were vaulted in stone and large widths could not be spanned without intermediate supports. The chapter house at Batalha, Portugal—one of the widest stone vaults in medieval architecture—collapsed twice during construction and was finally built by prisoners condemned to death.

Consider the large cathedral of Norwich, as it appeared in the twelfth century. In comparison with Little Tey, the rectangle of the nave has become much narrower; chapels have been added to the apse, and a transept runs perpendicular to the main axis. All these "adaptations" increase the ratio of external wall and window to internal volume. It is often stated that transepts were added to produce the form of a Latin cross. Theological motives may have dictated the position of such "outpouchings," but the laws of size required their presence. Very few small churches have transepts. Medieval architects had their rules of thumb, but they had, so far as we know, no explicit knowledge of the laws of size.

Large organisms, like large churches, have very few options open to them. Above a certain size, large terrestrial animals look basically alike—they have thick legs and relatively short, stout bodies. Large medieval churches are relatively long and have abundant outpouchings. The "invention" of internal organs allowed animals to retain the highly successful shape of a simple exterior enclosing a large internal volume; the invention of internal lighting and structural steel has permitted modern architects to design large buildings of essentially cubic form. The limits are expanded, but the laws still operate. No large Gothic church is wider than long; no large animal has a sagging middle like a dachshund.

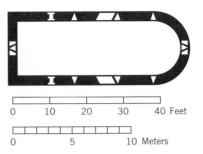

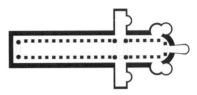

The great range of designs among medieval churches can be attributed partly to size. The twelfth-century parish church of Little Tey, Essex, England, was only 57 feet long and had a simple floor plan, top, while the floor plan for Norwich Cathedral, also twelfth century, shows adaptations—transept, chapels—required for the 450-foot-long building. The need for light and support dictated complex cathedral layouts.

I once overheard a children's conversation in a New York playground. Two young girls were discussing the size of dogs. One asked: "Can a dog be as large as an elephant?" Her friend responded: "No, if it were as big as an elephant, it would look like an elephant." How truly she spoke.

Elizabeth Cavicchi

Watching Galileo's Learning
Elizabeth Cavicchi

1   Introduction

By closely following Stillman Drake's biographies of Galileo, this essay interprets Galileo's free fall studies. Drake's biographies piece together how Galileo came to understand that the distance an object has fallen increases as the square of its descent time. He infers this story from calculations recorded among Galileo's working papers, including some that were not included in the definitive twenty volume set of Galileo's *Opera*, edited by Favaro in 1934.

Prior to Drake's studies, the interpretation of the historian Koyre prevailed among Galilean scholars. Koyre maintained that Galileo never made any observations of motion (except with the telescope); he says Galileo derived laws of motion through thought alone. For Koyre, Galileo's innovation in science lay in this introspective method of reasoning.

Koyre influenced the textbook treatment of Galileo's work. Many physics texts do not mention Galileo's free fall experiments. While some recent texts refer to his experiments, they simplify Galileo's process of learning by leaving out details and context which might assist readers in following how Galileo's experimenting and thinking developed. Texts also omit reference to the mathematics in use at the time, which were tools for Galileo's thought, and to the contemporary politics and thought, which Galileo's work challenged. Such simplifications and omissions are among the many ways textbooks distort their presentations of how people learned, and can learn, about the phenomena of nature. As a result, unable to imagine how Galileo could have come to the law of how things fall just by postulating it correctly, students may doubt their own potential to learn through exploring and questioning how things happen.

By comparison with the methods of analysis in use today, the mathematical tools and physical picture available to Galileo seem very limited. The task of stripping away our twentieth century sophistication, to better approximate the outlook and thinking of Galileo, is formidable. As a biographer tracking Galilean documents Drake acquired this outlook. However, he does not carefully provide clues that would facilitate a reader's understanding of Galileo's world view. My effort to understand Drake's argument seems parallel to Galileo's effort to understand motion. We begin from awareness of what we do not know, which deepens as we notice more confusion in our thinking and more complexity in the subject we are trying to understand. Through this deepening of what we do not know, we come to ways of making thoughts and questions that can take our understanding further. In watching Galileo's learning, I am repeatedly bewildered as my assumptions prevent me from seeing what Galileo did not know. I am also astonished by Galileo's creativity in using the tools he did have to find something new.

Elizabeth Cavicchi

Accelerated motion is subtle; Galileo devoted most of a lifetime to its study. We can re-experience some of the subtlety and complexity Galileo encountered by observing our own students as they try to make sense of it. I did this as a project with one student from the algebra course this reading is intended to supplement. I did not provide the student – Hazel Garland – with explanations of motion. Instead as we experimented and observed together, the experimenting itself became the beginnings of our further thought and experimenting.

Hazel and I released a weight, with a long paper tape attached, so that it fell freely through a gap in a spark-timer. At 1/60 s time intervals, a spark jumped through the gap, marking a dot on the paper tape. When we examined these paper tapes, Hazel found inconsistencies between the dot patterns on different tapes: this intrigued her. As I watched her developing analysis of these dots, I began to realize how ideas which seemed 'obvious' to me as a teacher – such as that the timer's periodicity was independent of the moment when we released the weight – were not at all obvious to Hazel. I saw how understandings of time intervals and motion developed through her careful and animated thought about what was on the paper tapes. As I tried to understand free fall from her perspective, I was repeatedly surprised by her creativity in working through what seemed – to me – to be limitations in what she knew. I came to see these 'limitations' as productive beginnings for what she came to know ever more deeply.

Through watching Hazel's experimenting, I came to appreciate how, by their notice of details in what natural phenomena do, learners can form new questions and work out ideas that develop what they know. While writing this essay, I realized that the engagement of learning through experimenting that I saw in Hazel's work was also evident in what Galileo did. I began to understand that the confusions and difficulties I encountered in interpreting Galileo's experimenting, and Drake's account of it, were integral to the work and process of how learning develops. These confusions cannot be simplified away or replaced by a neat logical sequence. The complexity of what happens in the learners' thoughts and work makes evident the depth of their response to the complexity of the physical motions they explore. It also makes possible their continued efforts at learning.

2   Falling Objects

When younger than we can now remember, we played by dropping a toy and noisily goading an attentive older companion to retrieve it. While the game now seems a test of patience, seeing something fall was a new experience for us, which we observed without asking either why or how things fall. When, as an older child, we asked what makes things fall, the answer was "gravity". When we asked what gravity was, we were told "it makes things fall". Gravity was an empty word, devoid of explanation.

Elizabeth Cavicchi

Better (or more complete, or less circular) answers to this question have evaded not only parents, but also people who explicitly study things and try to explain them. This simple question is too grand. Why does it take effort to get things to go where we want– lifting, dragging, carrying, bicycling– while they fall by themselves? No strings or other apparatus visibly pulls them down. Seeking a mechanism for falling is premature if we are not yet sure what is happening when something falls. Falling is so quick that it is not easy to say what is happening as something falls, even if it is watched carefully. Asking the question differently– "How do things fall?"– provides a clearer way to start towards an answer: by careful description of the phenomenon itself.

3   Early Explanations

The seeming simplicity of the "grand" question– why things fall– motivated early historical attempts to talk about falling in a way that made sense to ordinary people. The Greeks said the weight of an object makes it go towards earth's center. Things that have lightness, like fire, move away from the center. Weightier things fall faster than lighter ones. Aristotle (384-322 B.C.) extended this explanation with logic, to make many other statements about motion and change. His mostly incorrect view was considered authoritative for about a millennium.

People who looked carefully found that some of these ideas just did not match with nature. Two sixteenth century Italian instructors dropped objects differing only in weight from a high window or tower. The objects struck the ground almost at once. These observations were augmented when Galileo Galilei (1564-1642, Figure 1, [7]) began investigating what happens when something falls.

4   Galileo and the Inclined Plane

Galileo revised the question; he asked "how" things fall, instead of "why". This question suggested other questions that could be tested directly by experiment. By alternating questions and experiments, Galileo was able to identify details in motion no one had previously noticed or tried to observe. His questions about free fall suggested questions about other motions, such as motions Aristotle had classed as "natural". These motions start spontaneously without a push, such as a ball rolling down a slope, or a pendulum swinging. What he learned about one motion drove his study of the others.

In one experiment, performed near 1602, Galileo rolled balls down a ramp. He mounted lute strings crosswise to the motion, so that a distinct sound was produced each time a ball rolled over a string. He adjusted positions of the strings until time intervals between sounds were roughly equal. Perhaps Galileo

Figure 1: A portrait of Galileo, taken from his book on the sunspots, *Istoria e dimostrazioni intorno alle macchie solari*, published in Rome in 1613. By permission of the Houghton Library, Harvard University.

Elizabeth Cavicchi

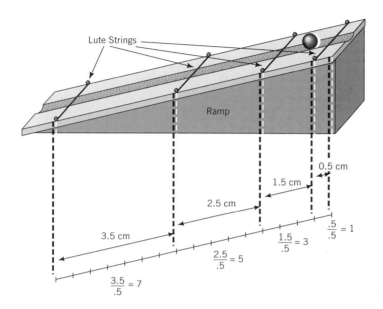

Figure 2: Whenever a ball rolling down a ramp bumps over a string, it makes a sound. The strings are positioned so the sounds mark out equal intervals in time. If we measure the distances between each pair of strings, the ratio of each distance to the first one makes a series of odd integers.

hummed a tune, with beats about a half-second apart, to estimate constant time intervals.

Once he had equalized the rolling time between each pair of strings, he measured the distance between each pair of adjacent strings. The distances between strings were unequal: each consecutive pair of strings was further apart than the preceding pair of strings (Figure 2). When he divided all the distances between pairs of strings by the first distance, the quotients were the odd integers 1, 3, 5, 7, ...

This discovery was revolutionary, as the first evidence that motion on Earth was subject to mathematical laws. Until this time, people assumed mathematics governed only motions in the heavens; terrestrial motions were considered disorderly. Galileo did not simply accept this belief, but tried to test it with direct observation.

Elizabeth Cavicchi

Galileo interpreted the separation between successive pairs of lute strings as a measure of the ball's speed while rolling between the two strings. The numerical pattern showed Galileo that the ball's speed increased, but it did not tell him how it increased. To understand this subtlety he again had to contradict physical theory of his contemporaries.

People already realized that one time or position measurement could change to another time or position by imperceptibly small, continuous increments. But speed was depicted as a motion that occurred in a perceptible time interval. While an object's speed could change between time intervals, they did not think speed could change instantaneously as we know it does today. Instantaneous change in speed was ill-defined for them, because they could not conceive of a speed without being able to see the motion directly. Galileo eventually understood and resolved this apparent contradiction, and concluded that speed changes continuously during free fall and natural motions. He wrote:

> I suppose (and perhaps I shall be able to demonstrate this) that the naturally falling body goes continually increasing its *velocità* according as the distance increases from the point which it parted[6].

5   Mathematical Tools

Galileo's ability to construct patterns from his measurements was limited by the mathematics available to him at the time. Although European mathematicians were then developing algebra beyond its Arabic and Hindu origins, and Galileo was aware of their work, he never used algebra in his physical studies. He never expressed any results using equations. His analysis of data and publications relied on a more classical training– popular at the time– in the logic, geometry, and numerical proportions of Euclid's *Elements* (fl. 300 B.C.).

One result of these mathematical limitations was that Galileo could not define speed the way we do today, as a distance divided by a time (for example miles *per* hour). In manipulating numbers according to the Euclidean theory of proportions, it was not legitimate to construct ratios of measurements with different units, such as dividing a distance by a time. Ratios could only be constructed by dividing one distance measurement by another like measurement (for example (20 meters)/(10 meters)), or one time by another time (for example (6 seconds)/(2 seconds)). The units of distance or time then canceled out, yielding a pure, dimensionless number.

Galileo's interpretation of speed as the separation between lute strings prevented him from seeing it as the distance traveled while the object rolled between the strings. If he had been able to identify that this one measurement conveyed data pertaining to two different physical quantities (speed and distance) at once, he might have worked out the relation inherring between distance and time two years earlier than he did. But, persisting with questions about what his observations revealed about motion, he continued experimenting by rolling balls

Elizabeth Cavicchi

down ramps of different lengths and of different vertical heights. He incorporated results from all these experiments into his mature, formal understanding of motion, but at the time he performed them, they did not lead him to the relation he sought.

6   Pendulum Swings

After failing to work out a lawful expression for motion along the incline, in 1604 Galileo began looking for it in the motion of pendulums. He had already experimented extensively with pendulums. He had already been the first person to demonstrate that a pendulum swings through a small arc in very nearly the same time that it swings through a wide arc.

Galileo timed pendulum swings with a water clock he designed. A water pipe was fed from a large elevated reservoir and plugged with a finger. When it was turned on by removing his finger from the pipe, water spurted from the pipe into a bucket. He unplugged the pipe for the duration of a pendulum swing and weighed the water that flowed into the bucket. As long as the flow rate was steady, weights of water were a measure of time; the longer the run, the more water, and hence the more weight (Figure 3). Galileo checked consistency of water flow through the pipe with a built-in timer: his own pulse. He used this clock to measure times for a wide variety of physical phenomena, such as free fall, balls rolling down ramps, and pendulum swings.

He measured one-quarter of a pendulum swing: from its moment of release to the moment when it was vertical. He mounted the pendulum so that when it hung vertically (like a plumb bob), the bob just touched a wall (Figure 4). He started his water clock when he released the pendulum and stopped it when he heard the bob bang against the wall. The sound helped him make measurements repeatably and reliably.

Since Galileo used the same clock for measurement of the pendulum and other natural motions, the times for all his measurements were expressed in the same 'units', which suggested that he could compare times for *different kinds* of natural motion, for example, pendulum swings and falling objects. Using a gear mechanism he designed, he tried to adjust the length of a pendulum string so that the time of its quarter swing exactly matched the time it took for an object to fall a given distance, by matching the weights of water that flowed during both motions. But his free fall measurement required too long a pendulum to be practical, so he tried matching the pendulum swing with half the free fall time he had measured, which corresponds to a fall of one meter. The length of the resulting pendulum string was a little less than one meter.

Galileo then made a pendulum twice as long as this one, and timed its quarter swing. He decided to double the pendulum length, rather than to try shorter lengths, because the mathematics of the time did not condone the construction of fractional lengths such as 1/2 or 1/4 of a given length. He found that swing

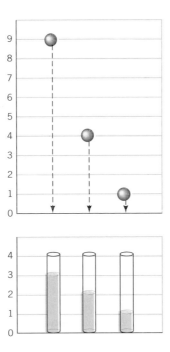

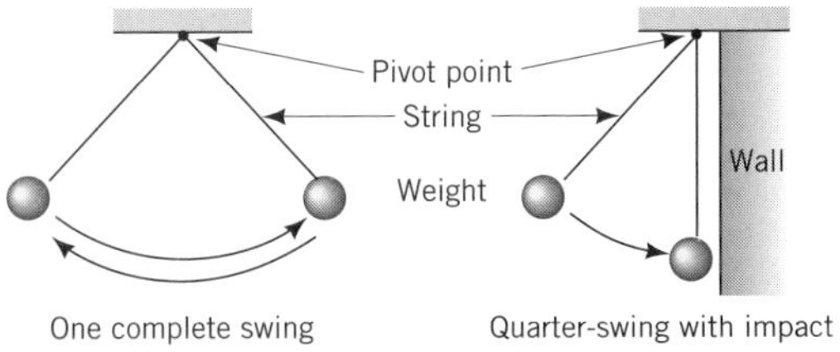

Figure 4: A complete pendulum swing and one quarter pendulum swing. When the bob completes one quarter swing, it hits a wall, making a sound.

Figure 3: The water clock was started when a weight was released, and stopped when it hit the ground. More water was collected during long falls than during short falls. The amount of water is a measure of the elapsed time.

time was proportional to a 'geometrical mean' involving pendulum length, which he computed using a straightedge and compass on a geometric drawing. This geometrical computation is equivalent to taking the square root of pendulum length. Today we say that a pendulum's period is proportional to the square root of its length. Galileo used his version of this rule to correctly predict the swing time of a thirty foot pendulum which he may have hung in the courtyard of the University of Padua where he worked.

Galileo linked the pendulum swing to the free fall by performing the experiment backward. He used the water clock to time a weight as it fell through the same height as the length of one of the pendulums he had already timed. He expressed the ratio of the two times (the free fall time over the pendulum time) as a ratio of the whole number weights 942/850. While he did not reduce this ratio to the decimal form 1.108, he was quite close to the true ratio $\pi/2\sqrt{2} = 1.1107$.

For Stillman Drake, Galileo's biographer, this ratio is the closest analogy in Galileo's work to something like a physical constant. Physical constants are an artifact of algebra. They drop out of ratios. Thus, although we now commonly say that Galileo discovered the constant acceleration due to gravity, g, he did not and could not. The mathematics, which he so carefully applied, and which revealed so much to him, masked numerical constancy.

Galileo combined his method of relating a pendulum's length to its swing

Elizabeth Cavicchi

time with the ratio he found between fall and pendulum times. He thus calculated the height a weight would have to fall from in order to match the swing time of one of his pendulums. When he analyzed the ratio calculation, he found that the falling time appeared twice. Time was multiplied by time; time was *squared*. In following the mathematical rules for combining ratios, he had constructed a new physical law: falling distance is proportional to the square of falling time. He had also constructed the new physical quantity 'time squared'. Until Galileo discovered this pattern in his calculation, no one had ever squared a physical quantity other than length, whose square is usually area.

7   Extending the Law

Galileo wondered if the same rule applied to the data collected in the inclined plane/lute string experiments. There he had computed the sequence of odd integers 1, 3, 5, 7, ... from ratios of the first distance to each successive distance covered in equal times. This pattern suggests to us that the link between distance and time involves a square: sums of the series of odd integers produce the series of squared integers $(1, 1 + 3 = 4, 1 + 3 + 5 = 9, 1 + 3 + 5 + 7 = 16)$. Galileo did not identify this pattern when he originally performed the experiment, because he interpreted the lute string separations as speeds rather than distances.

Galileo recorded the integer squares (1, 4, 9, 16, ...) beside his original distance measurements on the notebook page containing data from the lute string experiment. He found that the product of the first distance and each successive integer square (1, 4, 9, 16, ...) yielded the same number that he had originally measured as the ball's cumulative distance traveled from the start of motion. These squared integers behaved like the squares of times in free fall (Figure 5). Balls rolling down an incline and falling weights exhibited the same relation: distance traveled is proportional to the square of time. We can also say that the pendulum motion is somehow similar; the square root of its length is proportional to its swing time. All these 'natural' motions were somehow similar.

Galileo commented, though not in these modern words, that if you plot the total distance traveled for several consecutive and equal time intervals, the dots fall along a parabola (Figure 6). But to Galileo this comment was purely geometric; it did not provide him with a definition of speed or a characterization of projectile motion. For some time he questioned whether a object's speed in free fall could be represented by the total distance it had fallen, or by the distance it had fallen in the most recent interval of time. If the first choice were correct, the speeds of successive time intervals would also fall along a parabola, just like the distance measurements, while the second choice would place the speeds on a straight line of increasing slope (Figure 7).

Galileo tried to measure speed of a falling object by observing the effects of

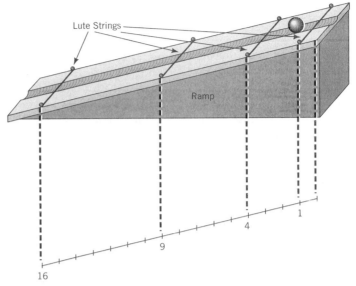

Figure 5: The total distance the ball had rolled after its release was proportional to the square of the integer identifying that equal time interval. The same law related distance and time in free fall motions.

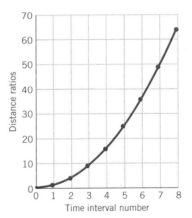

Figure 6: A plot of the total distance traveled in successive time intervals is a parabola.

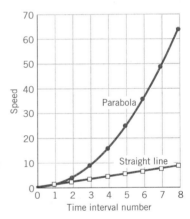

Figure 7: In the plot of the average speed in successive time intervals a parabola or a straight line?

Elizabeth Cavicchi

impact, measured, for example, by how much of a depression the object makes in the dirt when it hits. These estimates were misleading. The depth of the impact crater increased like a parabola with the total distance fallen. At first Galileo assumed that the depth represented the object's speed when it hit the ground. The results seemed to confirm this, but eventually he guessed that impact depth is proportional to the square of speed. He later constructed a geometric argument confirming the second choice, showing that speed is proportional to the total time, rather than the distance, of free fall.

8   Formal Publication

Galileo conveyed his new understanding of the proportionality between distance and time squared only in personal correspondence and in teaching students. It was finally incorporated into Galileo's last book, *Two New Sciences*, half a century after legend claims he first dropped weights from the Leaning Tower of Pisa.

By then, Galileo had been discovered by the Roman Inquisition, which had ruled that his work bordered on heresy by entertaining the possibility that the earth moved around the sun. He had been sentenced to life imprisonment (commuted to home incarceration) and prohibited from further publication. *Two New Sciences*, dedicated to a French ambassador who smuggled the manuscript out of Italy, was printed in the Protestant Netherlands by the Elsevier family in 1638.

But Galileo's long process of redesigning experiments and reworking data analyses was not suitable for description in a publication of the time. *Two New Sciences* had to present explanations formally derived from definitions and postulates. The work is in the form of a dialogue inspired by the writings of Plato, and includes discussions of long formal derivations by the characters. The proportionality between distance in free fall and the square of the falling time is proven geometrically as a theorem. The empirical data, through which Galileo first came to identify this law, is not mentioned.

It is easier to understand the classical presentation of *Two New Sciences* than to understand the process by which Galileo derived these results. The former is a treatise similar to the work of Euclid and Plato, while the latter represents a whole new way of thinking about motion. Galileo's actual paths of learning included stops, new starts, interconnections among phenomena and mathematical analogy.

9   Conclusion

While Galileo had described "how" things fall, first by direct measurement and then abstractly with the time-squared rule, he did not aim to explain "why"

Figure 8: Participants in Galileo's *Dialogo sopra i due Massimi Sistemi del mondo Tolemaico e Copernicano* published in Florence in 1632. By permission of the Houghton Library, Harvard University.

Elizabeth Cavicchi

things fall. Previous philosophers, in the tradition of Aristotle, regarded their task as finding causes for natural processes. Galileo diverged from this tradition in changing the question from "why do things fall" to "how do things fall". Galileo admits that he does not know the cause of free fall in an exchange between Simplicio (the Aristotelean) and Salviati (the speaker for Galileo) in the *Dialogue* which invoked the Inquisition's condemnation (Figure 8, [4]):

> Simplicio: The cause of this effect is well known; everyone is aware that it is gravity.
>
> Salviati: You are mistaken, Simplicio; what you ought to say is that everyone knows it is called 'gravity'. What I am asking for is not the name of the thing, but its essence, of which essence you know not a bit more than you know about the essence of whatever moves stars around... the name [*gravity*]... has become a familiar household word through the daily experience we have of it. But we do not really understand what principle or what force it is that moves stones downward, any more than we understand what moves them upward after they leave the thrower's hand, or what moves the moon around[9].

By revising experiments to make measurement practical, and analyzing those measurements with mathematics, Galileo composed the first accurate account of how things fall. He came to see numerical patterns in his data that he did not expect to see. These patterns were the first indication that terrestrial motions could be described by mathematics.

The outgrowth of that mathematical metaphor for nature is so tightly knitted into our contemporary physical understanding that it is almost impossible for us to see nature as freshly as he did. We translate physical quantities into algebraic abstractions; Galileo never did. By composing his measurements into ratios, he found a general relationship between distance and time. He worked with a few carefully made measurements; his work was not statistical. Galileo's method of learning from concrete, familiar examples, rather than from abstractions, may have something in common with that of students today.

Without knowing what he was going to find, or what methods might reveal it – and with the opposition of an authoritarian tradition that presumed to dictate how nature worked – Galileo extracted a regularity from falling motions that was new to understanding.

References

[1] *Aristotle's Physics*, W. D. Ross, trans., ed., (Oxford: Clarendon Press) 1966.

[2] L. N. H. Bunt, P. S. Jones, J. D. Bedient, *The Historical Roots of Elementary Mathematics*, (New York NY: Dover Pub., Inc.) 1988.

Elizabeth Cavicchi

[3] Elizabeth Cavicchi, "Halfway Through the Darkness: A Fieldwork Study of Time and Motions", Unpublished manuscript, May 1994.

[4] I. Bernard Cohen, *The Birth of a New Physics*, (New York NY: W. W. Norton & Co.) 1985.

[5] Bern Dibner, *Heralds of Science*, (Cambridge MA: MIT Press) 1969.

[6] Stillman Drake, *Galileo at Work: His Scientific Biography*, (Chicago IL: University of Chicago Press) 1978.

[7] Stillman Drake, *Galileo*, (New York, NY: Hill and Wang) 1980.

[8] Stillman Drake, "Galileo's physical measureents", *American Journal of Physics*, **54** 1985, 302-306.

[9] Stillman Drake, *Galileo: Pioneer Scientist*, (Toronto: University of Toronto Press) 1990.

[10] Howard Eves, *A Survey of Geometry*, (Boston MA: Allyn and Bacon, Inc.) 1972.

[11] Galileo Galilei, *Dialogue Concerning the Two Chief World Systems–Ptolemaic & Copernican*, Stillman Drake, trans., Albert Einstein, forward, (Berkeley CA: University of California Press) 1967.

[12] Galileo Galilei, *Discoveries and Opinions*, Stillman Drake, trans., (New York, NY: Doubleday) 1957.

[13] Galileo Galilei, *Two New Sciences Including Centers of Gravity and Force of Percussion*, Stillman Drake, trans., (Madison WI: University of Wisconsin Press) 1974.

[14] Morris Kline, *Mathematical Thought from Ancient to Modern Times*, (new York NY: Oxford University Press) 1972.

[15] A. Rupert Hall, *The Scientific Revolution 1500-1800: The Formation of the Modern Scientific Attitude*, (Boston MA: Beacon Press) 1966.

[16] David C. Lindberg, *The Beginnings of Western Science: The European Scientific Tradition in Philosophical, Religious, and Institutional Context, 600BC to AD 1450*, (Chicago IL: University of Chicago Press) 1992.

[17] George Sarton, *Ancient Science Through the Golden Age of Greece*, (New York NY: Dover Pub., Inc.) 1980.

[18] Thomas Settle, "Galileo and Early Experimentation" in *Springs of Scientific Creativity: Essays on Founders of Modern Science*, R Aris, H. T. Davis, R. H. Stuewer, eds., (Minneapolis MN: University of Minnesota Press) 1983.

Elizabeth Cavicchi

[19] Thomas Settle, "An Experiment in the History of Science", *Science*, Jan. 1961, v. 133, 19-23.

These are the texts which contributed to this reading on free fall. You might find others. This essay closely follows the accounts and approach of Stillman Drake.

Acknowledgements

This essay was written at the request of Judy Clark and Linda Kime, to accompany students' work with the free fall experiment in their course, "Explorations in College Algebra". Linda Kime invited me to observe her teaching this course in fall 1993, and encouraged my interviews of her students. From Linda and her students I began noticing how involved learning can be, and its fascination for students and teacher. Hazel Garland, a student in Linda's class, volunteered to meet me regularly in 1994, to experimentally explore motion and learning. Hazel's spontaneous work and perceptive insights opened me to see the possibilities, uncertainties, questions, and delights that can develop when student and teacher explore their understandings together. My own teachers and advisors, Eleanor Duckworth and Philip Morrison, made possible the questioning and thought about learning and physics that began emerging through my work with Hazel and reading about Galileo. Their responses changed my understandings. Courtney Cazden encouraged this essay, as did Irene Hall, my classmate in Cazden's course. Judy Clark consistently supported my efforts. Phylis Morrison advised the preparation of figures. Alva Couch worked with me on figure design and document preparation and shared my excitement at learning.

CHAPTER 1

Algebra Aerobics   1.2

1. $24\% + 24\% + 5\% + 14\% = 67\%$

    Note: If you calculated these by adding frequencies first, you would get $14/21 = 0.667$ or 67%

2. $11,645 + 10,423 + 10,582 = 32,650$. So $32,650/252,177 = 0.129$ or 13%

3. To find the relative frequency, divide the frequency count by the total count. To convert from a decimal to a percent, multiply by 100%.

Age	Frequency Count	Relative Frequency (%)
1–20	52	$52/137 = 0.38$ or 38%
21–40	35	$35/137 = 0.26$ or 26%
41–60	28	$28/137 = 0.20$ or 20%
61–80	22	$22/137 = 0.16$ or 16%
Total	137	$137/137 = 1.00$ or 100%

Algebra Aerobics   1.3

1.
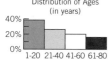

2. Table for the histogram in Figure 1.6

Age	Relative Frequency (%)	Frequency Count
1–20	20%	$(0.20)(1352) = 270$
21–40	35%	$(0.35)(1352) = 473$
41–60	30%	$(0.30)(1352) = 406$
61–80	15%	$(0.15)(1352) = 203$
Total	100%	1352

3. On the 20-year histogram in Figure 1.4, approximately 33% of the population is between 20 and 40 years old and approximately 22% of the population is between 40 and 60 years old. Thus, approximately 55% of the population was between 20 and 60 years old in 1991.

Algebra Aerobics   1.4

1. a. sum = \$8750, so mean = \$8750/9 = \$972.22; median = \$300; mode = \$300.

    b. sum = 4.7, so mean = 4.7/8 = 0.59; median = (0.4 + 0.5)/2 = 0.45

    The two modes are 0.3 and 0.4 (bimodal).

2. One of the values (\$6,000) is much higher than the others, which forces a high value for the mean. In cases like this, the median is generally a better choice for measuring central tendency.

3. Let t represent the unknown time. The sum of the seven reported times is 99.3. Therefore,

    $$(99.3 + t)/8 = 14.2 \rightarrow 99.3 + t = 113.6 \rightarrow t = 14.3 \text{ sec}$$

CHAPTER 2

Algebra Aerobics   2.1

1. a. The maximum median family income is \$36,062 in 1989. The coordinates of this point are (1989, 36062).

    b. The minimum is \$31,738 in 1982. The coordinates are (1982, 31738).

    c. The longest period of increase was from 1982 to 1987. The longest period of decrease was from 1978 to 1982.

2. a. Square the value of x, then reverse the sign, then add 1 (or, more elegantly, find the sum of the negative of the square of x, and 1)

    b. (0, 1), (1, 0), and (− 1, 0) are solutions

    c.

x	−3	−2	−1	0	1	2	3
y	−8	−3	0	1	0	−3	−8

3. a. Subtract 1 from the value of x, then square the result

    b. (0, 1) and (1, 0) are solutions

    c.

x	−3	−2	−1	0	1	2	3
y	16	9	4	1	0	1	4

Algebra Aerobics   2.2a

1. a. Tip = 15% of the cost of meal

    $$T = 0.15M$$

    Independent variable: M (price of meal)
    Dependent variable: T (tip)

The equation is a function since to each value of M there corresponds a unique value of T. The domain and range are positive numbers. The highest value for M that is believed to be "reasonable" varies from person to person. The highest value of T is 15% of the highest value selected for M.

b. $T = (0.15)(\$8) = \$1.20$

c. $T = (0.15)(\$26.42) = \$3.96$

In real life, we'd probably round this to the nearest nickel or dime.

2. a. Since the dosage depends on the weight, the logical choice is W as the independent, and D as the dependent variable.

b. In this formula, each value of W determines a unique dosage D, so D is a function of W.

c. Values for the domain must be larger than 0 and less than 10 kilograms. So we have

Domain = all values of W greater than 0 and less than 10 kilograms
= all values of W with $0 < W < 10$

As is often the case in the social sciences, the domain includes some questionable values (we don't expect a child to weigh close to 0 lb), but it spans all the appropriate ones.

Range = all values of D greater than 0 to a maximum of up to $10 \cdot 50 = 500$ mg
= all values of D with $0 < D < 500$

d. The following table and graph are representations of the function. Note that $(0, 0)$ and $(10, 500)$ are calculated only to help us draw the graph and are not actually included in the model:

Daily Ampicillin Dosage by Weight

W	D
Child's Weight (kg)	Daily Maximum Dosage (mg)
0	0
1	50
2	100
3	150
4	200
5	250
6	300
7	350
8	400
9	450
10	500

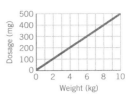

Maximum daily ampicillin dosage as a function of weight

3. The respiration rate (R) is not a function of the pulse rate (P), since a pulse rate of 75 determines two different respiration rates: 12 and 14.

Algebra Aerobics   2.2b

1. $f(1) = 5 \cdot 1 = 5.$ $f(10) = 5 \cdot 10 = 50$
$f(-2) = 5 \cdot (-2) = -10$

2. $g(0) = 2(0)^2 - (0) + 4$ $g(1) = 2(1)^2 - 1 + 4$
$= 0 - 0 + 4$ $= 2 - 1 + 4$
$= 4$ $= 5$

$g(-1) = 2(-1)^2 - (-1) + 4$
$= (2 \cdot 1) + 1 + 4$
$= 7$

3. $f(0) = 20;$ $f(20) = 0.$ When $x = 10$ or $x = 30$, then $f(x) = 10.$

4. $f(-4) = 2;$ $f(-1) = -1;$ $f(0) = -2;$ $f(3) = 1$

When $x = -2$ or $x = 2$, then $f(x) = 0$

Algebra Aerobics   2.2c

1. The domain and the range are now all the real numbers.

2. The domain and range are now all the real numbers. The table should now include negative values, and the graph will continue indefinitely in two directions.

We could write $f(W) = 50W$. Then $f(15) = (50)(15) = 750$, $f(-15) = (50)(-15) = -750$, and $f(15,000) = (50)(15,000) = 750,000$

Algebra Aerobics 2.3a

1. $f(-3) = 30; f(0) = 6; f(1) = 2$

2. $2x - 2 - 3y - 15 = 10 \Rightarrow y = (27 - 2x)/-3 = (2x - 27)/3$ or $y = \frac{2}{3}x - 9.$

So y is a function of x: $f(x) = \frac{2}{3}x - 9$ and the domain is all real numbers.

3. $y = \dfrac{1}{3}(x^2 + 2x + 4)$. So y is a function of x: $f(x) =$

$\dfrac{1}{3}(x^2 + 2x + 4)$ and its domain is all real numbers.

4. $3x + 2y = 6 \Rightarrow y = -\dfrac{3}{2}x + 3$. So y is a function of x:

$f(x) = -\dfrac{3}{2}x + 3$ and its domain is all real numbers.

5. $y = 4 - x - \sqrt{x - 2}$. So y is a function of x: $f(x) = 4 - x - \sqrt{x - 2}$ and the domain is all $x \geq 2$.

Algebra Aerobics   2.3b

1.   **a.** yes. **b.** When $x < 0$, the function is increasing. When $x > 0$, the function is decreasing. **c.** The graph appears to "blow up," meaning that the closer x gets to 0, the higher will be the value for y.

2.   Graphs (b) and (c) represent functions, while (a) and (d) do not represent functions since they fail the vertical line test.

3.   Neither is a function of the other. The graph fails the vertical line test, so weight is not a function of height. If we imagine reversing axes, so that weight is on the horizontal and height is on the vertical, the new graph will also fail the vertical line test.

4.   **a.** D is a function of Y, since each value of Y determines a unique value of D.

b. Y is not a function of D, since one value of D, \$2.7, yields two values for Y, 1993 and 1997.

Algebra Aerobics   2.4

1. $y = 3x$

2. $y = 3x - 2$

3. $y = -x^2$

CHAPTER 3

Algebra Aerobics   3.1

1. $(143 - 135)$ lb/5 yr = 1.6 lb/yr

2.   (\$27.5 − \$18.3) trillion/4 yr = \$2.3 trillion/yr

3.   **a.** −892 deaths/yr (or a decrease in the annual deaths by 892 deaths/yr)

b. −1578.4 deaths/yr or a decrease of 1578.4 deaths/yr

4.   **a.** Answers will vary.

Algebra Aerobics   3.2

1.   **a.**

Year	Total Population (in millions)	Average Annual Rate of Change (millions/yr)
1800	910	n.a.
1850	1130	$\dfrac{1130 - 910}{1850 - 1800} = \dfrac{220}{50} = 4.4$
1900	1600	9.4
1950	2510	18.2
1970	3702	59.6
1980	4456	75.4
1990	5293	83.7
1993	5555	87.3
1996	5771	72.0

b.

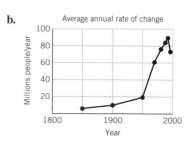

Average annual rate of change

c. For the time periods in Table 3.5, the largest average annual rate of change was from 1990 to 1993, which means that the line segment with the steepest slope connects (1990, 5293) and (1993, 5555).

d. According to these numbers, the average annual rate of change has been increasing until the time period from 1993 to 1996 when it started to decrease.

Algebra Aerobics   3.3

1.   1983–84; 1986–87; 1991–93

2.   84−85: −,   85−86: +,   86−87: −,   87−88: +, 88−89+,   89−90: +,   90−91: 0,   91−92: +,   92−93: −

3.   **a.** $(-5 - 3)/(4 - (-2)) = -8/6 = -4/3$;

b. $(500 - 500)/(7 - 3) = 0/4 = 0$

4.   Example 5: $\dfrac{-10 - 2}{-3 - 3} = \dfrac{-12}{-6} = 2$;

Example 6: $\dfrac{-7 - 3}{4 - (-6)} = \dfrac{-10}{10} = -1$

5.   Answers will vary.

Algebra Aerobics   3.4

1.   **a.** Social Security payments increased by \$285.980 − \$120.42 = \$165.508 billion

b. Social Security payments increased by $165.508/ ($120.472) = 1.37 or 137%

c. The average rate of change for Social Security payments between 1980 and 1992 was:

$$\frac{(\$285.980 \text{ billion} - \$120.472 \text{ billion})}{(1992-1980) \text{ yr}}$$

$$= \frac{\$165.508 \text{ billion}}{12 \text{ yr}} \approx 13.8 \text{ billion dollars/yr}$$

2. Answers will vary.

3. a. 1901–1910. According to Samuel Eliot Morison in *The Oxford History of the American People,* "unlimited and unrestricted immigration, except for Orientals, paupers, imbeciles, and prostitutes, had been national policy down to World War I." Postwar policies limiting immigration were instituted in the 1920s.

b. The largest change was between the 1920s and the 1930s—a decline of 3,579,000 immigrants. Morison writes that during the Great Depression immigration numbers actually became negative and "only rose after 1936 when fascism produced a new crop of refugees."

   i. −3,579,000

   ii. 3,579,000 represents (3,579,000)/(4,107,000) = 0.87 or 87% of 4,107,000. So the number of immigrants between the two decades decreased by 87%.

CHAPTER 4

Algebra Aerobics   4.1

1. a. $10 = 7.0 + 1.5(2) \Rightarrow (2, 10)$ satisfies the equation

b. $13.5 \neq 7.0 + 1.5(5) \Rightarrow (5, 13.5)$ does not satisfy the equation

2. From the equation, the exact weight is $W = 7.0 + 1.5(4.5) = 13.75$ lb.

3. $14 = 7.0 + 1.5A \Rightarrow 7.0 = 1.5A \Rightarrow A = 7.0/1.5 = 4.67$ months

4. For example, if we choose the two points $(-2, 4.0)$ and $(5, 14.5)$, then the rate of change $= \dfrac{14.5 - 4.0}{5 - (-2)} = \dfrac{10.5}{7} = 1.5$

5. For example,

$$A = -20 \Rightarrow W = 7 + 1.5(-20) = -23$$
$$A = 200 \Rightarrow W = 7 + 1.5(200) = 307$$

6. The slope or rate of change between $(-20, -23)$ and $(200, 307)$ (or any two points that lie on the line) should be 1.5. Calculating the slope we get $(307 - (-23))/(200 - (-20)) = 330/220 = 1.5$.

Algebra Aerobics   4.2a

1. a. $m = 5, b = 3$;   **b.** $m = 3, b = 5$;

   c. $m = 5, b = 0$;   **d.** $m = 0, b = 3$;

   e. $m = -1, b = 7.0$;   **f.** $m = -11, b = 10$

2. a. $m = (11 - 1)/(8 - 4) = 10/4 = 5/2$ or 2.5

   b. $m = (6 - 6)/(-3 - 2) = 0/-5 = 0$

   c. $m = (-1 - (-3))/(-5 - 0) = 2/-5 = -2/5$

3. a. $f(0) = 50 - 25 \cdot (0) = 50$

   b. $f(-2) = 50 - 25 \cdot (-2) = 50 + 50 = 100$

   c. $f(2) = 50 - 25 \cdot (2) = 50 - 50 = 0$

   d. Since the line passes through $(0, 50)$ and $(-2, 100)$, the slope must equal $\dfrac{100 - 50}{-2 - 0} = \dfrac{50}{-2} = -25$.

Algebra Aerobics   4.2b

1. 20,000 is in dollars (initial salary); 1,000 is in dollars/yr (annual increase)

2. dollars = dollars + (dollars/yr) (yr)

3. 15 is in dollars/person; 10 is in dollars

dollars $= \left(\dfrac{\text{dollars}}{\text{person}}\right) (\text{persons}) + \text{dollars}$.

4. When $A = 2, H = 30.0 + 2.5(2) = 30.0 + 5.0 = 35.0$
When $A = 12, H = 30.0 + 2.5(12) = 30.0 + 30.0 = 60.0$
So since H is an increasing function, the range is values of H such that $35.0 \leq H \leq 60$.

Algebra Aerobics   4.3a

1. $y = 1.2x - 4$

x	y
-6	-11.2
-3	-7.6
0	-4.0
3	-0.4
6	3.2

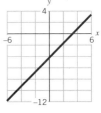

2. $y = 300 - 400x$

x	y
-4	1700
-2	1100
0	300
2	-500
4	-1300

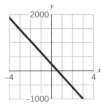

3. $S = 12,000 + 3,000x$

4. **a.** $C = 200,000 + 15x$

  b. The cost per item = $\dfrac{\text{total cost}}{\text{no. of items}} = \dfrac{200,000 + 15x}{x} =$
  $\dfrac{200,000}{x} + \$15.$ So as x, the number of items, increases, the cost per item will go down.

Algebra Aerobics   4.3b

1. $1 = (-4)(-3) + b \Rightarrow b = -11 \Rightarrow y = -4x - 11$

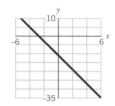

x	y
-2	-3
0	-11
4	-27

2. $8.5 = (1.6)(10) + b \Rightarrow b = -7.5 \Rightarrow S = 1.6Y - 7.5$ where S = salary and Y = years of education

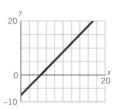

Y	S
0	-7.5
10	8.5
20	24.5

3. Substituting $(-3, -5)$ into $y = b + 4x$, we get $-5 = b + 4(-3) \Rightarrow -5 = b - 12 \Rightarrow b = 7.$

4. $m = (4 - 1)/(-2 - 3) = -0.6$
  $1 = (-0.6)(3) + b \rightarrow b = 2.8 \Rightarrow y = -0.6x + 2.8$

Algebra Aerobics   4.4

1. **a.** The line sketched into the figure seems a reasonable fit.

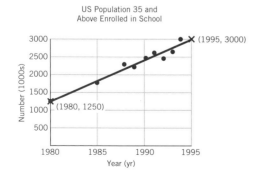

US Population 35 and Above Enrolled in School

b. The estimated coordinates of two points on the line are (1980, 1250) and (1995, 3000) where the first coordinate is the year and the second the number of students in the thousands. The approximate slope of the line is then:

$$(3000 - 1250)/(1995 - 1980) = 1750/15$$
$$\approx 117 \text{ thousand students/yr}$$

So each *year* between 1980 and 1995 approximately 117 thousand (or 117,000) more people 35 or older enrolled in school. Each *decade* 1170 thousand (or 1.17 million) more people 35 or older became students.

c. If x = number of years since 1980, then the point $(0, 1250)$ is the vertical intercept, so $b = 1250$. Hence the equation of the line is:

$$y = 1250 + 117x$$

where x = number of years since 1980, and y = number of students (in thousands) 35 years or older.

Algebra Aerobics   4.5a

1. $d = 5t$, so d is directly proportional to t. The d is in miles, t is in hours, so 5 is in mi/hr. This is more likely a walking than a driving pace.

2. Equation for Table 4.15: $y = -3x$ (directly proportional). Equation for Table 4.16 $y = 3x + 5$ (not directly proportional). Equation for Table 4.17: $C = 1.50\,G$ (directly proportional).

3. Let R = real geographic distance and M = distance on map. Since R is directly proportional to M, then $R = m \cdot M$. Then m = slope = 35 mi/1 in. so $R = 35\,M$.
  When $M = 4.5$, then $R = (35 \text{ mi/1 in.}) \cdot 4.5 \text{ in.} = 157.5$ mi

Algebra Aerobics   4.5b

1. **a.** $y = -5$; **b.** $y = -3$; **c.** $y = 5$; **d.** $y = 3$

2. $y = \$25,000$. The domain is 0 yr $\le x \le$ 10 yr. The range is the single value of $\$25,000$.

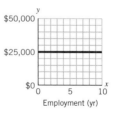

Employment (yr)

3. **a.** $x = 3$; **b.** $x = 5$; **c.** $x = -3$; **d.** $x = -5$

Algebra Aerobics   4.5c

1. $y = -x$

2. $m = 358.9$, so $1000 = (358.9)(4) + b \rightarrow b = -435.6$, and $y = 358.9x - 435.6$

3. a. $m = 1/3$; **b.** $m = -1$; **c.** $m = -10/31 = -0.32$

4. a. $m = -1/2 = -0.5, -5 = (-0.5)(3) + b \rightarrow -3.5 = b \rightarrow y = -0.5x - 3.5$

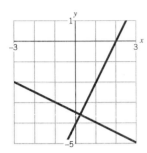

b. Any line parallel to $y = -0.5x - 3.5$, but not equal to it, will be perpendicular to $y = 2x - 4$, but not pass through $(3, -5)$. Two examples are $y = -0.5x + 4$ or $y = -0.5x - 7.5$.

c. The three lines are parallel.

Algebra Aerobics   4.6

1. and **2.**

   a. linear: $m = 2$; $b = 0$

   b. linear: $m = 0$; $b = 3$

   c. linear: Since $y = -11/34 - (41/34)x$, then $m = -41/34$; $b = -11/34$

   d. not linear: m is undefined; there is no vertical intercept

   e. not linear

   f. not linear (the rate of change is not constant)

   g. not linear

3. i. We could graph the data and verify that we have a straight line. Assuming that the freezing point depends upon salinity, we choose salinity as the independent variable and graph it on the horizontal axis.

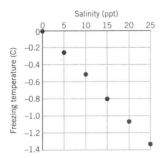

Graph of freezing temperature versus salinity.

The points appear to lie on a straight line with a negative slope, so the relationship may be described as linear.

ii. We could verify that a fixed change in the value of one variable corresponds to a fixed change in the value of the other variable. In Table 4.21 we can see that each 5 ppt increase in salinity corresponds to a decrease of approximately $-0.27°C$ in the freezing point. So we can consider the relationship linear.

iii. We could show that the average rate of change is constant. If we add a third column containing the average rate of change to the original table, we can see that the average rate of change remains fairly constant at about $-0.054°C/ppt$ (see table). Hence, we may consider the relationship as linear.

Salinity (ppt)	Freezing point (Celsius)	Average rate of change of freezing point with respect to salinity
0	0.000	n.a.
5	-0.266	-0.053
10	-0.534	-0.054
15	-0.804	-0.054
20	-1.076	-0.054
25	-1.350	-0.055

iv. We could find an m and b such that $y = b + mx$, where x = salinity (in grams of dissolved salts per 1000 g of water) and y = freezing point degrees centigrade? We can estimate a line by hand that would fit the data. Use any two points on the line, for example, $(0, 0.000)$ and $(20, -1.076)$, to find the equation of the line that passes through them. The slope, m, is

$$m = \frac{\text{change in freezing point}}{\text{change in salinity}}$$
$$= \frac{(-1.076 - 0)°C}{(20 - 0)\ \text{ppt}}$$
$$= \frac{-1.076°C}{20\ \text{ppt}}$$
$$\approx 0.054°C/\text{ppt}$$

Given our rate of change calculations in part iii, this is the value we expect. The equation of the line is then in the form $y = b - 0.054x$. The line goes through the origin $(0, 0)$, so $b = 0$. The equation of the line through the points $(0, 0.000)$ and $(20, -1.076)$ is

$$y = -0.054x \tag{1}$$

Do the other points in Table 4.21 satisfy this equation? Table 4.23 shows the given values for freezing point vs. the values predicted by Equation (1).

Salinity (ppt)	Actual freezing point (°C)	Predicted freezing point $y = -0.054x$
0	0.000	0.000
5	−0.266	−0.270
10	−0.534	−0.540
15	−0.804	−0.810
20	−1.076	−1.080
25	−1.350	−1.350

The results are very close, so Equation (1) gives a good representation of the relationship between x and y. The relationship may be described as linear.

v. We could find an A, B, and C such that: $Ax + By = C$. If we rewrite the equation $y = -0.054x$ as

$$0.054x + y = 0$$

then it is in the desired form $Ax + By = C$ where $A = 0.054$, $B = 1$, and $C = 0$.

CHAPTER 5

Algebra Aerobics   5.2

1. **a.**   **i.**   0.65

   ii.   0.68

   iii.   0.07

   iv.   0.70

   b. $|0.07|$; $|0.65|$; $|0.68|$; $|0.70|$

2. **a.**   The slope is 2120;
   The vertical intercept is −3460;
   The correlation coefficient is 0.76

   b.   a slope of $2120/yr. means that for every additional year of education the mean personal income increases by $2120.

   c. $2120 dollars for 1 yr

   $21,200 dollars for 10 yr

3. **i.**   The picture in the bottom right has the strongest linear relationship.

   ii.   The picture in the bottom left has the worst linear fit.

CHAPTER 6

Algebra Aerobics   6.1

1. Gas is the cheapest system from approximately 17.5 years of operation to approximately 32.5 years of operation.

Solar becomes the cheapest system after approximately 32.5 years of operation.

2. **a.**   (3, 1) is a solution since $4(3) + 3(-1) = 12 - 3 = 9$ and $5(3) + 2(-1) = 15 - 2 = 13$

   b. $(-1, 3)$ is not a solution since $4(-1) + 3(3) = -4 + 9 \neq 9$

3. **a.**
given	$12x - 9y = 18$
divide by 3	$4x - 3y = 6$
add $3y$ to each side of equation	$4x = 6 + 3y$

   Thus, the equations are equivalent.

   b.   There are an infinite number of solutions to the system of equations in part (a).

Algebra Aerobics   6.2a

1. **a.** $x = 1, y = 5$

   b. $x = 8, y = 15,100$

   c.   both C and $F = -40$

Algebra Aerobics   6.2b

1. **a.** $x = 2, y = 5$

   b. $w = 1/3, z = 2$

   c. $x = 1, y = 3$

   d. $s = 1/2, r = 6$

Algebra Aerobics   6.2c

1. **a.** $x = 1, y = 2$

   b. $x = 2, y = 5$

   c. $t = 2, r = 2$

   d. $x = 14,000, z = 3,600$

Algebra Aerobics   6.2d

1. **a.**   Infinitely many solutions—equations describe same line

   b.   No solution—parallel lines

   c.   Infinitely many solutions—equations describe same line.

2. One possible system of equations where there is no solution is

$$r = 5t + 10; r = 5t + 3$$

Note that the lines have the same slope, but different $y =$ intercepts, so they are parallel.

Algebra Aerobics   6.4a

1. **a.**   To estimate from the graph, try drawing dotted lines to find the amount of tax under each of the plans, as shown on the following graph.

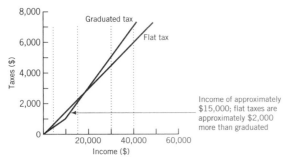

Flat Tax vs. Graduated Tax.

i. At an income of $5000, the flat tax would be about $200–$300 more.

ii. At an income of $15,000, the flat tax is also about $200–$300 more

iii. At an income of $40,000, the graduated tax is about $1,000 more.

b. **i.** $f(\$5000) = 0.15\,(\$5000) = \$750$
$g(\$5000) = 0.10\,(\$5000) = \$500$
$f(\$5000)$ is $250 larger than $g(\$5000)$

ii. $f(\$15,000) = 0.15\,(\$15,000) = \$2250$
$g(\$15,000) = \$1000 + 0.20\,(\$15,000 - \$10,000)$
$\quad\quad\quad\quad\quad = \$1000 + 0.20\,(\$5000)$
$\quad\quad\quad\quad\quad = \$2000$
$f(\$15,000)$ is $250 larger than $g(\$15,000)$

iii. $f(\$40,000) = 0.15\,(\$40,000) = \$6000$
$g(\$40,000) = \$1000 + 0.20\,(\$40,000 - \$10,000)$
$\quad\quad\quad\quad\quad = \$1000 + 0.20\,(\$30,000)$
$\quad\quad\quad\quad\quad = \$7000$
$g(\$40,000)$ is $1000 larger than $f(\$40,000)$

2. **a.** $g(i) = \begin{cases} 0.05i & \text{for } i \le \$50,000 \\ 2500 + 0.08(i - 50,000) & \text{for } i > \$50,000 \end{cases}$

b. $g(i) = \begin{cases} 0.06i & \text{for } i \le \$30,000 \\ 1800 + 0.09(i - 30,000) & \text{for } i > \$30,000 \end{cases}$

Algebra Aerobics   6.4b
First, estimate from the graph by drawing vertical lines at $30,000, at $60,000, and at $120,000, and by estimating the vertical distance between the two function graphs (see Algebra Aerobics 6.4a, problem 1(a).) Then compare your estimates with the following computations.

a. $f(\$30,000) = 0.0595\,(\$30,000) = \$1785$
$g(\$30,000) = 0.055\,(\$30,000) = \$1650$
$f(\$30,000)$ is $135 larger than $g(\$30,000)$

b. $f(\$60,000) = 0.0595\,(\$60,000) = \$3570$
$g(\$60,000) = \$2761 + 0.088\,(\$60,000 - \$50,200)$
$\quad\quad\quad\quad\quad = \$2761 + 0.088\,(\$9800)$
$\quad\quad\quad\quad\quad = \$2761 + \$862.40$
$g(\$60,000) = \$3623.40$
$g(\$60,000)$ is $53.40 larger than $f(\$60,000)$

c. $f(\$120,000) = 0.0595\,(\$120,000) = \$7140$
$g(\$120,000) = \$6263 + 0.098\,(\$120,000 - \$90,000)$
$\quad\quad\quad\quad\quad = \$6263 + 0.098\,(\$30,000)$
$\quad\quad\quad\quad\quad = \$6263 + \$2940$
$g(\$120,000) = \$9203$
$g(\$120,000)$ is $2063 larger than $f(\$120,000)$

CHAPTER 7

Algebra Aerobics   7.1
1. **a.** 10^{10} **b.** 10^{-14}

2. **a.** 0.000 000 01   **b.**   10,000,000,000,000

3. **a.** 10^{-9} or 0.000 000 001 sec   **b.** 10^{-1} or 0.1 m

c. 10^{9} or 1,000,000,000 bytes (a byte is a term used to describe a unit of computer memory)

4. **a.** $7 \cdot 10^{-2}$ or 0.07 m;   **b.** $9 \cdot 10^{-3}$ or 0.009 m;

c. $5 \cdot 10^{3}$ or 5000 m

Algebra Aerobics   7.2
1.   602,000,000,000,000,000,000,000

2. $3.84 \cdot 10^{8}$

3. $1 \cdot 10^{-8}$ cm

4.   0.000 000 002 m

5. **a.** $-705,000,000;$ **b.** $-0.000\,040\,3$

6. **a.** $-4.3 \cdot 10^{7};$ **b.** $-8.3 \cdot 10^{-6}$

Algebra Aerobics   7.3a
1. **a.** 10^{12} **b.** 8^{2} **c.** z^{-9}

d. cannot be simplified because bases are different.

e. $2 \cdot 7^{3}$ **f.** $(-5)^{2} = 25$, but $-5^{2} = -1 \cdot (5^{2}) = -25$.

2. $(3.3 \cdot 10^{-6}) \cdot (4.3 \cdot 10^{3}) = (3.3 \cdot 4.3) \cdot (10^{-6} \cdot 10^{3}) \approx 14 \cdot 10^{-3}$
$= 1.4 \cdot 10^{-2}$ or 0.014 sec. So it would take less than 2 hundredths of a second for the signal to cross the United States.

3. $x^{-2}(x^{5} + x^{-6}) = x^{-2}x^{5} + x^{-2}x^{-6}$
$\quad\quad\quad\quad\quad = x^{(-2+5)} + x^{(-2-6)}$
$\quad\quad\quad\quad\quad = x^{3} + x^{-8}$

Algebra Aerobics 7.3b

1. **a.** 10^{-2} or $1/10^2$; **b.** 8^{10}; **c.** 3^{-1} or $1/3$;

   d. cannot be simplified, different bases; **e.** $7^0 = 1$

2. **a.** $\dfrac{125.5 \cdot 10^6 \text{ people}}{152.5 \cdot 10^3 \text{ sq mi}} \approx 0.83 \cdot 10^3$ people/sq mi ≈ 830 people/sq mi in Japan.

   b. $\dfrac{263.8 \cdot 10^6 \text{ people}}{3620 \cdot 10^3 \text{sq mi}} \approx 0.073 \cdot 10^3$ people/sq mi ≈ 73 people/sq mi in the United States.

   c. Japan is much more densely populated, with more than 10 times as many people/sq mi as the United States.

3. (capacity of hard drive)/(capacity of diskette) = $2 \cdot 10^9$ bytes)/$(1.44 \cdot 10^6$ bytes) = $(2/1.44) \cdot (10^9)/(10^6) \approx 1.4 \cdot 10^3 = 1400$ diskettes.

Algebra Aerobics 7.3c

1. **a.** 10^{20} **b.** 10^{-20} **c.** 7^6 **d.** x^{20}

2. Surface area of Jupiter = $4\pi r^2 = 4\pi(7.14 \cdot 10^4 \text{ km})^2 \approx 641 \cdot 10^8 \text{ km}^2 \approx 6.41 \cdot 10^{10} \text{ km}^2$
   Volume of Jupiter = $4/3 \, \pi r^3 = 4/3 \, \pi \, (7.14 \cdot 10^4 \text{ km})^3 \approx 1520 \cdot 10^{12} \text{ km}^3 \approx 1.52 \cdot 10^{15}$

3. **a.** $t^{-3-(-12)} = t^9$ **b.** $v^{-3-(-6)}w^{7-(-10)} = v^3 w^{17}$

Algebra Aerobics 7.3d

1. **a.** $(3 \cdot 10^{-4})(4 \cdot 10^7) = 12 \cdot 10^3 = 1.2 \cdot 10^4 \approx 1 \cdot 10^4$

   b. $\dfrac{(5 \cdot 10^5)(2 \cdot 10^6)}{4 \cdot 10^{-3}} = \dfrac{5 \cdot 2}{4} \cdot \dfrac{10^5 10^6}{10^{-3}} = \dfrac{10}{4} \cdot \dfrac{10^{11}}{10^{-3}} = 2.5 \cdot 10^{14} \approx 3 \cdot 10^{14}$

2. Since only 3/7 of the farmable land is in use, we need to multiply the amount of land in the denominator by 3/7. Multiplying the denominator by 3/7 is equivalent to multiplying the whole fraction by 7/3. So (7/3) (450 people/sq mi) = 1050 people/sq mi. So Earth must sustain about 1050 people for each square mile of farmed land.

Algebra Aerobics 7.4

1. From Table 7.1 we have 1 km = 0.62 mi. So the conversion factor is 0.62 mi/1 km.
   Hence $9.46 \cdot 10^{12}$km $\cdot \dfrac{0.62 \text{ mi}}{1 \text{ km}} \approx 5.87 \cdot 10^{12}$ mi, which is very close to $5.88 \cdot 10^{12}$ mi.

2. 1 m = 100 cm, so the conversion factor for converting from cm to m is 1 m/100 cm
   So $1 \text{ Å} = 10^{-8} \text{ cm} \cdot \dfrac{1 \text{ m}}{100 \text{ cm}} = 10^{-8} \cdot \dfrac{1 \text{ m}}{10^2} = 10^{-8-2} \text{ m} = 10^{-10} \text{ m}$

3. 1 km = 0.62 mi, so the conversion factor for converting from miles to kilometers is 1 km/0.62 mi.
   Hence 135 mi $\cdot \dfrac{1 \text{ km}}{0.62 \text{ mi}} \approx 218$ km.

4. 1 km = 1000 m, so the conversion factor from kilometers to meters is 1000 m/1 km.
   Hence $7.8 \cdot 10^8$ km $\cdot \dfrac{1000 \text{ m}}{1 \text{ km}} = \dfrac{7.8 \cdot 10^8 \cdot 10^3 \text{ m}}{1} = 7.8 \cdot 10^{11}$ m.

5. 93,000,000 miles $\cdot$ 5,280 ft/mile $\cdot$ 2 dollars/ft $\approx$ $9.3 \cdot 10^7 \cdot 5.3 \cdot 10^3 \cdot 2$ miles $\cdot$ (ft/mile) $\cdot$ (dollars/ft) $\approx 99 \cdot 10^{10}$ dollars = $9.9 \cdot 10^{11}$ dollars or 990,000,000,000 dollar bills—almost a trillion dollars.

Algebra Aerobics 7.5a

1. **a.** 9; **b.** 12

2. **a.** $\sqrt{9x} = (9x)^{1/2} = 9^{1/2}x^{1/2} = 3x^{1/2}$;

   b. $\sqrt{\dfrac{x^2}{25}} = \left(\dfrac{x^2}{25}\right)^{1/2} = \dfrac{(x^2)^{1/2}}{25^{1/2}} = \dfrac{x^{2/2}}{5} = \dfrac{x^1}{5} = \dfrac{x}{5}$

3. **a.** $S = \sqrt{30 \cdot 60} = \sqrt{1800} \approx 42$ mph

   b. $S = \sqrt{30 \cdot 200} = \sqrt{6000} \approx 77$ mph

4. **a.** 5 and 6; **b.** 9 and 10

Algebra Aerobics 7.5b

1. **a.** 3; **b.** 2 **c.** $\frac{1}{2}$

2. **a.** 6.00; **b.** 72.0

3. $r = \sqrt[3]{\dfrac{3 \cdot 2 \text{ feet}^3}{4\pi}} = \sqrt[3]{\dfrac{3 \text{ feet}^3}{2\pi}} \approx \sqrt[3]{0.478 \text{ feet}^3} \approx 0.78$ feet

   Since 1 ft = 12 in., we can use a conversion factor of 1 = (12 in)/(1 foot) to get the radius of the balloon equal to 0.78 ft $\cdot$ (12 in)/(1 foot) or about 9.4 in.

Algebra Aerobics 7.5c

1. **a.** $2^{1/2}2^{1/3} = 2^{5/6}$ or $\sqrt[6]{32}$;

   b. $3^{1/2}9^{1/3} = 3^{1/2}3^{2/3} = 3^{7/6}$ or $\sqrt[6]{3^7} = 3\sqrt[6]{3}$

   c. $5^{1/2}5^{1/4} = 5^{3/4}$ or $\sqrt[4]{5^3} = \sqrt[4]{125}$

2. **a.** $\dfrac{2^{1/2}}{2^{1/3}} = 2^{1/2-1/3} = 2^{1/6}$ or $\sqrt[6]{2}$;

   b. $\dfrac{2^1}{2^{1/4}} = 2^{1-1/4} = 2^{3/4}$ or $\sqrt[4]{2^3} = \sqrt[4]{8}$

   c. $5^{1/4-1/3} = 5^{-1/12}$ or $\dfrac{1}{\sqrt[12]{5}}$

3. **a.** $c = 17.1 \cdot (0.25)^{3/8} = 17.1 \cdot (0.59) = 10.2$ cm

b. $c = 17.1 \cdot (25)^{3/8} = 17.1 \cdot (3.34) = 57.2$ cm

Algebra Aerobics 7.6a

1. Your salary is $1,000,000, and Henry's is $1,000.

2. **a.** Since $\dfrac{\text{radius of the Milky Way}}{\text{radius of the sun}} = \dfrac{10^{21}}{10^{9}} = 10^{21-9} =$ 10^{12}, the radius of the Milky Way is 12 orders of magnitude larger than the radius of the sun, or equivalently the radius of the sun is 12 orders of magnitude smaller than the radius of the Milky Way.

b. Since $\dfrac{\text{radius of a proton}}{\text{radius of the hydrogen atom}} = \dfrac{10^{-15}}{10^{-11}} =$ $10^{-15-(-11)} = 10^{-4}$, then the radius of the proton is four orders of magnitude smaller than the radius of the hydrogen atom, or equivalently the radius of the hydrogen atom is four orders of magnitude larger than the radius of a proton.

Algebra Aerobics 7.6b

1. The Armenian earthquake had tremors 10 times larger than those in Los Angeles.

2. The maximum tremor size of the Hawaiian earthquake of 1983 was 100 times smaller than the maximum tremor size of the largest earthquake.

Algebra Aerobics 7.6c

1. **a.** $6,370,000 \approx 10,000,000 = 10^7 \rightarrow$ plot it at 10^7

b. $0.000\,000\,7 = 7 \cdot 10^{-7} \approx 10 \cdot 10^{-7} = 10^{-6} \rightarrow$ plot it at 10^{-6}

2. **a.** plot at $10^7 \cdot 10^{-2} = 10^5$; **b.** plot at $10^{-6} \cdot 10^6 = 10^0$

Algebra Aerobics 7.7a

1. **a.** Since $10,000,000 = 10^7$, then $\log 10,000,000 = 7$.

b. Since $0.000\,000\,1 = 10^{-7}$, then $\log 0.000\,000\,1 = -7$.

c. Since $10,000 = 10^4$, then $\log 10,000 = 4$.

d. Since $0.0001 = 10^{-4}$, then $\log 0.0001 = -4$.

2. **a.** $100,000 = 10^5$; **b.** $0.000\,000\,01 = 10^{(-8)}$;

c. $10 = 10^1$

Algebra Aerobics 7.7b

1. **a.** Since $10^{0.4} \approx 2.511886$ and $10^{0.5} \approx 3.162278$, then $\log 3 \approx 0.5$. A calculator gives $\log 3 \approx 0.477121255$.

b. Since $10^{0.7} \approx 5.011872$ and $10^{0.8} \approx 6.30957$, then $\log 6 \approx 0.8$. A calculator gives $\log 6 \approx 0.77815125$.

c. Since $10^{0.8} \approx 6.30957$, then $\log 6.37 \approx 0.8$. A calculator gives $\log 6.37 \approx 0.804139432$

2. **a.** Write 3,000,000 in
scientific notation $\qquad$ $3,000,000 = 3 \cdot 10^6$
and substitute $10^{0.48}$ for 3 $\qquad$ $3,000,000 \approx 10^{0.48} \cdot 10^6$
then combine powers $\qquad$ $3,000,000 \approx 10^{6.48}$
and rewrite as a logarithm $\qquad$ $\log 3,000,000 \approx 6.48$

A calculator gives $\log 3,000,000 \approx 6.477121255$

b. Write 0.006 in scientific $\qquad$ $0.006 = 6 \cdot 10^{-3}$
notation
and substitute $10^{0.78}$ for 6 $\qquad$ $0.006 \approx 10^{0.78} \cdot 10^{-3}$
then combine powers $\qquad$ $0.006 \approx 10^{0.78-3} \approx 10^{-2.22}$
and rewrite as a logarithm $\qquad$ $\log 0.006 = -2.22$

A calculator gives $\log 0.006 \approx -2.22184875$.

3. **a.** $0.000\,000\,7 = 10^{\log 0.0000007} \approx 10^{-6.1549}$

b. $780,000,000 = 10^{\log 780,000,000} \approx 10^{8.892}$

c. We can only calculate logs of positive numbers, so $0.0042 = 10^{\log 0.0042} \approx 10^{-2.3768}$ and $-0.0042 \approx -10^{-2.3768}$

d. $-5,400,000,000 = -10^{\log (5,400,000,000)} = -10^{10.732}$.

CHAPTER 8

Algebra Aerobics 8.1a

1.

t	$N = 10 + 3t$	$N = 10 \cdot 3^t$
0	10	10
1	13	30
2	16	90
3	19	270
4	22	810

2.

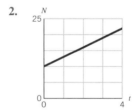

Since function is linear, the slope is constant ($m = 3$). The vertical intercept is (0, 10).

3.

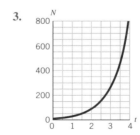

The slope is not constant, since for example, the average rate of change between 0 and 1 is $(30 - 10)/1 = 20$, and between 1 and 2 is $(90 - 30)/1 = 60$. The vertical intercept is $(0, 10)$.

4.

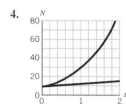

The exponential function dominates the linear function. It rises faster than the linear function, and the rate at which it rises also increases.

Algebra Aerobics 8.1b

x	$f(x) = 3^x$
-3	$3^{-3} = 1/27 \approx 0.037$
-1	$3^{-1} = 1/3 \approx 0.333$
0	$3^0 = 1$
1	$3^1 = 3$
3	$3^3 = 27$

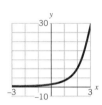

There is no value of x for which $f(x) \leq 0$. Hence, the function graph never lies below or touches the x-axis.

As $x \to +\infty, f(x) \to +\infty$. As $x \to -\infty, f(x) \, -\infty$.

c. exponential growth; $(0, 0.3)$

d. not exponential (linear)

e. exponential growth; $(0, 1)$

f. not exponential (quadratic)

2. The annual growth rate is 4.5% or 0.045 in decimal form. The annual growth factor is $1 + 0.045$ or 1.045. If n = number of years since 1996, and M = market size in billions of dollars, then \$34.8 billion is the initial amount (the sales when $n = 0$). An exponential model would be

$$M = 34.8 \cdot (1.045)^n$$

Since the predictions were only for 4 yr beyond 1996, we must restrict the domain to $0 \leq n \leq 4$. The year 2000 is 4 years beyond 1996, so $n = 4$, which lies within our domain. The projected market size M for the years between 1996 and 2000 are

Years since 1996	0	1	2	3	4
Market size (billions of \$)	\$34.8	\$36.4	\$38.0	\$39.7	\$41.5

3.

Exponential Function	Initial Value	Growth or Decay?	Growth or Decay Factor	Growth or Decay Rate
$A = 4(1.03)^t$	4	growth	1.03	0.03 or 3%
$A = 10(0.98)^t$	10	decay	0.98	0.02 or 2%
$y = 1000(1.005)^x$	1000	growth	1.005	0.005 or 0.5%
$y = 30(0.96)^x$	30	decay	0.96	0.04 or 4%
$A = 50{,}000(1.0705)^x$	\$50,000	growth	1.0705	0.0705 or 7.05%
$y = 200(0.51)^x$	200 g	decay	0.51	0.49 or 49%

Algebra Aerobics 8.2

1. a. growth (base of $4 > 1$); **b.** decay (base of $0.99 < 1$);

c. decay (base of $2/3 < 1$)

2. The graphs of $y = 3^x$ and $y = 5 \cdot 3^x$ both represent exponential growth. They have y intercepts of 1 and 5, respectively. The graph of $y = 5 \cdot 3^x$ lies above the graph of $y = 3^x$.

Algebra Aerobics 8.3

1. a. exponential growth; $(0, 100)$

b. exponential growth; $(0, 4)$

Algebra Aerobics 8.4a

1. a. $70/2 = 35$ yr; **b.** $70/0.5 = 140$ months

c. $70/8.1 = 8.64$ yr;

d. growth factor $= 106.5\% \to$ growth rate $= 6.5\%$
$70/6.5 = 10.77 \approx 11$ yr

2. a. $70/R = 10 \Rightarrow 70 = 10R \Rightarrow R \approx 70/10 = 7\%$/yr;

b. $R \approx 70/5 = 14\%$/min;

c. $R \approx 70/25 = 2.8\%$/sec

3. a. Since $a = 0.95$, $r = 0.05$, so $R = 5\%$. So $70/5 = 14$ months $=$ half-life

b. Since $a = 0.75$, $r = 0.25$, so $R = 25\%$. So $70/25 = 2.8$ sec $=$ half-life

c. $70/35 = 2$ yr $=$ half-life

Algebra Aerobics   8.4b

1. a. $70/3 \approx 23.3$ yr; **b.** $70/5 = 14$ yr; **c.** $70/7 = 10$ yr

2. a. $70/5 = 14\%$; **b.** $70/10 = 7\%$; **c.** $70/7 = 10\%$

3. a. $P = \$1000(1.04)^n$ **b.** $P = \$1000(1.11)^n$

c. $P = \$1000(2.10)^n$

4. If inflation is 10%/month, then what cost 1 cruzeiros this month would cost 1.10 cruzeiros next month. We have 1 cruzeiros $\approx$ 91% of 1.10 cruzeiros (since $1/1.1 \approx 0.91$), so a month later 1 cruzeiros would only be worth 0.91 cruzeiros or 91% of its original value. Thus, the decay factor is 0.91. So the exponential decay function $Q = 100\,(0.91)^n$, gives the purchasing power of 100 of today's cruzeiros at n months in the future.

When $n = 3$ months, then Q, the value of 100 of today's cruzeiros, will be $100\,(0.91)^3 \approx 100\,(0.75) = 75$ cruzeiros. When $n = 6$ months, then Q, the value of 100 of today's cruzeiros, will be $100(0.91)^6 \approx 100(0.57) = 57$ cruzeiros. When $n = 12$ months or 1 yr, then Q, the value of 100 of today's cruzeiros will be $100(0.91)^{12} \approx 100(0.32) = 32$ cruzeiros.

With a 10% monthly inflation rate, the value of a cruzeiros will shrink by more than two-thirds by the end of a year.

Algebra Aerobics   8.5

Judging from the graph, the number of *E. coli* bacteria grows by a factor of 10 (for example, from 10 to 100, or 100,000 to 1,000,000) in a little over 3 time periods. From the equation $N = 100 \cdot 2^t$, we know that every three time periods, the quantity is multiplied by 2^3 or 8. Every four time periods, the quantity is multiplied by 2^4 or 16. So the answers are consistent, since somewhere between three and four time periods the quantity should be multiplied by 10 (which is between 8 and 16).

CHAPTER 9

Algebra Aerobics   9.1

1. a. The surface area increases by a factor of 4 and the volume by a factor of 8.

b. The volume will grow faster than the surface area, so the ratio of surface area/volume will decrease as the radius increases.

Algebra Aerobics   9.2a

1. a. y is directly proportional to x^2;

b. not directly proportional;

c. y is directly proportional to x^{-2}

2. a. $g(2) = 40$; $g(4) = 320$; $g(4)$ is eight times larger than $g(2)$;

b. $g(5) = 625$; $g(10) = 5000$; $g(10)$ is eight times larger than g(5);

c. The value of $g(x)$ is eight times larger if x is doubled;

d. The value of $g(x)$ is eight times smaller if x is halved

3. a. y is equal to 3 times the fifth power of x;

b. y is equal to 2.5 times the cube of x;

c. y is equal to one-fourth times the fifth power of x

4. a. y is directly proportional to x^5 with a proportionality constant of 3;

b. y is directly proportional to x^3 with a proportionality constant of 2.5;

c. y is directly proportional to x^5 with a proportionality constant of 1/4

5. a. $R = \sqrt{\dfrac{P}{a}}$;

b. $h = \dfrac{3V}{\pi r^2}; r = \sqrt{\dfrac{3V}{\pi h}}$;

c. $Z = \dfrac{Y}{(a^2 + b^2)}; a = \sqrt{\dfrac{Y - Zb^2}{Z}}$

Algebra Aerobics   9.2b

1. a. $f(2) = 32$
$f(-2) = -32$
$f(s) = 4s^3$
$f(3s) = 108s^3$

b. $g(2) = -32$
$g(-2) = 32$
$g\left(\dfrac{t}{2}\right) = \dfrac{-t^3}{2}$
$g(5t) = -500t^3$

2. The domain in all cases is all real numbers.

a. $A = \pi r^2$ is a power function with $k = \pi$ and $p = 2$.

b. $y = z^5$ is a power function with $k = 1$ and $p = 5$.

c. $z = w^5 + 10$ is not a power function.

d. $y = 5^x$ is not a power function.

e. $y = 3x^5$ is a power function with $k = 3$ and $p = 5$

3.   a.

x	$f(x) = 4x^2$	$g(x) = 4x^3$
-4	64	-256
-2	16	-32
0	0	0
2	16	32
4	64	256

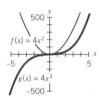

b.   As $x \to +\infty$, both $f(x)$ and $g(x) \to +\infty$.

c.   As $x \to -\infty$, $f(x) \to +\infty$, but $g(x) \to -\infty$.

d.   The domain of both functions is all the real numbers, the range of g is also all the numbers; however, the range of f is only all the nonnegative numbers.

e.   They intersect at the origin and at the point $(1, 4)$.

f.   All values greater than $x = 1$

Algebra Aerobics   9.3

1.   a.

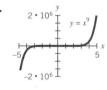

b.   no answer required

2.   i. $y = x$ and $y = 4x$ and $y = -4x$

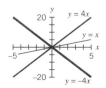

ii. $y = x^4$ and $y = 0.5x^4$ and $y = -0.5x^4$

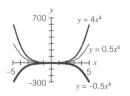

b.   check against above graphs

Algebra Aerobics   9.4

1.   a.   Graph of $y = 4^x$ and $y = x^3$

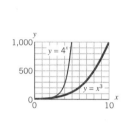

x	$y = 4^x$	$y = x^3$
0	1	0
1	4	1
2	16	8
3	64	27
4	256	64
5	1024	125
6	4096	216
7	16,384	343

$y = 4^x$ always dominates $y = x^3$

2.   a. $y = 2^x$ dominates eventually

b. $y = (1.000005)^x$ dominates eventually

Algebra Aerobics   9.5

1.   a.   degree 5 $f(-1) = -26$

b.   degree 3   when $x = -1, y = -26$

c.   degree 4 $g(-1) = -22$

d.   degree 2   when $x = -1, t = -9$

2.   a. $h(x) = x^4 - 3x^2 + 2x^2 + 4 = x^4 - x^2 + 4$

b. $h(0) = 4, h(2) = 16, h(-2) = 16$

Algebra Aerobics   9.6a

1.   The volume will increase by a factor of four.

2.

x	$r(x)$
0.01	600
0.25	24
0.50	12
1.00	6
2.00	3
5.00	1.2
10.00	0.6

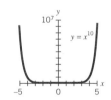

a. The function's value decreases as x increases.

b. The function blows up rapidly as x approaches zero.

c. No, the function is undefined at $x = 0$.

Algebra Aerobics   9.6b

1. a. i. $15/x^3$; **ii.** $-10/x^4$; **iii.** $3.6/x^1$

   b. i. $1.5x^{-2}$; **ii.** $-6x^{-3}$; **iii.** $\dfrac{-2}{3}x^{-2}$;

2. $t = kd^{-2}$ or $t = \dfrac{k}{d^2}$ for some constant k

3. The light is eight times as bright.

4. i. $g(x)$ is one-sixteenth its original value.

   ii. $g(x)$ is 16 times greater.

5. a.

x	$f(x) = 1/x^2$	$g(x) = 1/x^3$
0.01	10,000	1,000,000
0.25	16	64
0.50	4	8
1	1	1
2	0.25	0.125
10	0.01	0.001

b. The functions are not defined when x is zero. They are defined for negative values of x.

c. As $x \to +\infty$, both $f(x)$ and $g(x) \to 0$. When $x > 1$ as $x \to 0$, both $f(x)$ and $g(x) \to +\infty$.

Algebra Aerobics   9.7

1. a. $1/x^2$; value at 2 is 0.25

   b. $1/x^3$; value at 2 is 0.125

   c. $4/x^3$; value at 2 is 0.5

   d. $-4/x^3$; value at 2 is -0.5

2. a. $y = x^{-10}$

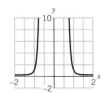

b. $y = x^{-11}$

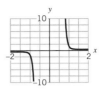

3. a. $y = x^{-2}$ and $y = x^{-3}$

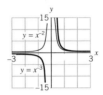

The graphs intersect at $(1, 1)$. As $x \to +\infty$, or $\to -\infty$, both graphs approach the x-axis. When $x > 0$, as $x \to 0$, both graphs "blow up" to $+\infty$. When $x < 0$, as $x \to 0$, the graph of $y = x^{-2}$ approaches $+\infty$, and the graph of $y = x^{-3}$ approaches $-\infty$.

b. $y = 4x^{-2}$ and $y = 4x^{-3}$

The graphs intersect at $(1, 4)$. As $x \to +\infty$ or $x \to -\infty$, both graphs approach the x-axis. When $x > 0$, as $x \to 0$, both graphs "blow up" and approach $+\infty$. When $x < 0$, as $x \to 0$, the graph of $y = 4x^{-2}$ approaches $+\infty$, and the graph of $y = 4x^{-3}$ approaches $-\infty$.

CHAPTER 10

Algebra Aerobics   10.2
See the table on page 611. The average velocity increases.

Time (sec)	Distance Fallen (cm)	Average Rate of Change (average velocity)
0.0000	0.00	N/A
0.0333	3.75	$\dfrac{(3.75 - 0.00)}{(0.0333 - 0.0000)} \approx 113$ cm/sec
0.0667	8.67	$\dfrac{(8.67 - 3.75)}{(0.0667 - 0.0333)} \approx 147$ cm/sec
0.100	14.71	$\dfrac{(14.71 - 8.67)}{(.1000 - .0667)} \approx 181$ cm/sec
0.1333	21.77	$\dfrac{(21.77 - 14.71)}{(.1333 - .1000)} \approx 212$ cm/sec
0.1667	29.90	$\dfrac{(29.90 - 21.77)}{(.1667 - .1333)} \approx 243$ cm/sec

Algebra Aerobics   10.3

1. **a.** $d = 1/2gt^2 + v_0t = 490t^2 + 125t$

   b. i. $490 + 125 = 615$ cm

   ii. $4410 + 375 = 4785$ cm

   c. $d = 490t^2 + 75t$

2. $d = (1/2)gt^2 + vt$
   cm = (cm/sec^2)(sec^2) + (cm/sec)(sec)
   cm = (cm) + (cm)
   cm = cm

3. m = (m/sec^2)(sec^2) + (m/sec)(sec)
   m = (m) + (m)
   m = m

Algebra Aerobics   10.4

1. **a.** $d = 1/2(980)t^2 + 50t$ or $d = 490t^2 + 50t$
   $v = 980t + 50$

   b.
time	distance	velocity
1 sec	540 cm	1030 cm/sec
2.5 sec	3187.5 cm	2500 cm/sec

2. **a.** $d = 1/2(32)t^2 + 20t$ or $d = 16t^2 + 20t$
   $v = 32t + 20$

   b.
time	distance	velocity
0.5 sec	14 ft	36 ft/sec
2 sec	104 ft	84 ft/sec

3. The formula is in meters as $4.9 = (1/2) 9.8$, which is g in meters.

Algebra Aerobics   10.5

1. Initial velocity of 500 meters per second in the same direction as gravity (downwards)

2. **a.** Initial height of 300 m

   b. Initial velocity of 50 m/sec in the opposite direction of gravity (i.e., upwards)

3. **a.** $t = 0.1$ $h = 200(0.1) - 4.9(0.1)^2 =$
   $200(0.1) - 4.9(0.01) = 19.951$ m

   $t = 2$ $h = 200(2) - 4.9(2)^2 = 200(2) - 4.9(4) =$
   380.4 m

   $t = 10$ $h = 200(10) - 4.9(10)^2 =$
   $200(10) - 4.9(100) = 1510$ m

   b.
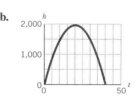

   c. approximately 41 sec; maximum height is approximately 2000 m

4.
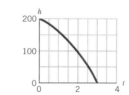

approximately 3.1 sec before hitting the ground

CHAPTER 11

Algebra Aerobics   11.1a

1. Vertex: $(0, 2)$; axis of symmetry is $x = 0$; concave down; two x-intercepts; the y-intercept is $(0, 2)$

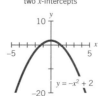

Vertex: (0, 2)
two x-intercepts

$y = -x^2 + 2$

Axis of symmetry is $x = 0$
and the y-intercept
is (0, 2)

2. vertex: $(-1, 0)$; axis of symmetry is $x = -1$; concave up; one x-intercept; the y-intercept is $(0, 1)$.

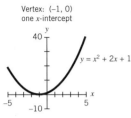

3. vertex: $(2, 1)$; axis of symmetry is $x = 2$; concave up; no x-intercepts; the y-intercept is $(0, 3)$.

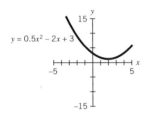

Algebra Aerobics   11.1b

1. a. min; narrower than $y = x^2$

   b. min; broader

   c. max; narrower

   d. max; broader

2. d, f, a, b, c, e

3. b is narrower than a; both open up

4. b is 6 units higher than a; both have the same shape and both open up

5. have same shape; a opens up; b opens down

6. b is much broader than a; both open down

7. Both are equally narrow. Both open down, (b) is 3 units higher than (a)

Algebra Aerobics   11.2

1. a. $(0, -4)$

   b. $(0, 6)$

   c. $(0, 1)$

2. a. vertex is $(0, 3)$

   b. vertex is $(0, 3)$

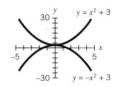

3. a. $(-1.5, -0.25)$

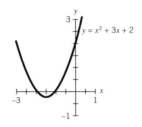

   b. $(1, 3)$

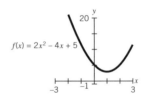

   c. $(-2, -3)$

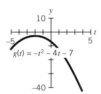

Algebra Aerobics   11.3a

1. a. $(0, 0)$; **b.** $(0, 5)$; **c.** $(0, -11/3)$

2. a. $(0, 80)$ initial position: 80 m

   b. $(0, 150)$ initial position: 150 cm

3.

	vertex	opens	number of x-intercepts
a.	$(-2, -11)$	up	2
b.	$(-0.25, 4.125)$	down	2
c.	$(0, 0)$	down	1
d.	$(0, -5)$	down	0

graph of (a):

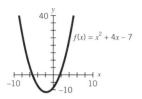

graph of (b):

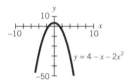

graph of (c):

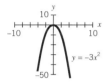

graph of (d):

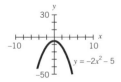

Algebra Aerobics   11.3b

1. **a.** discriminant $= 1 - 4(-5)(4) = 81 > 0 \Rightarrow$ two x-intercepts
   the function has two real zeros and hence the graph has two x-intercepts at $x = -1$ and $4/5$

   b. discriminant $= 784 - 4(4)(49) = 0 \Rightarrow$ one x-intercept
   the function has one real zero and hence the graph has one x-intercept at $x = 7/2 = 3.5$

   c. discriminant $= 25 - 4(2)(4) = -7 < 0 \Rightarrow$ no x-intercepts
   the function has two imaginary zeros at
   $$x = \frac{-5 \pm \sqrt{-7}}{4} = \frac{-5 \pm \sqrt{7}i}{4}$$

2. **a.** vertex is $(-0.1, 4.05)$; y-intercept $(0, 4)$

   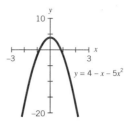

   b. vertex is $(3.5, 0)$; y-intercept $(0, 49)$

   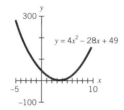

   c. vertex is $(-1.25, 0.875)$; y-intercept $(0, 4)$

   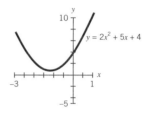

3. The first and third graphs (of $y = 2x^2 + 2$ and $y = -5x^2 + 3x - 1$) have no x-intercepts and hence those functions have no real zeros. The second graph (of $y = -x^2 + 2x + 3$) has 2 x-intercepts and hence the function has 2 real zeros. The fourth graph (of $y = 0.5x^2 + 5.0x + 12$) appears to have one x-intercept, and hence the function has one distinct real zero.

Algebra Aerobics   11.5

1. **a.** Since $c = 0$, the function can be factored easily as $y = t(-16t + 50)$ or equivalently as $y = -2t(8t - 25)$. In either case, $y = 0$. When $t = 0$ or $t = \dfrac{25}{8}$.

   b. $(y) = (x - 2)(x + 3)$. So $y = 0$ when $x = 2$ or $x = -3$.

   c. Since the discriminant is 41, which is not a perfect square, the function does not factor with integer coefficients. The zeros of the function given by the quadratic formula occur at $\dfrac{-1 \pm \sqrt{41}}{4}$. Since these are real numbers, they also represent the x-intercepts.

   d. $h(t) = 3(23 - 3t^2)$, but does not factor easily into two linear terms. To find the x-intercepts, we solve

$0 = 3(23 - 3t^2)$, giving $23 - 3t^2 = 0$ or $t^2 = \dfrac{23}{3}$ or
$t = \pm \sqrt{\dfrac{23}{3}} \approx \pm 2.77$.

e. $f(x) = -2(x^2 - 6x - 27)$ or $-2(x - 9)(x + 3)$. Setting $f(x) = 0$ to find x-intercepts, we get $x = 9$ or $x = -3$.

f. $g(x)$ does not factor easily. To find x-intercepts, we get $g(x) = 0$ and use the quadratic formula to get

$$x = \frac{-16 \pm \sqrt{256 - 1024}}{128} = \frac{-16 \pm \sqrt{-768}}{128}.$$

Since the discriminant, -768, is negative, there are no real roots and hence no x-intercepts.

Algebra Aerobics   11.6a

1. a. vertex is $(0, 0)$; **b.** vertex is $(-3, 0)$;

c. vertex is $(2, 0)$.

All vertices are on the x-axis. Vertex of (b) is vertex of (a) moved 3 units to the left. Vertex of (c) is vertex of (a) moved 2 units to the right.

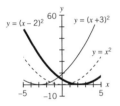

2. a. vertex is $(0, 0)$; **b.** vertex is $(1, 0)$;

c. vertex is $(-4, 0)$.

All vertices lie on the x-axis. The vertex of (b) is the vertex of (a) moved 1 unit to the right. The vertex of (c) is the vertex of (a) moved 4 units to the left.

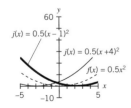

3. a. vertex is $(0, 0)$; **b.** vertex is $(-1.2, 0)$;

c. vertex is $(-0.9, 0)$.

All of the vertices are on the x-axis. The vertex of (b) is the vertex of (a) moved 1.2 units to the left. The vertex of (c) is the vertex at (a) moved 0.9 units to the right.

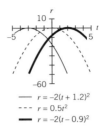

$$\text{---} \quad r = -2(t + 1.2)^2$$
$$\text{- - -} \quad r = 0.5t^2$$
$$\text{━━} \quad r = -2(t - 0.9)^2$$

Algebra Aerobics   11.6b

1. Notice how the forms where $k = 0$ the graphs in parts (b), (c) and (d) are simply the graph of the original function in part (a) shifted left or right. Notice also that when a k term is added on, the function is shifted up or down by that term similar to adding a c in the $a\,b\,c$ form.

In this case, the graph of part (b) is the graph of part (a) shifted 2 units to the right. The graph of part (c) is the graph of part (a) shifted 2 units to the right and 4 units up. The graph of part (d) is the graph of part (a) shifted 2 units to the right and down 3 units.

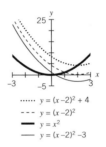

$$\cdots\cdots \quad y = (x - 2)^2 + 4$$
$$\text{- - -} \quad y = (x - 2)^2$$
$$\text{━━} \quad y = x^2$$
$$\text{---} \quad y = (x - 2)^2 - 3$$

2. The graph of part (b) is the graph of part (a) shifted to the left 3 units. The graph of part (c) is the graph of part (a) shifted 3 units to the left and down 1 unit. The graph of part (d) is the graph of part (a) shifted 3 units to the left and up 4 units.

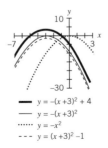

$$\text{━━} \quad y = -(x + 3)^2 + 4$$
$$\text{---} \quad y = -(x + 3)^2$$
$$\cdots\cdots \quad y = -x^2$$
$$\text{- - -} \quad y = (x + 3)^2 - 1$$

Algebra Aerobics   11.6c

1. a. $f(x) = (x + 1)^2 - 2$

b. $j(z) = 4z^2 - 8z - 6 = 4(z - 1)^2 - 10$

c. $h(x) = -3x^2 - 12x = -3(x + 2)^2 + 12$

2. a. $y = 2x^2 - 2x + 5.5$

b. $y = -\left(\dfrac{1}{3}\right)x^2 - \left(\dfrac{4}{3}\right)x + \dfrac{8}{3}$

3. a. $y = (x + 3)^2 - 2$

b. $y = 2(x + 1)^2 - 13$

4. a. vertex is $(-4, -5)$, vertical intercept $(0, 11)$; x-intercepts are at $-4 \pm \sqrt{5}$, i.e., at -1.764 and -6.236.

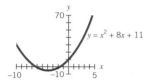

b. vertex is $\left(-\dfrac{2}{3}, -\dfrac{10}{3}\right)$ vertical intercept $(0, -2)$; x-intercepts are at $-\dfrac{2}{3} \pm \dfrac{\sqrt{10}}{3}$, i.e., at 0.387 and -1.721.

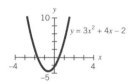

5. a. vertex is $(-5, -11)$ vertical intercept $(0, -8.5)$; x-intercepts are at $-5 \pm \dfrac{\sqrt{44}}{0.2}$, i.e., at 5.50 and -15.50.

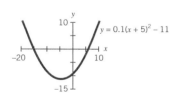

b. vertex is $(1, 4)$ vertical intercept $(0, 2)$; x-intercepts are at $1 \pm \sqrt{2}$, i.e., at 2.414 and -0.414.

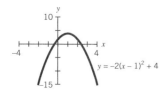

Algebra Aerobics 11.8a

1. a. Looking at the graph we can see that the graph only crosses the x-intercept once. It happens at about $x = 1.3$.

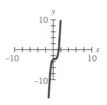

b. Clearly the graph does not intersect the x-axis.

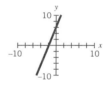

Algebra Aerobics 11.8b

1. a. degree is 1
y-intercept is $(0, 6)$
x-intercept is $(-2, 0)$

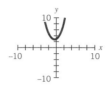

b. degree is 2
y-intercept is $(0, -4)$
$y = (x + 4)(x - 1)$
x-intercepts are $(1, 0)$ and $(-4, 0)$

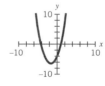

c. degree is 3
y-intercept is $(0, -75)$
$y = (x + 5)(x - 3)(2x + 5)$
so x intercepts are $(-5, 0)$, $(3, 0)$, and $(-2.5, 0)$

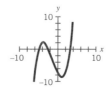

CHAPTER 12

Algebra Aerobics 12.1a

1. When $t = 0$, $M = 250$ When $t = 1$, $M = 750$. When $t = 2$, $M = 2,250$. When $t = 3$, $M = 6,750$.

2. **a.** $4 < t < 5$ **b.** $8 < t < 9$

3. **a.** Drawing a horizontal line across $y = 7$ until it hits the graph of the function, then from that point drawing a vertical line down until it hits the x-axis, we find an x-coordinate of approximately 1.8.

   b. Drawing a horizontal line across $y = 0.5$ until it hits the graph of the function, then from that point drawing a vertical line down until it hits the x-axis, we find an x-coordinate of approximately -0.5.

4. **a.** $2 < x < 3$ **b.** $-1 < x < 0$

Algebra Aerobics 12.1b

1. **a.** $\log (10^5/10^7) = \log (10^{-2}) = -2$ and $\log 10^5 - \log 10^7 = 5 - 7 = -2$.

   b. $\log (10^5 \cdot (10^7)^3) = \log (10^5 \cdot 10^{21}) = \log 10^{26} = 26$ and $\log (10^5) + 3 \log (10^7) = 5 + (3 \cdot 7) = 26$

2. $\dfrac{1}{2} [\log (2x - 1) - \log (x + 1)]$ 3. $\log \sqrt[3]{\dfrac{x}{x + 1}}$

4. **a.** $\log \dfrac{(x + 6)(x + 2)}{(x + 20)} = 0 \Rightarrow 10^0 = \dfrac{(x + 6)(x + 2)}{(x + 20)} \Rightarrow$ $x + 20 = x^2 + 8x + 12 \Rightarrow x = 1$

   b. $\log \dfrac{x + 12}{2x - 5} = 2 \Rightarrow 10^2 = \dfrac{x + 12}{2x - 5} \Rightarrow$ $200x - 500 = x + 12 \Rightarrow x = \dfrac{512}{199}$

5. $\log 10^3 - \log 10^2 = 3 - 2 = 1$. But $\dfrac{\log 10^3}{\log 10^2} = \dfrac{3}{2} = 1.5$.

Algebra Aerobics 12.1c

1. $6 = 2^t$ (dividing both sides by 10)
   $\log 6 = \log 2^t$ (taking the log of both sides)
   $\log 6 = t \log 2$ (using properties of logs)
   $t = \log 6/\log 2 \approx 0.778/0.301 \approx 2.58$

2. $7000 = 100 \cdot 2^t \Rightarrow 70 = 2^t \Rightarrow \log 70 = \log 2^t \Rightarrow \log 70 = t \log 2 \Rightarrow t = \log 70/\log 2 \approx 6.13$ time periods or approximately $(6.13)(20) = 122.6$ min for the bacteria count to reach 7000.
   $12{,}000 = 100 \cdot 2^t \Rightarrow 120 = 2^t \Rightarrow \log 120 = \log 2^t \Rightarrow \log 120 = t \log 2 \Rightarrow t = \log 120/\log 2 \approx 6.907$ time periods

or approximately $(6.907)(20) = 138.14$ min for the bacteria count to reach 12,000.

3. Using the rule of 70, since $R = 6\%$ yr, then $70/R = 70/6 \approx 11.7$ yr. More precisely, we have:

   $2000 = 1000(1.06)^t \Rightarrow 2 = 1.06^t \Rightarrow \log 2 = \log 1.06^t \Rightarrow$
   $\log 2 = t \log 1.06 \Rightarrow t = \log 2/\log 1.06 \approx 11.9$ yr

4. $1 = 100(0.5)^{t/28} \Rightarrow 0.01 = 0.5^{t/28} \Rightarrow \log 0.01 = \dfrac{t}{28} \log 0.5 \Rightarrow t = 28 \log 0.01/\log 0.5 \approx 186$ yr

Algebra Aerobics 12.2

1. **a.** $1000(1.085) = \$1{,}085$

   b. $1000 \left(1 + \dfrac{0.085}{4} \right)^4 = \$1{,}087.75$

   c. $1000 e^{0.085} = \$1{,}088.72$

2. **a.** $e^{0.04} = 1.0408 \Rightarrow 4.08\%$ is effective rate

   b. $e^{0.125} = 1.133 \Rightarrow 13.3\%$ is effective rate

   c. $e^{0.18} = 1.197 \Rightarrow 19.7\%$ is effective rate

3.

n	$1/n$	$1 + 1/n$	$(1 + 1/n)^n$
1	1	2	2
100	0.01	1.01	2.704813829
1000	0.001	1.001	2.716923932
1,000,000	0.000001	1.000001	2.718280469
1,000,000,000	0.000000001	1.000000001	2.718282052

The values for $(1 + 1/n)^n$ come closer and closer to the finite irrational number we define as e and are consistent with the value for e in the text of 2.71828.

Algebra Aerobics 12.3

1. **a.** 2;   **b.** -1;   **c.** 1/2;   **d.** -2;   **e.** 0

2. $\dfrac{1}{2} \ln (x + 2) - \ln x - \ln (x - 1)$

3. $\ln \dfrac{x}{(2x - 1)^2}$

4. $e^r = 1 + 0.064 = 1.064 \Rightarrow r = \ln 1.064 = 0.062 = 6.2\%$

5. $50{,}000 = 10{,}000 e^{0.078t} \Rightarrow 5 = e^{0.078t} \Rightarrow \ln 5 = 0.078t \Rightarrow t = \ln 5/0.078 \approx 20.6$ yr

6. **a.** $\ln e^{x+1} = \ln 10 \Rightarrow x + 1 = \ln 10 \Rightarrow x = -1 + \ln 10 \approx 1.30$

b. $\ln e^{x-2} = \ln 0.5 \Rightarrow x - 2 = \ln 0.5 \Rightarrow$
$x = 2 + \ln 0.5 \approx 1.31$

Algebra Aerobics   12.4

1. Since $\log x^2 = 2 \log x$ (by Property 3 of logarithms), the graphs of $y = \log x^2$ and $y = 2 \log x$ will be identical (assuming $x > 0$).

2. The graph of $y = -\ln x$ will be the mirror image of $y = \ln x$ across the x-axis.

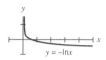

3. Acidic. If $4 = -\log [H^+]$, then $-4 = \log [H^+]$. So $[H^+] = 10^{-4}$. Since the pH is 3 less than pure water's, it will have a hydrogen ion concentration 10^3 or 1000 times higher than pure water's.

4. $N = 10 \log \left(\dfrac{1.5 \cdot 10^{-12}}{10^{-16}} \right) = 10 \log(1.5 \cdot 10^4) =$
$10 (\log 1.5 + \log 10^4) \approx 10 (0.176 + 4) = 10 (4.176) \approx 42$

5. Multiplying the intensity by $100 = 10^2$, corresponds to adding 20 to the decibel level.
Multiplying the intensity by $10,000,000 = 10^7$, corresponds to adding 70 to the decibel level.

Algebra Aerobics   12.5

1. **a.** decay; **b.** decay; **c.** growth

2. **a.** $e^x = 1.062 \Rightarrow x = \ln 1.062 \approx 0.060 \Rightarrow y = 1000 e^{0.06t}$

   b. $e^x = 0.985 \Rightarrow x = \ln 0.985 \approx -0.015 \Rightarrow y = 50 e^{-0.015t}$

3. Equation of the form $y = Ce^{rx}$, where $x =$ number of days, $C =$ initial amount of iodine-131, and $y =$ amount after x days. The points $(8, 2.40)$ and $(20, 0.88)$ satisfy the equation, so $2.40 = Ce^{8r}$ and $0.88 = Ce^{20r}$. So
$\dfrac{0.88}{2.40} = \dfrac{Ca^{20r}}{Ce^{8r}} \Rightarrow 0.367 = e^{12r} \Rightarrow \ln (0.367) =$
$\ln e^{12r} \Rightarrow -1.002 = 12 r \Rightarrow r = -0.084$, the decay rate.
Plugging this back in we get,
$2.40 = Ce^{-0.084 \cdot 8} \Rightarrow 2.40 = Ce^{-0.672} \Rightarrow$
$2.40 = C \cdot 0.511 \Rightarrow 2.40/0.511 = C \Rightarrow C = 4.70$ micrograms (μg)
So the exponential decay function is: $y = 4.70 e^{-0.084x}$.

Algebra Aerobics   12.6a

1. $y = 4x^3$ is a power function;
   $y = 3x + 4$ is a linear function;
   $y = 4 \cdot (3^x)$ is an exponential function.

2. and 4.

Function	Type of Plot on Which Graph of Function Would Appear as a Straight Line	Slope of Straight Line
$y = 3x + 4$	standard linear plot	$m = 3$
$y = 4 \cdot (3^x)$	semi-log	$m = \log 3$
$y = 4 \cdot x^3$	log-log	$m = 3$

Algebra Aerobics   12.6b

1. Note this is a log-log plot, so the slope of a straight line corresponds to the exponent of a power function
   Younger ages: slope $= 1.2 \Rightarrow$ original function is of the form $y = a \cdot x^{1.2}$, where $x =$ body height in cm, and $y =$ arm length in cm.
   Older ages: slope $= 1.0 \Rightarrow$ original function is of the form $y = a \cdot x^1$

2. **a.** exponential

   b. exponential

   c. power

Algebra Aerobics   12.7

1. Estimated surface area is about 200 cm^3. Computed area is $10 \cdot (70000)^{2/3} = 10 \cdot 1700 = 17,000 \text{ cm}^2$. Since $1 \text{ Kg} = 2.2 \text{ lb}$, then $70 \text{ Kg} = (70)(2.2) \text{ lb} = 154 \text{ lbs}$. Since $1 \text{ cm} = 0.394 \text{ in}, 1 \text{ cm}^2 = (0.394 \text{ in})^2 = 0.155 \text{ in}^2$. So $17,000 \text{ cm}^2 = (17000)(0.155) = 2,635 \text{ in}^2$.

2. **a.** power
   b. $y = ax^{-0.23}$
   c. When body mass increases by a factor of 10, heat rate is multiplied by $10^{-0.23}$ or about 0.215.

3. Estimated rate of heat production is somewhere between 10^3 and 10^4 kilocalories per day—say about 2,000–3,000 Kcal/day. Calculated value would be $151 (70)^{3/4} \approx 3,650$. These are roughly accurate values.

4. King Kong is a scaled up version of a gorilla. If it is 10 times larger, King Kong's weight would be 10^3 or 1,000 times more, but the ability to support the weight would only have increased by a factor of 10^2 or 100.

CHAPTER 1

3. a. Housing; 41.4%

   b. $5220.00

   c. Americans spend more on housing than anything else!

5. a. 4.0%, 2.5%, 1.75%, 2.2%, 12.9%

   b. In 2000 the largest concentration is in the 35–59 age range.

7. a. No

   b. During the final approach; during the load/unload phase.

   c. One cannot tell.

   d. The table does not tell about the kinds of planes that crash or the weather conditions under which planes crash.

11. He is correct, provided the person leaving has an I.Q. that is below the average I.Q. of the state left but above the average I.Q. of the state moved into.

13. a. $386/7 = 55.14$; 46; no mode.

   b. will not change the median of the list.

17. a. $\left(\sum_{i=1}^{5} x_i\right)/5$ **b.** $\left(\sum_{i=1}^{n} t_i\right)/n$

19. a. NE: 22.77; 9.20; MW: 19.85; 18.05

   b. Means are close; medians are far apart.

21. The mean net worth of a group, e.g., American families, is heavily biased upwards by the very high incomes of a relatively small subset of the group. The median net worth of such groups is not as biased. The two measures would be the same if net worths were distributed symmetrically about the mean.

23. Using midpoints: 34.987 years.; using "natural" values (i.e., 5, 15, . . . , 85): 35.464 years.

25. The population of China is substantially younger than that of the U.S.

27. a. There are far more people now than in 1971, and those who are most susceptible, the elderly, are now a greater proportion of the population than in 1971. Thus, it is not necessarily the case that proportionately twice as many

of them will have cancer. Also, better cancer detection tools could mean more frequent prevention of death. Lastly, not all cancers are fatal.

   b. The claims are not buttressed clearly enough by the data.

CHAPTER 2

1. a. 86–87, 89–90, 90–91, 91–92, 92–93, 93–94

   b. 87–88, 88–89, 93–94

   c. just below 12 billion dollars

   d. approximately 7.5 billion dollars

3. a. The shape depends a lot on the geographical location.

   b. Again the shape depends on the region.

   c. This would probably be a gradually rising and falling periodic curve.

   d. This is periodic and very dependent on latitude.

   e. The temperature of the coffee (according to Newton's Law of Cooling) goes down rapidly at first and then keeps going down but more slowly, eventually reaching room temperature.

   f. It is a mirror image of that for the cooling coffee with respect to the line representing a steady room temperature.

5. a. $(-1, 3)$ solves $y = 2x + 5$

   b. $(1, 0)$ and $(2, 3)$ solve $y = x^2 - 1$

   c. $(-1, 3)$ and $(2, 3)$ solve $x^2 - x + 1$

   d. $(1, 2)$ solves $y = 4/(x + 1)$

7. Juvenile arrests for murder:

   a. 1990–1991

   b. Approximately 1800 in 1988; 3500 in 1993.

   c. 1800 in 1988 and 3500 in 1993; $3500/1800 = 1.94$.

   d. The number of juvenile arrests has for the most part steadily risen from 1988 to 1993; but there is a noticeable dip going from 1993 to 1994.

9. a. $cst(gal) = 1.24 \cdot gal$ **b.** gal; cst.

c. function: each input of amount purchased gives a unique output in dollars.

d. $0 < \text{gal} \le 30; 0 < \text{cst} \le 37.20$

e. A small table and graph follow:

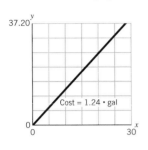

gal	cst
0	0.00
1	1.24
2	2.48
3	3.72
4	4.96
5	6.20
6	7.44
7	8.68

11. a. function

b. $\{1990, 1991, 1992, 1993, 1994, 1995\}$; $\{-0.5, 0, 1.2, 1.4, 2.3\}$.

c. 2.3; in 1991

d. increased: 1990–1991; 1993–1994. decreased from 1991–1993, 1994–1995.

e. not a function.

13. a. $5, -4, -3$ **b.** $x = \pm 2$

15. a. all real x **b.** $x \ge -2$ **c.** all real x

d. $x \ne -1$ **e.** all real x **f.** $x \ne \pm 2$

17. a. 2   **b.** -8 **c.** 10   **d.** -5

CHAPTER 3

1. a.

Year	Salary ($ millions)	Rate of Change over prior year
1987	0.41	N/A
1988	0.44	0.03
1989	0.50	0.06
1990	0.60	0.10
1991	0.85	0.25
1992	1.03	0.18
1993	1.08	0.05

b. The average rate was smallest in the 87–88. Its line segment slope is the least steep.

c. The average rate was greatest in the 90–91 period. The graph is steepest there.

3. a. A plot of the data is here.

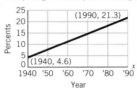

b. 16.7/50 or 0.334. Its units are percent completing 4 or more years of college education per year.

c. The average rate at which the percent of persons 25 years and older completing four years of college was 0.334 per year from 1940 to 1990.

5. a. 2   **b.** -0.5 **c.** 0

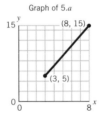
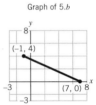
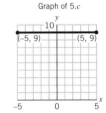

7. a. 0.47 years per calendar year.

b. Black females; statement like: Over the past 94 years the life expectancy of black females has increased on average far more than white females, white males, or black males.

9. a. 1.060; 1.178; 1.156

b. 89.7%; 77.98%; 91.96%

c. The rate of growth, e.g., is fastest for blacks.

d. 2010; 2019; 2007.

11. a. positive over $B < x < F$; negative over $F < X \le G$; 0 over $A \le x \le B$ and at $x = F$

b. positive over $B \le x \le C$ and $D \le x \le E$; negative over $E \le x \le G$; 0 over $A \le x \le B$ and $C \le x \le D$.

13. a. The number of newspapers published each year from 1915 to 1990 has gone down steadily, etc.

b. In 1920 there were 0.26 papers published per person. In 1990 there were 0.25 papers per person.

c.

Year	Avg. Rate for Newsp. Publ.	Avg. Rate for TV Stations
1960	−0.9	41.7
1970	−1.5	57.9
1980	−0.3	5.7
1990	−13.4	35.8

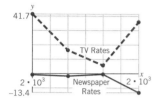

d. 1450 TV stations in 2000 if the growth rate from 1980 to 1990 prevailed.

e. Students might say something like the following: Fewer newspapers are surviving while the readership of these fewer goes up.

15. The time period at the right end of the scale: 1970–1974 is much shorter than the time periods for other awardings. The other periods counted over a 10-year period, not a 4-year period.

17. a. Visually, the rate of change of Medicare costs appears to be greater than the rate of change of doctor's bills during the period from 1963 to 1979.

b. But calculations show that the rate of change in Medicare costs was actually less than the rate of change in doctor's bills. For Medicare costs it was $(20 − 4)/16 = 1$ billion dollars per year. For doctor's bills it was $(48 − 9)/16 = 2.4$ billion dollars per year. This is more than twice the rate for Medicare costs.

19. a. and **b.** In the right column is the table on accumulated Federal Debt from 1945 to 1995. A third column shows the average rates of change of the accumulated Federal Debt over the previous interval and that table's line

graph. In simple terms, the third column of the table represents the average amount by which the debt has grown from year to year and not the debt amount itself. That amount is given in the second column. The rate itself has gone up and down and was actually negative for the first 5-year period.

Year	Billions of $	Annual Average Rate of Change of Accumulated Debt
1945	260	N/A
1950	257	−0.6
1955	274	3.4
1960	291	3.4
1965	322	6.2
1970	361	7.8
1971	408	47
1972	436	28
1973	466	30
1974	484	18
1975	541	57
1976	629	88
1977	706	77
1978	777	71
1979	829	52
1980	909	80
1981	994	85
1982	1137	143
1983	1371	234
1984	1564	193
1985	1817	253
1986	2120	303
1987	2346	226
1988	2601	255
1989	2868	267
1990	3206	338
1991	3599	393
1992	4003	404
1993	4410	407
1994	4644	234
1995	4961	317

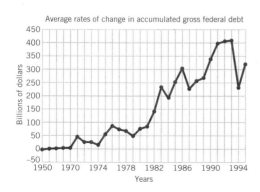

CHAPTER 4

1. The equation is: $y = 3x − 2$.
A small table is:

x	y
0	−2
1	1
2/3	0

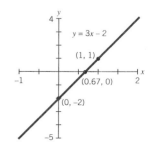

3. The equation is:
$y = 0x + 1.5$

x	y
0	1.5
−1	1.5
1	1.5

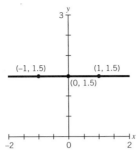

5. **a.** 9/5; 32

b. °F; °C

c.

C	F
0	32
20	68
100	212

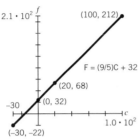

d. 1.8 or 9/5 is expected since the graph of the equation is a straight line with slope 9/5.

e. The graph is above to the right.

f. $(−4/5)C = 32$ or $C = 32 \cdot (−5/4) = −40°C$

7. If H is height in inches and W is the recommended weight in pounds, then a formula relating these two is:

$$W = 100 + 5(H − 60) \text{ or } W = 5H − 200.$$

A reasonable domain for this function is $54 \leq H \leq 78$ inches, assuming adult women. This takes into account very short and very tall women. The corresponding range is $70 \leq W \leq 190$.

Its graph with that domain is given on the right.

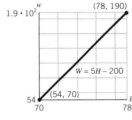

9. **A.** $m = 0$ $b = 4$ $y = 4$

B. $m = 1$ $b = 4$ $y = x + 4$

C. $m = 0.5$ $b = 0$ $y = 0.5x$

D. $m = −1$ $b = 4$ $y = −x + 4$

E. $m = 1$ $b = −4$ $y = x − 4$

F. $m = −1$ $b = −2$ $y = −x − 2$

13. Answers here are generic; student answers will be specific.

a. $y = mx + b_1$ and $y = mx + b_2$: same slope; different y-intercepts ($b_1 \neq b_2$)

b. $y = m_1x + b$ and $y = m_2x + b$: different slopes ($m_1 \neq m_2$); same y-intercept.

c. $y = m_1x$ and $y = m_2x$: different slopes ($m_1 \neq m_2$); y-intercepts are both 0.

d. $y = m_1x + b_1$ and $y = m_2x + b_2$, with $m_1 \cdot m_2 = −1$.

15. Graph **A**: different slopes; both slopes positive; same y-intercept.

Graph **B**: different slopes; one positive, one negative; same y-intercept.

Graph **C**: same slopes; both positive; different y-intercepts.

Graph **D**: different slopes; one positive, one negative; different y-intercepts.

17. **a.** $y = 4.20 − 1.62x$ **b.** $y = −2000 + 1400x$

c. $y = −386.7 + 0.2x$ **d.** $y = 12778.75 − 6.25x$

e. $y = 2.18 + 0.74x$ **f.** $y = 3.89 + 0.83x$

g. $y = 7.2$ **h.** not possible since its equation is $x = 275$

i. $y = −0.5 + 0.5x$

19. The line's equation is $y = −4 + 2x$. Its original graph (for comparison purposes) is given below. The graph on the left below it appears steeper than the original. The vertical scale was kept the same as in the original; the horizontal was enlarged. The graph below it on the right appears less steep than the original. The vertical scale was extended and the horizontal scale was kept the same as in the original.

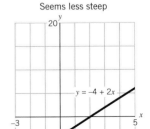

Seems more steep   Seems less steep

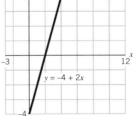
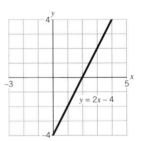

21. a. Best fit straight line for that data is: $y = -1.74 + 1.55x$ where x measures years from 1970. (See the graph below.); 1.55% per year.

b. Sometime in the year 2036.

c. It is likely that the percentage will level off instead of climbing continually, say, at 50%.

d. The distribution of dentists in the U.S.; the number of males and females enrolled in dental school currently; the percentage of dental graduates who are women; and the like.

Graph for 21a

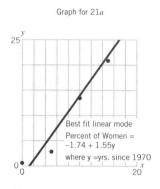

Graph for 21c

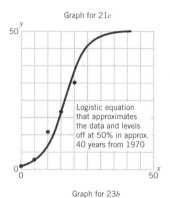

23.

Graph for 23a

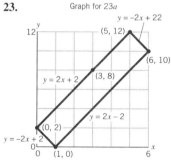

Graph for 23b

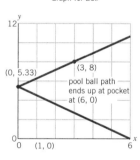

Graph for 23c

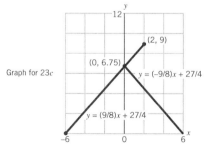

a. i. $(0, 2)$.

ii. $(1, 0)$; then $(6, 10)$; then $(5, 12)$.

iii. Yes; no; 2 is the ratio of the vertical length of the table to its horizontal length.

See the graphs above.

b. $(6, 0)$.

c. i. No

ii. No

iii. Trial and error should lead to points $(0, b)$ with $6 < b < 7$.

iv. Aim at the point whose coordinates are $(0, 6.75)$. See the graph.

CHAPTER 5

1. a. 2530

b. The rate of change of mean personal income with respect to years of education is $2530 per year.

c. $2350; $25,300.

d. Individual values and outliers are not well described by the regression line.

e. It deals only with mean total personal income and not with actual personal income; it is not really a very good predictor for those with very little education or for those with quite a bit (see, e.g., the outlier on the graph for 20 years of education).

f. Student answers will vary.

3.

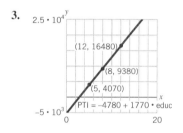

a. The rate of change of personal total income with respect to years of education.

b. $(5, 4070)$, $(8, 9380)$ and $(12, 16460)$; $(16460 - 9380)/(12 - 8) = 7080/4 = 1770$.

c. the same as that given in **a**.

d. The graph is given above.

e. A notable difference is the rate at which income increases for each; it is much more rapid for males.

5. a. It is the rate of change of the mean height of a son in inches for each increase of 1 inch in height of the father.

b. 66.754 in.; 71.398 in.

   c. There are only 17 different heights for fathers and for each the mean or average height of sons whose fathers have that height is given.

7. a. Using technology: for private colleges: $C = 6530 + 801 \cdot yr$ and for public colleges: $C = 1276 + 164 \cdot yr$; where yr = years from 1985. See the accompanying graphs below.

   b. The average rate of change of education cost per year for private colleges is approximately $800 and the same rate for public colleges is approximately $164.

   c. The estimated mean tuition cost for private college education in 2000 is: $18542 and for public college education that tuition cost is estimated at $3736. The figure at public colleges seems reasonable but recent rates of increase in tuition at private colleges make the figure for public colleges in 2000 seem rather low.

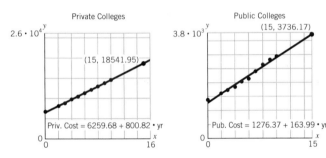

9. a. The equation of the best fit regression line is $y = 64.9 + 0.63x$. See the accompanying graph below.

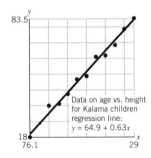

   b. x represents age in months; y the height in centimeters. 64.9 is the vertical intercept and represents the (extrapolated) height of a Kalama child at birth; the slope is 0.63 and represents the number of centimeters on average that a Kalama child would grow in one month. The correlation coefficient is 0.994 and thus the line is a good fit.

   c. 81.6 centimeters

11. a. See the scatter plot of data and graph of the regression line mentioned in **b** and **c** in the next column, where x = number of years since 1965 and y = % of population 18 and older that smokes.

i. $(24.2 - 42.4)/(28 - 0) = -0.65$

ii. $(24.2 - 25.5)/3 = 0.43$.

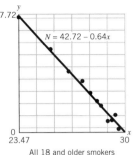

All 18 and older smokers regression line and data from 1965–1993

b. and c. The regression line's equation is $y = 42.72 - 0.64x$ and its graph is to the right. Its correlation coefficient is -0.993. Comparisons with hand generated lines will vary with each student.

d.

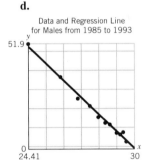

The regression line equation for males is $y = 51.1 - 0.89x$ and for females it is $y = 35.38 - 0.43x$; the correlation coefficient is -0.996 for males and -0.954 for females. In both, y measures years from 1985 and x is the % of the population being measured who smoke.

e. From 1985 to 1993: overall, among males and among females the % of smokers has been decreasing.

13. One way to check to see if there is hope of a linear relationship is to look at correlation coefficients of each possible pair of data columns. Those close to 1 in absolute value give promise. These are:

Life expectancy of males and the same for females has a cc of 0.947.

Life expectancy of males and infant mortality rates has a cc of -0.910.

Life expectancy of males and literacy % has a cc of 0.724.

Life expectancy of females and infant mortality rates has a cc of − 0.965.

Life expectancy of females and literacy % has a cc of 0.809.

Infant mortality and literacy % have a cc of − 0.855.

15.

Regression Line for 1880–1990 of Actual Farm Population

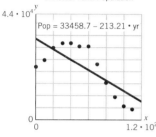

Regression Line for 1890–1990 of Farm Population as % of the Total Population

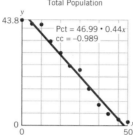

a. and b. If FP measures thousands of persons and FPP measures percentages of the whole population, then the regression line equations and correlation coefficients are: $FP = -213.21x + 33458.71$ (with a cc of -0.75) and $FPP = 49.99 - 0.44x$ (with a cc of -0.989). Note that x in both cases is measured in years since 1880.

c. Plotting from 1950 gives the regression line equation: $FP = -464.98x + 21107$ (with a cc of -0.97).

17. a. The regression line equation is $y = 171.59 - 1.48x$, where x measures years since 1972 and y represents the winning time for women in minutes; the cc is -0.83.

Graph for **17 a.** (1972 to 1996)

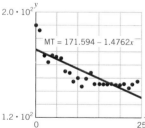

Graph for **17 c.** (1981 to 1996)

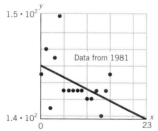

b. 115.35 minutes.

c. $y = 148 - 0.275x$, where x measures years since 1981. The cc is -0.44. It predicts that the winning time in 2010 is 137.55 minutes, which seems more reasonable than the prediction in **b**.

d. The Marathon times decrease slowly from 1970 to 1980. They rose a bit in the early '80s, peaking in '85. They have been more or less the same since.

19. a. $TempC = 100 - 0.001 \cdot x$, where x is feet above sea level. The correlation coefficient is -0.999. The plot of the data and the formula are given below.

b. $TempF = 212 - .002 \cdot x$, where x is feet above sea level. The correlation coefficient is -0.9998. Its plot is given below.

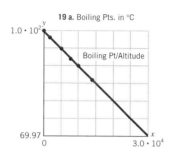

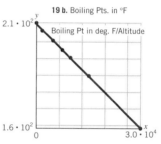

c. 171.36 °F; 214.85 °F. **d.** 56,700 ft. **e.** 100,000 ft.

CHAPTER 6

1. a. One: different slopes to straight-line graphs.

b. No solution: graphs are parallel lines.

c. One: different slopes to straight-line graphs.

3. a.

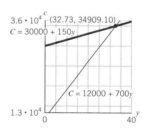

b. rate of change of total cost for gas heat in units of dollars per year since installation; rate of change in total cost for solar heat in the same units.

c. initial cost in dollars for installing the gas heating; initial cost in dollars for installing solar heating.

d. At $Y = 33$ and $C = 32900$; at 33 years the cost for both is the same.

e. $Y \approx 32.73$, $C \approx 34909.09$

f. Up to 32 years, 9 months and 12 days after installation solar heat is more expensive; after that gas heat is.

5. $1500 at 4%, $500 at 8%.

7. **a.** no; the starting salaries are the same; group B's salary increases at a faster rate.

b. no; the starting salary of group C is less than that of group A and its rate of growth per year is less than that of group A.

c. Those in group B will have the largest salary in 5 years; that salary will be $39,000.

9. **a.** $11x + 7y = 68$ (4)

b. $9x + 7y = 62$ (5)

c. $x = 3$ and $y = 5$ solves (4) and (5)

d. Thus $z = 2 \cdot 3 + 3 \cdot 5 - 11 = 10$

e. Thus the solution is $x = 3$, $y = 5$, $z = 10$ and the check is:
(1) $2 \cdot 3 + 3 \cdot 5 - 10 = 11$
(2) $5 \cdot 3 - 2 \cdot 5 + 3 \cdot 10 = 35$ and
(3) $1 \cdot 3 - 5 \cdot 5 + 4 \cdot 10 = 18$

11. **a.** Demand curve will shift to the left and the equilibrium point for the new demand curve will shift down to the left along the supply curve.

b. The supply curve shifts to the left as does the equilibrium point.

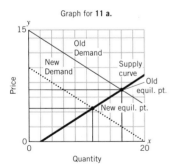

Graph for **11 a.**

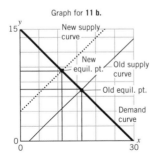

Graph for **11 b.**

13. **a.** The equilibrium point tends to move to the left and goes up.

b. The equilibrium point tends to move to the right and goes down.

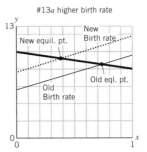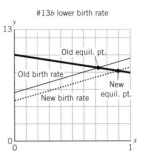

#13a higher birth rate   #13b lower birth rate

15. **a.** $f(x) = 1$ if $0 \leq x \leq 1$; and $f(x) = x$ if $1 < x \leq 3$.

b. $g(x) = -x + 1$ if $0 \leq x \leq 1$; and $g(x) = x - 1$ if $1 < x \leq 3$.

17. **a.** Table for f whose domain is all real x.

x	$f(x)$
0	5
3	5
8	5
10	5
15	15
20	25

b. Table for g whose domain is $-10 \leq t < 10$.

t	$g(t)$
-10	11
-5	6
1	0
3	3
5	5
10	10

19. Answers will vary from state to state. Check against the tax form itself.

21. **a. b.** and **c.**   Graph and table of the information asked for:

Income	Tax @ 8% rt.	Tax @ grad. rt.
0	0	0
20000	1600	0
50000	4000	1500
100000	8000	4000
150000	12000	9000
200000	16000	14000
300000	24000	24000

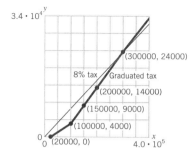

Time	D-Beg	D-Adv
0	0.00	0.00
1	0.06	0.06
2	0.12	0.13
3	0.18	0.19
4	0.23	0.25
5	0.29	0.31
6	0.35	0.38
7	0.41	0.44
8	0.47	0.50
9	0.53	0.56
10	0.58	0.63
11	0.64	0.71
12	0.70	0.80
13	0.76	0.89
14	0.82	0.98
15	0.88	1.06
16	0.93	1.15
17	0.99	1.24
18	1.05	1.33
19	1.11	1.41
20	1.17	1.50

d. $F(x) = 0.08x$ if $x \geq \$0$

e. $G(x) = 0$ if $\$0 \leq x \leq \$20,000$
$= 0.05(x - 20000)$ if $\$20,000 < x \leq \$100,000$
$= 4000 + 0.10(x - 100000)$ if $x > \$100,000$

f. and g. $x = \$300,000$, tax $= \$24,000$; below $\$300,000$ graduated tax is better; above $\$300,000$ flat tax is better.

h. It is likely that the new tax would be voted in, since more than half the people in that state would pay considerably less under the graduated tax.

23.

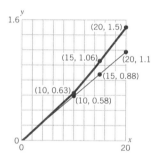

See the accompanying graphs for **a** and **b**. For **c** there is a table below for each function, for $T = 0, 1, \ldots, 20$

a. For $0 \leq T \leq 20$:

$$D_{beginner} = (3.5/60)T$$

since there is 1/60 of a hour in a minute; note that T is measured in minutes and $D_{beginner}$ is measured in miles.

b. For $0 \leq T \leq 10$:

$$D_{advanced} = (3.75/60)T$$

and for $10 < T \leq 20$:

$$D_{advanced} = 0.625 + (5.25/60)(T - 10)$$

c. The beginner and advanced function graphs do not intersect for $T > 0$. They do intersect at $T = 0$ and this point represents the start for each group.

CHAPTER 7

1. **a.** 10^6 **b.** 10^{-5}

 c. 10^9 **d.** 10^3

3. **a.** 10^{-1} **b.** $4 \cdot 10^3$

 c. $3 \cdot 10^{12}$ **d.** $6 \cdot 10^{-9}$

5. **a.** $2.9 \cdot 10^{-4}$ **b.** $6.54456 \cdot 10^2$

 c. $7.2 \cdot 10^5$ **d.** $1.0 \cdot 10^{-11}$

7. **a.** 723000   **b.** 0.000526

 c. 0.001   **d.** 1500000

9.

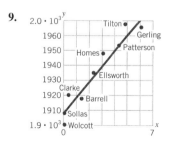

a. Wolcott (0.14, 1901); Sollas (0.2, 1908); and Clarke (0.23, 1921).

b. Barrell did it in 1918; the coordinates are: (2.4, 1918).

c. Using the coordinates of the points for Sollas (0.2, 1908) and Patterson, etc. (4.4, 1953), gives $m = 10.71$ and $b = 1907.8$; see accompanying graph.

d. One meaning for the slope is that for each increase of a billion years in the estimated age of Earth, time advances—on average—10.71 years.

11. a. $7.25 \cdot 10^{25}$ **b.** $7.25 \cdot 10^{-21}$ **c.** $1.3793 \cdot 10^{-26}$

    d. $-7.25 \cdot 10^{25}$ **e.** $-7.25 \cdot 10^{-21}$

13. a. 10^7 **b.** $11000 = 1.1 \cdot 10^4$

    c. $2000 = 2.0 \cdot 10^3$ **d.** x^{15}

    e. x^{50} **f.** $1.6409 \cdot 10^4$

    g. z^5 **h.** 1

    i. 3^{-1} **j.** 4^{11}

15. $1.4112 \cdot 10^{21}$ miles.

17. a. $3.72736 \cdot 10^{12}$ feet   **b.** $3.2736 \cdot 10^{-15}$ feet

19. a. $1.5 \cdot 10^8$ km and $1.08 \cdot 10^8$ km.   **b.** $7.2 \cdot 10^{-1}$

    c. 0.72 a.u.   **d.** 39.3 a.u.

21. a. and **b.** 27.79 kg/m^2; overweight.   **c.** Student answers will vary.   **d.** He is correct.

23. a. 4   **b.** -4 **c.** 1.414

    d. not defined   **e.** 2.8284   **f.** 1

25. a. between 3 and 4   **b.** between 4 and 5

    c. between 6 and 7

27. a. 31.342   **b.** 70.166   **c.** 1762.496

29. a. 5 orders of magnitude larger.

    b. 36 orders of magnitude larger

    c. 6 orders of magnitude larger

31. a. 10^{-7} cm   **b.** 10^8 cm   **c.** 15 orders of magnitude

    d. 10^3 Å   **e.** 10 Å

33. a. $\log_{10}(100) = 2$ **b.** $\log_{10}(10000000) = 7$

    c. $\log_{10}(0.01) = -2$ **d.** $10^1 = 10$

    e. $10^4 = 10000$ **f.** $10^{-4} = 0.0001$

35. and **36. a.** $1 < \log 11 < 2$ (since $10 < 11 < 100$ and $\log 11 \approx 1.0414$)

b. $4 < \log_{10} 12000 < 5$ (since $10000 < 12000 < 100000$ and $\log_{10} 12000 \approx 4.0792$)

c. $-1 < \log 0.125 < 0$ (since $0.1 < 0.125 < 1$ and $\log_{10} 0.125 \approx -0.9031$)

37. a. to the left 3 units

    b. to the left by approximately 0.3 units.

    c. moving to the right by 0.3 units

    d. moving to the right by 1 unit.

39. a. 7

    b. 2.85

    c. $3.16 \cdot 10^{-12}$

    d. A higher pH means a smaller hydrogen ion concentration.

    e. neutral, acidic, and acidic.

CHAPTER 8

1. a. **i.** For $y = 2^x$: $C = 1$ and $a = 2$; for $y = 5^x$: $C = 1$ and $a = 5$; for $y = 10^x$: $C = 1$ and $a = 10$

    ii. growth for all three; the first multiplies y by 2; the 2nd by 5; the 3rd by 10.

    iii.

x	2^x	x	5^x	x	10^x
-1	0.5	-1	0.2	-1	0.1
0	1	0	1	0	1
1	2	1	5	1	10
3	8	2	25	2	100

    iv. The graphs of these functions are illustrated on the same coordinate system below. Predictions will vary from student to student.

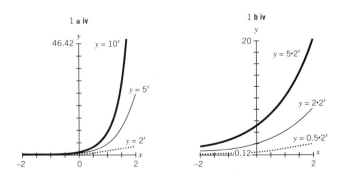

v. All have the same value for C, and thus all cross the y-axis at same spot; all have the same basic shape. Where they differ is in how fast they grow. The larger the base, the faster the function grows in y values as x grows.

b. i. For $y = 0.5 \cdot 2^x$: $C = 0.5$ and $a = 2$; for $y = 2 \cdot 2^x$: $C = 2$ and $a = 2$; for $y = 5 \cdot 2^x$: $C = 5$ and $a = 2$.

ii. growth for all three functions; y is multiplied by 2 in each case.

iii.

x	$0.5 \cdot 2^x$	x	$2 \cdot 2^x$	x	$5 \cdot 2^x$
-1	0.25	-1	1	-1	2.5
0	0.5	0	2	0	5
1	1	1	4	1	10
2	2	2	8	2	20

iv. The graphs of the three functions on the same coordinate system are illustrated on page 628, bottom right.

v. These functions have the same base but different positive values of C. In such a situation the larger the value of $|C|$, the faster the function grows when the bases are the same.

c. i. For $y = 3x$: $C = 1$ and $a = 3$; for $y = (1/3)x$: $C = 1$ and $a = 1/3$; for $y = 3 \cdot (1/3)x$: $C = 3$ and $a = 1/3$

ii. growth for the first and decay for the other two; y is multiplied by 3 in the first and by 1/3 in the other two.

iii.

x	3^x	x	$(1/3)^x$	x	$3 \cdot (1/3)^x$
-1	1/3	-1	3	-1	9
0	1	0	1	0	2
1	3	1	1/3	1	1

iv. The graphs for (a) and (b) and (c) are drawn on the same coordinate system in the figure.

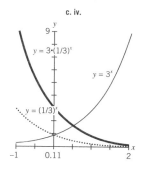

c. iv.

3. a. Answer is omitted.

b. $P = 736.99 \cdot 1.0099^t$ is the best fitting exponential, where t is measured in years since 1800 and P is measured in millions of people.

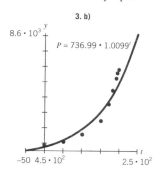

3. b)

c. and d. The 736.99 is the model's value for the world's population measured in millions in 1800. The base 1.0099 says that the growth rate for this model is 0.99%. This rate is measured in millions of persons per year. P's domain, should be $0 \le t \le 196$; P's range: $736.99 \le P \le 5077.50$.

e. i.

t	P
-50	450.44
120	2403.36
225	6755.66
250	8641.31

ii. during 1831, 1950, 1972 and 2043 respectively

iii. Approximately 70.39 yrs

5. 350 million; 490 million; in 2113.

7. Increasing 20% per day is not the same as doubling every 5 days since $(1.2)^5 = 2.488$, which is more than 2.

9. a. $V = 100 \cdot (1.04)^t$ **b.** $V = 1000 \cdot (1.04)^t$ **c.** Yes

11. a. at $x = 0$, $y = 1$.   **b.** at $x = 0$, $y = 5$.

13. a. $LN(t) = 4.2 - 0.5t$ **b.** $EN(t) = 4.2(0.880952)^t$

c. Linear in FY 1997; in exponential in FY 1998. These points are marked on the graphs.

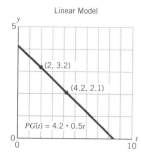

Linear Model

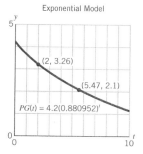

Exponential Model

d. $LN(2) = 3.2$ million; $EN(2) = 3.259521$ million recipients.

e. My source gave total dollar amounts and not the number of recipients.

15. a. $LS(t) = 100 - 10t$; after 5 min.

b. $ES(t) = 100(0.90)^t$; after 6.57 hrs.

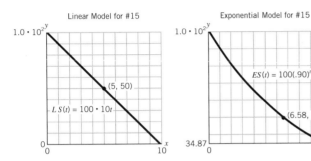

Linear Model for #15

Exponential Model for #15

$ES(t) = 100(.90)^t$

17. a. 3.125% is left (1/32)

b. **i.** Approximately 2.2 hours

ii. $A(t) = 100(1/2)^{t/2.2}$

iii. $5 \cdot 2.2 = 11$ hours and 3.125 milligrams.

19. a. for Quintana Roo: 11.07 years; for Baja California: 16.5 years

b. **i.** $M(t) = 21(1.05)^t$

ii. Here is a small table of values and its graph:

t	M(t)	t	M(t)
0	22.05	10	34.21
1	23.15	20	77.72
2	24.31	50	240.82

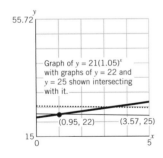

Graph of y = 21(1.05)ˣ with graphs of y = 22 and y = 25 shown intersecting with it.

(0.95, 22)   (3.57, 25)

iii. See the graph.

iv. The population would reach 22 million in approximately 0.95 years. It would reach 25 million in approximately 3.57 years.

21. a.

t	Aerospace	Bennington
0	$50,000	$35,000
1	$52,000	35,000 + 10%(35,000) = 35,000(1.1) = $38,500
2	$54,000	38,500 + 10%(38,500) = $42,350
3	$56,000	42,350 + 10%(42,350) = $46,585
4	$58,000	46,585 + 10%(46,585) = $51,243.50
5	$60,000	51,243.50 + 10%(51,243.50) = $56,367.85
6	$62,000	56,367.85 + 10%(56,367.85) = $62,004.64
7	$64,000	68,205.10
8	$66,000	75,025.61

b.

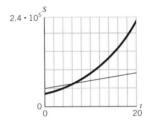

c. In the 7th year

d. Aerospace: $S = \$50,000 + \$2,000t$
Bennington: $S = \$35,000(1.1)^t$

e. Tenth year
Aerospace: $S = \$50,000 + \$2,000(10) = \$70,000$
Bennington: $S = \$35,000(1.1^{10}) = \$90,780.99$
Twentieth year
Aerospace: $S = \$50,000 + \$2,000(20) = \$90,000$
Bennington: $S = \$35,000(1.1^{20}) = \$235,462.50$

23. a. $M(n) = 10^n$.

b. 100,000 and 10,000,000,000.

c. The number grows very quickly.

25. a. This is a semi-log plot.

b. In 1929 it was 400; in 1933 it was 40; this is the depression era.

c. 1000 first in 1966; 3000 first in 1988; 6000 first in 1996.

d. A straight line on semi-log graph means that the growth is exponential because the equation of that straight line is $\log y = mx + b$. This means $y = 10^{mx+b}$ which can be written as $y = Ca^x$, where $C = 10^b$ and $a = 10^m$.

CHAPTER 9

1. a. 20, 20

b. 40, -40

c. $-20, -20$

d. $-40, 40$

3. a. $S = 1.256637 \cdot 10^{-19}$ m²

b. $V = 4.18879 \cdot 10^{-30}$ m³

c. $S/V = 3.0 \cdot 10^{10}$

5. a. $Y = kX^3$

b. $k = 1.25$

c. Increased by a factor of 125

d. Divided by 8

e. $X = (Y/k)^{1/3}$

7. a. $S(h) = 16\,h^2$; $V(h) = 4\,h^3$

b.
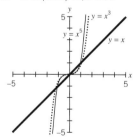

h	$S(h)$	$V(h)$
1	16	4
2	64	32
3	144	108
4	256	256

See the graph of this chart.

c. S is multiplied by 9; V is multiplied by 27.

d. It will decrease since it is $4/h$.

9. a. $d = kt^2$ **b.** 16   **c.** $t = \sqrt{d}/4$

11. a. Graphs of $y = x$, x^3 and x^5

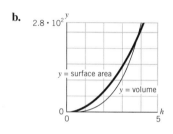

b. Graphs of $y = x^2$, x^4 and x^6

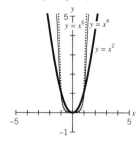

c. Graphs of $y = x^3$, $2x^3$ and $-2x^3$

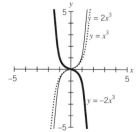

d. Graphs of $y = x^2$, $4x^2$ and $-4x^2$

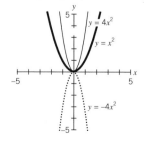

13.

$0 \le x \le 3$
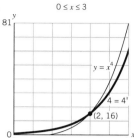

$3 \le x \le 6$
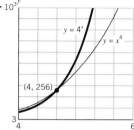

a. The two graphs intersect at (2, 4) and (4, 256). Thus, the functions are equal for $x = 2$ and $x = 4$.

b. The graphs of the two functions are in the two diagrams above. As x increases both functions grow. For $0 \le x < 2$ (see the diagram above on the left) we have that $x^4 < 4^x$; at $x = 2$ both have a y value of 16; from $2 < x < 4$ (see both diagrams) we have the $4^x < x^4$; for $x = 4$ they are again equal; for $x > 4$ we have that $x^4 < 4^x$ (see the diagram on the right); lastly, the relative sizes of the y values stay that way as x keeps on increasing.

c. Eventually the graph of $y = 4^x$ dominates: see the graph above on the right.

15. a. $y = 2x$ goes with Table #4

b. $y = x/2$ goes with Table #3

c. $y = 2/x$ goes with Table #1

d. $y = x^2$ goes with Table #5

e. $y = 2^x$ goes with Table #2

17. $3 \cdot 2^x < 3 \cdot x^2$, if $0 < x < 2$ and $3 \cdot 2^x > 3 \cdot x^2$, if $x > 2$; if $x = 2$ then $3 \cdot 2^2 = 3 \cdot 2^2$.

19. a. $1/8, -1/8$

b. $1/2, -1/2$

c. $-1/2, 1/2$

d. $-32, 32$

21. a.

x	$1/x$	x	$1/x^2$	x	$1/x^3$	x	$1/x^4$
-1.00	-1	1.00	1	-1.00	-1	-1.00	1
-0.50	-2	-0.50	4	-0.50	-8	-0.50	16
-0.25	-4	-0.25	16	-0.25	-64	-0.25	256
0.25	4	0.25	16	0.25	64	0.25	256
0.50	2	0.50	4	0.50	8	0.50	16
1.00	1	1.00	1	1.00	1	1.00	1

Graphs of $y = 1/x$ and $y = 1/x^2$

Graphs of $y = 1/x^3$ and $y = 1/x^4$

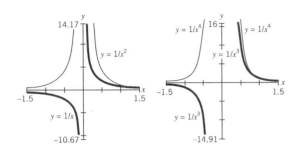

Graphs of $g(x) = 5x$ and $h(x) = x/5$

Graphs of $t(x) = 1/x$ and $f(x) = 5/x$

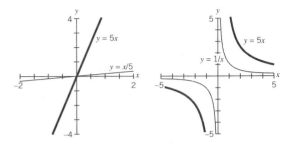

b. domain of $y = 1/x$: $x \neq 0$; range of $y = 1/x$: $y \neq 0$
domain of $y = 1/x^2$: $x \neq 0$; range of $y = 1/x^2$: $y > 0$
domain of $y = 1/x^3$: $x \neq 0$; range of $y = 1/x^3$: $y \neq 0$
domain of $y = 1/x^4$: $x \neq 0$; range of $y = 1/x^4$: $y > 0$

c. As $x \to \infty$ all four functions $\to 0$ from the positive side of the x axis.
As $x \to -\infty$ $1/x$ and $1/x^3$ approach 0 from the negative side of the x axis while $1/x^2$ and $1/x^4$ approach 0 from the positive side of the x axis.

d. As $x \to 0$ from the left, $y = 1/x \to -\infty$, i.e., gets very large negatively; as $x \to 0$ from the right, $y = 1/x \to \infty$, i.e. gets very large positively.
As $x \to 0$ from the left or right, $y = 1/x^2 \to \infty$, i.e., gets very large positively.
As $x \to 0$ from the left or right, $y = 1/x^3 \to$ behaves like $y = 1/x$.
As $x \to 0$ from the left or right, $y = 1/x^4 \to$ behaves like $y = 1/x^2$.

23. a. The volume becomes 1/3 of what it was.

b. It becomes $1/n$ of what it was.

c. The volume is doubled.

d. It become n times what it was.

25. a.

x	$5x$	x	$x/5$	x	$1/x$	x	$5/x$
-2	-10	-2	-0.4	-2	$-1/2$	-2	$-5/2$
2	10	2	0.4	2	1/2	2	5/2
-1	-5	-1	-0.2	-1	-1	-1	-5
1	5	1	0.2	1	1	1	5
0	0	0	0.0	0.5	2	0.5	10

b. Omitted

27. $I(d) = k/d^2$ and thus $I(4)/I(7) = 49/16 = 3.06$—it will 3.06 times as great.

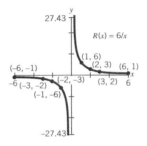

29. A table for r is given below.

x	$r(x)$
1	6
2	3
3	2
6	1
-1	-6
-2	-3
-3	-2
-6	-1

The domain of the abstract function is $x \neq 0$. For $x > 0$, as $x \to 0$ we have $r(x) \to +\infty$ and for $x < 0$ as $x \to 0$ we have that $r(x) \to -\infty$.

31. a. Let $L =$ wavelength of wave, $v =$ velocity of wave, and $t =$ time between waves. We have that $t = L/(k\sqrt{L}) = (1/k) \cdot \sqrt{L}$ and thus time is directly proportional to the square root of the wavelength of the wave.

b. 4 times the distance.

33. a. at $x = 0$ and at $x = 1/2$ **b.** at $x = 0$ and at $x = 1$

c. at $x = \sqrt{(1/2)}$ **d.** at $x = 1$

e. at $x = 1$.

CHAPTER 10

1. No answer is given.

3. a. 1.7 is the initial velocity of the object falling; it is measured in meters per sec.; 4.9 is half the gravitational constant when it is measured in (meters/sec)/sec.

b.

t	d
0.0	0.000
0.1	0.219
0.2	0.536
0.3	0.951

c. and **d.** The shape of the graph is that of a quadratic.

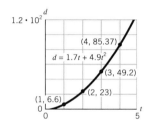

5. a. $d = 16t^2 + 12t$:

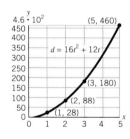

t	d
0	0
1	28
2	88
3	180
5	460

b. $v(t) = 12 + 32t$

c. See the graph below.

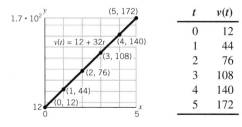

t	$v(t)$
0	12
1	44
2	76
3	108
4	140
5	172

7. For $h = 85 - 490t^2$:

a. 85 is the height in centimeters of the falling object at the start; -490 is half the gravitational constant when measured in (cm/sec)/sec; it is negative in value since h measures height above the ground and the gravitational constant is connected with pulling objects down.

b. The initial velocity is 0 cm/sec.

c.

t	h
0.0	85.0
0.1	80.1
0.2	65.4

d. Below is the graph of the function with the table entries marked on it.

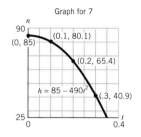

9. a. The initial velocity is positive

b. $h = 50 + 10t - 16t^2$.

11. a. The first graph shows that the child's weight increases with age but eventually levels off.

b. The second graph shows that the rate at which weight increases is slow at first, then very rapidly increases for a while, then goes down rapidly after peaking, and then tapers off so that there is almost no change.

c. The third graph shows that the acceleration of growth peaks early, then decreases quickly, then slows down until it bottoms out, then rises rapidly and finally levels off.

13. For $h = 4 + 64t - 16t^2$. See the tables and accompanying graph.

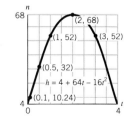

t	h	t	h
0.0	4.00	1	52
0.1	10.24	2	68
0.5	32.00	3	52

a. $(32 - 4)/0.5 = 28/0.5 = 56$ ft/sec; rising

b. $(10.24 - 4)/0.1 = 6.24/0.1 = 62.4$ ft/sec; rising

c. $(52 - 4)/1 = 48$ ft/sec; rising

d. $(68 - 52)/1 = 16$ ft/sec.; rising

e. $(52 - 52)/2 = 0$ ft/sec; both

f. $(3.3584 - 4)/0.01 = -64.16$ ft/sec; falling

15. **a.** $\sqrt{d}/4$ **b.** $d \geq 0$ **c.** $t = \sqrt{d}/4$

d.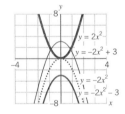

t	d
0	0.0
16	1.0
32	1.4
64	2.0
96	2.5

17. **a.** $d = -(1/2)gt^2 + 146.67t$ **b.** 2.36 m

19. **a.** 110 cm/sec; 660 cm/sec; $v(t) = 60 + 10t$ cm/sec.

b. 85 cm/sec.

21. **a.** $v(t) = 200 + 60t$ m/sec.   **b.** $d(t) = 200t + 30t^2$ m

23. For any $n > 0$: $(n + 1)^2 - n^2 = 2n + 1$, an odd integer.

CHAPTER 11

1. **a.** **i.** $-8, 12$ **ii.** $10, 14$ **iii.** $2, -14$

b. **i.** $0, 2$ **ii.** $-9, -7$ **iii.** $2, 14$

3. The graphs with their labels are given here.

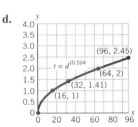

5.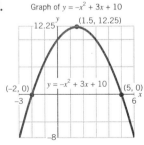

All such equations whose graphs turn down will have two horizontal intercepts. The example given here that faces up has no horizontal intercepts. Others could have one horizontal intercept, such as $y = 10(x - 1)^2$, or two horizontal intercepts, such as $y = (x - 5)(x - 2)$.

7. **a.** $y = x^2 + x - 12$

vertex at $(-0.5, -12.25)$
roots at $x = -4$ and 3 (by factoring)
rough sketch: $a = 1 > 0$ and thus the graph opens upward, goes through $(-4, 0)$ and $(3, 0)$; y-intercept at -12

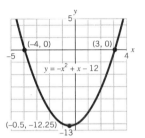

b. $y = x^2 + 3x + 7.2$

vertex at $(-1.5, 4.95)$
no real roots: discriminant $= -21.8$
rough sketch: $a = 1 > 0$ and thus the graph opens upward, vertex at $y = 4.95$ means no x-intercepts; y-intercept at 7.2

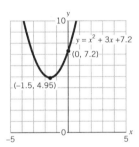

c. $y = -x^2 + 5x - 6.25$

vertex at $(2.5, 0)$
double root at $x = 2.5$;
rough sketch: $a = -1 < 0$ and thus the graph opens down, with double root at 2.5; y-intercept at -6.25

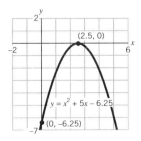

d. $y = x^2 + 8x + 15$

vertex at $(-4, -1)$;
roots: $x = -3$ and $x = -5$; (factoring)
rough sketch: $a > 0$ and thus the graph opens up;
y-intercept $= 15$; x-intercepts -3 and -5

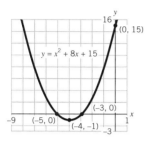

e. $y = x^2 + 3x - 11$

vertex at $(-1.5, -13.25)$;
roots at $x = [-3 \pm \sqrt{(53)}]/2$;
rough sketch: $a = 1 > 0$ and thus the graph opens up;
y-intercept at -11;
roots at 2.14 and -5.14

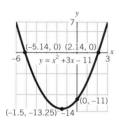

9. a. $6t^2 - 7t - 5 = 0$ and thus $t = (7 \pm \sqrt{169})/12 = (7 \pm 13)/12$ or $t = 5/3$ or $-1/2$.
[*Note* that $6t^2 - 7t - 5 = (3t - 5)(2t + 1)$.]

b. $9x^2 - 12x + 4 = 0$ and thus $x = (12 \pm \sqrt{0})/18$ or $x = 2/3$. [*Note* that $9x^2 - 12x + 4 = (3x - 2)^2$]

c. $3z^2 - z - 9 = 0$ and thus $z = (1 \pm \sqrt{109})/6$ or $z \approx 1.91$ or -1.57 (rounded to 2 decimal places)

d. $x^2 + 6x + 7 = 0$ and thus $x = (-6 \pm \sqrt{8})/2$; thus $x \approx -1.59$ or -4.41 (rounded to 2 decimal places)

11. a. At $t = 0, h = 4$ ft.

b. $h = 0$, by the quadratic formula, when $t = (-50 \pm \sqrt{2756})/(-32) = 3.20$ or -0.08 and the latter is rejected because h is not defined for such a value of t.

c. $h = 30$ ft at $t = 0.659$ sec and at $t = 2.466$ sec (via quadratic formula); it is never 90 feet high since the discriminant in the resultant quadratic formula is -3004.

d. It reaches its maximum height at $t = -50/(-32) = 1.56$; and the height at that moment is 43.06 ft.

13.

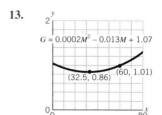

a. and **b.** See the accompanying graph. The minimum gas consumption rate suggested by the graph occurs when $M = 32.5$ mph and it is approximately 0.85 gph. Computation on a calculator gives the same M but the gas consumption rate is 0.86 when rounded off.

c. In 2 hours 1.72 gallons will be used and one will have traveled 65 miles.

d. If $M = 60$ mph then $G = 1.01$ gph. It takes 1.083 hours to travel 65 miles at 60 mph and one will have used 1.094 gallons.

e. Clearly traveling at the speed that minimizes the gas consumption rate does not conserve fuel if the trip lasts only 2 hours.

f. and **g.**

mph	gph	gpm	mpg
0	1.07	x.xxxxx	0.0
10	0.96	0.09600	10.4
20	0.89	0.04450	22.5
30	0.86	0.02867	34.9
40	0.87	0.02175	46.0
50	0.92	0.01840	54.3
60	1.01	0.01684	59.4
70	1.14	0.01629	61.4
80	1.31	0.01638	61.1

For **f.** we have:
$G/M = (0.0002M^2 - 0.013M + 1.07)/M$.
Eyeballing gives the minimum y at $M \approx 73$ mpg.

For **g.** we have:
$M/G = M/(0.0002M^2 - 0.013M + 1.07)$.
Eyeballing gives the maximum y at the same M. This is expected since max $= 1/\min$.

15.

a. $y = (x + 4)(x + 2)$

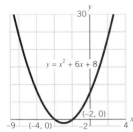

$y = x^2 + 6x + 8$

b. $z = 3(x - 3)(x + 1)$

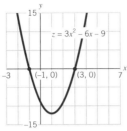

$z = 3x^2 - 6x - 9$

c. $f(x) = (x - 5)(x + 2)$

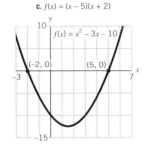

$f(x) = x^2 - 3x - 10$

d. $w = (t - 5)(t + 5)$

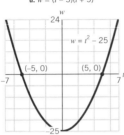

$w = t^2 - 25$

e. $r = 4(s - 5)(s + 5)$

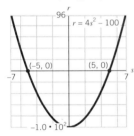

$r = 4s^2 - 100$

f. $3(x + 1)(x - 4/3)$

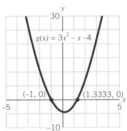

$g(x) = 3x^2 - x - 4$

17. Since the vertex is at $(-1, 4)$ we have that $y = a(x + 1)^2 + 4$. Since the y-intercept is at $(0, 2)$ we have that $2 = a(0 + 1)^2 + 4$ or $a = -2$ and thus $y = -2x^2 - 4x + 2$. [Check: the roots are at $x = -1 \pm \sqrt{2} \approx 2.4$ or -1.4—which values agree with the graph—and substitution gives $-2(-1)^2 - 4(-1) + 2 = 4$ and $-2(0)^2 - 4(0) + 2 = 2$.]

21. a. $y = a(x - 2)^2 + 4$ and $7 = a(1 - 2)^2 + 4$ gives $3 = a$ and thus $y = 3(x - 2)^2 + 4 = 3x^2 - 12x + 16$

b. Concave up: $y = a(x - 2)^2 - 3$ if $a > 0$; concave down: $y = a(x - 2)^2 - 3$ if $a < 0$.

23. a.

Time	Distance	Velocity	Time	Distance	Velocity
1	16	32	2.5	100	80
1.5	36	48	1.87	56	59.87
2	64	64			

b. If $t = 3$ then $d = 16 \cdot 9 = 144$ ft. and $v = 32 \cdot 3 = 96$ ft/sec.

c. The graphs from the table given in a. are found below. The distance graph is given on the left and the velocity graph is given on the right.

Distance Graph   Velocity Graph

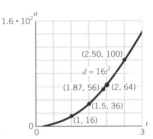

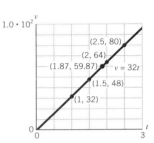

d. Time the drop of the pebble in seconds and use $d = 16t^2$ to find d.

25.

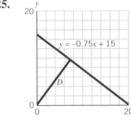

$y = -0.75x + 15$

a. The graph is illustrated on the left, with x and y marked off in miles, and with a typical point (x, y) marked on the highway along with the straight line to that point.

b. The highway goes through the points $(0, 15)$ and $(20, 0)$ and thus has the equation $y = -.75x + 15$.

c. and d. $DS^2 = x^2 + y^2 = x^2 + (15 - .75x)^2 = 1.5625x^2 - 22.5x + 225$

e. Let DS = distance squared. Then $DS(x) = 1.5625x^2 - 22.5x + 225$ and the minimum occurs at the vertex, which is at $x = 22.5/(2 \cdot 1.5625) = 7.2$. The minimum for DS is 144 and thus the minimum distance is $\sqrt{144}$ or 12 miles.

f. The coordinates of the point of shortest distance from $(0, 0)$ are $(7.2, 9.6)$

27.

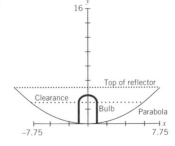

a. A schematic of the lightbulb with a semi-circular end of radius 1 in. is given in the illustration on the previous page. The focus is marked at the 3 in. mark and the parabola is drawn.

b. 3 in. (The bulb is 4 inches long and the filament is 1 in. from one end and thus 3 in. from the other end.)

c. With the vertex at the origin its equation is $y = ax^2$; the focal length $= 3$ and thus $3 = 1/(4a)$ and thus $a = 1/12$. The equation is $y = (1/12)x^2$.

d. Since the reflector is to be 1 in. above the glass end of the bulb, we solve $5 = (1/12)x^2$. This gives $x = \pm\sqrt{60}$ and thus the diameter of the reflector 5 in. above the base is $2 \cdot \sqrt{60} = 15.4$ in. (See the top dotted horizontal line in the diagram on the previous page at the 5 in. level.)

e. The center of the bulb is 3 in. from the base and thus to get the distance from the center of the bulb to the reflector's side we solve $3 = (1/12)x^2$. This gives $x = 6$. Since the bulb has a 1 in. radius, the clearance from the outside of the bulb to the side of the reflector is 5 in.

29. a. Algebraically one finds the values of x that satisfy the two equations:

$$2x^2 - 3x + 5.1 = -4.3x + 10 \qquad \text{or}$$
$$2x^2 + 1.3x - 4.9 = 0 \qquad \text{or}$$
$$x = [-1.3 \pm \sqrt{[1.3^2 - 4(2)(-4.9)]}]/4 \qquad \text{or}$$
$$= [-1.3 \pm \sqrt{40.89}]/4 \qquad \text{or}$$
$$= [-1.3 \pm 6.39]/4 \qquad \text{or}$$
$$= 1.27 \text{ or } -1.92$$

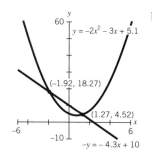

b. The graphs are given in the diagram; they show the points of intersection at $x = -1.92$, $y = 18.27$ and at $x = 1.27$, $y = 4.52$

31.

x	y	average rate of change of y with regard to x	avg rate of change of avg rate of change
-3	-3	N/A	N/A
-2	1	4	N/A
-1	3	2	-2
0	3	0	-2
1	1	-2	-2
2	-3	-4	-2
3	-9	-6	-2

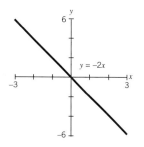

The graph of the average rate has a constant slope of -2 and thus is linear. The function to be graphed is the function $y = -2x$.

33. Generic answers are given here:

a. $f(x) = (x - a) \cdot (x - b) \cdot (x - c) \cdot g(x)$ with a, b, and c distinct and $g(x)$ any polynomial.

b. $h(x) = a(x + 1)(x - 3)(x - 10)$, for any $a \neq 0$.

c. $p(x) = ax^3 + bx^2 + cx + 4$ for $a \neq 0$ and any b and c.

d. $q(x) = ax^5 + bx^3 + cx^2 + dx - 4$, for any $a \neq 0$ and any b, c, and d.

35. a. $y = 3x^3 - 2x^2 - 3$ has only one real root at $x \approx 1.28$.

b. $y = x^2 + 5x + 3$ has two real roots: $x = -2.5 \pm 0.5\sqrt{13} \approx -4.30$ or -0.70

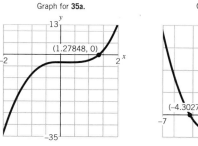
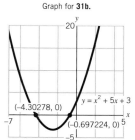

Graph for 35a.

Graph for 31b.

37. a. Goes with Graph 3: quadratic, face up; 1 is y-intercept

b. Goes with Graph 4: quadratic, face down; one root.

c. Goes with Graph 2: cubic, positive leading coefficient; negative y-intercept

d. Goes with Graph 5: cubic, negative leading coefficient; negative y-intercept

e. Goes with Graph 6: quartic; positive leading coefficient; two roots;

f. Goes with Graph 1: exponential; horizontal asymptote; increasing; positive y-intercept.

CHAPTER 12

1. Student estimates will vary for each interest rate. The ones given here are based on rough approximations of the position of 300 between the two relevant data entries.

   a. Approximately 37   **b.** Approximately 22 years

   c. Approximately 15 years

3. **a.** 0.001   **b.** 1,000,000   **c.** 2.1544

   d. 1   **e.** 10   **f.** 0.1

5. **a.** $\log(25) = \log(5^2) = 2 \cdot \log(5) \approx 1.398$

   b. $\log(1/25) = \log(25^{-1}) = -\log(25) \approx -1.398$

   c. $\log(10^{25}) = 25 \cdot \log(10) = 25$

   d. $\log(0.0025) = \log(25 \cdot 10^{-4}) \approx 1.398 - 4 = -2.602$

7. **a.** $\log\left(\dfrac{K^3}{(K + 3)^2}\right)$ **b.** $\log\left(\dfrac{(3 + n)^5}{m}\right)$

9. Let $w = \log(A)$. Then $10^w = A$ and thus $A^p = 10^{wp}$ and $\log(A^p) = wp = p \cdot w = p \cdot \log(A)$

11. Rewriting $f(t) = 100(1/2)^{t/2}$ makes things a bit easier:

    a. **i.** $0.6 = (0.5)^{t/2}$ or
    $t/2 = \log(0.6)/\log(0.5) = 0.7370$ or
    $t \approx 1.47$ hours

    ii. $0.4 = (0.5)^{t/2}$ or
    $t/2 = \log(0.4)/\log(0.5) = 1.3219$ or
    $t \approx 2.64$ hours

    iii. $0.2 = (0.5)^{t/2}$ or
    $t/2 = \log(0.2)/\log(0.5) = 2.3219$ or
    $t \approx 4.64$ hours

    b. Its half life is 2 hours as can be seen by using the rewritten formula.

13. In general, $A_k(n) = 10000 \cdot (1 + 0.085/k)^{kn}$ gives the value of the $10,000 after n years if interest is compounded k times per year and $A_c(n) = 10000 \cdot e^{0.085n}$ gives that value if the interest is compounded continuously.

    a. annually: $(1 + 0.085/1)^1 = 1.085$ and thus the effective rate is 8.50%

    b. semi-annually: $(1 + 0.085/2)^2 \approx 1.0868$ and thus the effective rate is 8.68%.

    c. quarterly: $(1 + 0.085/4)^4 \approx 1.0877$ and thus the effective rate is 8.77%

    d. continuously: $e^{0.085} \approx 1.0887$ and thus the effective rate is 8.87%

15. **a.** $U(x) = 10 \cdot (1/2)^{x/5}$ where x is measured in billions of years.

    b. $x = 5 \cdot \log(0.1)/\log(0.5) = 16.6096$ billion years.

17. **a.** $t = \log(2)/\log(1.12) \approx 6.12$ years

    b. $t = \log(2)/\log((1 + .12/4)^4) \approx 5.86$ years

    c. $t = \ln(2)/0.12 \approx 5.78$ years.

19. **a.** $10^n = 35$ **b.** $N = N_0 e^{-kt}$

21. Let $w = \ln(A)$ and $z = \ln(B)$ and thus $e^w = A$ and $e^z = B$ and therefore $A \cdot B = e^{w+z}$ and thus $\ln(A \cdot B) = w + z = \ln(A) + \ln(B)$

23. The half life is $\ln(1/2)/(-r) = \ln(2)/r = 100 \cdot \ln(2)/R = 69.3147/R$ which is approximately $70/R$.

25. **a.** $\ln\left(\dfrac{R^3}{\sqrt{P}}\right)$ **b.** $\ln\left(\dfrac{N}{N_0^2}\right)$

27.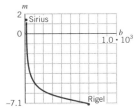

    a. At $B = B_0$.

    b. See the rough sketch of its graph. The scale is very spread out. The positions of Rigel and Sirius are indicated on it.

    c. As the brightness increases the magnitude decreases.

    d. If the brightness goes from B to $5B$ then the magnitude M becomes $2.5 \cdot \log(5) \approx 1.75$ units less bright.

29. $db = 10\log(I/I_0) = 28$ implies that $I/I_0 = 10^{2.8}$ and thus $I = 10^{-13.2}$ watts/cm^2 and 92 dB has $I/I_0 = 10^{9.2}$ and thus $I = 10^{-6.8}$ watts/cm^2.

31. If the combined intensity is called Cm and the individual intensity is In, then $Cm = 5 \cdot In$; thus $dB(Cm) = 10\log(5 \cdot In/I_0) = 10\log(5) + 10\log(I/I_0) \approx 6.99 + dB(In)$.

33. Answers of students will vary. Look for a curve that goes through roughly the middle of each cluster. Without the data table it is not possible to give the best fit exponential with any precision.

35. **a.** $N = 10e^{0.044017t}$ **b.** $Q = 5 \cdot 10^{-7} \cdot e^{-2.631089A}$

37. **a.** The population density is directly proportional to the -2.25 power of the length of the organism.

b. The population density of the organism is directly proportional to the -0.75 power of the body mass of the organism.

c. From **a** and **b** we have that $d \cdot m^{-0.75} = c \cdot x^{-2.25}$ or that $m = (d/c) \cdot x^3$

Note that $\log(p)$ is not directly proportional to $\log(x)$ or to $\log(m)$, since the $\log(p)$ intercept is not 0 in either graph. However, the change in $\log(p)$ is directly proportional both to the change in $\log(x)$ and to the change in $\log(m)$.

39. In this context, the general equation for a straight line ℓ is: $\log(y) = mx + \log(b)$ and if we solve for y we get: $y = b \cdot 10^{mx}$

a. $y = b_1 \cdot 10^{mx}$ and $y = b_2 \cdot 10^{mx}$ with $b_1 \neq b_2$. Thus they have the same power of 10 but differ in their y-intercepts.

b. $y = b_3 x^{mx}$ and $y = b_4 x^{mx}$ with $b_3 \neq b_4$. Thus they have the same power of x but differ in their y-intercepts.

Index

Page references followed by italic *n* indicate footnotes.

Limited Use License Agreement

This is the John Wiley & Sons, Inc. (Wiley) limited use License Agreement, which governs your use of any Wiley proprietary software products (Licensed Program) and User Manual(s) delivered with it.

Your use of the Licensed Program indicates your acceptance of the terms and conditions of this Agreement. If you do not accept or agree with them, you must return the Licensed Program unused within 30 days of prior receipt or, if purchased, within 30 days, as evidenced by a copy of your receipt, in which case, the purchase price will be fully refunded.

License: Wiley hereby grants you, and you accept, a non-exclusive and non-transferable license, to use the Licensed Program and User Manual(s) on the following terms and conditions:

a. The Licensed Program and User Manual(s) are for your personal use only.

b. You may use the Licensed Program on a single computer, or on its temporary replacement, or on a subsequent computer only.

c. You may modify the Licensed Program for your use only, but any such modifications void all warranties expressed or implied. In all respects, the modified programs will continue to be subject to the terms and conditions of this Agreement.

d. A backup copy or copies may be made only as provided by the User's Manual(s), but all such backup copies are subject to the terms and conditions of this Agreement.

e. You may not use the Licensed Program on more than one computer system, make or distribute unauthorized copies of the Licensed Program or Users Manual(s), create by decompilation or otherwise the source code of the Licensed Program or use, copy, modify, or transfer the Licensed Program, in whole or in part, or User's Manual(s), except as expressly permitted by this Agreement.

If you transfer possession of any copy or modification of the Licensed Program to any third party, your license is automatically terminated. Such termination shall be in addition to and not in Lieu of any equitable, civil, or other remedies available to Wiley.

Term: This License Agreement is effective until terminated. You may terminate it at any time by destroying the Licensed Program and User Manual(s), and any copies made (with or without authorization).

This Agreement will also terminate upon the conditions discussed elsewhere in this Agreement, or if you fail to comply with any term or condition of this Agreement. Upon such termination, you agree to destroy the Licensed Program, User Manual(s), and any copies made (with or without authorization) of either.

Wiley's Rights: You acknowledge that the Licensed Program and User Manual(s) are the sole and exclusive property of Wiley. By accepting this Agreement, you do not become the owner of the Licensed Program or User Manual(s), but you do have the right to use them in accordance with the provisions of this Agreement. You agree to protect the Licensed Program and User Manual(s) from unauthorized use, reproduction, or distribution.

Texas Instruments Graph Link™ License Agreement

By downloading the software and/or documentation you agree to abide by the following provisions.

1. License: Texas Instruments Incorporated ("TI") grants you a license to use and copy the software program(s) and documentation from the linked web page ("Licensed Materials").

2. Restrictions: You may not reverse-assemble or reverse-compile the software program portion of the Licensed Materials that are provided in object code format. You may not sell, rent or lease copies that you make.

3. Copyright: The Licensed Materials are copyrighted. Do not delete the copyright notice, trademarks or protective notice from any copy you make.

4. Warranty: TI does not warrant that the program or Licensed Materials will be free from errors or will meet your specific requirements. The Licensed Materials are made available "AS IS" to you or any subsequent user.

5. Limitations: TI makes no warranty or condition either express or implied, including but not limited to any implied warranties of merchantability and fitness for a particular purpose, regarding the Licensed Materials.

In no event shall TI or its suppliers be liable for any indirect, incidental or consequential damages, loss of profits, loss of use or data, or interruption of business, whether the alleged damages are labeled in tort, contract or indemnity.

Some states do not allow the exclusion or limitation of incidental or consequential damages, so the above limitation may not apply.